109

N	ν	Nu
Ξ	ξ	Xi
O	ο	Omicron
Π	π	Pi
P	ρ	Rho
Σ	σ	Sigma
T	τ	Tau
Υ	υ	Upsilon
Φ	φ	Phi
X	χ	Chi
Ψ	ψ	Psi
Ω	ω	Omega

Statistics for Business and Economics

JOSEPH G. VAN MATRE
University of Alabama in Birmingham

GLENN H. GILBREATH
Virginia Commonwealth University

Statistics for Business and Economics

1980
BUSINESS PUBLICATIONS, INC.   Dallas, Texas   75243
Irwin-Dorsey Limited   Georgetown, Ontario   L7G 4B3

© BUSINESS PUBLICATIONS, INC., 1980

ISBN 0-256-02276-3
Library of Congress Catalog Card No. 79–54422
Printed in the United States of America

2 3 4 5 6 7 8 9 0 MP 7 6 5 4 3 2 1 0

To Mindy, Kelly, and Sandy
JGV

To my Mother
GHG

Preface

This book is primarily designed for undergraduate students in business, economics, and management; further, the material covered is suitable for some courses at the MBA level. The text's material is sufficient for a two-semester or two-quarter course. However, it may be used for shorter courses by omitting certain topics. Suggested coverage for varying lengths of courses is given below.

This course provides the sole exposure to statistics for many students, while others may continue study in this and related areas such as operations research or econometrics. To meet the needs of both groups, this book emphasizes a conceptual approach with both mathematical and verbal explanations. Selected derivations are presented, often in footnotes. Illustrations and exercises are plentiful; the amount of "busy" work is kept to a minimum.

We assume that the reader's mathematical maturity is at the college algebra level. Students with a knowledge of elementary calculus will benefit from the supplementary explanations given for certain concepts. However, calculus-based material is relegated to footnotes or optional sections so that no loss of continuity occurs if this material is omitted. Optional sections which contain calculus, selected derivations, or advanced concepts are clearly marked with asterisks. A brief review of notation, symbols, and summation operations is given in Appendix A.

Many features to help stimulate student interest and enhance the learning process have been incorporated into the book. Much attention has been paid to the selection of illustrations and exercises throughout the text. Real data and topical situations are used extensively. Some of the situations examined are airline overbooking, radiation leakage at nuclear power plants, wage and salary discrimination, electricity demand, risks of heart attack, auditing accounts payable, bank account confirmation programs, insurance charges and claims, and stock portfolios and security analysis.

Each chapter is introduced with a quotation that expresses an important point within the chapter. Biographical sketches of major and minor figures in statistics are included to provide a historical frame of reference and to give some human interest to a subject that, at times, seems too methodical. Each chapter concludes with a summary of the major concepts; a convenient list of important terms, symbols, and formulas; true-false and fill-in questions; and a generous number of carefully designed exercises. Answers to the odd-numbered exercises are given at the end of the book. A detailed solutions manual for all exercises is available to instructors.

One of the unique features of this book is the devotion of a full chapter to the actual use of statistics in today's business world. Chapter 16, Statistics in Business, provides a look at how some of the basic concepts are used in accounting, banking, finance, marketing, personnel management, and other areas of application. The intent of the chapter is to help show the reader why there is a need for statistical concepts and to provide a brief look at how these concepts are used in decision making. This material can be used in at least three ways. The entire chapter, or selected applications, could be read as a motivational tool in an introduction to the course, or as a topic is introduced. Second, study of one or more methods in a given chapter could be followed by a look at a specific application. Finally, the entire chapter could serve as a capstone to the course.

All topics under the general heading of "descriptive statistics" are presented in a single chapter, Chapter 2. This includes frequency distributions, averages, and measures of variation. The result is a more unified and complete treatment of these closely related topics. Index numbers are presented as descriptive measures in Chapter 3. However, this topic stands alone and can be placed at any point in the sequence of topics for a given course. This material is organized so that the topic may be covered only briefly or in some depth.

Probabilistic concepts are essential for an understanding of statistical inference. A careful presentation of the basic concepts is given in Chapter 4, while probability mass functions and probability density functions are given in Chapter 5. Sampling and sampling distributions, covered in Chapter 6, complete the preparation for estimation and testing. For hypothesis testing, the modern approach to reporting the *p-value* is provided, in addition to discussions of the traditional decision rules. The use of various decision rules is illustrated throughout the text.

Separate chapters are devoted to nonparametrics, multiple regression, chi-square, and analysis of variance. Chapter 7 covers decision theory through normal priors and the two-action problem without extended digressions into utility theory and the assessment of prior distributions. An expanded coverage of time series analysis, Chapter 14, includes exponential smoothing and a discussion of quadratic and exponential trend equations. The use of widely available computer routines for performing the more laborious computations is illustrated throughout the text. Yet we caution the reader that the use of a computer routine is not a substitute for sound statistical reasoning; it is merely a computational aid.

The book may be used for courses of varying lengths. For two quarters or two semesters, we suggest that the book be covered much in its entirety. Several chapters stand alone, however, and may be considered optional. These include Chapter 3, Index Numbers; Chapter 7, Statistical Decision Theory; Chapter 14, Time Series Analysis; and Chapter 15, Nonparametric Methods. A one quarter or one semester course could be built around the chapters remaining after excluding these optional chapters. Chapters 9, 10, 11, and 13 may also be considered optional for these shorter courses. Note that Chapter 12, on Simple Linear Regression, does *not* require any material from Chapters 10 and 11. Optional sections of chapters are marked with asterisks. More detailed suggestions may be found in the solutions manual.

In addition to our own ideas, a book of this nature is a product of interactions with our students, colleagues, and consulting clients. We have benefited from the ideas and helpful suggestions of many. Our consulting work provided us with the opportunity to work with real problems, and the examples and exercises of the book often reflect these experiences. We are particularly indebted to the readers of the entire manuscript: Ronald S. Koot (Pennsylvania State University), Jean D. Gibbons (University of Alabama), Ernest Houck (Virginia Polytechnic Institute and State University), and Richard Haase (Drexel University). Their comments have done much to improve the content of this work. Of course, any faults that remain are the sole responsibility of the authors. We appreciate the suggestions and encouragement of our colleagues at the University of Alabama in Birmingham: Al Kvanli, Jim Dilworth, M. Gene Newport, Carol Pfrommer, and Herb Tsang. The following graduate assistants provided help in the problem solutions and proofreading: Carla Grogan, Julie Hurd, David Spaulding, Rick Burns, and Karen Squires. In the business community, special thanks to Alan Barton. Finally, for patience and care in typing and retyping the manuscript, we thank Donna Pruet, Betty Smith, and Sandy Van Matre.

Several colleagues at Virginia Commonwealth University provided valuable suggestions. George C. Canavos, Maurice V. Gilbert, Walter S. Griggs, Jr., and Elbert G. Miller, Jr., were always willing to discuss an issue and lend their support. Several persons assisted in the typing of various versions of the manuscript. In particular, Carolyn England, Jamie Froelich, and Ann Mc-Farland shared the major effort.

We are grateful to the Literary Executor of the late Sir Ronald A. Fisher, F.R.S., to Dr. Frank Yates, F.R.S. and to Longman Group Ltd., London, for permission to reprint Table III from their book *Statistical Tables for Biological, Agricultural and Medical Research* (6th ed., 1974). We also thank the other authors and publishers cited throughout the text who allowed us to reprint various figures, tables, and excerpts from articles.

Finally, many others have assisted us in various ways; they go unnamed but not forgotten. We trust our colleagues will find that this book meets their classroom needs and would appreciate their comments and suggestions.

January 1980   ***Joseph G. Van Matre***
                                           Glenn H. Gilbreath

Contents

confidence

EXAM II   *EXAM III*   *Final Exam*   *midt*

** Optional material.

** Optional material.

Statistics:
An Overview

1 STATISTICS: AN OVERVIEW

> ...'I wonder what Latitude or Longitude I've
> got to?' (Alice had not the slightest idea what
> Latitude was, or Longitude either, but she
> thought they were nice grand words to say).
>
> Lewis Carroll
> *Alice's Adventures in Wonderland*

1.1 INTRODUCTION

As the student begins this study of statistics, he must be apprised that many new concepts lie ahead. One can understand these concepts only by "speaking the language" of statistics. This is an accepted fundamental for an introductory course in any subject, because the particular terms involved must be mastered before they can be used. Leafing through this text, the student will observe a variety of new symbols (e.g., σ, μ) and words (e.g., median, regression). The Greek alphabet printed on the inside front cover indicates how these symbols are pronounced (σ: sigma and μ: mu), but the statistical meaning of these symbols must be learned. We trust that when a student completes this book, his statistical knowledge cannot be characterized by Alice's use of longitude and latitude.

1.2 STATISTICS DEFINED

The word statistics calls forth different things to different people. For example, statistics are Tony Dorsett's yards per carry and season touchdowns, or the price-earnings ratio of a stock. However, as we shall see, the study of statistics possesses great utility in capacities other than providing a means of description, although this aspect should not be ignored.

A somewhat traditional definition of the discipline is:

> *statistics:* a body of methods dealing with the collection, description, analysis, and interpretation of information that can be given in numerical form.

Since the definition above is rather broad, we consider the subject in more detail by discussing the major divisions within the field; namely, descriptive statistics, inferential statistics, and statistical decision theory.

Descriptive Statistics

Descriptive statistics are measures that are used to characterize (describe) a mass of numerical data. For example, suppose one were given the SAT scores of the 300 freshmen of the School of Business of some university, and then later were asked "How do the B-school students' entrance scores look?" Probably, no one would reply by beginning to read the 300 scores. An obvious possible descriptor is what the statistical layman calls the average. This and other descriptive measures will be the subject of the next chapter. Art Buchwald's column of September 16, 1971, in the *Washington Post* provides another illustration on the use (or misuse?) of descriptive statistics:

There was good news out of Washington last week. According to Attorney General John Mitchell, President Nixon's war on crime has been successful, and the results of the administration's monumental efforts have been so great that "fear is being swept from the streets of some—though not all—American cities." . . .

The reason for the euphoria in the Justice Department is that FBI statistics for 1970 indicated that the rate of increase of crime had gone down from 12 percent in 1969 to 11.3 percent in 1970.

This sounded terrific, until I read that the same statistics revealed that 566,700 more crimes had been committed in 1970 than in 1969.

Admittedly confused, I sought out my friend Professor Heinrich Applebaum, the great Justice Department statistician, whose definitive book *Do Decimal Points Have a Sex Life?* is used in every math class in the country.

"Professor Applebaum, the Justice Department reports that the rate of crime has gone down in the country under President Nixon. Yet the same report says there has been a million more crimes in the last two years. How can that be?"

"It's quite simple," said Applebaum. "Percentagewise crime has gone down, crimewise it's gone up."

"But where does that leave the average person?"

"It depends whether you're a Republican or a Democrat. If you're a Republican you have nothing to fear walking the streets of our American cities. But if I were a Democrat, I'd stay home."

"Are you saying that the Republicans are trying to take the crime issue out of the 1972 campaign?"

"They have," Applebaum said. "The last year the Democrats were in office crime had gone up 13.8 percent. When the Republicans took over in 1969 it only went up 12.0 and last year 11.3 percent. The Democrats can't argue with that."

"But still more people were robbed, mugged, murdered, and raped in 1969 and 1970 than they were in the previous years."

"We're not talking about people," Applebaum said, irritated. "We're talking about percentages. You can't think about the people who were molested in 1969 and 1970, you have to think about the ones who weren't mugged . . . this year thanks to President Nixon's leadership."

"It's hard to think in those terms," I admitted.

"That's because you're not running for election next year. You must understand the reporting of crimes is a very serious business, and can cause great conflict. J. Edgar Hoover, in order to prove he is doing his job, has to show that crime is going up in the country. At the same time the administration has to prove that crime is going down."

"The Attorney General has solved the problem by reporting the percentages, which are lower, and Hoover, by reporting the crimes, which are higher. That's the beauty of statistics. It makes everyone feel better."[1]

---

[1] Reprinted with permission.

The preceding levity serves a useful purpose in addition to the provision of humor. Statistical information is sometimes used as a prop for conclusions unwarranted by the data. After studying this text, the reader should have an improved awareness and ability to detect improper use of statistical methods.

Inferential Statistics

Inferential statistics is a body of techniques used in drawing conclusions concerning some group on the basis of incomplete knowledge. The group of interest is usually termed the population or universe:

population: the set of elements on which information is desired.

The population under study can be relatively small (e.g., the 250 salaried employees of the Normal Company) or very large (e.g., the 70,000 industrial customers served by Midstate Power Company). Of course, the population in a particular study should be clearly identified, or the aims of the investigation may not be realized. Sometimes the desired information on a population is obtained by means of a census.

census: the individual observation of the entire population.

One example of a census is the decennial Census of Population and Housing directed by the U.S. Bureau of the Census, but note that the definition above would encompass much more than federal efforts. Measures computed on the basis of a census are termed *parameters* and are usually symbolized by Greek letters.

Pragmatism frequently dictates that information concerning a population be obtained in a manner other than the census. Time and monetary considerations are important factors in such a decision. Or occasionally, one may encounter measurements that can be obtained only through destructive testing; for example, the life of a light bulb can only be determined by recording the time until burn-out. In such a case, a census is obviously out of the question. Hence, information or conclusions concerning some populations will be based on incomplete knowledge (i.e., something less than census information). For the statistician, this partial knowledge will typically be obtained via sampling.

sample: that portion of the population available for analysis.

A measure computed on the basis of sample data is termed a *statistic* (note the singular) and usually symbolized by a Roman letter. A major portion of this text is concerned with sampling and the use of sample data to reach conclusions concerning the population.

Decision Theory

Statistical decision theory, in the elementary approach taken in Chapter 7, is concerned with the choice of an action from among several possible courses of action. A unique feature of decision theory is the explicit consideration of monetary consequences, or payoffs. For example, suppose Students, Inc., a profit-maximizer, has the program concession for the upcoming football game, and program sales are a function of attendance. The game is sold out; however, the number attending is greatly influenced by the game-day weather. Weather forecasts are available, but, of course, a weather forecast is of the form, "Partly cloudy, with a 40 percent chance of rain." In view of this uncertainty concerning the weather, and hence attendance, how many programs should be printed? Solutions to problems of this type are found in Chapter 7.

1.3 STATISTICS IN BUSINESS

Statistical methods have been applied to problems ranging from business to medicine to agriculture. A superficial examination of the professional literature in almost any field will substantiate the pervasiveness of statistical analysis. Particularly in business disciplines, the observer notes an increasing quantification of knowledge in both theory and practice. A major factor in this phenomenon is the computer. Computers have increased the quantity and complexity of data available to decision makers, and meaningful analyses of large volumes of data are almost impossible without the utilization of statistical methods. Thus, decisions involving such matters as plant expansion, product change, investment actions, and personnel policies are employing statistical information more and more. The business graduate, regardless of major, will likely be of more value to an employer and enhance his or her career prospects if familiar with statistical concepts. Some specific examples should help illustrate our point.

Marketing

Statistical analyses are frequently used in providing information for marketing decisions. *Measuring Markets: A Guide to the Use of Federal and State Statistical Data*[2] contains several case studies illustrating the application of statistics to marketing problems. The data, available from governmental agencies such as the Bureau of the Census, is often reported in such terms as median (see Chapter 2) family income. The problems, e.g., location of a supermarket or the establishment of sales quotas for television sets, involve indexing (Chapter 3), regression analysis (Chapters 12 and 13), and other techniques.

Marketing research is particularly concerned with the gathering and analyzing of data. In a 1975 survey, marketing research managers were asked what specific courses a recruit should have taken. Of the respondents, 97 percent said some statistical training, and 59 percent desired work in advanced statistics.[3] These individuals should be familiar with sample surveys, experimental designs, time series analysis, and other techniques to be introduced in the text.

Management

The analysis of training programs, performance appraisals, and manpower planning are examples of management problems frequently utilizing statistical

---

[2] U.S. Department of Commerce, *Measuring Markets: A Guide to the Use of Federal and State Statistical Data* (Washington: U.S. Government Printing Office, 1974).

[3] A. B. Blankenship, "What Marketing Research Managers Want in Trainees," *Journal of Advertising Research* (vol. 15, 1975), pp. 7–14.

analyses. The Equal Pay Act, Title VII of the Civil Rights Act of 1964, and other laws relating to discrimination in employment have further increased the need for data analysis in personnel departments. For example, the law requires that employment tests be significantly related to job performance; such test validation requires techniques such as regression analysis. Further, many sex discrimination lawsuits are initiated and largely decided on the basis of statistical evidence. To quote Fuentes, "Courts, recognizing the difficulty of establishing discrimination on an individual basis, have placed particular emphasis on the use of statistics to prove discriminatory patterns and practices."[4] Today's managers must detect and rectify such problems because of their social responsibilities as well as their desire to avoid the expense of litigation.

Finance

Financial managers have routine contact with information in numerical form. Financial forecasts, breakeven analyses, and investment decisions under uncertainty are but part of their activities. Since the late 1950s, sophisticated models dealing with inventories, cash balances, etc., have been developed and applied. These models frequently involve regression analysis, probability concepts, variances, expected values, and other topics of statistics.[5] The area of security analysis is highly quantitative; for example, time series data are utilized to detect trends. In common stocks, one of the measures frequently studied is a stock's beta coefficient. This statistic measures how a given stock moves with the stock market as a whole. Beta coefficients can be obtained through regression analysis.

Accounting

Statistical methods are also employed in accounting. In particular, the auditing function makes frequent application of statistical sampling and estimation procedures, and the cost accountant uses regression analysis. The importance of these methods is underscored by their appearance on the semiannual Uniform CPA Examinations. McKenzie's article notes that questions relating to statistical sampling appeared on all eight CPA exams given during 1974–77; questions pertaining to regression analysis were absent from only one of these eight exams. In the Accounting Practice part of the CPA exam, time allocated to questions concerning quantitative methods (statistics and other subjects such as linear programming) represented 6.9 percent of the total time of that part of the 1974–77 exams.[6]

This text introduces the student to statistical methods for use by today's managers. After the reader becomes familiar with these concepts, specific

---

[4] Sonia Pressman Fuentes, "The Law Against Sex Discrimination in Employment and its Relationship to Statistics," *The American Statistician* (vol. 26, April 1972), p. 18.

[5] Peruse a managerial finance text and observe the applications of statistics; for example, look at J. Fred Weston and Eugene F. Brigham, *Managerial Finance* (Hinsdale, Ill.: The Dryden Press, 1978).

[6] Patrick B. McKenzie, "Quantitative Methods on the CPA Examination: An Update," *Collegiate News and Views*, vol. 32, no. 1 (Fall 1978), pp. 10–13.

applications may be cited. The last chapter discusses in detail the applications of statistics in a variety of real business problems.

1.4 HISTORICAL REVIEW

Before beginning with descriptive statistics in the following chapter, a brief historical review of the discipline should be of interest. Probability is often, and correctly, termed the foundation of statistics. Around the 15th century, probability became a subject of interest primarily in connection with the study of various games of chance. The history of games, however, goes much further back in time. For example, games employing dice had existed many centuries prior to the introduction of cards, which themselves appeared in Europe around 1350 A.D. (We note that loaded dice seem to have been known in Roman times.)

One of the older problems considered (*circa* 1500) was of the following form:

> *Problem:* A and B each put $5 in a "pot" and proceed to toss a coin. If six heads are observed first, A wins; if six tails, B wins. After nine tosses, they have observed five heads and four tails. At this point the game is interrupted. What is a fair division of the $10 pot?

The requisite theorems for a solution to the above problem were somewhat slow in developing, and several decades passed before the correct solution was obtained. As a matter of fact, the first book on probability theory did not appear until 1657; the author was a Dutch astronomer, Christianus Huygens. Huygens' work stimulated interest in the field, and major advancements were soon made by such individuals as Jacques Bernoulli (1654–1705), Pierre Montmort (1678–1719), and Abraham de Moivre (1667–1754). (De Moivre in particular should be noted for his development of the normal distribution and the central limit theorem in the early eighteenth century; both topics appear in later chapters of this text.) In 1812, one of the greatest single works on probability was published: Laplace's *Analytical Theory of Probabilities*. This 500-page volume brought together and systematized the existing body of knowledge concerning probability. (Laplace, one of our most interesting personalities, had close friends who were guillotined during the French Revolution, yet was nimble enough to be Napoleon's minister of the interior and then be made a marquis by the restored monarchy.)

A selective mentioning of other work that shall be the subject of subsequent chapters would include the development of regression by Sir Francis Galton (1822–1911). A man of varied interests, Galton also did pioneering work in fingerprint identification. Karl Pearson (1857–1936) developed the chi-square goodness-of-fit test. Finally, Sir Ronald Fisher (1890–1962) made major contributions in the field of experimental design. Galton, Pearson, and Fisher all made significant advances other than those cited above.

We would be remiss if, at some point, we did not mention modern data processing equipment. The history of computational aids would begin with the development of the bamboo rod abacus in China circa 500 B.C. In 1642 Pascal invented the first adding machine, followed by Leibniz's mechanical calculator in 1671. Another major advancement occurred in the 1880s, when Herman

Hollerith developed the punched card and associated devices. The first large-scale electromechanical computer, the Mark I, became operational in 1944, and the initial version of today's electronic computer was constructed in 1946.

We noted earlier the effects of the computer as concerns the quantities of information available to today's manager. But computers have also had a tremendous effect on the nature of statistical analyses in all fields. For example, the requisite theory for applications of a statistical technique such as multiple regression (further discussed in Chapter 13) had been developed many years before real-world applications occurred in any significant number. Why the delay? Because the computational burden presented an almost insurmountable task until the development of the computer. Particularly in the area of multivariate statistics, many advances in theory and applications are taking place.

1.5 STATISTICS COURSES

The preceding discourse has primarily dealt with statistics in general. However, any university catalogue is replete with such courses as Mathematical Statistics I, Psychological Statistics, Statistics for Industrial Engineering, and, of course, Statistics for Business. What is the distinguishing factor among such offerings? Quite simply, it is one of emphasis and level. Each discipline selects statistical topics that are of particular relevance to the field. For example, the industrial engineering student would spend significant time on quality control and acceptance sampling. The psychology course would devote far more time than we to such subjects as experimental design and nonparametrics. The mathematics department would strive for a strong theoretical base with rigorous proofs and an emphasis on probability.

The topics presented in this text were similarly chosen for their relevance to business applications. As we have noted, the accountant employs sampling and estimation concepts in his auditing procedures. The marketing staff will frequently use time series or regression analysis, or both, in an effort to improve sales forecasts, and the personnel manager may choose analysis of variance as an aid in the evaluation of training programs. These and other topics will be introduced as we proceed through the text. Before beginning Chapter 2, the student should review Appendix A on notation, symbols, and summation operations.

1.6 TERMS

Statistics

Population

Census

Parameters

Sample

Statistic

QUESTIONS

1. A company auditor has examined *all* expense account vouchers for the month of June; the average amount per voucher was $230.17. When is this number considered a parameter? A statistic?

2. Distinguish between statistic (singular) and statistics (plural).

3. Define a population that might be encountered by
   a. The marketing manager.
   b. The personnel manager.
   c. An accountant.

4. Describe a business problem for which a census would be impractical because of:
   a. Time constraints.
   b. Money constraints.
   c. Destructive testing.

REFERENCES

Dudycha, Arthur L., and Dudycha, Linda W. "Behavioral Statistics: An Historical Perspective." In Kirk, Roger E., ed., *Statistical Issues, A Reader for the Behavioral Sciences.* Belmont, Cal.: Brooks/Cole Publishing Co., 1972.

Fairley, William B., and Mosteller, Frederick *Statistics and Public Policy.* Reading, Mass.: Addison-Wesley Publishing Co., 1977.

Gerald, Curtis F. *Computers and the Art of Computation.* Reading, Mass.: Addison-Wesley Publishing Co., 1972.

Goodman, A. F. "The Interface of Computer Science and Statistics: An Historical Perspective." *The American Statistician,* vol. 22, no. 3 (June 1968), pp. 17–20.

Kendall, M. G. "Studies in the History of Probability and Statistics." *Biometrika,* vol. 43 (June 1956), pp. 1–13.

Tanur, Judith M., et al., eds. *Statistics: A Guide to the Unknown.* San Francisco: Holden-Day, Inc., 1972.

Descriptive Statistics

2

2 DESCRIPTIVE STATISTICS

Statisticians, this is your problem:
count, measure, compare. . . .

Jean-Jacques Rosseau
The Social Contract (1762)

2.1 INTRODUCTION   The first chapter introduced descriptive statistics as computed measures that are used to characterize (describe) a mass of numerical data. Descriptive statistics enable us to represent concisely a large group of numerical values on the basis of a few summary measures. The definition and usage of such measures are the major subjects of this chapter.

Suppose a manager in a local business firm asks an assistant to provide the key salary data for a report on 150 clerical employees. What salary information should the assistant provide? Perhaps an alphabetical list of employees, showing their salaries, is available. Typically, without considerable study, one could elicit little information from the raw data. Furthermore, users of such information (for example, upper management) are not inclined to bury themselves in data analyses. Rather they prefer, and rightfully, that the data be summarized in a manner that will present the relevant characteristics. Key characteristics might include the smallest and largest salaries, a typical salary, and some comparison of different salaries.

In addition to computed measures, various tables and charts can be used to describe different characteristics of data. Company sales of a conglomerate can be described by a table showing sales in each product line. A chart may be used to depict the rise of the Consumer Price Index over the last two decades. Median family income is a measure used to describe family well-being in your city. These tables, charts, and computed measures describe a large body of data, whether it be sales by various product lines, prices for a large number of items, or incomes of all families in your city.

In summary, managers require information for decision making purposes. In order for them to assimilate information efficiently into the decision making process, data must be available in a convenient, easy-to-read form. This allows the manager quickly to identify the key characteristics of the data. The effective manager, as well as the statistician, must "count, measure, and compare . . ." The descriptive measures discussed in this chapter are used for precisely this purpose. First, frequency distributions and certain charts are examined as methods for summarizing and presenting data. Then various computed measures useful for describing different characteristics of data are discussed. Computed measures are classified as measures of central tendency and measures of variation.

2.2 DATA SOURCES

Data for a statistical study can be obtained through *original investigation* (a census or sample) or from sources of *published data*. The latter are usually less expensive and more efficient to use, provided that the necessary data are accessible. Published data are available from *primary* and *secondary sources*. A primary source is the organization that originally collects and publishes the data. Two private organizations that serve as primary sources are the American Institute for Public Opinion and the A. C. Nielsen Company. They regularly collect and publish data generally referred to as the *Gallup Poll* and the *Nielsen Survey*, respectively. The U.S. Bureau of the Census is a public source of such primary data as the *Census of Population* and the *Census of Agriculture*.

Many governmental and private organizations collect data from various primary sources, organize the information in a different fashion, and then republish it. Such sources are secondary. The annual *Statistical Abstract of the U.S.* is a secondary source of data published by the U.S. Bureau of the Census. It contains data originally collected for the *Census of Population* as well as data obtained from the Interstate Commerce Commission, American Telephone and Telegraph Company, Institute of Life Insurance, etc.

The reference by Goeldner and Dirks at the end of the chapter contains a comprehensive list of data sources.

2.3 FREQUENCY DISTRIBUTIONS

Data collected by original investigation are generally recorded in the form of *raw scores*. For example, the price at which a certain stock closed trading today on the New York Stock Exchange is a raw score measuring price. In order to determine the average price today of any list of stocks, such as utility stocks, it is first necessary to obtain a raw score for each. A collection of raw scores is referred to as *raw data* or *ungrouped data*. The raw scores of 200 freshman males on the mathematics portion of the Scholatic Aptitude Test (SAT) are given in Table 2.1. By glancing at these data, can you determine the typical score for the 200 students? What is the largest and smallest score? It is difficult to answer these questions when the data are presented in their original, ungrouped form. Hence there is a need to summarize and classify the data in a manner that facilitates the determination of their important characteristics.

The first step in classifying and summarizing ungrouped data is to make an *array*. An array is a listing of the data in order of numerical magnitude. The ordered list may range from the smallest raw score to the largest, or vice versa. An array of the 200 SAT-Mathematics scores is given in Table 2.2. A glance at the array shows immediately that the scores range from a low of 318 to a high of 748. One might guess that some score in the neighborhood of 500 is a typical score, but this is only a guess. Further summarization of the data is needed to provide information efficiently. The array would be suitable only if the number of raw scores is small, say 10 to 20.

The fundamental method for tabular summarization of a large volume of data is the *frequency distribution*. A frequency distribution shows the number of raw scores that have numerical values within each of a series of non-overlapping intervals. Table 2.3 is a frequency distribution of the SAT-Mathematics scores. Each interval of scores is termed a *class* of the frequency distribution. The

TABLE 2.1:
SAT-MATHEMATICS
SCORES FOR 200 MALE
FRESHMEN

412	546	359	477	476	599	409	387
506	576	478	543	420	498	591	514
644	491	574	675	625	533	477	611
376	699	533	366	520	337	525	383
574	537	420	462	479	485	566	511
672	472	332	460	468	464	485	463
508	572	477	501	509	599	364	639
408	411	456	422	366	521	641	464
567	521	506	611	524	618	532	318
503	416	475	460	466	400	462	465
356	500	472	359	424	470	552	426
526	457	566	565	542	697	414	613
422	625	527	422	627	472	596	523
561	529	389	486	480	339	645	437
476	470	473	495	521	491	459	509
425	537	540	560	374	480	479	429
515	326	489	440	565	617	681	606
460	638	453	483	454	492	544	533
586	530	531	597	530	748	350	649
418	529	491	381	578	516	562	456
601	518	488	625	503	651	543	538
505	690	586	546	496	428	618	647
436	376	508	435	712	512	542	406
550	546	640	502	546	597	457	539
487	581	521	613	473	600	507	570

TABLE 2.2: ARRAY OF
SAT-MATHEMATICS
SCORES

318	414	460	479	507	531	565	613
326	416	460	480	508	532	566	617
332	418	462	480	508	533	566	618
337	420	462	483	509	533	567	618
339	420	463	485	509	533	570	625
350	422	464	485	511	537	572	625
356	422	464	486	512	537	574	625
359	422	465	487	514	538	574	627
359	424	466	488	515	539	576	638
364	425	468	489	516	540	578	639
366	426	470	491	518	542	581	640
366	428	470	491	520	542	586	641
374	429	472	491	521	543	586	644
376	435	472	491	521	543	591	645
376	436	472	495	521	544	596	647
381	437	473	496	521	546	597	649
383	440	473	498	523	546	597	651
387	453	475	500	524	546	599	672
389	453	476	501	525	546	599	675
400	456	476	502	526	550	600	681
406	456	477	503	527	552	601	690
408	457	477	503	529	560	606	697
409	457	477	505	529	561	611	699
411	459	478	506	530	562	611	712
412	460	479	506	530	565	613	748

TABLE 2.3: FREQUENCY
DISTRIBUTION OF
SAT-MATHEMATICS
SCORES

Scores	Number of students
300 and under 350.	5
350 and under 400.	14
400 and under 450.	23
450 and under 500.	50
500 and under 550.	52
550 and under 600.	25
600 and under 650.	22
650 and under 700.	7
700 and under 750.	2
Total	200

range of scores included in a class is defined by *class limits*. The first class has class limits of 300 and 350. The smaller value 300 is the *lower limit*—and *is* a member—of the class, while the larger value 350 is the *upper limit*—but is *not* a member—of the class. The frequency distribution shows that five of the 200 students have scores equal to or greater than 300 but less than 350.

Two other important characteristics of a frequency distribution are the *class interval* and *midpoint*. The *class interval* represents the width of a class. This may be determined by subtracting the lower limit of one class from the lower limit of the following class (alternatively, successive upper limits may be used). For example, the lower limits of the classes "300 and under 350" and "350 and under 400" are 300 and 350, respectively. Thus the class "300 and under 350" has a class interval of (350 − 300), or 50. Verify that each of the nine classes in this frequency distribution has a class interval of 50.

We define the *midpoint* of a class as the sum of two successive lower limits divided by two. In Table 2.3, the midpoint of the class "300 and under 350" is computed as (300 + 350)/2 = 325. Likewise, the midpoint of the following class is 375.

Does the frequency distribution help us to determine a typical score? A glance at the frequency distribution indicates that more than 100 of the 200 scores are concentrated in the two classes "450 and under 500" and "500 and under 550." Our earlier guess of 500 as a representative score appears reasonable.

2.4 PREPARING A FREQUENCY DISTRIBUTION

A frequency distribution is a compact summary of raw scores that groups the scores by magnitude into a series of classes. Such a table quickly conveys much information about the data and can also be used for further analysis of the data. However, the identity of the exact raw scores is unavailable in a frequency distribution. As an example, what is the numerical value of the smallest raw score in the frequency distribution? This cannot be determined from Table 2.3. We only know that the smallest value is not below 300. The array in Table 2.2 shows 318 as the smallest value. Unless the frequency distribution is carefully prepared, characteristics of the data determined from the frequency distribution may not accurately reflect the raw scores.

A sequence of steps can be followed to assist in preparing a frequency distribution that provides a good representation of the raw scores. These steps are:

1. Array the ungrouped data.
2. Determine the number of classes.
3. Decide on the width of each class.
4. Choose class limits.
5. Record class frequencies.

Some considerations in this process are worthwhile. As a general rule, the number of classes should be between five and twelve. A larger number of scores requires more classes for proper summarization than a small number of scores. Use of less than five classes tends to over-summarize the data, while use of more than twelve usually fails to provide an adequate summary. Equal class intervals, as in Table 2.3, are generally desirable.

Establishing a proper frequency distribution is, in part, a trial-and-error process. Given an initial decision for the width and number of classes, we should examine the resulting frequency distribution. As a result, we may decide to try a different number of classes, width of classes, or a different starting point (lower limit of the first class) for the list of classes. To illustrate, assume we have taken the raw scores in Table 2.1 and developed the array in Table 2.2. How are the number of classes and width of each class determined?

The approximate class interval and number of classes may be examined by using the following relation.

$$\text{Class interval} = \frac{\text{Largest score} - \text{Smallest score}}{\text{Number of classes}} \tag{2.1}$$

The largest score in the array is 748 and the smallest is 318. If we considered a frequency distribution with five classes, the approximate class interval is

$$\text{Class interval} = \frac{748 - 318}{5}$$

$$= \frac{430}{5} = 86.$$

Now the class interval should be some number that is computationally convenient. Rather than a class interval of 86, we might be more comfortable with classes of size 75, 80, or perhaps 100. Thus Equation (2.1) provides only a guide for the class interval, given a certain number of classes.

The next step in constructing the frequency distribution is to choose class limits for one of these intervals. Using class intervals of 100 and starting with a value of 300, the frequency distribution in Table 2.4 is established. Class frequencies are determined by counting the number of raw scores in each class from the array of scores in Table 2.2.

The trial-and-error process of establishing a proper frequency distribution requires us to consider alternatives to Table 2.4. Might class intervals smaller

TABLE 2.4: FREQUENCY
DISTRIBUTION OF
SAT-MATHEMATICS
SCORES (class intervals
of 100)

Scores	Number of students
300 and under 400.	19
400 and under 500.	73
500 and under 600.	77
600 and under 700.	29
700 and under 800.	2
Total	200

75% of data in 2 classes — ∴ condensed too much

than 100, along with more classes, provide a better portrayal of the 200 raw
scores? Using ten classes, Equation (2.1) provides an approximate class interval
of 43. The frequency distribution in Table 2.3 is close to this, having nine classes
with intervals of 50. If we consider 12 classes, the approximate class interval is
35.83. A frequency distribution of the SAT-Mathematics scores with 12 classes
and intervals of 40 is given in Table 2.5.

TABLE 2.5: FREQUENCY
DISTRIBUTION OF
SAT-MATHEMATICS
SCORES (class intervals
of 40)

Scores	Number of students
300 and under 340.	5
340 and under 380.	10
380 and under 420.	13
420 and under 460.	21
460 and under 500.	43
500 and under 540.	42
540 and under 580.	26
580 and under 620.	19
620 and under 660.	13
660 and under 700.	6
700 and under 740.	1
740 and under 780.	1
Total	200

too much detail

Which of the three distributions provides the best tabular summary of the
SAT-Mathematics scores? Table 2.4, with five classes, tends to over-summarize
the data. More than half of the 200 scores are in two adjacent classes, "400 and
under 500" and "500 and under 600." Very little information about the differences in scores is conveyed by this distribution. On the other hand, Table 2.5
retains too many characteristics of the raw scores. It is more difficult to glance
over 12 classes and easily determine data characteristics. Thus Table 2.3
provides a compromise between the distributions having "too few" and "too
many" classes. This distribution with nine classes quickly conveys characteristics
of the data and provides a good summary while retaining the observed differences in the grouped scores.

In conclusion, there is usually no one *best* frequency distribution for a set of
data, although some are better than others. The construction of a frequency
distribution leaves much to the subjective judgment of the analyst.

2.5 TYPES OF DATA AND FORMS OF FREQUENCY DISTRIBUTIONS

Data for statistical summary and analysis exist in many different forms. For our purposes here, the different types of data may be classified as follows:

1. Quantitative data.
   a. Discrete.
   b. Continuous.
2. Qualitative data.

Any data that possess the characteristic of numerical magnitude are *quantitative*. Your score on an examination, the SAT-Mathematics scores in Table 2.1, an executive's salary, and the hourly production rate of a factory worker are all examples of data that are characterized by numerical magnitude. Numerical magnitude further implies order, which allows us to classify data into an ordered sequence of classes as a frequency distribution. Since quantitative data are characterized by different numerical values, the data represent the values of a *variable*. The quantitative data in Table 2.3 represent 200 values of the variable "SAT-Mathematics Scores."

Qualitative data are characterized by exhaustive and distinct categories that do not possess magnitude. For example, all employees of a small business might be classified into an exhaustive list of three distinct categories identified as managerial, clerical, or production. These three categories possess no numerical characteristics. The categories are exhaustive since all employee functions are accounted for in these three job descriptions.

An example of a frequency distribution with qualitative data is given in Table 2.6. These findings result from a 1976 survey of 74,000 executives of leading U.S. businesses that was designed, in part, to determine where executives received their undergraduate and graduate degrees. The categories, or classes of the frequency distribution, do not have numerical characteristics. For qualitative data, some characteristic other than magnitude must be used to determine a sequence for listing the categories that represent classes. Alphabetical order may be appropriate in some cases. Here the schools are ranked by the number of executives receiving degrees from each school.

Quantitative data may be further classified into one of two types, discrete or continuous. The term *discrete data* refers to quantitative data that are limited

TABLE 2.6: THE TOP TEN SCHOOLS IN EDUCATION OF LEADING EXECUTIVES OF U.S. CORPORATIONS IN 1976

Institution	Number of Executives
Harvard University	5,017
New York University	2,502
Yale University	2,271
University of Pennsylvania	1,863
University of Michigan	1,762
Columbia University	1,712
Northwestern University	1,468
City College (CUNY)	1,454
Princeton University	1,404
University of Wisconsin	1,308

Source: Appeared originally in *The Chronicle of Higher Education.* Reprinted with permission.

to certain numerical values of a variable. Examples of discrete variables include population, personal weekly earnings, and class enrollments. Values of variables such as population and class enrollment are limited to whole numbers. A state population may be 3,205,665 but not a fractional value such as 3,205,665.8. A university class may enroll 30 or 31 students but not 30.5. Personal weekly earnings of a worker paid an hourly wage of $6.26 may total $250.40 but not an amount of $250.425. An example of a frequency distribution containing discrete data, the number of production workers employed at various plant locations of an international corporation, is shown in Table 2.7.

TABLE 2.7: NUMBER OF PRODUCTION EMPLOYEES AT PLANTS OF UCOR, INC.

Number of Workers	Number of Plants
50–59	5
60–69	11
70–79	22
80–89	9
90–99	3
Total	50

Notice the difference in the form of the class intervals in Figures 2.3 and 2.7. The class "50–59" in Table 2.3 is read as 50 through 59. If the form of classes used in Table 2.3 is used here, the first class would read "50 and under 60." In Table 2.7 the upper limit of any class is one unit less than the lower limit of the following class. This form of class statement is often, but not always, used for discrete data because the "space" between class limits implies that the variable is limited to certain values.

Continuous data can take on all values of the variable. The time required for an assembly operation in a factory is a continuous variable. Time may be recorded as one minute, 1.1 minute, 1 minute and 4 seconds, etc. The variable "time" is not limited to certain numerical values. Other examples of continuous variables include weight, distance, and volume. A frequency distribution for continuous data, time required for ticketing at an airline service counter, is shown in Table 2.8.

Notice in Table 2.8 that the "and under" form of frequency distribution is preferred for continuous data. There is no "space" between class limits, which

TABLE 2.8: TIME REQUIRED FOR TICKETING AT A SERVICE COUNTER OF BLUEBIRD AIRLINES

Time (minutes)	Number of Customers
2 and under 6	9
6 and under 10	15
10 and under 14	28
14 and under 18	21
18 and under 22	6
22 and over	1*
Total	80

open-ended class (determined by values of data)

* This customer required 42 minutes.

implies that the data may take on all continuous values of the variable. A time of 5.9975 minutes could be recorded in the first class while a time of 6.0 minutes belongs to the second class.

If the form of classes given in Table 2.7 is used for continuous data, then one must recognize that the actual data may have values outside the stated class limits. If, for example, continuous data is recorded in a class stated as 50–59, the data may take on values between 49.5 and 59.5. The latter values are "real" class limits, which should be used in any analysis of the frequency data.

The last class in Table 2.8, "22 and over," is called an *open-end class.* This class is "open" at the upper end, i.e., it has no upper limit. Open-end classes may be used to include those values of a variable in a frequency distribution that are extreme in relation to the other values. In Table 2.8, 79 of the 80 customers took less than 22 minutes for ticketing, and one required 42 minutes. The time of 42 minutes is extremely large relative to the other times. To incorporate all observations, a frequency distribution is prepared with equal class intervals for the smallest 79 items, and an open-end class is used for the extreme item. Extremely small items could be included in a class that is "open" at the lower end. Suppose all the scores in Table 2.3 remain the same except that the smallest is 150. This score could be included in an open-end class "under 300." In general, equal class intervals are preferred. However when extreme values exist, unequal class intervals may be more suitable.

2.6 CUMULATIVE AND RELATIVE FREQUENCIES

It is often useful to express class frequencies in a different fashion. Rather than listing the actual frequency opposite each class, it may be appropriate to list either cumulative frequencies, relative frequencies, or cumulative relative frequencies. In Table 2.9, two columns are added to the original distribution of SAT-Mathematics Scores from Table 2.3. Column (3) lists the number of students who scored less than the upper limit of each class. For example, 92 of the 200 students scored less than 500. The number of students with scores equal to or more than the lower limit of each class is given in Column (4). As an illustration, 56 students have scores of at least 550, i.e., equal to or more than 550. The frequencies given in Columns (3) and (4) are termed, respectively,

TABLE 2.9: CUMULATIVE DISTRIBUTION OF SAT-MATHEMATICS SCORES

(1) Scores	(2) Number of Students	(3) Less Than Upper Class Limit	(4) Equal to or More Than Lower Class Limit
300 and under 350	5	5	200
350 and under 400	14	19	195
400 and under 450	23	42	181
450 and under 500	50	92	158
500 and under 550	52	144	108
550 and under 600	25	169	56
600 and under 650	22	191	31
650 and under 700	7	198	9
700 and under 750	2	200	2
Total	200		

less-than cumulative frequencies and *equal-to-or-more-than cumulative frequencies*. Such cumulative frequencies conveniently provide cumulative totals relative to the class limits.

A *relative frequency* distribution presents the proportion of the total number of observations that belongs to each class. Relative frequencies for the SAT-Mathematics Scores are shown in Column (3) of Table 2.10. The proportion of students scoring "450 and under 500" is 50 out of 200, or 0.250. Since a percentage figure is a proportion multiplied by 100, we may also say that 25 percent of the students scored "450 and under 500."

TABLE 2.10:
RELATIVE
FREQUENCY
DISTRIBUTION OF
SAT-MATHEMATICS
SCORES

(1) Scores	(2) Number of Students	(3) Proportion of Students
300 and under 350	5	0.025
350 and under 400	14	0.070
400 and under 450	23	0.115
450 and under 500	50	0.250
500 and under 550	52	0.260
550 and under 600	25	0.125
600 and under 650	22	0.110
650 and under 700	7	0.035
700 and under 750	2	0.010
Total	200	1.000

× 100 = %

Occasionally we may wish to examine the proportion or percentage of observations larger than or smaller than some amount. The proportion of students with SAT-Mathematics Scores less than the upper limit of each class is given in Column (2) of Table 2.11. These are *cumulative relative frequencies*. Likewise, the proportion of students scoring equal to or more than the lower limit of each class is given in Column (3). The cumulative relative frequencies may be determined from either the list of cumulative frequencies in Table 2.9 or the list of relative frequencies in Table 2.10. Using Table 2.11, we see that 79 percent of the students scored at least 450. What proportion of students scored less than 700?

TABLE 2.11:
CUMULATIVE RELATIVE
FREQUENCIES OF
SAT-MATHEMATICS
SCORES

(1) Scores	(2) Proportion Less Than Upper Class Limit	(3) Proportion Equal to or More Than Lower Class Limit
300 and under 350	0.025	1.000
350 and under 400	0.095	0.975
400 and under 450	0.210	0.905
450 and under 500	0.460	0.790
500 and under 550	0.720	0.540
550 and under 600	0.845	0.280
600 and under 650	0.955	0.155
650 and under 700	0.990	0.045
700 and under 750	1.000	0.010

52% made 500 or more

Source: Table 2.9.

All three variations of frequency lists, the cumulative frequency, relative frequency, and cumulative relative frequency, simply express the data in a different fashion. Such variations accommodate the many different uses for descriptive summaries of data.

2.7 CHARTS OF FREQUENCY DISTRIBUTIONS

Graphical displays are excellent methods for presenting data. A visual portrayal of a frequency distribution enables one quickly to discern points of concentration as well as differences within the observations. The four principal charts of frequency distributions are the bar chart, histogram, frequency polygon, and ogive.

Bar charts are useful for portraying frequency distributions of qualitative data. Figure 2.1 is a *horizontal bar chart* for the qualitative data in Table 2.6. The length of each bar represents the number of executives receiving degrees from a particular school.

FIGURE 2.1: THE TOP TEN SCHOOLS IN EDUCATION OF LEADING EXECUTIVES OF U.S. CORPORATIONS IN 1976

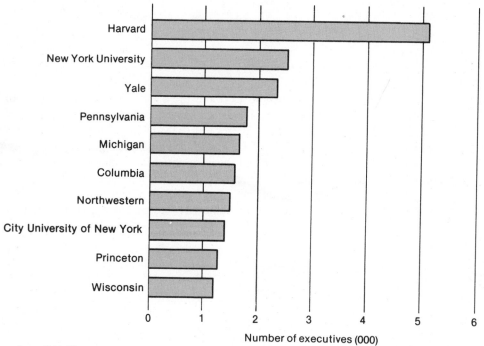

Source: Table 2.6.

A chart for quantitative data is based on a two-dimensional coordinate system. The scale for the horizontal axis is constructed from the class limits of the frequency distribution. Class frequencies represent the vertical dimension. The chart portraying discrete quantitative data is called a *histogram*. A histogram uses vertical bars drawn between class limits so that the height of each bar represents the class frequency. A histogram for the number of production employees at plants of UCOR, Inc., is shown in Figure 2.2. The histogram is especially suited for discrete data since the bars clearly distinguish among classes, implying that the variable measured is discontinuous.

FIGURE 2.2:
DISTRIBUTION OF
PRODUCTION
EMPLOYEES AT PLANTS
OF UCOR, INC.

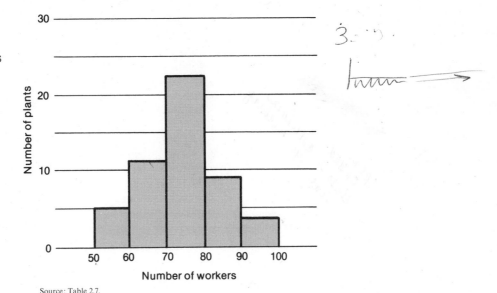

Source: Table 2.7.

A line chart called a *frequency polygon* is used to represent a frequency distribution with continuous data.[1] A frequency polygon of the ticketing time for 80 customers at an airline service counter is shown in Figure 2.3. The polygon is formed by connecting points by a series of straight lines. The horizontal and vertical coordinates of the points are, respectively, the class midpoint and class frequency.

Notice in Figure 2.3 that a footnote is used to identify the ticketing time for the customer included in an open-end class. A similar procedure could be used for open-end classes in a histogram.

As the number of observations in a frequency distribution increases, the frequency polygon often becomes more of a smooth curve. If the ticketing times in Figure 2.3 represent a sample of 80 customers, then the times for the population of customers may approach the smooth curve in Figure 2.4. Smoothed curves of data are used extensively throughout statistical analysis. Our study of statistical concepts in the remainder of the text utilizes many different shapes of smoothed curves.

Ogives are line charts of cumulative frequency distributions. Either relative cumulative frequencies or the actual cumulative frequencies may be used. Both

---

[1] Even though the histogram is usually associated with discrete data and a frequency polygon is appropriate for continuous data, this distinction is not always followed in practice. Subjective considerations play a role in selecting the type of chart useful for portraying a set of data.

FIGURE 2.3: TIME
REQUIRED FOR
TICKETING AT A
SERVICE COUNTER OF
BLUEBIRD AIRLINES

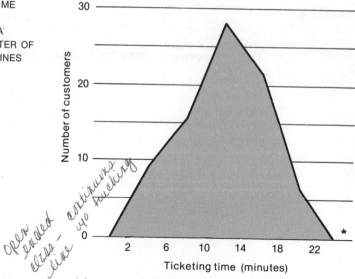

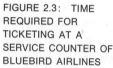

* One customer required 42 minutes.
Source: Table 2.8.

FIGURE 2.4:
POPULATION OF TIMES
REQUIRED FOR
TICKETING AT A
SERVICE COUNTER OF
BLUEBIRD AIRLINES

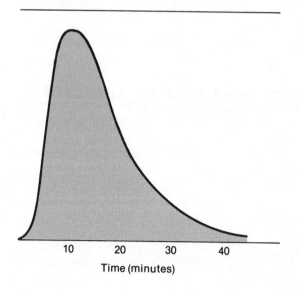

are used in Figure 2.5 for the data on SAT-Mathematics scores. The vertical axis on the left side represents the number of students, and that on the right side represents the percentage of students. For the *less-than ogive,* the less-than cumulative frequencies are plotted opposite the upper class limits. The equal-to-or-more-than cumulative frequencies are plotted opposite the lower class limits to determine the *equal-to-or-more-than ogive*. From the ogives, we see that 75 percent of the students, or 150, scored at least 460 (approximately). Likewise, 92 students scored less than 500. Fifty percent of the students scored equal to or more than what amount?

FIGURE 2.5:
CUMULATIVE
DISTRIBUTIONS OF
SAT-MATHEMATICS
SCORES

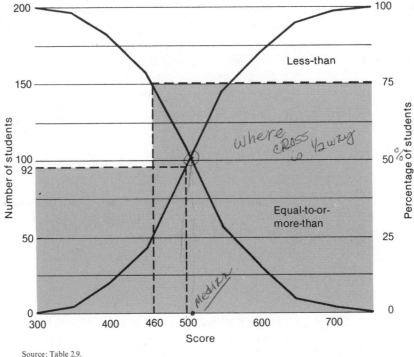

Source: Table 2.9.

The frequency distribution is shown in the previous sections as the primary method for summarizing raw scores in tabular form. Frequency distributions may provide actual, relative, or cumulative frequencies. Graphical displays of a frequency distribution may be histograms, frequency polygons, or ogives. The following sections present various computed measures used to describe statistical characteristics of data. The measures are computed from both raw scores and frequency distributions.

2.8 MEASURES OF CENTRAL TENDENCY

A *measure of central tendency* is a single number used to represent a group of data. In Section 2.3 an SAT-Mathematics score of "about 500" was called a typical score. In other words, the value 500 is a single number used to represent or typify the scores for all 200 students. The most representative single value of a group of data is sought as a measure of central tendency. Since different data sets possess different numerical characteristics, one measure of central tendency may better represent a certain group of data than another. The three principal measures of central tendency are the arithmetic mean, the median, and the mode.

2.9 ARITHMETIC MEAN

If a retail store manager is asked "How is business?" he may reply "Fairly good; average daily sales are about $750." An investor buying shares of a certain stock at three different prices (on different dates) may pay an average price per share of $43.25. A student's average score on all examinations in a course may be 86.4. In all three instances, the term "average" is used as a typical or representative amount. This general use of the term "average" is a layman's expression for the *arithmetic mean*, a statistical measure of central tendency.

Ungrouped Data

> The *arithmetic mean* of a group of raw scores is the sum of the raw scores divided by the number of scores.

The arithmetic mean of a sample of n observations, called the *sample mean*, is

$$\bar{X} = \frac{\sum_{i=1}^{n} X_i}{n}. \tag{2.2}$$

The symbol for the sample mean, $\bar{X}$, is read as "X-bar".[2] If the set of data is not a sample, but rather the entire population of N items, the *population mean*, μ (Greek letter "mu"), is:

$$\mu = \frac{\sum_{i=1}^{N} X_i}{N}.$$

The subscripts may be omitted with the understanding that the summation is taken over all values of the index i. This practice is followed throughout the text except where inclusion of subscripts is helpful. Consequently, the sample and population means are written as

$$\bar{X} = \frac{\sum X}{n}$$

and

$$\mu = \frac{\sum X}{N}.$$

EXAMPLE. A sample of last year's personnel records shows the number of days' sick leave (X) for each of five employees as 4, 3, 8, 7, 3. The arithmetic mean, using Equation (2.2), is:

$$\bar{X} = \frac{\sum X}{n}$$

$$= \frac{X_1 + X_2 + X_3 + X_4 + X_5}{n}$$

$$= \frac{4 + 3 + 8 + 7 + 3}{5}$$

$$= \frac{25}{5} = 5.$$

---

[2] The reader unfamiliar with summation notation, i.e., uses of the symbol $\sum$, should review Appendix A before proceeding further.

**Mathematical
Properties**

The arithmetic mean possesses certain mathematical properties that will prove useful in later sections. First, the sum of the deviations of the raw scores about the mean is zero.[3] That is, $\sum(X - \bar{X}) = 0$.

Second, the sum of the squared deviations of the raw scores about the mean is a minimum, i.e., $\sum(X - \bar{X})^2$ is a minimum. Conceptually, this second property indicates that the arithmetic mean, rather than some other measure of central tendency, is the measure that minimizes the sum of squared deviations.[4] The usefulness of this property will be apparent in later discussions.

Third, if a new variable Y is defined by a linear transformation on the original variable X, then the mean of Y is given by the same transformation on the mean of X. Symbolically, if

$$Y = a + bX,$$

then

$$\bar{Y} = a + b\bar{X},$$

where a is the additive constant and b is the multiplicative constant.[5]

---

[3] This may be demonstrated algebraically:

$$\begin{aligned} \sum(X - \bar{X}) &= \sum X - \sum \bar{X} \\ &= \sum X - n\bar{X} \\ &= n\bar{X} - n\bar{X} \\ &= 0. \end{aligned}$$

[4] To show that $\bar{X}$ is the quantity A that minimizes the sum of squared deviations (SSD), let

$$\text{SSD} = \sum(X - A)^2.$$

The derivative of SSD with respect to A is

$$\frac{d(\text{SSD})}{dA} = -2\sum(X - A).$$

Set the above equal to zero and solve for A;

$$\begin{aligned} -2\sum(X - A) &= 0 \\ \sum X - nA &= 0 \\ A = \frac{\sum X}{n} &= \bar{X}. \end{aligned}$$

[5] Proof:

$$\bar{Y} = \frac{\sum Y}{n} = \frac{\sum(a + bX)}{n}$$

$$= \frac{1}{n}\left(\sum a + b\sum X\right)$$

$$= \frac{1}{n}\left(na + b\sum X\right)$$

$$= \frac{na}{n} + \frac{b\sum X}{n}$$

$$= a + b\bar{X}.$$

EXAMPLE. Weekly production (X) of an item for a sample of four weeks is given in Column 1 of Table 2.12. Find average weekly production ($\bar{X}$), $\sum(X - \bar{X})$, and $\sum(X - \bar{X})^2$. Additionally, the production cost Y (in dollars) is related to units produced (X) by the equation $Y = 500 + 2X$. Determine $\bar{Y}$ first by finding Y for all X, and second, by the relation $\bar{Y} = a + b\bar{X}$. A tabular format is used in Table 2.12 for the solution. The results show $\bar{X} = 21$, $\sum(X - \bar{X}) = 0$, and $\sum(X - \bar{X})^2 = 98$. The values of Y, shown in Column (4), have a mean, $\bar{Y}$, of 542. Further, $\bar{Y} = 500 + 2(21) = 542$. According to the third property, $\bar{Y}$ may be found from a knowledge of the linear function and $\bar{X}$; the individual Y values, the production cost per week, are not required.

TABLE 2.12: WEEKLY PRODUCTION DATA

(1) Units Produced per Week (x)	(2) (x − x̄)	(3) (x − x̄)²	(4) Production Cost per Week (y)
23	+2	4	546
18	−3	9	536
28	+7	49	556
15	−6	36	530
Total 84	0	98	2,168
Mean 21	—	—	542

Weighted Arithmetic Mean

The arithmetic mean, as discussed in the previous section, gives equal importance (or weight) to each of the raw scores in a set of data. In some cases, all raw scores do not have the same importance. A weighted mean is used to assign any degree of importance to each raw score by choosing an appropriate set of weights. The *weighted arithmetic mean*, $\bar{X}_w$, of n items is:

$$\bar{X}_w = \frac{\sum wX}{\sum w},\qquad (2.3)$$

where the w are the weights assigned to the raw scores.

EXAMPLE. A student's "average" grade (the mean) in History 101 is determined by counting each of three hour quizzes equally and counting the final examination twice as much as an hour quiz. Harry's scores, X, on the three hour quizzes are 72, 81, 75, and his final exam score is 88. Harry's mean grade, using Equation (2.3), is

$$\bar{X}_w = \frac{\sum wX}{\sum w}$$

$$= \frac{1(72) + 1(81) + 1(75) + 2(88)}{1 + 1 + 1 + 2}$$

$$= \frac{404}{5} = 80.8.$$

The choice of the weights is arbitrary as long as the relative weights remain the same. Weights for the hour quizzes could be 20 percent each with 40 percent for the final exam, and the weighted mean would be the same. Note that the mean grade with equal weights for the four scores, from Equation (2.1), is 79; weighting may mean the difference in a letter grade of B or C for Harry.

A variation of Equation 2.3 arises when a set of raw scores, some of which are equal in numerical value, are grouped according to the distinct numerical values. Suppose a fund drive for a local charity is under way and the eight employees in the administrative office of a local firm contribute a total of $110 in the following amounts: $15, $10, $5, $25, $30, $10, $5, $10. The mean contribution per employee, using Equation (2.2), is $13.75. The contributions may be grouped according to the different amounts as shown in Table 2.13.

TABLE 2.13: EMPLOYEE CHARITABLE CONTRIBUTIONS

(1) Amount of Contribution (x)	(2) Number of Employees (f)	(3) (fx)
5	2	10
10	3	30
15	1	15
25	1	25
30	1	30
Total	8	110

The first two columns show the number of employees contributing various amounts. This is a *frequency count* of raw scores. Column (3), determined as the product of the values in Columns (1) and (2), is summed to yield the total contribution of $110. Again, the mean is $110/8 = $13.75. The mean contribution per employee can be expressed as

$$\bar{X} = \frac{\sum fX}{n},\tag{2.4}$$

where X is a raw score, f is the frequency of a raw score, and n is the total number of raw scores (which is the same as $\sum f$). The above formula is essentially the same as Equation (2.3), except w is replaced by f. However, Equation (2.4) is not a weighted mean; it provides the same result as Equation (2.1) when there is a frequency count of raw scores.

Grouped Data

The arithmetic mean of a frequency distribution is computed from a slight modification of Equation (2.4). This is given by Equation (2.5),

$$\bar{X} = \frac{\sum fX}{n},\tag{2.5}$$

where the f is a class frequency, X is a class midpoint, and n is the total number of observations ($\sum f$).[6]

---

[6] So called "short-cut" methods for calculating various measures were popular before the widespread use of electronic calculators and computing equipment. The reader interested in these may consult the references to Shao and to Stockton and Clark at the end of the chapter.

EXAMPLE. Find the arithmetic mean of the SAT-Mathematics Scores given in Table 2.14. Using Equation (2.5) and the computations in Table 2.14,

$$\bar{X} = \frac{\sum fX}{n} = \frac{102,000}{200} = 510.$$

Recall our "guess" of a typical score in Section 2.3 was 500.

TABLE 2.14:
COMPUTATION OF
ARITHMETIC MEAN FOR
SAT-MATHEMATICS
SCORES

(1) *Scores*	*(2)* *Number of* *Students* *(f)*	*(3)* *Midpoint* *(x)*	*(4)* *(fx)*
300 and under 350	5	325	1,625
350 and under 400	14	375	5,250
400 and under 450	23	425	9,775
450 and under 500	50	475	23,750
500 and under 550	52	525	27,300
550 and under 600	25	575	14,375
600 and under 650	22	625	13,750
650 and under 700	7	675	4,725
700 and under 750	2	725	1,450
Total	200		102,000

Source: Table 2.3.

Is the mean of a frequency distribution a good approximation to the mean of the original ungrouped data? In the application of Equation (2.5) to a frequency distribution, the midpoint of each class, X, is used to represent all the items in a class. If for all classes, the arithmetic mean of the raw scores in a class equals the class midpoint, then $\sum f$ will equal the original total of all the raw scores used to develop the frequency distribution. Hence the arithmetic mean of the frequency distribution and the arithmetic mean of the raw scores will be identical. However, it is difficult to have the midpoint equal the average of the raw scores for all classes. Even so, the differences tend to balance one another out. Adherence to the rules in Section 2.4, combined with reasonable judgment, produces a frequency distribution that adequately summarizes raw scores for either presentation or computational purposes.

How is the arithmetic mean computed if a frequency distribution contains an open-end class? In order to compute this, the raw scores included in an open-end class (or their mean) must be known. The mean of the raw scores in the open-end class is used as the midpoint, X, for the class.

EXAMPLE. Using the data in Table 2.15, compute the arithmetic mean of the time required for ticketing by 80 customers at a service counter of Bluebird Airlines. The computations in Table 2.15 yield:

$$\bar{X} = \frac{\sum fX}{n} = \frac{990}{80} = 12.375 \text{ minutes.}$$

TABLE 2.15:
COMPUTATIONS OF
ARITHMETIC MEAN FOR
BLUEBIRD AIRLINES

(1) Time *(minutes)*	*(2)* Number of Customers *(f)*	*(3)* Midpoint *(x)*	*(4)* *(fx)*
2 and under 6.	9	4	36
6 and under 10.	15	8	120
10 and under 14.	28	12	336
14 and under 18.	21	16	336
18 and under 22.	6	20	120
22 and over	1*	42	42
Total	80		990

* This customer required 42 minutes.

JACQUES QUETELET (1796–1874)

During his early life in Ghent, Belgium, Quetelet pursued his interest in art and poetry while teaching mathematics in a secondary school. Turning to the study of advanced mathematics, he received a doctorate from the University of Ghent at age 23. He was active as an author, mathematician, physicist, public lecturer, and university teacher. Quetelet is best known for the descriptive studies of social characteristics of the Belgian population he made while serving as head of the Belgian Statistical Commission. His concept of *homme moyen*, or "average man," embodied the idea that an average may be used to represent a group of differing numerical values. Because of this work, he is known as the "father of descriptive statistics."

2.10 MEDIAN

A second measure of central tendency is the median. The median is widely used for summarizing income data. For example, the Bureau of the Census reports that median family income in the United States rose from $9,867 in 1970 to $13,719 in 1975. When data are listed in order by rank, the median is also a useful descriptive measure.

Ungrouped Data

The *median* is the middle value in an array of raw scores.

Hourly wage rates for seven employees are given below along with an array of the wages. With seven raw scores, the middle score in the array is the fourth, $6.65, which is the median wage.

Hourly Wage	Array	
$6.65.	$3.92	
9.21.	4.40	
5.50.	5.50	
4.40.	6.65	(median)
3.92.	6.80	
6.80.	7.25	
7.25.	9.21	

If the number of raw scores is odd, as above, the median is equal to one of the original observations. If the number of raw scores is even, the median is the arithmetic mean of the two middle scores in the array.

An investment firm purchased the same stock at six different times during the last year, paying the following prices per share (a price of \$15.125 is often quoted as $15\frac{1}{8}$):

Price/Share	Array
\$17.750	\$10.500
12.250	12.250
15.125	14.625
14.625	15.125
10.500	16.000
16.000	17.750

The two middle items in the even-numbered array are \$14.625 and \$15.125. The median price per share is (\$14.625 + \$15.125)/2 = \$14.875.

Mathematical Property

The important mathematical property of the median is that the sum of the absolute deviations about the median, Md, is a minimum, i.e., $\sum |X - Md| =$ a minimum. Recall that the arithmetic mean minimizes the sum of *squared* deviations. The median is the measure of central tendency that minimizes the sum of the *absolute* deviations.

EXAMPLE. For the six security prices (X) given previously, compare the sum of squared deviations and sum of absolute deviations about both the arithmetic mean (\$14.375) and the median (\$14.875).

(1) X	(2) $|X - \bar{X}|$	(3) $|X - Md|$	(4) $(X - \bar{X})^2$	(5) $(X - Md)^2$
\$17.750	3.375	2.875	11.391	8.266
12.250	2.125	2.625	4.516	6.891
15.125	0.750	0.250	0.562	0.062
14.625	0.250	0.250	0.062	0.062
10.500	3.875	4.375	15.016	19.141
16.000	1.625	1.125	2.641	1.266
Total	12.000	11.500	34.188	35.688

The sum of the absolute deviations about the median is smaller than the sum of the absolute deviations about the mean (Column [3] versus Column [2]). Comparing Columns (4) and (5), we see that the sum of squared deviations about the mean is less than the sum of squared deviations about the median.

Grouped Data

The *median of a frequency distribution* is defined as the value such that half of the observations are numerically less than or equal to this value.

The following formula is used.

$$Md = L + i\left(\frac{n/2 - F}{f}\right),$$ (2.6)

where

L = lower limit of the median class,
i = class interval of the median class,
n = total number of observations ($\sum f$),
F = cumulative frequency less than the median class, and
f = actual frequency of the median class.

TABLE 2.16: LESS-THAN CUMULATIVE FREQUENCIES OF SAT-MATHEMATICS SCORES

(1) Scores	(2) Number of Students	(3) Less Than Upper Class Limit
300 and under 350	5	5
350 and under 400	14	19
400 and under 450	23	42
450 and under 500	50	92
500 and under 550	52	144
550 and under 600	25	169
600 and under 650	22	191
650 and under 700	7	198
700 and under 750	2	200
Total	200	

Source: Table 2.9.

Less-than cumulative frequencies for the distribution of SAT-Mathematics Scores are shown in Table 2.16. The first step in computing the median is to locate the median class, which is the class that contains the $(n/2)$nd item. For Table 2.16, $n/2 = 200/2 = 100$. Since there are 92 scores less than 500 and 144 scores less than 550, the 100th score in numerical order is contained in the class "500 and under 550." Equation (2.6) is expressed in terms of this class, the median class. Substituting,

$$Md = 500 + 50\left(\frac{100 - 92}{52}\right)$$

$$= 500 + 7.69$$

$$= 507.69.$$

The result indicates that half of the 200 scores are less than or equal to 507.69.

The median may be approximated graphically from a chart of cumulative frequencies. Less-than cumulative frequencies for the SAT-Mathematics Scores are plotted in Figure 2.6. To locate the median, determine $n/2$ as in Equation (2.6), which is 200/2, or 100 students. From this value on the vertical axis, draw a horizontal line to intersect with the curve of less than cumulative frequencies; at that point, a vertical line drawn to the horizontal scale approximates the median, 507.69. The accuracy of the approximation depends on the accuracy

FIGURE 2.6: LESS-THAN
CUMULATIVE
FREQUENCIES FOR
SAT-MATHEMATICS
SCORES

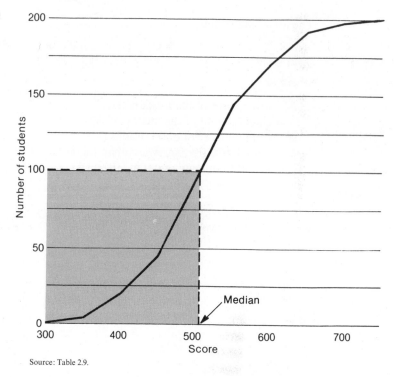

Source: Table 2.9.

of the chart. Note also in Figure 2.5 that a vertical line drawn from the intersection of cumulative distributions to the horizontal axis locates the same point.

2.11   MODE

While the term *mode* may not be familiar, interpretations of this measure of central tendency are encountered daily. Accountants, attorneys, and physicians often charge so-called "usual and customary" fees for their services. Likewise, a prevailing rate for an automobile tune-up exists in most localities. Usual, customary, and prevalent are common interpretations of the mode.

> The *mode* is the most frequently occurring value in a group of raw scores.

Quarterly dividends for each share of a stock paid during the last two years are (in cents), in order of payment:

$$15, 18, 20, 15, 15, 20, 25, 15.$$

The mode is 15 cents per share since this dividend was paid in four of the eight quarters, more frequently than any other amount. However, the mode is not necessarily a unique value. Eight quarterly dividends (in cents) for a second company are:

$$10, 12, 10, 10, 8, 12, 12, 14.$$

This set of data is said to be *bimodal* with modes of 10 and 12 cents, each occurring three times. In fact, the mode may not exist in some groups of raw scores. If there is not a value that occurs more frequently than others, there is no mode. The mode is of limited use in statistics and is not discussed further.

2.12 PERCENTILES   Examination scores such as SAT scores are often expressed in terms of percentiles. For example, a score of 582 is the 80th percentile of the distribution of SAT-Mathematics Scores given in Table 2.16.

> The *j*th *percentile* is defined as the value such that *j* percent of the scores have numerical values less than or equal to this amount.

Percentiles are particularly useful descriptive measures for large data sets where the data are given by a frequency distribution. The *j*th percentile, P_j, of a frequency distribution is:

$$P_j = L + i\left(\frac{\frac{jn}{100} - F}{f}\right) \tag{2.7}$$

where

    j = the desired percentile,
    L = lower limit of the desired percentile class,
    i = class interval of the desired percentile class,
    n = total number of observations ($\sum f$),
    F = cumulative frequency less than the desired percentile class, and
    f = actual frequency of the desired percentile class.

    EXAMPLE.   Verify with Equation (2.7) that the 80th percentile of the SAT-Mathematics Scores given in Table 2.16 is 582. Solution: First locate the 80th percentile class by computing $(jn)/100 = (80)(200)/100 = 160$. The 80th percentile class is the class containing the 160th item in numerical value. Using the less-than cumulative frequencies in Table 2.16, this is the class "550 and under 600." Substituting into Equation (2.7),

$$P_{80} = 550 + 50\left(\frac{160 - 144}{25}\right)$$

$$= 550 + 32$$

$$= 582.$$

Eighty percent of the SAT-Mathematics scores are less than or equal to 582. This may be verified graphically by examining Figure 2.5.

    The similarity between Equation (2.7) and that for the median, Equation (2.6), is not accidental. The median is the 50th percentile. Essentially, the median divides a distribution into two parts whereas percentiles divide a distribution into 100 parts. Related measures include quartiles, which divide the data into four parts, and deciles, which provide ten parts. All of these measures may be determined from Equation (2.7).

2.13 MEASURES OF VARIATION

Descriptive statistical measures are used to summarize the key characteristics of data. Measures of *central tendency* describe typical or representative values. *Variation*, a second characteristic of data, is used to measure differences in the data. Compare the following two groups of raw scores.

Group I: 52, 56, 60, 64, 68.

Group II: 40, 50, 60, 70, 80.

For both groups, the arithmetic mean and median are 60, yet the two groups of scores are obviously different in terms of variability. The scores in Group II are more varied, or dispersed, than the Group I scores. This difference in variability is illustrated by the following data plots.

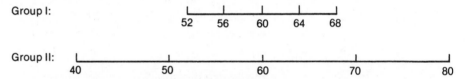

Group I:

52   56   60   64   68

Group II:

40   50   60   70   80

Measures of variation are divided into two categories. *Measures of absolute variation* are expressed in terms of the original data whereas *measures of relative variation* are expressed as percentages or ratios. The measures of absolute variation are the range, average deviation, variance, and standard deviation. The measures of relative variation are the coefficient of variation and skewness.

2.14 RANGE

The range is the simplest measure of absolute variation.

> The *range* is the difference between the largest and smallest value in a group of scores.

For the raw scores in Group I above, the range is $68 - 52 = 16$. The range of scores in Group II is $80 - 40 = 40$. Thus the range provides a numerical measure of distance between extremes.

For a frequency distribution, the range is found as the difference between the upper limit of the last class and the lower limit of the first class. The range of SAT-Mathematics Scores in Table 2.16 (Section 2.10) is $750 - 300 = 450$. All possible scores on the SAT, as stated by the College Entrance Examination Board, "range" from 200 to 800.

The simplicity of the range is its chief disadvantage as a measure of variation. The range is determined by only two values, the extremes. A more useful measure of variation takes into account all scores in a data set.

2.15 AVERAGE DEVIATION

The average deviation is a measurement of the absolute distances of raw scores from a measure of central tendency. Specifically,

> The *average deviation* is the arithmetic mean of the absolute deviations of all raw scores from either their mean or median.

The average deviation from the mean is:

$$AD_{\bar{X}} = \frac{\sum |X - \bar{X}|}{n}. \tag{2.8}$$

The average deviation from the median is:

$$AD_{Md} = \frac{\sum |X - Md|}{n}. \tag{2.9}$$

2.9 will always be < 2.8 — due to mathematical properties of 2.9

EXAMPLE. A sample of last year's personnel records provides the number of days' sick leave (X) for each of five employees as given below. The arithmetic mean, determined in Section 2.9, is 5. Find: (*a*) the median; (*b*) average deviation from the mean; (*c*) average deviation from the median.

Solution. (*a*) From the array, the median is 4.

X	$|X - \bar{X}|$	$|X - Md|$
3	2	1
3	2	1
4 (*Md*)	1	0
7	2	3
8	3	4
Total	10	9

3 − 5 =   − 2

(*b*) $AD_{\bar{X}} = \dfrac{10}{5} = 2;$ (*c*) $AD_{Md} = \dfrac{9}{5} = 1.8.$

The choice between using either $AD_{\bar{X}}$ or AD_{Md} depends on the choice of a measure of central tendency. If the arithmetic mean is used as a typical score, than $AD_{\bar{X}}$ would be preferred. Likewise, AD_{Md} is preferable when the median is chosen as the measure of central tendency.

As a measure of variation, the average deviation is of limited use. Even though the measure is easy to understand, the measures presented in the following section possess more desirable arithmetic properties.

2.16 VARIANCE AND STANDARD DEVIATION

The two most important measures of variation are the variance and standard deviation. Algebraically, they are superior to the average deviation, and they are the principal measures of variation used in statistical inference.

Ungrouped Data

> The *variance* is the arithmetic mean of the squared deviations of the raw scores about their mean.

The variance of a sample of n observations, called the *sample variance*, is

$$S^2 = \frac{\sum(X - \bar{X})^2}{n}. \tag{2.10}$$

If the data constitutes the entire population of N items, rather than a sample, the *population variance* is:

$$\sigma^2 = \frac{\sum(X - \mu)^2}{N}. \tag{2.11}$$

The symbol σ is the lower-case Greek letter sigma, so that the population variance is read as "sigma-squared."

> The *standard deviation* is the square root of the variance.

Alternatively, the *standard deviation* may be expressed as the square root of the arithmetic mean of the squared deviations of the raw scores about their mean. The standard deviation of a sample of n observations, the *sample standard deviation*, is

$$S = \sqrt{\frac{\sum(X - \bar{X})^2}{n}}. \tag{2.12}$$

The population standard deviation of N items, σ (called "sigma"), is

$$\sigma = \sqrt{\frac{\sum(X - \mu)^2}{N}}. \tag{2.13}$$

EXAMPLE. Using the data in the previous example, find the variance and standard deviation of the number of days' sick leave (X) for a sample of five employees.

Solution. The mean of the five observations (3, 3, 4, 7, 8) is 5. Using Equation (2.10), the sample variance is

$$S^2 = \frac{\sum(X - \bar{X})^2}{n}$$

$$= \frac{(3 - 5)^2 + (3 - 5)^2 + (4 - 5)^2 + (7 - 5)^2 + (8 - 5)^2}{5}$$

$$= \frac{4 + 4 + 1 + 4 + 9}{5}$$

$$= \frac{22}{5} = 4.4.$$

The standard deviation, $S = \sqrt{4.4}$, is 2.10 days.

Note that the variance is measured in squared units while the standard deviation is measured in the units of the original observations. Because of this, the standard deviation is the more widely used measure.

Ungrouped Data: Alternative Method

An alternative, and often computationally more efficient, formula for the sample variance is found by expanding Equation (2.10) and simplifying:

$$S^2 = \frac{\sum(X - \bar{X})^2}{n}$$

$$= \frac{\sum(X^2 - 2X\bar{X} + \bar{X}^2)}{n}$$

$$= \frac{\sum X^2}{n} - \frac{2\bar{X}\sum X}{n} + \frac{n\bar{X}^2}{n}$$

$$= \frac{\sum X^2}{n} - \bar{X}^2$$

or

$$= \frac{\sum X^2}{n} - \left(\frac{\sum X}{n}\right)^2. \tag{2.14}$$

Likewise, the sample standard deviation, by the alternate method, is

$$S = \sqrt{\frac{\sum X^2}{n} - \left(\frac{\sum X}{n}\right)^2}. \tag{2.15}$$

Using the data on the number of days' sick leave, the calculations necessary for determining the variance and standard deviation by the alternate method are given in Table 2.17. Substituting from Table 2.17,

$$S^2 = \frac{\sum X^2}{n} - \left(\frac{\sum X}{n}\right)^2$$

$$= \frac{147}{5} - \left(\frac{25}{5}\right)^2$$

$$= 29.4 - 25$$

$$= 4.4.$$

**TABLE 2.17:
UNADJUSTED AND
ADJUSTED SUM OF
SQUARES**

(1) Sick Leave (days) (x)	(2) (x²)	(3) (x − x̄)²
3	9	4
3	9	4
4	16	1
7	49	4
8	64	9
Total 25	147	22

Sums of Squares

The two methods for calculating the variance and standard deviation introduce *sums of squares*, a component of many statistical computations. The numerator of Equation (2.10), $\sum(X - \bar{X})^2$, the sum of squared deviations about the mean, is often referred to as the *adjusted sum of squares*. Using this terminology, the variance represents the arithmetic mean of the adjusted sums of squares. The numerator of the first term in Equation (2.15), $\sum X^2$, is called the *unadjusted sum of squares*. For the variable X (number of days' sick leave), Table 2.17 provides: (1) the sum of the original values of the variable, 25; (2) the unadjusted sum of squares, 147; and (3) the adjusted sum of squares, 22. The utility of these summations will be apparent in the following chapters.

Mathematical Property

If a variable Y is a linear transformation of the variable X so that

$$Y = a + bX,$$

then the variance of variable Y, S_Y^2, may be expressed in terms of the variance of variable X, S_X^2, by the relation[7]

$$S_Y^2 = b^2 S_X^2, \tag{2.16}$$

where a is the additive constant and b is the multiplicative constant in the linear relationship. The relationship between standard deviations is

$$(S_Y = bS_X. \quad \text{↗ std. dev.}$$

One use of this property is illustrated below.

EXAMPLE. A local telephone company has a base rate for residential customers of $6 per month plus a charge of ten cents for each directory-assistance call. The number of directory-assistance calls per month (X) made by one customer over a sample of five months is shown in Column (1) of Table 2.18. From Columns (1) and (2), the arithmetic mean and variance of the number of calls are:

$$\bar{X} = \frac{\sum X}{n} = \frac{40}{5} = 8;$$

$$S_X^2 = \frac{\sum(X - \bar{X})^2}{n} = \frac{46}{5} = 9.2.$$

---

[7] Proof:

$$S_Y^2 = \frac{\sum(Y - \bar{Y})^2}{n} = \frac{\sum Y^2}{n} - \bar{Y}^2$$

$$= \frac{\sum(a + bX)^2}{n} - (a + b\bar{X})^2$$

$$= \frac{\sum(a^2 + 2ab + b^2 X^2)}{n} - (a^2 + 2ab\bar{X} + b^2 \bar{X}^2)$$

$$= \frac{na^2}{n} + \frac{2ab\sum X}{n} + \frac{b^2 \sum X^2}{n} - a^2 - 2ab\bar{X} - b^2 \bar{X}^2$$

$$= b^2 \left(\frac{\sum X^2}{n} - \bar{X}^2 \right).$$

TABLE 2.18:
NUMBER OF
DIRECTORY-ASSISTANCE
CALLS AND MONTHLY
PHONE COST

(1) Number of Directory-Assistance Calls x	(2) $(x - \bar{x})^2$	(3) Monthly Charges y	(4) $(y - \bar{y})^2$
9	1	$ 6.9	0.01
5	9	6.5	0.09
10	4	7.0	0.04
4	16	6.4	0.16
12	16	7.2	0.16
Total 40	46	$34.0	0.46
Mean 8		$ 6.8	

Let the variable Y represent total monthly charges (\$) to the customer (excluding any toll call charges), where $Y = 6 + 0.10X$. According to the above mathematical property, the variance of Y, S_Y^2, is

$$S_Y^2 = b^2 S_X^2$$
$$= (0.10)^2(9.2)$$
$$= (0.01)(9.2)$$
$$= 0.092.$$

Verify this result by finding all values of Y for $Y = 6 + 0.10X$ and compute the variance of Y using Equation (2.10). Note that the additive constant 6 is not a factor in relating the variance of monthly charges to the variance of directory-assistance calls. Adding a constant to all scores in a data set does not affect the variance of the resulting scores.

Solution. The values of Y and their mean are given in Column (3) of Table 2.18. From Column (4),

$$S_Y^2 = \frac{\sum (Y - \bar{Y})^2}{n}$$

$$= \frac{0.46}{5} = 0.092.$$

Since the standard deviation of the number of directory assistance calls, $S_X = \sqrt{9.2}$, is 3.03, the standard deviation of the total monthly charges, S_Y, is

$$S_Y = b S_X$$
$$= (0.10)(3.03)$$
$$= 0.303.$$

Note further that $\bar{Y} = 6 + 0.10\,\bar{X}$, a property of the arithmetic mean discussed in Section 2.9.

Grouped Data

The variance is computed from a frequency distribution by the following formula:

$$S^2 = \frac{\sum f X^2}{n} - \left(\frac{\sum f X}{n}\right)^2, \tag{2.17}$$

where f is a class frequency, X is a class midpoint, and n is the total number of observations ($\sum f$). The standard deviation of frequency-distribution data is

$$S = \sqrt{\frac{\sum fX^2}{n} - \left(\frac{\sum fX}{n}\right)^2}. \tag{2.18}$$

EXAMPLE. Find the variance and standard deviation of the SAT-Mathematics Scores given in Table 2.19.

TABLE 2.19:
COMPUTATION OF
VARIANCE FOR
SAT-MATHEMATICS
SCORES

(1)	(2) Number of Students	(3) Midpoint	(4)	(5)
Scores	(f)	(x)	(fx)	(fx²)
300 and under 350	5	325	1,625	528,125
350 and under 400	14	375	5,250	1,968,750
400 and under 450	23	425	9.775	4,154,375
450 and under 500	50	475	23,750	11,281,250
500 and under 550	52	525	27,300	14,332,500
550 and under 600	25	575	14,375	8,265,625
600 and under 650	22	625	13,750	8,593,750
650 and under 700	7	675	4,725	3,189,375
700 and under 750	2	725	1,450	1,051,250
Total	200		102,000	53,365,000

Source: Table 2.3.

Solution. The class midpoints are given in Column (3) of Table 2.19. The products of the values in Columns (2) and (3) provide Column (4), which gives $\sum f X$ as the total. Column (5), obtained by the product of values in Columns (3) and (4), yields $\sum f X^2$ as its total. Substituting in Equation (2.17),

$$S^2 = \frac{53,365,000}{200} - \left(\frac{102,000}{200}\right)^2$$

$$= 266,825 - (510)^2$$

$$= 6,725.$$

The standard deviation, $\sqrt{6725}$, is 82.01.

Estimating the
Standard Deviation

Sometimes a quick estimate of the standard deviation is desired. In some instances an estimate, rather than the exact value, may be adequate. The following rule is helpful in such situations.

> The range of a set of data is approximately four to six times the standard deviation.

Thus the standard deviation is estimated as $\frac{1}{4}$ to $\frac{1}{6}$ of the range. The factor of $\frac{1}{4}$ provides a better estimate for a data set containing a small number of observations, while $\frac{1}{6}$ is more appropriate for a larger number of observations. An estimate of the standard deviation of the number of days' sick leave (X)

for five employees, Table 2.17, is $(\frac{1}{4})(5)$, or 1.25 days. The true standard deviation from Equation (2.12) is 2.10 days. An estimate of the standard deviation of the 200 SAT scores in Table 2.19 is $(\frac{1}{6})(450)$, or 75. The standard deviation from Equation (2.18) is 82.01. This method not only provides a rapid estimate of the standard deviation, but also serves as a check on the calculated value of the standard deviation. The rationale for this rule will become apparent in a later chapter.

2.17 COEFFICIENT OF VARIATION

The absolute measures of variation discussed previously are generally inadequate for comparing the variability among several groups of data. The average deviation, variance, and standard deviation are all expressed in terms of distances about a measure of central tendency, either the arithmetic mean or median. If, for example, the means and/or the units of measurement are different, then the standard deviations are not comparable. The coefficient of variation is a measure of relative variation that facilitates these comparisons.

> The *coefficient of variation*, V, expresses the standard deviation as a percentage of the arithmetic mean:
>
> $$V = \frac{S}{\bar{X}} \cdot 100. \qquad (2.19)$$

EXAMPLE. The mean closing price of Stock A over the past year is \$58 with a standard deviation of \$15. For Stock B, the mean is \$27 and the standard deviation is \$9. Which stock varies more in price? The coefficient of variation for Stock A, V_A, is $(\frac{15}{58}) \cdot 100 = 25.9$ percent. For Stock B, $V_B = (\frac{9}{27}) \cdot 100 = 33.3$ percent. Thus Stock B fluctuates in price (about its mean) more than Stock A.

In financial analysis, both the standard deviation and coefficient of variation are used as measures of risk. From above, Stock A is less risky than Stock B.

2.18 SKEWNESS

The shape of a frequency distribution also provides descriptive information about the data. Curves of frequency distribution are either symmetrical or skewed. Three distinct shapes of frequency curves are shown in Figure 2.7. In a *symmetrical* distribution the mean, median, and mode are equal, as Curve A shows.

Curve B, a distribution *skewed left*, contains some values that are much smaller than the majority of observations. The mode, unaffected by extreme items, is the point of greatest frequency. The value of the arithmetic mean is equally affected by all items and is "pulled" in the direction of the extreme values. Therefore the mean is the smallest of the three measures in a distribution skewed left. The median, less affected by extreme items, is located between the mode and mean.

When the extreme items are large in value relative to majority of other observations, the distribution is *skewed right* (Curve C). The mean is the largest of the three measures of central tendency in this case.

FIGURE 2.7: SHAPES OF
FREQUENCY
DISTRIBUTIONS

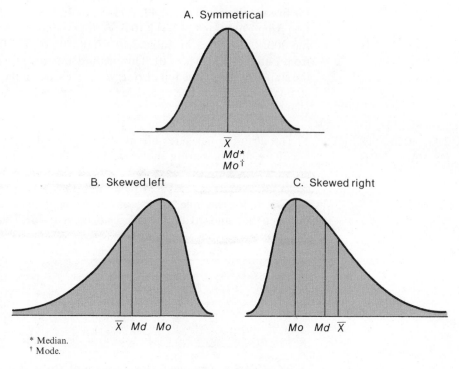

A. Symmetrical

$\overline{X}$
$Md*$
$Mo^\dagger$

B. Skewed left

$\overline{X}$ Md Mo

C. Skewed right

Mo Md $\overline{X}$

* Median.
† Mode.

Karl Pearson, an important contributor to statistical methodology, found that in moderately skewed distributions, the median is generally located about two-thirds of the distance from the mode to the mean. The *Pearsonian measure of skewness*, a relative measure, is

$$Sk = \frac{3(\overline{X} - Md)}{S}.$$ (2.20)

This measure is positive for distributions skewed right, resulting in the term *positive skewness*. For distributions skewed left, the sign is negative; hence *negative skewness*.

EXAMPLE. Is the distribution of SAT-Mathematics Scores, first given in Table 2.3, skewed?

$$Sk = \frac{3(510 - 507.69)}{82.01} = +0.08.$$

Sketch a curve of the distribution given in Figure 2.3 to verify that it is slightly positively skewed.

The arithmetic mean and median are equal in a symmetrical distribution. In such cases, Equation (2.20) shows that Sk has a value of zero. If a distribution is not symmetrical, then Equation (2.20) shows that Sk has a nonzero value. Hence, skewness is a measure of the lack of symmetry in a distribution.

Knowledge of skewness is useful in describing the shape of a distribution and in choosing the most representative measure of central tendency. If the distribution of outstanding balances on credit card statements mailed by an oil company in 1977 has skewness of $+1.5$, then we know the distribution is skewed similarly to Figure 2.7(C). This implies that the bulk of the distribution shows relatively small balances, though there are a few large balances. Further, the mean balance is larger than the median balance since the mean is more affected by the extreme (large) balances. If the distribution of outstanding balances on statements mailed in 1979 has skewness of $+0.6$, the distribution is still positively skewed but less so than in 1977. There should be less difference in the mean and median balance in 1979 than in 1977.

The choice of the most representative measure of central tendency depends in part on what one wishes to convey about the data. The shape of a distribution also influences this choice. Since the mean, median, and mode are equal in a symmetrical distribution, any of the three measures may be used. The mean is usually chosen because of its familiarity and its usefulness in other statistical procedures. In the case of a skewed distribution, the median better conveys the typical value. Further considerations in selecting representative values are given in the next section.

2.19 COMPARISON OF MEASURES

Choosing a descriptive measure that best summarizes the desired characteristic for a group of data involves careful consideration. The choice of a measure of central tendency should be based on (1) the distribution of the data, (2) the characteristics of the available measures, and (3) the intended use of the measure.

The arithmetic mean, frequently termed the average, is the most familiar and widely used measure of central tendency. The mean gives equal weight to all values; consequently it is affected by extreme values more than the other measures. The mathematical properties of the arithmetic mean make it particularly useful in further computations. If a distribution is symmetrical or only slightly skewed, the arithmetic mean is generally best.

Airlines use the arithmetic mean as a typical weight of passengers. The distribution of weight is approximately symmetrical for a large number of passengers. Moreover, the mean can be used to estimate the total weight of all passengers; this is necessary to determine the capacity and range of the plane, baggage allowance, fuel requirements, etc. The arithmetic mean is the only measure of central tendency that can be used to estimate a total.

The median is a simple measure; it is easy to determine and it is easily understood. Because it is less affected by extreme values than the arithmetic mean, the median is widely used to characterize skewed distributions. Income data is generally skewed right; hence the U.S. Bureau of the Census uses the median as a typical value (median family income in 1975 was $13,719). The arithmetic mean would be affected by the relatively small number of extremely large incomes and therefore misleading as a typical measure. The distribution of time required for many services, such as airline ticketing (Figure 2.4), is often positively skewed.

Care must be exercised in using the mode. Not every distribution has a mode; some distributions have more than one mode. The mode is best used in those situations where there is clearly a "customary and usual" value. Retailers make extensive use of the mode in placing orders for merchandise. In what size typing paper should an office supply firm maintain its largest inventory? The most frequently requested (modal) size, $8\frac{1}{2}$ by 11 inches, should be stocked in the largest quantities. What denomination of paper currency should be printed next year for circulation? Again, the modal value is of interest.

Sometimes there is no one "best" measure of central tendency. What is the typical salary of the starting backfield of a National Football League team if the quarterback earns $1 million and the three other backs each earn $100,000? The arithmetic mean is $325,000, which is not typical of any of the players' salaries. Either the median or mode, $100,000, is preferable to the mean. Occasionally, it may be helpful to report all three measures.

In choosing between measures of variation, the range provides a simple measure that is easily obtained and understood. A disadvantage is that the range is based on only two items, the extreme values, and therefore may be unrepresentative. The average deviation provides a measure of variability based on all observations, but it lacks the advantageous arithmetic properties of the variance and standard deviation. The standard deviation is the most widely used measure of variation. Even though it is more difficult to calculate and interpret, its mathematical properties make it ideal for further analysis. For comparing variability among different groups of data, the coefficient of variation is superior.

The arithmetic mean and standard deviation are the principal descriptive measures. The concepts of statistical inference presented later rely heavily on these measures and their extensions.

2.20 USE OF COMPUTERS

Computing facilities are available to all but the smallest of businesses today, in the form of either a minicomputer or a large-scale computer, full-time or on a time-sharing basis. Additionally, a wide variety of computer programs and "packages" of computer routines are available for performing statistical computations. Computer-calculated statistical measures may be obtained through the use of packaged programs simply by specifying what analysis is desired. The manager may, through judicious use of standard computer routines, more efficiently analyze the data available in order to promote sound decision making.

Illustrations of two widely available computer routines are provided here. In Figure 2.8, a computer-generated frequency distribution and histogram of the SAT-Mathematics scores is given. This is obtained by using the 200 scores in Table 2.1 and a procedure in the MINITAB statistical package.[8] The classes of the frequency distribution are identified by the class midpoints. There are ten classes with class intervals of 50. Compare this with the frequency distribu-

---

[8] Described in Thomas A. Ryan, Brian L. Joiner, and Barbara F. Ryan, *MINITAB Student Handbook* (North Scituate, Mass.: Duxbury Press, 1976).

FIGURE 2.8: MINITAB FREQUENCY DISTRIBUTION AND HISTOGRAM

HIST 1

COLUMN 1   MAXIMUM 748.0000   MINIMUM 318.0000

MIDDLE OF INTERVAL	NUMBER OF OBSERVATIONS	
300.	1	*
350.	12	************
400.	22	**********************
450.	33	*********************************
500.	51	***
550.	39	***************************************
600.	24	************************
650.	12	************
700.	5	*****
750.	1	*

tion in Table 2.3. In the histogram, class frequencies are represented by horizontal lines of asterisks.

The descriptive statistics of the SAT-Mathematics scores given in Figure 2.9 were obtained by use of a procedure in the *Statistical Package for the Social Sciences* (SPSS).[9] These measures were computed from the raw scores given in Table 2.1. In previous sections of the chapter, these measures were computed from the frequency distribution in Table 2.3. Review these to see how closely the measures computed from the frequency distribution approximate the measures computed from the raw scores.[10] This comparison reveals how accurately the frequency distribution summarizes the raw scores.

FIGURE 2.9: SPSS DESCRIPTIVE STATISTICS

VAR01   SAT MATHEMATICS SCORES

MEAN	509.180	MEDIAN	506.500	STD DEV	84.691
VARIANCE	7172.645	SKEWNESS	0.139	RANGE	430.000
MINIMUM	318.000	MAXIMUM	748.000		
VALID CASES	200	MISSING CASES	0		

Computer-generated statistical measures are illustrated throughout the text. Keep in mind, however, that the computer merely acts as a computational device. Computer routines cannot substitute for sound statistical reasoning. The burden of selecting the appropriate analysis, interpreting the results, and using the information in decision making remains with the user. The manager can effectively utilize statistical analysis and the power of the computer only after obtaining a thorough understanding of statistical principles.

---

[9] Described in Norman H. Nie et al., *Statistical Package for the Social Sciences*, 2d. ed. (New York: McGraw-Hill Book Co., 1975).

[10] The computer routine uses a slightly altered formula for the standard deviation and variance. The denominator, n, of Equations (2.10) and (2.12) is replaced by $(n - 1)$. The purpose of this adjustment is discussed in Chapter 8.

2.21 SUMMARY

Statistical description of data is provided through tables, charts, and computed measures. A frequency distribution is the principal tabular summary of either discrete or continuous data. The frequency distribution may show actual, relative, or cumulative frequencies. Actual and relative frequencies may be charted as either a histogram (a bar chart) or a frequency polygon (a line chart). Two charts of cumulative frequencies are the less-than ogive and the equal-to-or-more-than ogive. Data classified into a frequency distribution are referred to as grouped data. Ungrouped data consist of the individual raw scores.

Measures of central tendency and variation are the primary computed measures used to describe the statistical characteristics of data. Any of the various measures of central tendency may be chosen as the most representative or typical measure. The arithmetic mean is widely used and understood as a measure of central tendency. The median is generally a more representative measure in skewed distributions. The mode should be used when there is a "customary and usual" value.

The range, average deviation, variance, and standard deviation are the chief measures of absolute variation. Variability is a measure of distance between observations or distance of observations from a measure of central tendency. The variance and standard deviation provide measures of distance from the arithmetic mean. Measures of relative variation are useful for comparing different groups of data. The coefficient of variation expresses the standard deviation as a percentage of the mean. Skewness measures the lack of symmetry in a distribution.

Many of these descriptive statistics are important measures in statistical inference. Thus a sound understanding of these concepts is an important prerequisite to later chapters.

2.22 TERMS AND FORMULAS

Absolute variation	Ogive
Adjusted sum of squares	Open-end class
Arithmetic mean	Percentile
Bar chart	Primary source
Bimodal	Qualitative data
Central tendency	Quantitative data
Class interval	Range
Class limits	Raw score
Continuous data	Relative frequency
Cumulative frequency	Relative variation
Discrete data	Secondary source
Frequency distribution	Skewness
Frequency polygon	Standard deviation
Grouped data	Unadjusted sum of squares
Median	Ungrouped data
Midpoint	Variance
Mode	Variation

Arithmetic mean of a
  frequency distribution

$$\bar{X} = \frac{\sum f X_{mp}}{n}$$

Average deviation from the mean

$$AD_{\bar{X}} = \frac{\sum |X - \bar{X}|}{n}$$

Average deviation from the median

$$AD_{Md} = \frac{\sum |X - Md|}{n}$$

Class interval

$$\frac{\text{Largest score} - \text{Smallest score}}{\text{Number of classes}}$$

Coefficient of variation

$$V = \frac{S}{\bar{X}} \cdot 100$$

Median of a frequency distribution

$$Md = L + i \left(\frac{n/2 - F}{f} \right)$$

Pearsonian measure of skewness

$$Sk = \frac{3(\bar{X} - Md)}{S}$$

Percentile (jth) of a frequency
  distribution

$$P_j = L + i \left(\frac{\frac{jn}{100} - F}{f} \right)$$

Population arithmetic mean

$$\mu = \frac{\sum X}{N}$$

Population standard deviation

$$\sigma = \sqrt{\frac{\sum (X - \mu)^2}{N}}$$

Population variance

$$\sigma^2 = \frac{\sum (X - \mu)^2}{N}$$

Sample arithmetic mean

$$\bar{X} = \frac{\sum X}{n}$$

Sample standard deviation

$$S = \sqrt{\frac{\sum (X - \bar{X})^2}{n}} \text{ or}$$

$$\sqrt{\frac{\sum X^2}{n} - \left(\frac{\sum X}{n} \right)^2}$$

Sample variance

$$S^2 = \frac{\sum (X - \bar{X})^2}{n} \text{ or}$$

$$\frac{\sum X^2}{n} - \left(\frac{\sum X}{n} \right)^2$$

Standard deviation of a
  frequency distribution

$$S = \sqrt{\frac{\sum f X^2}{n} - \left(\frac{\sum f X}{n} \right)^2}$$

Variance of a frequency distribution

$$S^2 = \frac{\sum fX^2}{n} - \left(\frac{\sum fX}{n}\right)^2$$

Weighted arithmetic mean

$$\bar{X}_w = \frac{\sum wX}{\sum w}$$

Linear transformations

$$\bar{Y} = a + b\bar{X}$$

$$S_Y^2 = b^2 S_X^2$$

QUESTIONS

A. True or False

_F_ 1. In a distribution skewed left, one generally finds mean > median > mode.

2. The mode may not exist for some sets of ungrouped data.

_T_ 3. A frequency distribution of the student enrollment in all U.S. colleges and universities contains discrete data.

4. Relative frequencies are actual frequencies expressed in percentage terms.

_F_ 5. A nationwide real estate firm sold 250 single family dwellings in 1979 at prices ranging from $15,000 to $135,000. A reasonable estimate of the standard deviation of selling prices is $2,000.

6. The measure of central tendency most easily approximated from a histogram is the median.

_T_ 7. The variance is always nonnegative.

8. Addition of a constant to each of a group of scores adds the same constant to the arithmetic mean of the scores.

_F_ 9. One-fourth of the scores are less than the 75th percentile while three-fourths are larger.

10. The sum of the squared deviations about the mean is more than the sum of the squared deviations about other measures of central tendency.

B. Fill in

1. The average of the squared deviations about the arithmetic mean is the ___variance___.

2. The measure of central tendency most affected by extremely large observations is the ___mean___.

3. A chart presenting cumulative frequencies is an ___ogive___.

4. The sum of the actual deviations from the arithmetic mean equals __0__ ___.

5. Any frequency curve that has a large number of observations at one end of the scale and relatively few observations at the other end is said to be ___skewed___.

6. The measure of central tendency that divides the sum of the scores by the number of scores is the ___mean___.

7. The ___c of variation___ is a measure of relative variation expressed in percentage terms.

8. _____ refers to the lack of symmetry in a distribution.

9. Data assembled or collected from records already published is called _____ data.

10. The most frequently occurring value in a group of scores is referred to as the _____.

11. If the standard deviation of a group of measurements made in feet is 0.20, the standard deviation of the measurements made in inches would be ___2.4___.

12. If all the values in a set of data are identical, the variance equals _____.

EXERCISES

1. The credit office of a department store mailed statements for payment due to 40 customers today. Construct a frequency distribution of the balances due as given below.

$33.70	$56.32	$10.97	$45.09
57.05	39.89	50.12	59.49
9.97	62.54	20.15	42.16
75.93	21.47	9.96	34.42
48.67	36.05	63.75	18.56
35.26	17.76	32.76	68.10
11.49	82.70	53.91	39.73
5.97	30.06	15.00	79.01
21.25	50.10	41.72	27.18
94.86	19.92	25.01	51.45

2. A recent survey of some of the largest banks in the United States provides information on the number of branches each bank operates. The number of branch banks operated by 60 of these banks is given below. Construct a frequency distribution of these data.

110	17	3	83
71	14	14	75
28	30	8	80
19	16	2	97
30	14	188	54
12	10	215	54
11	7	46	48
4	135	36	104
39	68	27	74
43	78	24	70
32	53	32	88
19	70	32	38
9	10	27	61
5	15	34	3
75	10	80	3

3. Plot a histogram of the frequency distribution constructed in Question 2. Estimate the modal number of branches per bank from the histogram.

4. A survey of 50 shoppers at a nearby shopping mall reveals the distribution of ages tabulated below.
   a. Plot a frequency polygon of the distribution.
   b. Determine a relative frequency distribution for the data.
   c. Find the cumulative less than relative frequencies and plot as a less-than ogive.
   d. Estimate the median age from the ogive.
   e. Estimate, from the ogive, the 80th percentile of the distribution of ages.
   f. What percentage of the shoppers is under 35?

Age	Number of Shoppers
10 and under 20	3
20 and under 30	25
30 and under 40	10
40 and under 50	7
50 and under 60	4
60 and under 70	1
Total	50

5. Compute the following measures from the distribution of ages given in Question 4:
   a. Arithmetic mean.
   b. Median.
   c. Variance.
   d. Standard deviation.

6. Examine your frequency polygon of the distribution of ages in Question 4a. Is the distribution skewed? Verify your answer by computing the coefficient of skewness. Describe what the skewness means in terms of the age distribution of the shoppers.

7. A frequency distribution of the length of telephone calls monitored at the switchboard of an office is given below.
   a. Compute the arithmetic mean and standard deviation of the calling time.
   b. Determine, and explain the meaning of, the 90th percentile of the distribution.

Length of Calls (minutes)	Number of Calls
0 and under 2	10
2 and under 4	25
4 and under 6	20
6 and under 8	40
8 and under 10	5
Total	100

8. A real estate firm advertised, in the Sunday newspaper, lakefront lots for sale in a vacation retreat. The number of inquiries per day about the lots was recorded for the next five business days: 6, 16, 9, 7, and 2.
   a. Determine the arithmetic mean and median number of inquiries per day.
   b. Compute the average deviation from both the mean and median.

9. Use the data on the number of inquiries per day (X) in Question 8.
   a. Calculate the unadjusted sum of squares, $\sum X^2$.
   b. Calculate the adjusted sum of squares, $\sum (X - \bar{X})^2$.
   c. Show how both sums of squares can be used to calculate the standard deviation of the number of inquiries per day.

10. The total of tuition and fees charged in 1976–77 to in-state undergraduate students in six major Virginia colleges and universities is given below. Find the mean, median, mode, variance, and standard deviation of these charges.

Institution	Tuition Plus Fees
College of William and Mary	$958
James Madison University	804
Old Dominion University	620
University of Virginia	734
Virginia Commonwealth University	826
Virginia Polytechnic Institute and State University	657

Source: *Richmond Times-Dispatch*, January 2, 1977, p. C-2.

11. The Environmental Protection Agency provides estimates of fuel consumption (miles per gallon) in highway driving for the following 1977 models of U.S. compact automobiles equipped with automatic transmissions.

Model	MPG
Gremlin	24
Pacer	23
Skyhawk	26
Nova	23
Vega	28
Mustang II	29
Pinto	32
Bobcat	32
Starfire	28
Ventura	29

Find the
   a. Arithmetic mean.
   b. Median.
   c. Mode.
   d. Variance.
   e. Standard deviation of gasoline mileage.

12. The ten largest U.S. defense contractors, in terms of the value of contracts awarded by the Department of Defense, for the fiscal year ending June 30, 1976, are listed below. Determine the arithmetic mean and median of the contract values. Which measure of central tendency is most representative of contracts awarded to this group of contractors? Why?

Contractor	Value of Contract ($ billions)
McDonnell Douglas Corp.	$2.460
Lockheed Aircraft Corp.	1.510
Northrop Corp.	1.480
General Electric Corp.	1.350
United Technologies Corp.	1.230
Boeing Co.	1.180
General Dynamics Corp.	1.070
Grumman Corp.	0.982
Litton Industries, Inc.	0.978
Rockwell International Corp.	0.966

Source: *The Wall Street Journal*, December 21, 1976, p. 2.

13. The ten largest holdings of common stock in the investment portfolio of the Virginia Supplemental Retirement System, the retirement program for state employees, as of June 30, 1976, are given below.
   a. What is the market value per share of IBM stock in the portfolio?
   b. What is the market value of the typical share in the entire portfolio of stocks? (Find the mean market value per share.)
   c. What is the mean cost per share for the portfolio of ten stocks?

Company	Number of Shares	Market Value ($000)	Cost ($000)
IBM	65,400	18,099	16,267
Phillip Morris	183,600	9,410	7,264
CBS	160,000	9,400	6,944
Dow Chemical	176,200	8,502	4,271
Eastman Kodak	80,900	8,101	8,594
Caterpillar Tractor	85,000	7,746	5,096
General Electric	128,400	7,335	6,059
Continental Oil	180,000	6,908	3,994
Citicorp	185,100	6,688	6,278
Schering-Plough	122,000	6,558	7,012
Total		88,747	71,779

Source: *Richmond Times-Dispatch*, December 24, 1976, p. A-10.

14. The prizes awarded during a recent contest sponsored by a national magazine are given below. Determine the
   a. Arithmetic mean.
   b. Median.

    c.   Mode of the amounts awarded.

    d.   What is the most typical amount awarded?

Prize	Number Awarded	Amount
First	1	$25,000
Second	1	20,000
Third	2	10,000
Fourth	5	5,000
Fifth	50	500
Sixth	400	25

15. Information on the ages of three persons out of a group of four is provided below. Find the missing age of individual C.

Individual	Age	Deviation from $\bar{X}$
A.	17	−8
B.	32	+7
C.	—	—
D.	21	−4

16. If the mean weight for 20 packages mailed today by the shipping department of a manufacturing company is 60 kilograms, what is the total weight of all the packages?

17. A manufacturing firm periodically uses a machine to test the quality of output from one production process. A CPA firm, currently auditing last year's records, wishes to verify the manufacturer's statement of operating expenses for the machine. The auditor knows that monthly operating cost can be expressed as a fixed cost of $50 plus a variable cost of $8 for each hour of operation. The auditor selects a sample of five month's records from the past year and finds that the hours per month the machine was operated were 3, 6, 9, 4, and 8.

    a.   Find the arithmetic mean hours per month the machine was used in the sample of five months.

    b.   Use your answer in (*a*) to compute the arithmetic mean of monthly operating cost, and use this to estimate operating cost for the entire year.

18. The top ten money winners in men's and women's tennis in 1976 are given below. Winnings are based on tournaments and exclude bonuses, challege matches, and endorsements.

    a.   Compare the winnings of the ten men with the ten women by computing the arithmetic mean, median, and standard deviation of each group.

    b.   Which measure of central tendency, the arithmetic mean or median, provides the most representative measure of the typical winnings for a member of each group? Why?

c. Compare the variation in winnings for men and women in both absolute and relative terms.

Men			Women		
Rank	Name	Winnings	Rank	Name	Winnings
1.	Jimmy Connors	$303,335	1.	Chris Evert	$289,165
2.	Raul Ramirez	253,442	2.	Evonne Goolagong	173,285
3.	Manuel Orantes	205,884	3.	Virginia Wade	124,880
4.	Guillermo Vilas	201,226	4.	Martina Navratilova	94,535
5.	Bjorn Borg	198,420	5.	Rosemary Casals	87,185
6.	Harold Solomon	193,182	6.	Betty Stove	85,025
7.	Roscoe Tanner	178,906	7.	Sue Barker	69,660
8.	Wojtek Fibak	176,539	8.	Francoise Durr	63,830
9.	Eddie Dibbs	171,571	9.	Mima Jausovec	45,905
10.	Ilie Nastase	165,205	10.	Billie Jean King	42,970

Source: Reprinted courtesy of *Tennis* Magazine from the February 1977 issue. Copyright © 1977, Tennis Features, Inc., U.S.A.

19. Using the top ten men's tennis winnings (X) in question 18 above, subtract $200,000 from all the original observations, creating a new variable $Y = X - \$200,000$.

   a. Compute the arithmetic mean of variable Y.

   b. Show that $\bar{Y}$ may also be obtained directly from $\bar{X}$ (Question 18a) by use of the relation $\bar{Y} = \bar{X} - \$200,000$.

   c. Determine the standard deviation of variable Y from your knowledge of the standard deviation of variable X (Question 18a).

   d. What does your answer imply about the effect on the standard deviation of subtracting a constant from all observations?

20. The capacity, in kilowatts (X), of the ten largest nuclear power generating plants in the U.S. in 1976 is given below. Define a new variable Y, thousand kilowatts, so that $Y = X/1000$.

   a. Compute the mean, variance, and standard deviation of variable Y.

   b. Using the mathematical properties of linear transformation, find the mean, variance, and standard deviation of variable X from the results given in (a).

Location	Operating Utility	Capacity (kilowatts)	Year Operative
Columbia County, Ore.	Portland General Electric Co.	1,130,000	1975
Decatur, Ala. (Unit 1)	Tennessee Valley Authority	1,065,000	1974
Decatur, Ala. (Unit 2)	Tennessee Valley Authority	1,065,000	1974
Peach Bottom, Pa. (Unit 2)	Philadelphia Electric Co.	1,065,000	1974
Peach Bottom, Pa. (Unit 3)	Philadelphia Electric Co.	1,065,000	1974
Zion, Ill. (Unit 1)	Commonwealth Edison Co.	1,050,000	1973
Zion, Ill. (Unit 2)	Commonwealth Edison Co.	1,050,000	1974
Bridgman, Mich. (Unit 1)	Indiana & Michigan Power Electric Co.	1,050,000	1975
Clay Station, Calif.	Sacramento Municipal Utility District	913,000	1975
Seneca, S.C. (Unit 1)	Duke Power Co.	886,000	1973

Source: Nuclear Regulatory Commission.

21. A real estate broker is in charge of home sales in two new subdivisions. The mean selling price of new homes in Ridge Run is $62,500 with a standard deviation of $14,000. In Broad Meadows the mean price is $35,900 and the standard deviation is $9,600. In relative terms, which subdivision has the larger variation in home prices?

22. Ben Budget keeps careful records of the amounts spent on evenings out. He finds a coefficient of variation of 50 percent and a variance of $16. What is his average (mean) expenditure per evening?

23. The following measures from the grade distribution on a recent test in statistics are available: mode = 83, arithmetic mean = 74, median = 77. From your knowledge of the general relationships between these measures, sketch a curve indicating the general shape of this distribution of grades. Indicate the approximate position of the three values and label the curve as to positive or negative skewness. Where are the extreme values located in a distribution of this shape?

24. A portfolio manager buys the following stocks: 150 shares of General Motors at $65 per share, 100 shares of International Business Machines at $250 per share, and 200 shares of Shell Oil at $75 per share. What is the mean price per share purchased, excluding brokerage fees?

25. *a.* Use a computer program available in your university computer center to prepare a frequency distribution of the data in Question 2. Compare this with the frequency distribution you prepared in Question 2.
    b. Use a computer program to determine descriptive measures of central tendency and variation from the data in Question 2.

26. United University awarded 200 M.B.A. degrees last year. The starting salary for these graduates is given below.

Annual Salary ($ thousands)	Number of Graduates
10 and under 12	15
12 and under 14	35
14 and under 16	55
16 and under 18	45
18 and under 20	30
20 and under 22	15
22 and under 24	5
	200

Determine the following:
a. Mean.
b. Median.
c. Variance.
d. The Pearsonian measure of skewness, *Sk*.

REFERENCES

Croxton, Frederick E.; Cowden, Dudley J., and Klein, Sidney. *Applied General Statistics*, 3d ed. Englewood Cliffs. N.J.: Prentice-Hall, Inc., 1967.

Doane, David P. "Aesthetic Frequency Classifications." *The American Statistician*, vol. 30, no. 4 (November 1976), pp. 181–83.

Goeldner, C. R., and Dirks, Laura M. "Business Facts: Where to Find Them." *MSU Business Topics*, vol. 24, no. 3 (Summer 1976), pp. 23–36.

Shao, Steven P. *Statistics for Business and Economics*, 3d ed. Columbus, Ohio: Charles E. Merrill Publishing Co., 1976.

Stockton, John R., and Clark, Charles T. *Introduction to Business and Economic Statistics*, 5th ed., Cincinnati, Ohio: South-Western Publishing Co., 1975.

3

Index Numbers

3 INDEX NUMBERS

The cost of living has gone up
another dollar a quart.

W. C. Fields

3.1 INTRODUCTION

Index numbers are descriptive measures used as indicators of change in the magnitude of business activity over time. Index numbers for measuring change in prices, income, production, employment, as well as other variables, are abundant. The most widely known and used index is the Consumer Price Index, which is compiled by the U.S. Department of Labor's Bureau of Labor Statistics. Popular uses of this index include the monitoring of changes in retail prices, measuring the purchasing power of a dollar, and as an indicator of the "cost of living." This index, like all indexes, provides a measure of change relative to a fixed point in time, called the *base period* of the index. Thus an index is a way to express a percentage change from a base period.

The major classifications of index numbers, according to the type of business activity measured, are price, quantity, and value. *Price indexes* measure changes in some price characteristic; for example, the Consumer Price Index measures changes in retail prices paid for goods and services purchased by urban wage earners, clerical, professional, and self-employed workers, as well as unemployed and retired persons. *Quantity indexes* measure changes in some volume characteristic, such as the Index of Industrial Production, which measures changes in the physical volume of output in the industrial sector of the economy.[1] *Value indexes* measure change in some criterion of value; the Index of Total Construction Value provides indicators of change in the dollar value of new construction contracts awarded. A brief description of several major indexes is provided in Table 3.1.

All of the major indexes are designed for special purposes, and each has its own unique formula. Nevertheless, the special methods employed are based on some standard methods of index number construction. The types of index numbers, classified according to the method of construction, are:

1. Simple index
2. Composite index
   a. Relative of aggregates

---

[1] During the depression, many people could not afford the regular cuts of meat, and one restaurant owner began offering rabbit stew. One day a regular customer told the owner that the rabbit stew didn't taste right. "Yes," the owner replied, "rabbits are so scarce I had to put in some horse meat, but it's fifty-fifty." After another bite, the customer inquired: "Are you sure it's fifty-fifty?" "Yes," said the owner, "one rabbit to one horse."

TABLE 3.1: CHARACTERISTICS OF SOME MAJOR INDEXES

Index	Compiler/Publisher	Measures	Current Base Period	Value in 1976	1977
1. Consumer Price Index	Department of Labor, Bureau of Labor Statistics	Retail prices paid for goods and services purchased by urban wage earners, clerical, professional, and self-employed workers, and by unemployed and retired persons.	1967	170.5	181.5
2. Producer Price Index (formerly Wholesale Price Index)	Department of Labor, Bureau of Labor Statistics	Prices of all commodities sold in primary markets, i.e., prices paid by the first large-volume buyer.	1967	183.0	194.2
3. Index of Industrial Production	Federal Reserve System Board of Governors	Physical volume of output in the industrial sector of the economy (manufacturing, mining, and gas and electric utilities).	1967	129.8	137.1
4. Daily Index of Spot Market Prices	Department of Labor, Bureau of Labor Statistics	Price movements in 22 basic commodities whose markets are among the first to be influenced by changes in economic conditions (metals, livestock and farm products, fats and oils).	1967	201.0	209.6
5. Index of Prices Received by Farmers	Department of Agriculture	Prices on all crops and livestock.	1910–14	464	456
6. Index of Average Hourly Earnings	Department of Labor, Bureau of Labor Statistics	Hourly earnings for production or nonsupervisory workers in private nonagricultural industries (includes overtime in manufacturing).	1967	183.0	196.8
7. Index of Help-Wanted Advertising	The Conference Board	Volume of classified advertising in 51 major U.S. newspapers.	1967	95	118
8. Index of Exports	Department of Commerce, Bureau of Economic Analysis	Value of exported U.S. merchandise, excluding Department of Defense shipments of military supplies and equipment.	1958	202.1	211.8
9. Standard & Poor's Composite Index of Stock Prices	Standard & Poor's Corporation	Price of 500 common stocks.	1941–43 = 10	102.01	98.20
10. Index of Total Construction Value	F. W. Dodge Corporation	Changes in the dollar value of new construction contracts awarded in 50 states. Includes residential, commercial, and institutional building plus non-building construction such as highways and public works.	1967	199	252

Source: These indexes are published regularly in one or more of the following: *Business Conditions Digest, Economic Indicators, Economic Report of the President*, and *Survey of Current Business*.

b. Average of relatives
c. Relative of weighted aggregates
d. Average of weighted relatives

The construction of a price index by these methods is now examined. The same methods are useful for quantity and value indexes.

3.2 SIMPLE INDEX

A *simple index* measures the relative change from the base period for a *single* item. To determine a simple index of price in any given time period, express the price in the given time period, P_g, as a ratio to the price in the base period, P_b, and multiply by 100. The simple index of price for period g with a base period b is:

$$I_{g/b} = \frac{P_g}{P_b} \cdot 100. \tag{3.1}$$

EXAMPLE. The average retail price of an item is $4 in 1977, $5 in 1978, and $6 in 1979. Determine the percentage change in price from 1977 to 1978 and 1979 and determine a simple index of price for 1978 and 1979, using 1977 as the base.

Solution. The change in price from 1977 to 1978 is ($1/$4) · 100 = 25 percent. For 1979, the change from 1977 is ($2/$4) · 100 = 50 percent. A simple index of price in 1977 with a 1976 base is

$$I_{78/77} = \frac{P_{78}}{P_{77}} \cdot 100 = \frac{\$5}{\$4} \cdot 100 = 125.$$

For 1979, the index is

$$I_{79/77} = \frac{P_{79}}{P_{77}} \cdot 100 = \frac{\$6}{\$4} \cdot 100 = 150.$$

The results are summarized below.

Year	Price	Percentage Change from 1977	Index (1977 = 100)
1977	$4	0	100
1978	5	25	125
1979	6	50	150

Three basic characteristics of index numbers are illustrated by the above example. First, the index for the base period is 100.[2] The statement "1977 = 100" is used to identify the base. Second, the change in the value of the index from the base period to any given period is simply a measure of percentage change

---

[2] It is standard practice to use 100 as the value of an index in the base period. For an exception, see Standard & Poor's Composite Index of Stock Prices in Table 3.1.

from the base period. The difference in the 1978 index, 125, and the 1977 index, 100, is 25, which indicates that price increased 25 percent. Alternatively, the 1978 index of 125 percent indicates that the 1978 price is 125 percent of the base period (1977) price. Likewise, the 1979 price is 150 percent of the 1977 price. Simple indexes of price are also called *price relatives*. Third, the change in the value of an index for two periods does *not* indicate percentage change unless one time period is the base period. For example, the difference in the 1979 and 1978 indexes (150 − 125) does not indicate a 25 percent increase in price from 1978 to 1979. An index number provides a measure of change *from the base period only*. There is a 20 percent increase in price from 1978 ($5) to 1979 ($6). The result may be obtained indirectly from the indexes; a change from 125 to 150 in the index is a 20 percent increase.

3.3 COMPOSITE INDEX

A *composite index* measures relative change from the base period for a group of closely related items. For example, the Producers Price Index, formerly called the Wholesale Price Index, includes price measurements on over 2,300 items. The major indexes listed in Table 3.1 are all composite indexes. The four basic forms of composite indexes were outlined previously. Two are considered unweighted indexes while two are weighted.

3.4 RELATIVE OF AGGREGATES INDEX

The *relative of aggregates* (RA) price index is the total price of a group of items in a given period divided by the total price in the base period, multiplied by 100. The relative of aggregates price index for period g with base period b, $RA_{g/b}$, is

$$RA_{g/b} = \frac{\sum P_g}{\sum P_b} \cdot 100. \tag{3.2}$$

EXAMPLE. Advances in electronic circuitry during the 1970s provided substantial price reductions for electronic calculators. DATACHIP manufactures and sells three calculator models: the Basic, providing only four-function arithmetic features; the Financial, with functions for financial analysis; and the Scientific, which is programmable and has sophisticated mathematical functions. Retail prices of these models for three years follow. Using 1971 as the base, the relative of aggregates price indexes for 1975 and 1979 are:

$$RA_{75/71} = \frac{\sum P_{75}}{\sum P_{71}} \cdot 100 = \frac{445}{1145} \cdot 100 = 38.86,$$

and

$$RA_{79/71} = \frac{\sum P_{79}}{\sum P_{71}} \cdot 100 = \frac{193}{1145} \cdot 100 = 16.86.$$

Since the index in the base period, 1971, is 100, the difference in the indexes for 1971 and 1975, (100 − 38.86 = 61.14), indicates that the average price of the three models declined by 61.14 percent. Alternatively, the average price in 1975 is 38.86 percent of the 1971 average price. The decline in price from 1971 to 1979 is 83.14 percent.

	Retail Price ($)		
Model	*1971*	*1975*	*1979*
Basic.	95	15	8
Financial	350	180	60
Scientific	700	250	125
Total	1145	445	193
RA Index	100	38.86	16.86
(1971 = 100)			

FRANCIS YSIDRO EDGEWORTH (1845–1926)

Edgeworth is responsible for introducing statistical analysis into economics and social science. Born in Edgeworthstown, Ireland, he studied classics at Trinity College in Dublin and Oxford University in England. After leaving Oxford, he studied commercial law and was called to the bar in 1877, but found little satisfaction in the legal profession. He undertook a program of self-study in mathematics and, as a lecturer on logic at King's College, London, became fascinated with economic science after reading W. Stanley Jevons' *Theory of Political Economy*. Statistics, at that time, had been applied primarily to astronomy and geodesy. Edgeworth's primary aim became to develop and apply statistical theories to social and economic data. He devoted much effort to determining the best "index numbers" to measure prices and other economic data. Responsible for introducing the phrase "coefficient of correlation," he often related shapes of distributions to the outline of hats. A normal curve was a French gendarme's hat, a skewed curve was a hat blown to one side, and extreme values were located on the rim of the hat.

3.5 AVERAGE OF RELATIVES INDEX

The *average of relatives* (*AR*) price index is the arithmetic mean of the price relatives, $(P_g/P_b) \cdot 100$, for a group of items. The average of relatives price index for period g with base b is

$$AR_{g/b} = \frac{\sum\left(\dfrac{P_g}{P_b} \cdot 100\right)}{n}, \tag{3.3}$$

where n is the number of items. Average of relatives price indexes for the data in the previous example are given in Table 3.2. Each price relative is computed by Equation (3.1), using a 1971 base. For example, the 1975 price relative for the financial model is $(P_{75}/P_{71}) \cdot 100 = 51.43$.

After the relatives are determined, the indexes are

$$AR_{75/71} = \frac{\sum\left(\dfrac{P_{75}}{P_{71}} \cdot 100\right)}{n} = \frac{102.93}{3} = 34.31,$$

Model	1971		1975		1979	
	Price	Relative	Price	Relative	Price	Relative
Basic	95	100	15	15.79	8	8.42
Financial	350	100	180	51.43	60	17.14
Scientific	700	100	250	35.71	125	17.86
Total		300		102.93		43.42
AR Index		100		34.31		14.47
(1971 = 100)						

and

$$AR_{79/71} = \frac{\sum\left(\dfrac{P_{79}}{P_{71}} \cdot 100\right)}{n} = \frac{43.42}{3} = 14.47.$$

According to the average of relatives method, the average price for the calculators declined by 65.59 percent from 1971 to 1975 and by 85.53 percent from 1971 to 1979.

As evidenced here, the two price index methods generally produce different results. The relative of aggregates method indicates a 61.14 percent decrease ($RA_{75/71} = 38.86$) in the average price of the calculators from 1971 to 1975, while the average of relatives method gives a 65.69 percent decrease ($AR_{75/71} = 34.31$). The results differ because of the implicit weighting systems involved. The relative of aggregates method is a composite of absolute prices, with each item counted equally. However, higher-priced items will influence the total price ($\sum P$) more than lower-priced items. Thus there is a built-in bias toward higher-priced items. Different units of measurement for the items (e.g., dozen eggs, pound of coffee, quart of milk) further influence the results.[3] The average of relatives method overcomes these difficulties by focusing on the relative change in price, rather than the absolute price, of the items. But this method is not free of all bias. Those items with the larger relative change in price exert the most influence on the average of relatives index. An advantage of the method is that it provides a mechanism for examining relative change in individual items (price relatives) as well as an aggregate measure of change.

Explicit weighting systems may be used to assign any degree of importance to each item in a composite index. In determining a price index for grocery items, we may wish to weight milk more than margarine because the typical family consumes several quarts of milk for each pound of margarine. The following weighted indexes result from the application of weighting systems to the two previous methods.

---

[3] Consider a price index for meat where the prices are average per animal. Suppose in 1975 the prices for cattle, hogs, and chickens were $400, $200, and $1 while in 1976 the prices were $400, $200, and $3. Although the price of chickens has tripled, the RA index would show virtually no change.

3.6 RELATIVE OF WEIGHTED AGGREGATES INDEX

The *relative of weighted aggregates* (*RWA*) price index uses the quantity purchased as a weight for the price of each item. The rationale for this is that the most frequently purchased items should influence the price index more than the items purchased in smaller amounts. Various measures of quantity purchased are used as weights in this method. The most popular is the quantity purchased in the base period. The relative of weighted aggregates price index for period g with base b and base period quantity weights Q_b is

$$RWA_{g/b} = \frac{\sum(P_g Q_b)}{\sum(P_b Q_b)} \cdot 100. \tag{3.4}$$

EXAMPLE. The 1973, 1975, and 1977 average prices of four beverages are given in Columns (1), (2), and (3) of Table 3.3. The average quantity purchased per week in 1973 by a typical family of four is given in Column (4). Compute an index of beverage prices for each year by the relative of weighted aggregates method. Let 1973 be the base period and use base period quantity weights.

TABLE 3.3: BEVERAGE PRICE INDEX BY RELATIVE OF WEIGHTED AGGREGATES METHOD (1973 = 100)

	(1)	(2) Price	(3)	(4)	(5)	(6) Weighted Price	(7)
				Quantity			
Beverage	1973	1975	1977	1973	$P_{73}Q_{73}$	$P_{75}Q_{73}$	$P_{77}Q_{73}$
Coffee (lb.)	$1.70	$2.44	$3.10	$\frac{1}{2}$	$ 0.85	$ 1.22	$ 1.55
Milk (qt.)	0.34	0.40	0.45	4	1.36	1.60	1.80
Orange juice	0.35	0.37	0.40	2	0.70	0.74	0.80
(12 oz. concentrate)							
Soft drinks	0.89	1.10	1.29	2	1.78	2.20	2.58
(six-carton)							
Total					$ 4.69	$ 5.76	$ 6.73
RWA Index (1973 = 100)					100	122.8	143.5

Solution. The price of each item in each time period is weighted by the quantity purchased in 1973, Q_{73}. These weighted prices are given in Columns (5), (6), and (7). For example, the weighted 1975 price of coffee is $P_{75}Q_{73} = (\$2.44)(\frac{1}{2}) = \$1.22$. The beverage price indexes for 1975 and 1977 are:

$$RWA_{75/73} = \frac{\sum(P_{75}Q_{73})}{\sum(P_{73}Q_{73})} \cdot 100 = \frac{\$5.76}{\$4.69} \cdot 100 = 122.8;$$

$$RWA_{77/73} = \frac{\sum(P_{77}Q_{73})}{\sum(P_{73}Q_{73})} \cdot 100 = \frac{\$6.73}{\$4.69} \cdot 100 = 143.5.$$

A relative of weighted aggregates index with base-period-quantity weights is also known as the *Laspeyres* index. The Laspeyres price index measures the relative change in price that must be paid for the base-year bill of goods in the given period. In the beverage example, it costs 22.8 percent more in 1975 than in 1973 to purchase the 1973 quantities of beverages. The 1973 beverage purchases cost 43.5 percent more in 1977 than in 1973. The interpretation of the relative of weighted aggregates index differs slightly if an alternative quantity weighting system is used.

Many of the major indexes, such as the Consumer Price Index and Wholesale Price Index, are computed by modifications of the Laspeyres formula.

3.7 AVERAGE OF WEIGHTED RELATIVES INDEX

The weights used for the *average of weighted relatives* (*AWR*) price index are called *value weights*, the product of price and quantity. If we are willing to purchase three units of an item at a price of $4 per unit, then (3)($4), or $12, is a measure of the value of this item relative to other purchases. Value weighting systems may use base-period prices and quantities, given-period prices and quantities, or some variation of these. The average of weighted relatives price index for period g with base b and base-period value weights ($P_b Q_b$) is:

$$AWR_{g/b} = \frac{\sum \left(\frac{P_g}{P_b} \cdot 100 \cdot P_b Q_b \right)}{\sum (P_b Q_b)}. \tag{3.4}$$

EXAMPLE. Determine the 1975 and 1977 average of weighted relatives price index for the beverages in Table 3.4 using 1973 as the base period and base-period-value weights.

TABLE 3.4: BEVERAGE PRICE INDEX BY AVERAGE OF WEIGHTED RELATIVES METHOD (1973 = 100)

	(1)	(2)	(3)	(4)	(5)	(6)	(7)	(8)	(9)
		Price			Value	Price Relatives		Weighted Relatives	
				Quantity	Weight				
Beverage	1973	1975	1977	1973	$P_{73}Q_{73}$	1975	1977	1975	1977
Coffee (lb.)	$1.70	$2.44	$3.10	$\frac{1}{2}$	$0.85	143.5	182.4	$122	$155
Milk (qt.)	0.34	0.40	0.45	4	1.36	117.6	132.4	160	180
Orange juice . . . . (12 oz. concentrate)	0.35	0.37	0.40	2	0.70	105.7	114.3	74	80
Soft drinks . . . . (six-carton)	0.89	1.10	1.29	2	1.78	123.6	144.9	220	258
Total					$4.69			$576	$673
AWR Index (1973 = 100)								122.8	143.5

Solution. Base period value weights, $P_{73}Q_{73}$, for each beverage are computed as the product of Column (1) and Column (4). Price relatives of each item for 1975 and 1977 are given in Columns (6) and (7). The weighted price relative for coffee in 1975 is $(P_{75}/P_{73}) \cdot 100 \cdot P_{73}Q_{73} = (143.5)(\$0.85) = \$122$. The average of weighted relatives price indexes are:

$$AWR_{75/73} = \frac{\sum \left(\frac{P_{75}}{P_{73}} \cdot 100 \cdot P_{73}Q_{73} \right)}{\sum (P_{73}Q_{73})} = \frac{\$576}{\$4.69} = 122.8;$$

and

$$AWR_{77/73} = \frac{\sum\left(\dfrac{P_{77}}{P_{73}} \cdot 100 \cdot P_{73}Q_{73}\right)}{\sum(P_{73}Q_{73})} = \frac{\$673}{\$4.69} = 143.5.$$

Notice that the numerical results obtained in Table 3.4 are exactly those previously determined by the relative of weighted aggregates method. The relative of weighted aggregates index with base-period-quantity weights always produces the same result as the average of weighted relatives index with base period value weights. In fact, algebraic simplification of Equation (3.4) produces Equation (3.3):

$$AWR_{g/b} = \frac{\sum\left(\dfrac{P_g}{P_b} \cdot 100 \cdot P_b Q_b\right)}{\sum(P_b Q_b)} = \frac{\sum(P_g Q_b)}{\sum(P_b Q_b)} \cdot 100 = RWA_{g/b}.$$

The two methods do not produce identical results if other weighting systems are employed, however. The importance of distinguishing the two methods centers on the usefulness of the average of weighted relatives index as a mechanism for updating an index. The value weights, $P_b Q_b$, remain constant as a new price relative, $(P_g/P_b) \cdot 100$, is determined for each item in the most recent time period. The price relatives provide measures of relative change for each item. Price relatives are then weighted by the constant-value weights in order to obtain a composite index for the most recent time period. The relative of weighted aggregates index, while numerically equivalent, is not as convenient to update, and it does not provide a measure of relative change for each item included in the composite index.

3.8 REVISION OF INDEXES

Index numbers are occasionally revised to take into account changes in the economy effected by technology, consumer tastes, and spending patterns. A comprehensive revision of the Consumer Price Index (CPI) was introduced in April 1977. Prior to the revision, the CPI measured the change in price of goods and services for all wage earners and clerical workers. The restriction of the index to this specific group of consumers raised questions as to the usefulness of the index for all segments of the economy. But in fact, the index had become so popular that it was being used for purposes other than that for which it was designed. Thus it was felt that a new index was needed. The earlier revision of the CPI was based on a 1960–61 Consumer Expenditure Survey covering 66 areas of the country. To develop the new index and update the existing one, a Consumer Expenditure Survey was conducted in 1972–73 to measure the "market basket" of goods and services purchased by some 40,000 families in 216 areas of the country. The updated version of the earlier index is now termed the Consumer Price Index for Wage Earners and Clerical Workers, Revised. This is abbreviated as CPI-W in current government publications. The newly developed, more comprehensive index is called the Consumer Price Index for All Urban Consumers, and is abbreviated CPI-U. Unlike the old index, it includes purchasing patterns for professional and self-employed workers, and unemployed and retired persons. The CPI-U covers expenditures for 80 percent of the noninstitutional population versus 35 to 40 percent under the old index.[4] Publication of the new index began in January 1978. For times before that, both CPI series are given the same value. Values of the series are given below for three months in 1978 for comparison purposes.[5] The current base period for both CPI series is 1967.

Month 1978	CPI-W	CPI-U
January	187.1	187.2
June	195.3	195.3
October	200.7	200.9

3.9 CHANGE OF BASE PERIOD

The base period of many major indexes is changed occasionally in order to reflect current trends and economic activity. Since introducing the Consumer Price Index in 1917, the Bureau of Labor Statistics has successively changed the base period to $1935-39 = 100, 1947-49 = 100, 1957-59 = 100$, and $1967 = 100$. In addition to the above reason, a change of base period may be desirable for measuring changes from a fixed time period, other than the base period, and for comparing indexes that do not have the same base period.

The Consumer Price Index ($1967 = 100$) and Standard & Poor's Composite Index of Stock Prices ($1941-43 = 10$; includes 500 common stocks) are given for the years 1972–76 in Columns (1) and (3) of Table 3.5. If an investor uses these indexes to compare changes since 1972 in stock prices with changes since 1972 in the Consumer Price Index, several problems are encountered. Neither index has a 1972 base; the two do not even have the same base period. Further, Standard & Poor's Composite Index uses a base value 10, rather than the common value of 100 for the base period. In order to compare changes in the two indexes since 1972, change the base period of each index to 1972 with a base value of 100. The index for any *given* period g with a *new base* period n, $I_{g/n}$, is determined by dividing the index for period g with *old base* period b, $I_{g/b}$, by the old index for time period n with base b, $I_{n/b}$, and multiplying the result by the *desired base value*. This is:

$$I_{g/n} = \frac{I_{g/b}}{I_{n/b}} \cdot \text{(desired base value)}. \tag{3.5}$$

The Consumer Price Index for 1974 with a new base period of 1972 and a base value of 100 is:

$$I_{74/72} = \frac{I_{74/67}}{I_{72/67}}(100) = \frac{147.7}{125.3}(100) = 117.88.$$

---

[4] See Shiskin (1974) for further discussion of the revision.
[5] *Survey of Current Business.*

TABLE 3.5: CHANGING
THE BASE OF THE
CONSUMER PRICE INDEX
AND STANDARD &
POOR'S COMPOSITE
INDEX OF STOCK
PRICES TO 1972 = 100

	(1)	(2)	(3)	(4)
	Consumer Price Index		*Standard & Poor's Composite Index of Stock Prices*	
Year	*(1967 = 100)*	*(1972 = 100)*	*(1941–43 = 10)*	*(1972 = 100)*
1972	125.3	100	109.20	100
1973	133.1	106.23	107.43	98.38
1974	147.7	117.88	82.85	75.87
1975	161.2	128.65	86.16	78.90
1976	170.5	136.07	104.20	95.42

Source: Council of Economic Advisors, *Economic Indicators* (December 1976).

Likewise, Standard & Poor's Composite Index for 1974 with a new base period of 1972 and a base value of 100 is:

$$I_{74/72} = \frac{I_{74/41-43}}{I_{72/41-43}} (100) = \frac{82.85}{109.20} (100) = 75.87.$$

The results of changing the base of both indexes to 1972 = 100 are given in Columns (2) and (4) of Table 3.5. This shows that Standard & Poor's Composite Index declined 4.58 percent over the years 1972 to 1976, while the Consumer Price Index rose 36.07 percent. What do these results indicate about the stock market as an investment over this time?

3.10 USE OF CONSUMER PRICE INDEX

The Consumer Price Index is often referred to as a "cost of living" index. In this sense, the index is widely used to compare price changes with changes in income, wage rates, retail sales, and the like. A basic concern of every individual is whether income keeps pace with the cost of living. To illustrate, per capita personal income for the years 1972 to 1976 is given in Column (2) of Table 3.6. The increase from 1972 to 1976 is $2,017, or 44.5 percent. But retail prices increased substantially over this period, as shown by the Consumer Price Index, given in Column (1). The effect of changing prices on a dollar series, such as per capita personal income, may be removed by *deflating* the series expressed in *current dollars* by an appropriate price index to produce a measure expressed in terms of *real dollars*. This relationship is

$$\text{Real dollars} = \frac{\text{Current dollars}}{\text{Price index}} (100). \tag{3.6}$$

The measure of real dollars, also called *constant dollars*, is expressed in terms of the price level at the time of the base period of the index. Real per capita personal income, expressed in terms of 1967 prices, for 1974 is ($5486/147.7)(100) = $3,714. This means that the 1974 income of $5,486 could purchase $3,714 worth of goods and services at 1967 prices, as measured by the consumer price index. Real per capita personal income for 1972 to 1976 is given in Column (3) of Table 3.6. The increase in real dollars from 1972 to 1976 is 6.2 percent, much less than the 44.5 percent increase in current dollars.

TABLE 3.6: USING THE CONSUMER PRICE INDEX

Year	(1) Consumer Price Index (1967 = 100)	(2) Per Capita Personal Income Current Dollars	(3) Per Capita Personal Income Real Dollars (1967 = 100)	(4) Purchasing Power of a Dollar (1967 = 100)
1972	125.3	$4,537	$3,621	0.7981
1973	133.1	5,049	3,793	0.7513
1974	147.7	5,486	3,714	0.6770
1975	161.2	5,902	3,661	0.6203
1976	170.5	6,554	3,844	0.5865

Source: Bureau of Labor Statistics.

The *purchasing power* of a dollar expresses the value of a dollar received in a current time period relative to base-period prices. At the retail level, the Consumer Price Index is used to measure purchasing power. The *purchasing power of a retail dollar* in period g relative to base period b, $PP_{g/b}$, is

$$PP_{g/b} = \frac{1}{CPI_{g/b}} (100). \qquad (3.7)$$

The purchasing power of dollar in 1974 is

$$PP_{74/67} = \frac{1}{CPI_{74/67}} (100) = \frac{1}{147.7} (100) = 0.677.$$

Thus a dollar received in 1974 is worth $0.677 compared to a dollar received in 1967. In other words, the 1974 dollar is valued at 67.7 percent of a 1967 dollar,

Reprinted by permission *The Wall Street Journal*

a decline in purchasing power of 32.3 percent. A measure in real-dollar terms can also be obtained from a measure of purchasing power:

$$\text{Real dollars} = (\text{Current dollars})(\text{Purchasing power}). \qquad (3.8)$$

Real per capita personal income in 1974, relative to 1967 prices, is $(\$5,486)(0.677) = \$3,714$.

3.11 SUMMARY

Index numbers measure relative change, usually over time. Specifically, an index measures percentage change from the base period. The major classifications of indexes are price, quantity, and value. A simple index measures the relative change over time for a single item. Composite indexes measure the relative change in a group of closely related items. The major methods for computing index numbers are the relative of aggregates, average of relatives, relative of weighted aggregates, and average of weighted relatives.

The Consumer Price Index, prepared by the Bureau of Labor Statistics, is probably the most widely referenced index. It is used to compare changes in retail prices with changes in other variables such as income and retail sales. Since January 1978, there have been two Consumer Price Index series. The Consumer Price Index for Wage Earners and Clerical Workers is a revision of the original index. The newer Consumer Price Index for All Urban Consumers is more comprehensive and representative of a larger segment of the population. A frequent application of the Consumer Price Index is deflating the values of a retail dollar series expressed in current dollars. The results are the values of the series in constant or real dollars.

3.12 TERMS AND FORMULAS

Base period	Price index
Composite index	Purchasing power
Constant dollars	Quantity index
Current dollars	Real dollars
Deflation	Simple index
Laspeyres index	Value index

Average of relatives price index

$$AR_{g/b} = \frac{\sum\left(\dfrac{P_g}{P_b} \cdot 100\right)}{n}$$

Average of weighted relatives price index

$$AWR_{g/b} = \frac{\sum\left(\dfrac{P_g}{P_b} \cdot 100 \cdot P_bQ_b\right)}{\sum(P_bQ_b)}$$

Change of base period

$$I_{g/n} = \frac{I_{g/b}}{I_{n/b}} \cdot (\text{desired base value})$$

Price relative

$$\frac{P_b}{P_g}(100)$$

Purchasing power of a retail dollar $\qquad PP_{g/b} = \dfrac{1}{\text{CPI}_{g/b}} (100)$

Real dollar measures $\qquad\qquad \text{Real dollars} = \dfrac{\text{Current dollars}}{\text{Price index}} (100)$

$\qquad\qquad\qquad\qquad\qquad\qquad$ Real dollars =
$\qquad\qquad\qquad\qquad\qquad\qquad$ (Current dollars)(Purchasing power)

Relative of aggregates price index $\qquad RA_{g/b} = \dfrac{\sum P_g}{\sum P_b} \cdot 100$

Relative of weighted aggregates
price index $\qquad\qquad\qquad\qquad RWA_{g/b} = \dfrac{\sum (P_g Q_b)}{\sum (P_b Q_b)} \cdot 100$

Simple price index $\qquad\qquad\qquad I_{g/b} = \dfrac{P_g}{P_b} \cdot 100$

QUESTIONS

A. True or False

1. A price relative is the ratio of prices for a single item at two points in time, multiplied by 100.

2. A quantity index has a value of 64 in 1978 and a value of 59 in 1979. This indicates that quantity declined 5 percent from 1978 to 1979.

3. The purchasing power of a dollar, based on the CPI (1967 = 100), is 0.5845 in 1976. Thus a 1976 dollar is worth 58.45 percent of a 1967 dollar.

4. Deflation is the process of converting current dollars into constant dollars.

5. The Consumer Price Index is published by the Consumers' Union (a nonprofit organization).

6. If a price index has a value of 100 in 1977, the base period for the index must be 1977.

7. The CPI is an unweighted index.

8. A simple index measures the relative change from the base period for a single item.

9. The relative of weighted aggregates index with base-period-quantity weights always produces the same result as the average of weighted relatives index with base-period-value weights.

B. Fill in

1. The _____ price index expresses the total price for a group of items in a given period as a ratio to total price for the same items in the base period, multiplied by 100.

2. Retail sales in current dollars that have been adjusted for changes in the Consumer Price Index are expressed in terms of _____ dollars.

3. A department store reports 1979 sales of $756,000; sales in real dollars are $504,000. The index used to determine real-dollar sales has a 1979 value of _____ .

4. A simple index of quantity is 120 in 1974 and 130 in 1975. If the actual quantity in 1975 is 520, the quantity in 1974 is _____ .

5. Index numbers can be classified according to the method of construction as simple or _____ .

6. A relative of weighted aggregates index with _____ period weights is also known as the Laspeyres index.

7. The _____ price index is the arithmetic mean of the price relatives (multiplied by 100) for a group of items.

8. _____ weights are used in the average of weighted relatives price index.

EXERCISES

1. The average price for four food items in 1973, 1975, and 1977 is given below, along with the quantity of each purchased weekly by a typical family of four.

| | Price | | | Quantity |
Food	1973	1975	1977	1973
Bread (lb.)	$0.35	$0.42	$0.48	2
Eggs (doz.).	0.59	0.83	0.95	1
Ground beef (lb.)	0.69	0.89	0.97	3
Potatoes (lb.)	0.45	0.98	1.70	2

Using 1973 as the base period, determine and interpret indexes for 1975 and 1977 by:
a. The relative of aggregates method.
b. The average of relatives method.
c. The relative of weighted aggregates method.
d. The average of weighted relatives method.

2. A retailer's advertising expenditures by major media for 1978 and 1979 are listed below. Using 1978 as the base period, compute for 1979:
a. The relative of aggregates.
b. The average of relatives index of advertising expenditures.

| | Expenditure | |
Medium	1978	1979
Newspapers	$8,001	$8,442
Magazines	1,504	1,465
Radio.	1,837	2,025
Television	4,851	5,272
Direct mail.	3,986	4,155

3. The portfolio of stocks for a small investor is given below along with the average price of each stock in recent years.

Company	Shares Purchased in 1977	Average Price/Share		
		1977	1978	1979
Best Bank	50	15	19	12
Good Foods	250	40	44	50
Recycled Oil	100	25	24	30

Determine the following indexes of price per share for the portfolio, using 1977 as the base period:
a. Relative of aggregates.
b. Average of relatives.
c. Relative of weighted aggregates (use shares purchased in 1977 as weights).
d. Average of weighted relatives with base-period-value weights.

4. Did the investor's portfolio in Question 3 outperform the market between 1977 and 1979? Answer this by comparing the relative of weighted aggregates index with the changes in the Dow-Jones Industrial Average (in *The Wall Street Journal, Economic Indicators*, etc.) for these years.

5. The manager of the local processing center for a national credit card service reports that the total of sales drafts for the Thanksgiving-to-Christmas period (November 26 to December 24) was 30 percent larger in 1976 than in 1975. Realizing this period was one working day longer in 1976, the manager computes average daily sales to obtain a more representative comparison: $910,776 in 1976 and $737,515 in 1975, a 23.5 percent increase. For further analysis, the manager will express average daily sales in real dollars. The Consumer Price Index (1967 = 100) is 161.2 in 1975 and 170.5 in 1976. What is the percentage change in real average daily sales from 1975 to 1976?

6. A Congressional opponent of a recently proposed increase in the U.S. defense budget argues that defense spending has traditionally increased too rapidly, and produces the Department of Defense expenditures for 1970–76 given below. A proponent argues that defense spending, measured

Year	Expenditures of U.S. Department of Defense (millions of dollars)	Consumer Price Index (1967 = 100)
1970	78,360	116.3
1971	79,922	121.3
1972	76,679	125.3
1973	75,000	133.1
1974	79,307	147.7
1975	87,471	161.2
1976	90,160	170.5

Source: U.S. Department of Treasury, Survey of Current Business.

in constant dollars, actually declined in almost every year over this period. Express the Department of Defense expenditures in terms of constant dollars using the Consumer Price Index. Comment on the significance of the two arguments.

7. An economist wishes to determine real disposable personal income per capita (PIPC) in the United States for the period 1965–72. The economist has ascertained PIPC in current dollars for these years. The Consumer Price Index will be used to convert PCPI in current dollars to real dollars. The CPI is available for 1965–69 with a 1957–59 base and for 1969–72 with a 1967 base.

    a. Convert the 1965–68 CPI values in Column (2) to a 1967 base.
    b. Use the resulting values and those of Column (3) to determine real disposable PIPC (in 1967 dollars) for the years 1965–72.
    c. What percentage increase in real disposable PIPC occurred between 1965 and 1972?

Year	(1) U.S. Disposable Personal Income per Capita	(2) Consumer Price Index (1957–59 = 100)	(3) Consumer Price Index (1967 = 100)
1965	$2,436	109.9	—
1966	2,604	113.1	—
1967	2,749	116.3	—
1968	2,945	121.2	—
1969	3,130	127.7	109.8
1970	3,376	—	116.3
1971	3,605	—	121.3
1972	3,843	—	125.3

Source: *Economic Report of the President*, 1970 and 1975.

8. Three measures of common stock prices are given below for the period 1972–76. The New York Stock Exchange Index includes all stocks (more than 1,500) listed on the New York Stock Exchange. The Dow-Jones Industrial Average is based on 30 stocks, while the Standard & Poor's Index is based on 500. Which of the three measures has the largest percentage change between 1972 and 1976?

Year	New York Stock Exchange Composite Index (December 31, 1965 = 50)	Dow-Jones Industrial Average	Standard & Poor's Composite Index (1941–43 = 10)
1972	60.29	950.71	109.20
1973	57.42	923.88	107.43
1974	43.84	759.37	82.85
1975	45.73	802.49	86.16
1976	54.46	993.20	102.01

Source: *Economic Indicators*, December 1976, p. 31.

9. Information on total personal income in the United States for 1970–75 is given below.
   a. How is the information in Column (2) determined from the data in Column (1)?
   b. Using this information, estimate the percentage change in the cost of living from 1970 to 1975.

Year	(1) Personal Income in Current Dollars* (billions)	(2) Personal Income in Real Dollars (1967 = 100) (billions)
1970.	801.3	689.0
1971.	859.1	708.2
1972.	942.5	752.2
1973.	1,052.4	790.7
1974.	1,153.3	780.8
1975.	1,249.7	775.2

* Source: *Economic Indicators*, December 1976, p. 5.

10. Both the relative of aggregates and average of relatives indexes have a built-in bias. Explain how this bias arises. How can this bias be overcome?

11. In 1967, Joe Max started work with his present company at a salary of $7,100 per year. Because his annual salary ten years later is $11,500, Joe feels proud of the regular salary increases he has received. A colleague of Joe's says he should ask his employer for more money, since the 1977 CPI is 181.5 (1967 = 100). Has Joe's salary kept up with the cost of living, as measured by the CPI?

12. Data are given below for two indicators of construction activity. The F. W. Dodge Corporation provides an Index of Total Construction Value, including residential, commercial, and institutional building, plus highways and public works. E. H. Boeckh and Associates prepares an Index of Residential Construction Cost.
   a. Which index has the largest percentage change from 1976 to 1977?
   b. Compare the changes in these construction indicators with changes in the Consumer Price Index.

Year	Dodge (1967 = 100)	Boeckh (1972 = 100)	CPI (1967 = 100)
1976	199	136.2	170.5
1977	252	148.7	181.5

Source: *Survey of Current Busines*, Federal Housing Administration.

13. The Federal Housing Administration (FHA) operates loan insurance and subsidy programs designed to help stabilize the home mortgage market, improve construction standards, and facilitate sound financing. Amounts

of mortgage and other loans insured by the FHA are given below for 1975–77. Using 1975 as the base period, construct:

a. The relative of aggregates.

b. The average of relatives indexes of FHA-insured loans.

Loan Type	FHA-Insured Mortgage and Other Loans Made ($ millions)		
	1975	1976	1977
Home mortgages.	6,166	6,362	8,841
Project mortgages	976	2,314	2,818
Property improvement loans.	739	942	1,341

Source: Federal Housing Administration.

14. The number of public and private housing starts, by region of the United States, is given below for 1975–77. Using 1975 as the base period, construct indexes of housing starts by

a. The relative of aggregates method.

b. The average of relatives method.

c. Which region has the largest percentage increase in housing starts from 1975 to 1976? From 1975 to 1977?

Region	Private and Public Housing Starts (thousands)		
	1975	1976	1977*
Northeast	149.9	169.5	201.4
North Central.	295.4	400.7	465.8
South.	448.3	574.9	784.2
West	277.7	402.5	537.8
Total U.S.	1,171.3	1,547.6	1,989.2

* Preliminary.

Source: Bureau of the Census.

15. The 1975 *Economic Report of the President* lists the average gross weekly earnings in private nonagricultural industries for 1973 as $145.43 in current dollars and $109.26 in 1967 dollars (based on the CPI). Use this to determine the 1973 Consumer Price Index.

16. Select one of the indexes from Table 3.1 and use your library to research and summarize the development and use of the index. Include the purpose of the index, method of construction, data collection procedures, use of the index, revisions, etc.

17. A leading newspaper reported in 1978 that the new Consumer Price Index for all urban consumers (CPI-U) increased by two percentage points from May 1978 to June 1978. Does this mean that the average price, for the

items included in the index, increased two percent? (Note: The CPI-U was 193.3 and 195.3 in May and June, 1978, respectively.)

REFERENCES

The Consumer Price Index: Concepts and Content Over the Years. Report 517, Bureau of Labor Statistics, 1977.

Croxton, Frederick E.; Cowden, Dudley J.; and Klein, Sidney   *Applied General Statistics*, 3d ed. Englewood Cliffs, N.J.: Prentice-Hall, Inc., 1967.

Shao, Steven P.   *Statistics for Business and Economics*, 3d. ed. Columbus, Ohio; Charles E. Merrill Publishing Co., 1976.

Shiskin, Julius   "Updating the Consumer Price Index: An Overview." *Monthly Labor Review* (July 1974), pp. 3–20.

Stockton, John R., and Clark, Charles T.   *Introduction Business and Economic Statistics*, 5th ed. Cincinnati, Ohio: South-Western Publishing Co., 1975.

Wallace, William H., and Cullison, William E.   *Measuring Price Changes: A Study of the Price Indexes*, 3d ed. Richmond, Va.: Federal Reserve Bank of Richmond, 1976.

"Which Price Indicator Do You Believe?" *Business Week* (May 14, 1979), pp. 120–22.

Introduction to Probability

4 INTRODUCTION TO PROBABILITY

I think therefore it would be useful,
not only to gamesters but to all men in
general, to know that chance has rules
which can be known, and that through
not knowing these rules they make faults
every day.

Montmort, 1708

4.1 MEASURING UNCERTAINTY

Every day we make decisions in situations characterized by uncertainty. You may decide to bet a dollar with a friend that you will beat him or her in your next tennis match because you feel that the chances of winning are in your favor. Many people attend college partly because they feel that a college education improves their chances for a higher lifetime income. You may decide to forgo studying tonight for tomorrow's history class because you feel there is little chance that your instructor will give an unannounced quiz. These are all personal decisions that are made in the face of uncertainty. You may be uncertain, in varying degrees, about who will win the next tennis match, whether a college education is right for you, or whether an unannounced quiz will be given.

Probability is a measure of the chance that an uncertain event will occur.

A fundamental understanding of probability is not only helpful in understanding how we make personal decisions, but more important for our purposes, it is the foundation of inferential statistics.

The early development of probability was spurred by an interest, sometimes pecuniary, in games of chance, particularly those involving dice. Card games also provided a source for analysis of chance events. However cards were not introduced until *circa* 1350 and took centuries to displace dice as the major element in gaming. In fact, Montmort (quoted earlier) was one of the first to study both dice and card games. These early studies, however motivated, did establish that "chance has rules which can be known . . . ." The objective of this chapter is to introduce probability and its rules and to learn their applicability "not only to gamesters, but all men in general . . . ."

4.2 EXPERIMENTS AND EVENTS

In order to use probability as a measure of uncertainty, a given situation must be expressed in the language of probability. Some of the terms in this language are now introduced. Your playing a tennis match, as an example situation, may be called an *experiment*. The experiment has two *outcomes:* either you win the match or you lose. The playing of a tennis match is equivalent to conducting an experiment when you are uncertain about which outcome will occur.

> An *experiment* is any well-defined situation or procedure that results in one of two or more possible outcomes.

In analyzing any situation, it is always helpful to determine the possible outcomes, as two outcomes are possible for the playing of a tennis match. How many outcomes are possible when a single six-sided die (one of a pair of dice) is tossed against a wall and allowed to come to rest? There are six possible outcomes, or faces of the die that may face upward, after the die comes to rest. These faces are usually identified as 1, 2, 3, 4, 5, and 6. These outcomes may also be called the possible *events* that may occur. But an event may also refer to more than one distinct possible outcome. Suppose we divide the six faces of the die into two groups called "even-numbered face" and "odd-numbered face." Thus the *event* "even-numbered face" consists of three outcomes, the faces 2, 4, and 6. The outcomes 1, 3, and 5 are associated with the event "odd-numbered face." If the die is tossed and the outcome is the 6 facing up, the event that occurred is an even-numbered face.

> An *event* identifies one or more outcomes of an experiment.

The discussion above refers to the hierarchy of events in the measurement of probability. An *elementary event* expresses individual outcomes in their simplest terms, that is, in the most *elementary* form. *Compound events* are expressed in terms of a group of outcomes or a group of *elementary events*. In the toss of a single six-sided die, there are six elementary events. The events "even-numbered face" and "odd-numbered face" are compound events, each made up of three elementary events. These are illustrated in Table 4.1.

TABLE 4.1:
ELEMENTARY AND
COMPOUND EVENTS FOR
A SIX-SIDED DIE

Six Elementary Events	Two Compound Events
1 3 5	Odd-numbered face
2 4 6	Even-numbered face

In practice, the terms elementary events and compound events are used rather loosely. You may find that the word "event" is used to refer to either individual outcomes or a group of outcomes. This causes no confusion however. If we speak of the *event* "even-numbered face," then this obviously is a compound event since it refers to more than one outcome. Likewise, a reference to the *event* "four" in tossing a single six-sided die immediately indicates an elementary event.

A list of events is *exhaustive* if the list identifies all possible outcomes in a situation. The two events "win" and "lose" are exhaustive for the playing of a tennis match, since no other event is possible: one of these two must occur. In tossing a single six-sided die, the events 1, 2, 3, 4, 5, and 6 are exhaustive. Are the events even-numbered face and odd-numbered face exhaustive?

When all possible outcomes of an experiment are grouped into two categories or events, either event may be expressed as the *complement* of the other event. When a single die is tossed, either an even-numbered face or an odd-numbered face must occur. The event "odd-numbered face" can be referred to as the complement of the event "even-numbered face." In the playing of a tennis match, the event "lose the match" is the complement of the event "win the match."

An experiment may consist of one or more *trials*. If a die is tossed ten times to observe the outcome of each toss, each toss is called a *trial* and the experiment consists of ten trials. If you play a tennis match once a week with the same person for a year, this may be considered an experiment over one year made up of 52 trials.

In the toss of a single die, consider the following events: the four, the five, and an even-numbered face. Notice that the event "even-numbered face" includes the outcomes 2, 4, and 6. If the die was tossed and a four occurred, the result could be expressed as the elementary event "four" or as the compound event "even-numbered face." If we bet some money on the occurrence of *either* a four or an even-numbered face, it would be important to know that the event "four" is also the event "even-numbered face." However, if we bet on the occurrence of either a five or an even-numbered face, we would recognize that the event "five" is not an "even-numbered face." In such a case, we say that the events "five" and "even-numbered face" are *mutually exclusive* events, because they cannot occur together on a single trial. The events "four" and "even-numbered face" are *not mutually exclusive*, as they can occur together on a single toss of the die.

CHRISTIANUS HUYGENS (1629–95)

Lord Huygens was born in the Netherlands into one of that country's leading families. He studied law and mathematics at various European universities before receiving a law degree from the Protestant University at Angers, France, in 1655. On a visit to Paris he became interested in the work of his contemporaries on such topics as the historical problem mentioned in Chapter 1, and subsequently prepared a treatise on dice games, *De Ratiociniis in Aleae Ludo*, which was published in 1657. In this work, Huygens says "I do not treat here a simple game of chance, but I have thrown out the elements of a new theory, both deep and interesting." Indeed his treatise, the first book on probability theory, was used as a standard reference for some 50 years. Particularly well developed in his work was the concept of mathematical expectation (a topic to be pursued in a later chapter). Huygens is today considered the father of probability theory.

> Two or more events that cannot occur together on a single trial of an experiment are *mutually exclusive* events.

4.3 COUNTING

Counting the number of ways in which events may occur in an experiment plays a major role in probability. Some rules for counting that are valuable in the study of probability are presented in this section. The first of these rules is called the *fundamental principle of counting*.

> The *fundamental principle of counting* specifies that if one event can occur in n_1 ways and another event can occur in n_2 ways, the two events can occur together in $n_1 n_2$ ways.

Apply the fundamental principle of counting to the following situation. A six sided die and a coin are tossed simultaneously. How many events are possible? The toss of the die may result in any one of six events, the faces numbered 1, 2, 3, 4, 5, and 6. The toss of the coin produces one of two events, either a head or a tail. The two events can occur together in $(6)(2) = 12$ ways.

Two different experiments are involved in the above example. The toss of the die is one, the toss of the coin is another. The fundamental principle of counting determines the number of possible outcomes for the joint experiment. Define a a particular outcome of one experiment and a particular outcome of another experiment as a *joint event*. If the toss of the die results in the face numbered "4" and the toss of the coin produces a "head," then the joint event is called "4 and head," which may be written "4, head." The fundamental principle of counting specifies that there are 12 distinct joint events that may occur as a result of the two experimental situations. All 12 joint events are identified in Table 4.2, where H denotes head and T denotes tail.

The fundamental principle of counting holds for any number of experiments, not just two as illustrated previously. The number of joint events possible is simply the product of the number of outcomes possible in each experimental situation.

TABLE 4.2: JOINT EVENTS FOR TOSSING A DIE AND A COIN

1, H	1, T
2, H	2, T
3, H	3, T
4, H	4, T
5, H	5, T
6, H	6, T

EXAMPLE. A compulsive gambler makes a bet in each of four uncertain situations: the toss of a six-sided die, the toss of a coin, the winner of a Monopoly game with five participants, and the winner of an NFL playoff game. The total number of possible joint events is 120, the product of 6, 2, 5, and 2, the number of outcomes possible in the four experimental situations.

Permutations

Other important counting rules pertain to the arrangement of items with regard to the order of the items. Consider a three-digit number such as 527. If the order of the digits were switched, would the same three-digit number exist? Certainly not: switching the order of the 5 and 2 gives the number 257. The order or sequence in which the digits appear is exactly what determines the numerical value of the three-digit number, or for that matter, any number.

> *Permutations* are groups of items where both the composition of the groups and the order within a group are important.

The number 527 is a different permutation from 257 because of order. How many different permutations of the digits 5, 2, and 7 are possible? First, determine the answer to this question by examining how a three-digit number is constructed. Set up a rectangle with three spaces inside, as shown below. If we start filling spaces from the left with one of the three digits, any one of the three may be

chosen. Since any one of three digits may be selected for this first space, place a 3 in that position.

3		

Assuming a digit is chosen for this first position, how many digits remain as we decide which to place in the next position? Two out of three remain, and a 2 is written in the position to the right of the 3.

3	2	

After selecting two out of the three available digits, only one can be chosen for the final position. Thus the spaces in the rectangle show that any one of three digits can be placed in the first position,

3	2	1

any one of two digits in the second, and only one in the last position. Appealing to the fundamental principle of counting, the number of ways all three positions can be filled is the product of 3, 2, and 1. Thus there are six ways of arranging the three distinct digits to form different three-digit numbers. A general rule for the number of permutations follows, assuming that each item can occur only once in each arrangement.

The *number of permutations*[1] of n distinct items arranged x at a time is

$$_nP_x = \frac{n!}{(n-x)!},$$ (4.1)

where $n!$, read n factorial, is

$$n! = n(n-1)(n-2)\ldots(1).$$

The number of permutations of the three digits arranged three at a time is

$$_3P_3 = \frac{3!}{(3-3)!} = \frac{3!}{0!} = (3)(2)(1) = 6.$$

The six permutations are 527, 257, 725, 572, 275, and 752.

How many two-digit numbers can be formed from the three digits 5, 2, and 7? This is a permutation of 3 items arranged 2 at a time, which gives:

$$_3P_2 = \frac{3!}{(3-2)!} = \frac{(3)(2)(1)}{1} = 6.$$

The six permutations are 52, 25, 57, 75, 27, and 72.

EXAMPLE. Six people have been nominated as officers in an organization in which you are a member. Three offices, president, vice president, and secretary/treasurer, must be filled. Each of the six nominees may serve in any office. How many different arrangements of officers may be chosen from the six candidates? Order of selection is important since Joe as president, Mary as vice president, and Tom as secretary/treasurer constitutes a different arrangement of officers from Mary as president, Tom as vice president, and Joe as secretary/treasurer. Additionally, any of the six candidates may be chosen. Using the formula, a total of 120 arrangements is possible.

$$_6P_3 = \frac{6!}{(6-3)!} = \frac{6!}{3!}$$

$$= \frac{6 \cdot 5 \cdot 4 \cdot 3!}{3!} = 6 \cdot 5 \cdot 4$$

$$= 120.$$

A different kind of permutation problem exists when not all of the items are distinct. There are 3! or 6 permutations of three distinct items arranged three at a time (recall the permutations of 527). If all items are not distinct, the number of possible permutations would be reduced. Suppose the three items are the integers 5, 2, and 2. How many different three-digit numbers may be formed? Only three: 522, 252, and 225. In the arrangement 225, we cannot distinguish between the first "2" and the second "2," i.e., switching the order

---

[1] If $x = n$ here, the denominator $(n-x)!$ is 0! or 1 (by definition, 0! = 1), and $_nP_n = n!$.

of these first two digits would not produce a different three-digit number. If all three digits were distinct, switching any two digits would yield a different three-digit number. The following permutation rule applies to these situations.

> For n items that can be divided into k categories where x_1 of the items are of one kind, x_2 of the items are of another kind, . . . , and x_k of the items are of another kind so that $x_1 + x_2 + \cdots + x_k = n$, the number of *permutations* of all n items is
>
> $$_nP_{x_1, x_2, \ldots, x_k} = \frac{n!}{x_1! x_2! \cdots x_k!}. \qquad (4.2)$$

Applying the above rule to determine the possible permutations of 522,

$$_3P_{1,2} = \frac{3!}{1!2!} = 3.$$

EXAMPLE. An accounting firm has four employees who are Certified Public Accountants (CPAs), five accountants without certificates, and two clerks. How many ways can the group of 11 employees be assigned to eleven one-person projects today if we recognize only the distinction between the CPA, the non-CPA accountant, and the clerk?

$$_{11}P_{4,5,2} = \frac{11!}{4!5!2!} = 6,930.$$

In the preceding examples, we have not allowed an integer or item to be repeated within a given sequence. Suppose we allow each of the integers in the three-digit number 527 to be used more than once. If any integer could be used up to three times in the three digits, numbers like 522, 775, 555, and 222 are possible. Refer to the three-rectangle scheme used earlier and determine the possible number of permutations of three-digit numbers if the integers can be duplicated in the three-digit number.

3	3	3

There could be any of three integers in the first position, any of three in the second, and any of three in the third position, which gives 27 possible permutations. Allowing duplications greatly increases the number of permutations possible. Similarly, the number of permutations of three integers from the ten integers 0, 1, . . . , 9, where duplication is allowed is $(10)(10)(10)$ or 1,000. The following rule may be used to determine the number of permutations where duplications are possible.

> The number of *permutations* of n distinct items that may be used to occupy r positions where each of the n items may be duplicated is n^r.

EXAMPLE. A state Division of Motor Vehicles uses a four-letter prefix on automobile license plates. If any four of the 26 letters from the alphabet may be used on a license plate, including the use of letters more than once on the same license plate, how many different four-letter prefixes are available?

Solution. $n^r = (26)^4 = 456,976.$

Combinations

Permutations concern ways in which both order and composition are important. But in some situations order is insignificant. Suppose your instructor assigns Problems 1, 5, and 8 at the end of this chapter for homework and asks you to turn in your solutions tomorrow. Would your homework assignment be different if your instructor said "Let's see, work Problem 8, Problem 1, and, oh yes, also Problem 5?" The order in which the problems are assigned is different but you still have the same homework assignment. Order is not important here.

> *Combinations* are groups of items where the composition of the group, but not order, is important.

Refer to Table 4.2, which lists the joint events possible when a die and coin are tossed. Would changing the order in which the outcomes are listed, e.g., from 3, *H* to *H*, 3, constitute a different joint event? No, the joint events are combinations of outcomes; order is not important.

The six permutations of any two digits from the three digits 5, 2, and 7 are repeated below.

52	25
57	75
27	72

How many combinations of two digits can be formed from these three? Only three. Observe that, in each row, the two digits in the left column appear in a different order in the right column. If order is not important, each row contains the same combination of two digits. The following rule may be used for combinations.

> The *number of combinations*[2] possible by selecting x out of n distinct items is
>
> $$_nC_x = \frac{n!}{x!(n-x)!}. \qquad (4.3)$$

---

[2] The symbol for combinations, $_nC_x$, is also written as C_x^n and $\binom{n}{x}$. The relationship between the permutation and combination formulas is: (1) there are $_nC_x$ ways to comprise a group of x from n without regard to order; (2) given a group of x there are $x!$ ways to arrange these items; (3) hence the number of permutations of x items from a group of n is given by (using the fundamental principle of counting) the product of (1)(2) or $_nP_x = (_nC_x)(x!)$.

Use this formula to verify the combinations of two digits from those in the number 527.

$$_3C_2 = \frac{3!}{2!(3-2)!}$$

$$= \frac{3!}{2!1!} = 3.$$

How many combinations of all three digits used at once are possible using those in the number 527?

$$_3C_3 = \frac{3!}{3!(3-3)!}$$

$$= \frac{3!}{3!0!} = 1.$$

Zero factorial appears in the denominator above. By definition, $0! = 1$. The illustration points out another rule to remember.

> Only *one combination* is possible when n items are selected without regard to order out of a group of n distinct items.

Values of $_nC_x$ are given in Appendix C for $n \le 20$.

EXAMPLE. How many ways can an instructor select five students for a group project out of a class of 12? Since order is not important, the number of possible groups is

$$_{12}C_5 = \frac{12!}{5!(12-5)!} = \frac{12!}{5!7!}$$

$$= \frac{12 \cdot 11 \cdot 10 \cdot 9 \cdot 8 \cdot 7!}{5 \cdot 4 \cdot 3 \cdot 2 \cdot 1 \cdot 7!} \quad = 120$$

$$= 792.$$

The various rules for counting are used extensively in probability. The following section provides some fundamental concepts in probability.

4.4 PROBABILITY FUNDAMENTALS

A statement of probability is useful in any situation characterized by uncertainty. An uncertain situation is any in which we are unsure about the outcome. We may know the possible events that can occur but we may not know exactly which will occur. Or we may be uncertain about the time of occurrence for some event. Events that occur in uncertain situations are called *random events*.

If Pete and Bill regularly play tennis and we know that Bill wins about two-thirds of all games, we still do not know who will win the next game played. You might wish to make a bet that Bill will win the next game. But you are not *certain* of this, for the events occur in a *random* manner. Denote the two possible events for each game as B (Bill wins) and $\bar{B}$ (Bill loses, in which case Pete wins). Suppose the next six games played produce the following events: B, $\bar{B}$, B, B, $\bar{B}$, B. Given the information that Bill wins two-thirds of all games and the outcomes for the last six games, can you now predict with certainty who will win the next game? Of course not. The two possible events B and $\bar{B}$ still occur randomly. Our best prediction would be to state that Bill wins about two-thirds of the time. The numerical value $\frac{2}{3}$ is a statement about the chances of Bill winning. Thus $\frac{2}{3}$ is a measure of the uncertainty associated with Bill winning the next game, the event B. Probability is used as a measure of uncertainty, so that the value $\frac{2}{3}$ represents the probability of the event B. The notation used to represent the probability of event B is simply $P(B)$. Likewise, $P(\bar{B})$ represents the probability that Pete wins. The definition of probability, stated earlier, follows.

Probability measures the likelihood or chance that an uncertain event will occur.

Equally Likely Events

Occasionally all events in an experimental situation have the same chances of occurrence. Refer to Table 4.1, which lists the six elementary events that may occur when a six-sided die is tossed. A die is called fair if each of the six faces has the same chance of occurrence. Events that have the same chance of occurring are called *equally likely events*. If a die is fair, then the face "2" is just as likely to occur as the face "5" or any other face. Since each of the compound events "even-numbered face" and "odd-numbered face" is made up of three elementary events, the compound events are also equally likely if the die is fair. Are the joint events in Table 4.2 equally likely if both the die and coin are fair?

Axioms of Probability

The assignment of a number to represent the probability of an event is based on a set of axioms which were first formalized by the Russian mathematician A. N. Kolmogoroff and published in 1933. Given a set of n mutually exclusive and collectively exhaustive events $E_1, E_2, \ldots, E_i, \ldots, E_n$, where $P(E_i)$ represents the probability that event E_i will occur, the axioms may be stated as follows. → *list all outcomes*

Axiom 1: $0 \leq P(E_i) \leq 1$. The probability that an event E_i will occur is assigned a number between 0 and 1 inclusive, with 0 representing complete certainty that the event will not occur and 1 representing complete certainty that it will occur.

Axiom 2: $\sum P(E_i) = 1$. The sum of the probabilities assigned to any set of mutually exclusive and collectively exhaustive events is 1.

Axiom 3: $P(E_1 \text{ or } E_2) = P(E_1) + P(E_2)$. The probability that either of two (or more) mutually exclusive events occurs is the sum of the individual probabilities that the two (or more) events occur separately.

The following illustrations show how these axioms are used to assign probabilities to events. First, a fair coin is tossed once and we wish to assign a number as the probability of each of the two possible outcomes, H (a head) and T (a tail). The number assigned to represent the probability of each outcome must be between zero and one (Axiom 1), and the sum of the two probability numbers must be one (Axiom 2). Algebraically, Axiom 2 requires $P(H) + P(T) = 1$. By definition of a fair coin, the outcomes are equally likely, i.e., $P(H) = P(T)$. Substituting $P(H)$ in place of $P(T)$ in the equation of Axiom 2, we find:

$$P(H) + P(T) = 1,$$
$$P(H) + P(H) = 1,$$
$$2P(H) = 1,$$
$$P(H) = \tfrac{1}{2}.$$

Substituting the result into the equation for Axiom 2,

$$P(H) + P(T) = 1,$$
$$\tfrac{1}{2} + P(T) = 1,$$
$$P(T) = \tfrac{1}{2}.$$

Thus the only number that can be used to represent the probabilities of the outcomes in the toss of a fair coin that also satisfies the axioms of probability are $P(H) = \tfrac{1}{2}$ and $P(T) = \tfrac{1}{2}$. We may also summarize these probabilities as $P(H) = 0.50$ and $P(T) = 0.50$. The probability of an event should be expressed as either a whole fraction in the simplest terms or as a decimal fraction. Notice that Axiom 3 is illustrated also:

$$P(H \text{ or } T) = P(H) + P(T) = \tfrac{1}{2} + \tfrac{1}{2} = 1.$$

Again refer to Table 4.1. If a fair die is tossed, what are the probabilities of the six elementary events? Following the same line of reasoning as above, each mutually exclusive event is equally likely and has a probability of $\tfrac{1}{6}$. In the same example, what is the probability of the compound event "even-numbered face"? Since the two compound events are equally likely, each has a probability of $\tfrac{1}{2}$. Following a different approach, Axiom 3 may also be used to find the probability of the event "even-numbered face." An even-numbered face occurs for the elementary events 2, 4, 6. Applying Axiom 3, the probability that either of the three mutually exclusive events occurs is

$$P(2, 4, \text{ or } 6) = P(2) + P(4) + P(6)$$
$$= \tfrac{1}{6} + \tfrac{1}{6} + \tfrac{1}{6}$$
$$= \tfrac{3}{6}$$
$$= \tfrac{1}{2}.$$

The two illustrations above indicate a convenient way to assign a number for probability when the events are equally likely. This procedure also involves counting both the number of events that may occur and the number of ways in which specified events may occur, a subject discussed previously in Section 4.3. When a fair coin is tossed, how many events are possible? Two: H and T. In how many ways can the event H (a head) occur? Only one, when the coin lands head-up after the toss. What is the ratio of the number of ways H can occur to the number of events that may occur? The ratio is $\frac{1}{2}$, which is also the $P(H)$ as determined previously. Using the same procedure, determine the probability of an even-numbered face as a result of the toss of a fair die. How many elementary events are possible? Six. How many ways (how many elementary events) can an even-numbered face occur? Three ways: either the 2, 4, or 6 occurs. The ratio of the number of ways an even-numbered face can occur to the number of events possible is $\frac{3}{6}$ or $\frac{1}{2}$, the same result previously determined another way. The procedure applied here is stated as follows.

> If all possible events are *equally likely*, the *probability* that a specified event will occur equals the number of ways in which the specified event can occur expressed as a ratio to the number of ways in which any event can occur.

EXAMPLE. A person deals a single card from a well-shuffled deck of 52 playing cards. What is the probability that the card is an ace? The effect of shuffling the cards is to insure that each of the 52 elementary events is equally likely. The probability of an ace is the number of elementary events in the deck that can be identified as an ace, expressed as a ratio to the total number of elementary events, 52. There are four aces, so the probability ratio is 4/52, or 1/13. Verify that the probability of dealing a heart is 1/4.

Odds

The concept of *odds* is closely related to probability and can be expressed in terms of the basic axioms. The odds in favor of an event or the odds against an event are ways of summarizing a measure of uncertainty that is equivalent to stating the probability of the event. Suppose in the next National Football League Super Bowl Game the Pittsburgh Steelers play the Dallas Cowboys. Prior to the game your newspaper carries a story stating the odds as 3 to 2, written 3:2, in favor of Pittsburgh. This is equivalent to a statement that the odds against Pittsburgh are 2:3. What do these odds imply about the probability of Pittsburgh winning? The relationship between odds and probability is described as follows:

> If the *odds* in favor of an event are $f:a$, the *probability* of the event occurring is $f/(f + a)$ and the probability that the event will not occur is $a/(f + a)$.

If the odds in favor of Pittsburgh are 3:2, the probability of Pittsburgh winning is $3/(3 + 2) = \frac{3}{5} = 0.60$. The probability that Pittsburgh will lose, which is the probability that Dallas wins, is $2/(3 + 2) = \frac{2}{5} = 0.40$. Observe that the probabilities obtained conform to the axioms of probability.

**Probability of
Complement Event**

The previous illustration also points out how we may use the idea of the *complement* of an event, as discussed in Section 4.2, in probability computations. The complement of the event "Pittsburgh wins" is the event "Dallas wins." From Axiom 2 the sum of the probabilities of these events must equal one, since the two events are the only possible outcomes. If the P(Pittsburgh wins) $= 0.60$ and we know that

$$P(\text{Pittsburgh wins}) + P(\text{Dallas wins}) = 1,$$

then

$$0.60 + P(\text{Dallas wins}) = 1,$$

Solving,

$$P(\text{Dallas wins}) = 1 - 0.60$$
$$= 0.40.$$

This shows that if all outcomes in an experiment can be grouped into two events E_1 and E_2 so that E_1 is the complement of E_2, then Axiom 2 specifies that $P(E_1) + P(E_2) = 1$. From this, we may solve for $P(E_2)$ by finding $1 - P(E_1)$. Evaluating the probability of one event by finding the probability of the complement event is a useful rule. The general rule states that $P(E) = 1 - P(\bar{E})$, where $\bar{E}$ is the complement of event E.

Our further investigation into the computation of probability will be enhanced by examining some different interpretations of probability measures.

**4.5   THEORY OF
PROBABILITY**

The mathematical properties of probability described by the three axioms provide a basis for establishing the probability of an event. The assignment of a number as a probability may be viewed as a process of making deductions based on observed phenomena. After playing tennis several times with a person, you may have observed enough play to establish that you feel you have a 75 percent chance of winning each time you play this person. This does not imply that you will always win while playing with this person; it indicates that you expect to win three-quarters of the times you two play. In fact, you may decide to state formally that your probability of winning is 0.75. Exactly how you use observed phenomena, such as past play with this person, to evaluate probability has led to some debate in the statistical profession. Despite a diversity of views, sufficient similarity exists to classify the theories of probability into two broad categories.

The most widely held theory of probability is the *objective* or *frequency approach*. In practice, the frequency approach implies that the probability of an event is the *relative frequency* with which an event occurs in a large number of trials. We previously determined that the probability of a head in the toss of a fair coin is 1/2. According to the frequency approach, the value 1/2 represents the relative frequency with which heads occur in a large number of tosses. In formal mathematical terms, the frequency approach identifies probability as the limit of a relative frequency. This limit is often referred to as the *Venn limit*, from an interpretation given by the English mathematician John Venn (1834–1923).

Precisely stated, if f is the number of occurrences of event E during n trials, the probability of event E, $P(E)$, approaches the ratio of f/n as the number of trials grows infinitely large. This can be stated as

$$P(E) = \lim_{n \to \infty} \frac{f}{n}.$$

as n gets large the f/n will stabilize n? & #.

Of course we cannot observe an infinitely large number of trials. The application of this theoretical construct is interpreted to mean that as n becomes larger, the relative frequency becomes stable (a phenomenon called *statistical regularity*). Probability is then simply the ratio of the number of occurrences of an event to the total number of trials in a long-run experiment. Repetition of an experiment under identical conditions is the key to probability estimation in this concept. The accuracy of the estimate of probability is increased by observing a larger number of trials. According to this, an estimate of the probability of a head in a coin toss based on 1,000 trials would be better than an estimate based on 50 trials. If we know the coin is fair, then $\frac{1}{2}$ is the expected long-run relative frequency of heads. The idea of determining probability by identifying equally likely events conforms to the frequency approach.

The frequency approach to probability is also called *objective* since it is based entirely on empirical evidence, i.e., what is observed during repeated trials. The *subjective approach* to probability, in opposition to the former, denies that probability statements are empirical statements at all. Under the subjective approach, information about a probability is not limited to objective information gathered in an empirical investigation. Instead, information prior to an experiment, whether of an objective or subjective nature, is relevant to the determination of a numerical probability.

Probability, according to the subjective approach, represents the *degree of belief* an individual has about an uncertain situation. Because different individuals may have differing degrees of belief about a numerical probability, the subjective concept yields what is often called a *personal probability*. Subjective probability is not merely psychological, however. The theory is subjective or *personalistic* in the sense that a person can have any degree of belief whatever in any given statement based on any evidence, provided only that his other degrees of belief have suitable values. In other words, they must conform to the generally accepted laws of probability. Consider the following question. What is the probability that a woman will be elected president of the United States no later than 1984? We cannot perform a large number of trials of an experiment to determine this probability. Past history is of little help since all U.S. presidents have, to date, been men. Does this imply that the probability of electing a woman no later than 1984 is zero? Most agree that the likelihood of this event is closer to zero than to one, but that still does not pinpoint a numerical probability. The assessment of subjective probability has attracted much interest in recent years. Several research efforts currently under way have a goal of improving assessment techniques.

One of the basic ideas in subjective probability assessment is to express a degree of belief in terms of bets or wagers. The bet or wager is used to help verify an individual's degree of belief. If a construction contractor proposes to bid on a new contract on the basis of his belief that the odds in favor of obtaining the contract are 2:1, this means he should be willing to bet $200 against $100 that he will obtain the contract. If he considers this bid unfair, then he does not agree with the odds and should adjust them until a *fair bet* is determined. Or, if he has convictions against gambling, then he must agree that the bet is fair for a gambling individual. Provided he agrees with these odds and would be willing to make this wager (or consider it fair), then we say he has a personal probability, or degree of belief, of $200/(200 + 100) = \frac{2}{3}$ that he will be awarded the contract.

Both of the approaches to probability have considerable merit. Fortunately, most of the debate over these two approaches centers upon philosophical ideas and not on application of the laws of probability. Practitioners make use of both approaches, depending on the nature of the problem situation. Your ability to determine and interpret numerical probabilities will certainly be enhanced by relying on both approaches.

4.6 APPLICATION OF BASIC CONCEPTS

How are the concepts of probability applied in the analysis of various problem situations? First, you need a good understanding of the concepts presented up to this point in the chapter. Second, you need to possess a willingness to experiment with these concepts in order to learn, sometimes through trial and error, the significance of each in different problem situations. Each problem has certain unique characteristics that you must recognize. It is frequently helpful to identify, at least in your mind, all possible outcomes in each situation. Next determine which of these outcomes are associated with the event(s) for which a probability measure is desired. Finally, approach the determination of a numerical probability in a logical manner while being guided by the basic axioms and concepts. Frequently, the solution may be found in more than one way. When this is possible, arriving at the same solution through more than one procedure provides verification for your reasoning and increases your potential for solving other problems. A few examples are presented now to illustrate this process.

A single card is drawn from a well-shuffled deck of 52 playing cards. What is the probability that the card is a diamond? This is similar to an illustration in Section 4.4. Obviously there are 52 elementary events possible when a card is drawn. If the cards are shuffled well, we may assume that there is no particular order to the way the cards are contained in the deck. In other words, the cards are randomly distributed in the deck so that each card (elementary event) has an equal chance of being selected. How many elementary events are associated with the event "diamond"? Thirteen, since the deck is divided into four suits of thirteen cards each. Since all elementary events are equally likely, we can use the ratio described in Section 4.4 to determine the probability of a diamond.

The ratio of diamonds to all cards is 13 to 52. Letting D represent the event "diamond," $P(D) = \frac{13}{52} = \frac{1}{4}$.

Now that we have established that the probability of a diamond is 1/4, is there another way that this result can be verified? Consider that the 52 equally likely outcomes can be grouped in four suits of thirteen cards each. The four suits are also equally likely. The event "diamond" is one of the four suits. Therefore $P(D) = \frac{1}{4}$, which again utilizes the ratio approach. Can you think of other ways to verify this result?

Continuing the same example, what is the probability that the card drawn is the ace of hearts? This is one of the 52 elementary outcomes. Since all outcomes are equally likely, the probability of the ace of hearts is $\frac{1}{52}$.

Using the same example, what is the probability that the card drawn is either a diamond or the ace of hearts? To answer this, determine how many cards in the deck are either diamonds or the ace of hearts. Fourteen out of 52 elementary events are either diamonds or the ace of hearts, so that the probability is $\frac{14}{52} = \frac{7}{26}$. We can verify this result by referring to Axiom 3. This axiom states that the probability of either of two mutually exclusive events equals the sum of the probabilities of the events. Are the events diamonds and the ace of hearts mutually exclusive? Yes, we cannot select a single card that has both characteristics. Then Axiom 3 may be used to verify this probability. If D represents the event "diamonds" and AH represents the event "ace of hearts," Axiom 3 provides the following result.

$$P(D \text{ or } AH) = P(D) + P(AH)$$

$$= \frac{13}{52} + \frac{1}{52}$$

$$= \frac{14}{52}$$

$$= \frac{7}{26}.$$

Axiom 3 may be used in this manner for any situation involving mutually exclusive events.

General Rule of Addition

As a final reference to the same example, determine the probability that the card is either an ace or a heart. Suppose the following solution is offered. Let A represent the event "ace" and H represent "heart." Then

$$P(A \text{ or } H) = P(A) + P(H)$$

$$= \frac{4}{52} + \frac{13}{52}$$

$$= \frac{17}{52}.$$

Do you agree with this? The alert reader will examine the numerical result and try to verify it by asking, "Are there 17 cards in a deck that are either aces or hearts?" The answer is, of course, no. There are only 16. The ace of hearts has been counted twice, once as an ace and once as a heart. The correct numerical result should be 16/52. Axiom 3 holds only for mutually exclusive events. Aces and hearts are *not mutually exclusive* events since they can occur together (the ace of hearts). Axiom 3 has been used when the appropriate conditions are not met, and therefore the result is incorrect. The following *General Rule of Addition* can be used in such cases.

> Given events E_1 and E_2, the probability that either occurs is
>
> $$P(E_1 \text{ or } E_2) = P(E_1) + P(E_2) - P(E_1 \text{ and } E_2). \qquad (4.4)$$

This indicates that the probability of two nonmutually exclusive events equals the sum of the probabilities of the individual events minus the probability that the two events occur together. This rule is applicable to both mutually exclusive and nonmutually exclusive events. In the case of the former, $P(E_1 \text{ and } E_2) = 0$, one obtains the third axiom. Applying this,

$$P(A \text{ or } H) = P(A) + P(H) - P(A \text{ and } H)$$

$$= \frac{4}{52} + \frac{13}{52} - \frac{1}{52}$$

$$= \frac{16}{52}$$

$$= \frac{4}{13}.$$

The last illustration is included to show the importance of logically examining numerical results in evaluating probability statements. Axiom 3 was inappropriate because the events were not mutually exclusive. In order to make a correct probability deduction, it is necessary to identify the type of events in a problem situation.

EXAMPLE. A local accounting firm currently employs 20 people and boasts of the following qualifications within the firm: 12 Certified Public Accountants (CPAs), 5 holders of master's degrees, 2 employees who have earned both the master's degree and CPA certification, 1 university co-op student without other qualifications, and 4 secretaries. Note all events are not mutually exclusive since $12 + 5 + 2 + 1 + 4 = 24$, and the firm size is only 20. An employee is selected at random for a task (assuming all are capable and have an equal chance to be selected). What is the probability that the person selected holds either CPA certification or a master's degree? Using the general rule of addition, with $A = \text{CPA}$ and $M = \text{master's degree}$,

$$P(A \text{ or } M) = P(A) + P(M) - P(A \text{ and } M)$$

$$= \frac{12}{20} + \frac{5}{20} - \frac{2}{20}$$

$$= \frac{15}{20} = 0.75.$$

Thus three-fourths of the employees in the firm hold either a master's degree, CPA certification, or both.

Venn Diagrams

A diagram showing the *sample space* for CPAs and holders of master's degrees is shown in Figure 4.1. Such a diagram is called a *Venn diagram*. The sample space of an experiment is a representation of all possible outcomes. In the language of set theory, Figure 4.1 represents the *union* of CPAs and holders of master's degrees. The two employees holding both CPA certification and master's degree are represented by the *intersection* of the two sets.

FIGURE 4.1: VENN DIAGRAM: NON-MUTUALLY EXCLUSIVE EVENTS

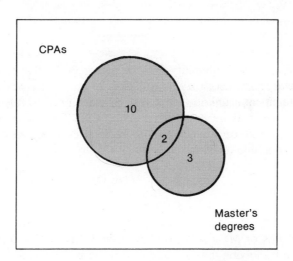

Continuing the accounting firm illustration, what is the probability that the person selected at random for the task is either a CPA or a co-op student? These events are mutually exclusive. Designating $A = \text{CPA}$, $C = \text{co-op student}$, and using Equation (4.4) (although Axiom 3 is applicable):

$$P(A \text{ or } C) = P(A) + P(C) - P(A \text{ and } C)$$

$$= \frac{12}{20} + \frac{1}{20} - 0$$

$$= \frac{13}{20} = 0.65.$$

The result indicates that 65 percent of the employees are either CPAs or co-op students. A Venn diagram for these mutually exclusive events is shown in Figure 4.2.

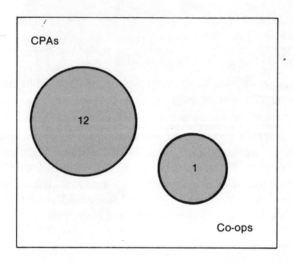

What is the probability that the person selected at random for the task is a secretary? The necessary information for immediately evaluating this probability statement is unknown but may be determined by reference to Axiom 2. The employees may be classified as either A = CPAs, M = holders of master's degrees, C = co-op students, or S = secretaries. From Axiom 2, the probabilities of all possible outcomes may be written as:

$$P(A \text{ or } M) + P(C) + P(S) = 1.$$

Since $P(A \text{ or } M) = 0.75$ from above and $P(C) = \frac{1}{20} = 0.05$, $P(S)$ may be determined by solving the equation for this term. The result is:

$$P(S) = 1 - P(A \text{ or } M) - P(C)$$
$$= 1 - 0.75 - 0.05$$
$$= 1 - 0.80 = 0.20.$$

This indicates that 20 percent of the firm's employees are secretaries. The procedure used here for determining $P(S)$ from Axiom 2 is actually an extension of the use of the complement of an event. A diagram illustrating the sample space for all employees in this situation is given in Figure 4.3.

The previous examples serve as illustrative guidelines for analyzing a situation and verifying the accuracy of numerical results. Expertise in problem analysis improves with experience, which may be gained from exercises at the end of this chapter. Some additional concepts are now introduced.

FIGURE 4.3: VENN
DIAGRAM FOR
ACCOUNTING FIRM

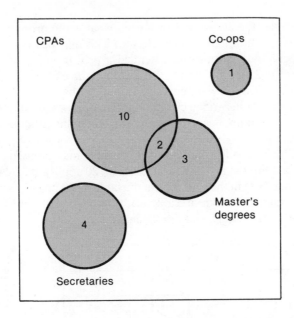

4.7
CLASSIFICATION OF
PROBABILITIES

Consider the following situation. A large firm in your state is about to hire a recent college graduate to fill a desirable trainee position. After interviewing many applicants, the firm has identified 20 applicants that it considers, for all practical purposes, about equally qualified for the position. The tabulation below shows that 15 of the 20 are graduates of a large state university while 5 are graduates of a small private university.

	State	Private	Total
Male	10	2	12
Female	5	3	8
Total	15	5	20

Further, there are 12 male and 8 female graduates. If the firm selects an applicant from these 20 so that each has an equal chance of being selected, the following probabilities are immediately obvious.

$$P(\text{male graduate}) = \frac{12}{20} = 0.60$$

$$P(\text{female graduate}) = \frac{8}{20} = 0.40$$

$$P(\text{state university graduate}) = \frac{15}{20} = 0.75$$

$$P(\text{private university graduate}) = \frac{5}{20} = 0.25.$$

Joint Probability

Under the above circumstances, what is the probability that a male private university graduate will be selected? This may be interpreted as the chance that the graduate is male *and* attended a private university. Examining the table, there are 2 out of 20 who possess both of these characteristics, giving a probability of 0.10. Such a statement is a joint probability statement. *Joint probability statements* refer to the chance that two or more events occur together. Examine the table of applicants to verify that the joint probability of female and state university is $\frac{5}{20} = 0.25$.

The format for the table of applicants may be used to summarize the joint probabilities in any situation. Since all 20 applicants are equally likely, dividing the numbers in the four cells of the table by 20 produces a *joint probability table*. This table is shown in Table 4.3, where S = state university, P = private university, M = male, and F = female.

TABLE 4.3: JOINT PROBABILITY TABLE FOR JOB APPLICANTS

	S	P	$Total$
M	0.50	0.10	0.60
F	0.25	0.15	0.40
Total	0.75	0.25	

Marginal Probability

Notice in Table 4.3 that the row totals and column totals provide the measures of probabilities first mentioned in this section. The total of the row labeled M indicates that the probability of selecting a male applicant is 0.60. The table shows that the 0.60 probability is the sum of two joint probabilities;

$$P(M \text{ and } S) + P(M \text{ and } P) =$$
$$0.50 \quad + \quad 0.10 \quad = 0.60.$$

> A probability that can be expressed as a sum of joint probabilities is a *marginal probability.*

Is the probability of selecting a private university graduate, $P(P)$, a marginal probability? Yes,

$$P(P) = P(M \text{ and } P) + P(F \text{ and } P)$$
$$= \quad 0.10 \quad + \quad 0.15$$
$$= \quad 0.25.$$

All the row totals and column totals in Table 4.3 represent marginal probabilities. In fact, the probabilities found in the *margins* of any joint probability table are *marginal probabilities*, since they represent sums of joint probabilities.

Conditional Probability

Suppose an applicant is selected by the firm, and we are told only that the individual graduated from a state university. We still do not know whether the candidate selected is male or female. Earlier, we determined $P(M) = 0.60$ and $P(F) = 0.40$. Should the new information, that a state university graduate is selected, cause us to change the probability of either male or female? The answer is yes, as explained below. Prior to knowing the type of school, each of the 20 candidates had an equal chance of being selected. Twelve out of 20 were male and 8 out of 20 were female so that probabilities of 0.60 and 0.40 were appropriate. Now that we know a private university graduate was not selected, the person selected must be one of the 15 state university graduates. In effect, the sample space of the experiment is reduced to these 15. The proportion of state university graduates who are male and female is $\frac{10}{15} = \frac{2}{3} = 0.67$ and $\frac{5}{15} = \frac{1}{3} = 0.33$, respectively. Thus the probabilities of male and female have changed with the added information regarding the type of school. The new probability of selecting a male candidate is written as $P(M \mid S) = 0.67$. Any probability statement of the form $P(E \mid A)$ is called a *conditional probability*. The vertical line separating E and A is read as the word "given," so that the entire statement is read as "the probability that E occurs, given that A has already occurred." The conditional probability $P(E \mid A)$ answers the question: What is the probability of E subject to the condition that A has occurred? In the same manner, the conditional probability of female given state university graduate is $P(F \mid S) = 0.33$. A formal statement of conditional probability follows.

> The *conditional probability* of event E given event A, is the ratio of the joint probability of the two events to the marginal probability of A. For $P(A) \neq 0$,
> $P(E \mid A) = P(E \text{ and } A)/P(A)$. $\hspace{2em}$ (4.5)

Use this definition to verify that $P(M \mid S) = 0.67$ and $P(F \mid S) = 0.33$, According to the definition,

$$P(M \mid S) = \frac{P(M \text{ and } S)}{P(S)}$$

Substituting these probabilities from Table 4.3,

$$P(M \mid S) = \frac{0.50}{0.75} = 0.67.$$

Likewise,

$$P(F \mid S) = \frac{P(F \text{ and } S)}{P(S)}$$

$$= \frac{0.25}{0.75} = 0.33.$$

Even though it is not necessary to rely on the definition to determine $P(M \mid S)$ and $P(F \mid S)$, the illustration demonstrates some important relationships between joint, marginal, and conditional probabilities. There are some other important relationships here.

Computing Joint Probability

A method for computing joint probabilities can be obtained from this definition. Conditional probability is defined as

$$P(E|A) = \frac{P(E \text{ and } A)}{P(A)}.$$

Multiply both sides of the equation by $P(A)$.

$$P(A) \cdot P(E|A) = P(E \text{ and } A).$$

> The *joint probability* of two events E and A is computed as $P(E \text{ and } A) = P(A) \cdot P(E|A)$. (4.6)

Use this rule to verify that the joint probability of a male state university graduate (from above) is 0.50.

$$P(M \text{ and } S) = P(S) \cdot P(M|S) \qquad (4.7)$$

$$= (0.75)\left(\frac{2}{3}\right)$$

$$= 0.50.$$

Statistical Independence

From the joint probability table in Table 4.3, $P(M) = 0.60$ and $P(M|S) = 0.67$. This kind of comparison is used to determine whether events are statistically independent or dependent.

> Two events are *statistically independent* if $P(E) = P(E|A)$. Events E and A are *dependent* if $P(E) \neq P(E|A)$.

This correctly implies that the probability of event E is unchanged by the knowledge that event A occurred if events E and A are independent. In the job applicant situation the events male and state university graduate are *dependent* since $P(M) \neq P(M|S)$. In other words, knowledge that an applicant is a state university graduate affects the probability that the applicant is male.

Consider a different situation. A fair coin is tossed twice. What is the probability of two heads occurring? Since the coin is fair, the events head and tail are equally likely on each toss with probability 1/2. Now Equation (4.6) can be used to compute the probability of the joint events H_1, head on the first toss, and H_2, head on the second toss. From Equation (4.6),

$$P(H_1 \text{ and } H_2) = P(H_1) \cdot P(H_2|H_1). \qquad (4.8)$$

But since the probability of a head on each toss is $\frac{1}{2}$, $P(H_2) = P(H_2|H_1)$. The events H_1 and H_2 are independent. The occurrence on the first toss does not affect the probability of a head on the second toss. Substituting $P(H_2)$ in place of $P(H_2|H_1)$ in Equation (4.8) and solving,

$$P(H_1 \text{ and } H_2) = P(H_1) \cdot P(H_2)$$

$$= \left(\frac{1}{2}\right) \cdot \left(\frac{1}{2}\right)$$

$$= \frac{1}{4}.$$

The result indicates that a special joint probability rule exists for independent events. If E and A are independent events,

$$P(E \text{ and } A) = P(A) \cdot P(E). \tag{4.9}$$

EXAMPLE. Recall the example of the accounting firm in Section 4.7. The employees are classified by CPA certification and master's degree in the table below.

	Master's Degree	No Master's Degree	Total
CPA	2	10	12
Not CPA	3	5	8
Total	5	15	

Are the events *CPA* and master's degree independent? Determine this by examining the table to find $P(M) = \frac{5}{20} = 0.25$ and $P(M|CPA) = \frac{2}{12} = 0.17$. Since $P(M) \neq P(M|CPA)$, the events are not independent.

EXAMPLE. A person deals three cards from a well-shuffled deck of 52 cards. What is the probability that all three cards are aces? Are the events A_1, ace on first card; A_2, ace on second card; and A_3, ace on third card; independent? No, after dealing each card the size of the deck changes. Further, if either the first card, the second card, or both cards are aces, the number of aces remaining in the deck is reduced. To compute this probability, expand Equation (4.6) to three events.

$$P(A_1 \text{ and } A_2 \text{ and } A_3) = P(A_1) \cdot P(A_2|A_1) \cdot P(A_3|A_1 \text{ and } A_2)$$

$$= \left(\frac{4}{52}\right)\left(\frac{3}{51}\right)\left(\frac{2}{50}\right)$$

$$= \left(\frac{1}{13}\right)\left(\frac{1}{17}\right)\left(\frac{1}{25}\right)$$

$$= \frac{1}{5525} = 0.00018.$$

The result shows the usefulness of the multiplicative laws.

PIERRE-SIMON de LAPLACE (1749–1827)

Born at Beaumont, Normandy, France, Laplace went to Paris at age 18 with a letter of introduction to the mathematician D'Alembert. The mathematician secured Laplace a post at the Paris Military Academy, where Laplace served as a teacher of mathematics and examiner for cadets. There Laplace examined Napoleon Bonaparte for a second lieutenant's commission in 1785. Laplace held several government positions during the French revolution. In 1799 Napoleon named Laplace minister of interior, a post he held for only six months before becoming chancellor. As minister, Laplace designated July 14 as a national holiday, Bastille Day. He wrote many papers on probability, chance, and mathematics. His *Theorie Analytique des Probabilities*, published in 1812, has been called the greatest single work on probability; much of the later development of probability is based on it.

4.8 PROBABILITY TREES

Probability trees are diagrams useful for performing joint probability computations. Suppose an entrepreneur applies for financial support from three different sources (A, B, and C) in order to undertake a new venture. He decides to proceed with the venture if financial support is obtained from at least one source, even though the support requested from all three is preferable. He estimates the probability that source A will provide support is $\frac{1}{3}$; source B, $\frac{1}{2}$; and source C, $\frac{1}{4}$.

A probability tree for this situation is given in Figure 4.4. The two outcomes for source A, approve or deny financial support, are shown as branches at the

FIGURE 4.4:
PROBABILITY TREE FOR
ENTREPRENEUR

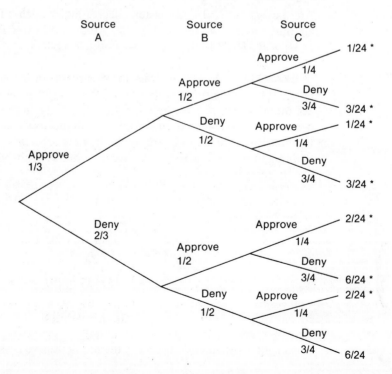

beginning (left) of the diagram. Moving right, the diagram indicates that, for either outcome of source A, there are two possible outcomes for source B. Likewise, for each outcome of source B, two outcomes are possible for source C. The possible outcomes for each source and associated probabilities are labeled on the branches.

Joint probability computations are obtained from the diagram by following a sequence of branches from left to right. The uppermost sequence of branches represents approval from all three sources. The joint probability of approval from all three sources is the product of probabilities along that sequence, $(\frac{1}{3})(\frac{1}{2})(\frac{1}{4})$, or $\frac{1}{24}$. Similarly, the joint probability of approval from sources A and B and denial from source C is $(\frac{1}{3})(\frac{1}{2})(\frac{3}{4})$, or $\frac{3}{24}$. These and other joint probabilities are given at the end (right) of each sequence of branches. The probability that the request for support will be approved by at least one of the sources is the sum of the joint probabilities that are marked by asterisks on the diagram; this is 18/24 or 0.75.

Probability trees are helpful devices for performing joint probability calculations. Note that the joint probabilities at the end points of a probability tree must sum to one. In the present illustration, this allows us to make use of the complement rule of probability. If E represents the event "at least one source approves support," then the complement, $\bar{E}$, is the event "all sources deny support." Therefore,

$$P(E) = 1 - P(\bar{E}) = 1 - \frac{6}{24} = \frac{18}{24} = 0.75.$$

4.9 THE BIRTHDAY PROBLEM

An illustration using several of the concepts presented up to this point in the chapter concerns the likelihood of finding duplicate birthdays in a group of people of any size. To examine this, first consider the probability of finding *no* identical birthdays in a group of five people (assume the people are randomly chosen).

Given the birthday of the first, what is the probability that the second has a different birthday? Assuming a 365-day year, the second could have any 364 of 365 birthdates that are different from the first, i.e., $P(\text{different}) = \frac{364}{365}$. The third person's birthday could be any 363 of 365 without coinciding with the birthday of either of the first two. This probability is $\frac{363}{365}$. Continuing this line of reasoning, the probability that no two of the five birthdays coincide is

$$\left(\frac{364}{365}\right)\left(\frac{363}{365}\right)\left(\frac{362}{365}\right)\left(\frac{361}{365}\right) = 0.973.$$

Appealing to the complement of this event, the probability that at least two out of the five have the same birthday, $1 - 0.973$, is 0.027. This result is shown in Table 4.4 along with the probabilities for groups of up to 100 individuals. Use Table 4.4 to determine the smallest group needed to make the probability of coincident birthdays larger than $\frac{1}{2}$. Is the result surprising?

TABLE 4.4: THE PROBABILITY THAT AT LEAST TWO OUT OF A GROUP OF SIZE N HAVE THE SAME BIRTHDAY

N	Probability	N	Probability	N	Probability	N	Probability
1	0.0	26	0.59824082	51	0.97443199	76	0.99977744
2	0.00273973	27	0.62685928	52	0.97800451	77	0.99982378
3	0.00820417	28	0.65446147	53	0.98113811	78	0.99986095
4	0.01635591	29	0.68096854	54	0.98387696	79	0.99989067
5	0.02713557	30	0.70631624	55	0.98626229	80	0.99991433
6	0.04046248	31	0.73045463	56	0.98833235	81	0.99993311
7	0.05623570	32	0.75334753	57	0.99012246	82	0.99994795
8	0.07433529	33	0.77497185	58	0.99166498	83	0.99995965
9	0.09462383	34	0.79531686	59	0.99298945	84	0.99996882
10	0.11694818	35	0.81438324	60	0.99412266	85	0.99997600
11	0.14114138	36	0.83218211	61	0.99508880	86	0.99998159
12	0.16702479	37	0.84873401	62	0.99590957	87	0.99998593
13	0.19441028	38	0.86406782	63	0.99660439	88	0.99998928
14	0.22310251	39	0.87821966	64	0.99719048	89	0.99999186
15	0.25290132	40	0.89123181	65	0.99768311	90	0.99999385
16	0.28360401	41	0.90315161	66	0.99809570	91	0.99999537
17	0.31500767	42	0.91403047	67	0.99844004	92	0.99999652
18	0.34691142	43	0.92392286	68	0.99872639	93	0.99999740
19	0.37911853	44	0.93288537	69	0.99896367	94	0.99999806
20	0.41143838	45	0.94097590	70	0.99915958	95	0.99999856
21	0.44368834	46	0.94825284	71	0.99932075	96	0.99999893
22	0.47569531	47	0.95477440	72	0.99945288	97	0.99999922
23	0.50729723	48	0.96059797	73	0.99956081	98	0.99999942
24	0.53834426	49	0.96577961	74	0.99964864	99	0.99999958
25	0.56869970	50	0.97037358	75	0.99971988	100	0.99999969

Note: These values are based on a 365-day year. The probability that at least two have the same birthday is computed as one minus the probability that no two have the same birthday.

4.10 BAYES THEOREM

In Table 4.3, joint probabilities were given for the chance that certain university graduates would be hired by a firm, where S = state university, P = private university, M = male, and F = female. Suppose the placement director at the state university is confident that a state university graduate will be hired for the position and wishes to determine the probability that the successful applicant is male, i.e., determine $P(M|S)$. The placement director knows only that there are 60 percent males and 40 percent females in the group of applicants, that $\frac{5}{6}$ of the males are state university graduates and that $\frac{5}{8}$ of the females are state university graduates. Thus $P(M) = 0.60$, $P(F) = 0.40$, $P(S|M) = \frac{5}{6}$, and $P(S|F) = \frac{5}{8}$.

The conditional probability $P(M|S)$ is defined as

$$P(M|S) = \frac{P(M \text{ and } S)}{P(S)}. \tag{4.10}$$

The joint probability $P(M \text{ and } S)$ may be calculated as either

$$P(S) \cdot P(M|S)$$

or

$$P(M) \cdot P(S|M). \tag{4.11}$$

Since $P(M|S)$ is unknown, substitute Equation (4.11) for the joint probability in Equation (4.10).

$$P(M|S) = \frac{P(M) \cdot P(S|M)}{P(S)}. \tag{4.12}$$

The denominator of Equation (4.12), a marginal probability, can be expressed as a sum of joint probabilities.

$$P(S) = P(M \text{ and } S) + P(F \text{ and } S).$$

Substituting for $P(S)$ in Equation (4.10),

$$P(M|S) = \frac{P(M \text{ and } S)}{P(M \text{ and } S) + P(F \text{ and } S)}. \tag{4.13}$$

The result shows that a conditional probability can be written as the ratio of a joint probability to a sum of joint probabilities. The joint probability $P(M$ and $S)$, which appears in both the numerator and denominator of Equation (4.13), was written as the product $P(M) \cdot P(S|M)$ in Equation (4.11). Likewise, the joint probability $P(F$ and $S)$ can be written as the product $P(F) \cdot P(S|F)$. Substituting these expressions for joint probabilities in Equation (4.13), the result is

$$P(M|S) = \frac{P(M) \cdot P(S|M)}{P(M) \cdot P(S|M) + P(F) \cdot P(S|F)}. \tag{4.14}$$

Now all terms of the original conditional statement in Equation (4.12) have been expanded to show that the conditional probability $P(M|S)$ can be calculated in terms of other conditional statements. To verify, substitute the appropriate probabilities into Equation (4.14).

$$P(M|S) = \frac{(0.60)(5/6)}{(0.60)(5/6) + (0.40)(5/8)}$$

$$= \frac{0.50}{0.50 + 0.25} \tag{4.15}$$

$$= \frac{0.50}{0.75} = 0.67.$$

Compare this result with that in the previous section. Also note that the joint probabilities obtained in Equation (4.15) conform to Equation (4.13) and can be verified from the joint probability table in Table 4.3.

We did not need to develop Equation (4.14) in order to solve for $P(M|S)$ in this problem. But in other situations, a similar procedure could be required. Your understanding of the algebra of relationships among joint, marginal, and conditional probabilities is extremely valuable in mastering the framework of probability analysis. The basis of Equation (4.14), an expanded conditional probability statement, is called *Bayes theorem*.

> Given one event A and the n mutually exclusive and exhaustive events E_1, $E_2, \ldots, E_n$, *Bayes theorem* states that the probability of any event E_i for $i = 1, 2, \ldots, n$ is
>
> $$P(E_i|A) = \frac{P(E_i) \cdot P(A|E_i)}{P(E_1) \cdot P(A|E_1) + P(E_2) \cdot P(A|E_2) + \cdots + P(E_n) \cdot P(A|E_n)}.$$
>
> (4.16)

The denominator of Bayes theorem may be written more compactly with summation notation as

$$P(E_i|A) = \frac{P(E_i) \cdot P(A|E_i)}{\sum_{i=1}^{n} P(E_i) \cdot P(A|E_i)}. \tag{4.17}$$

EXAMPLE. Consider the possibility of a dangerous radioactive leakage occurring at a nuclear power generating plant. Suppose that a device designed to warn against leakage detects a problem in sufficient time to take corrective action with probability 0.98. Also the probability that the device indicates action should be taken when no problem exists, i.e., the device overreacts, is 0.10. An environmentalist claims that the chance of leakage is 0.03 at any given plant. What is the probability that a leakage problem actually exists if the warning device now indicates this? Let L and $\bar{L}$ represent the events leakage and no leakage, respectively, and D represent the device indicating a problem. The available information is $P(D|L) = 0.98$, $P(D|\bar{L}) = 0.10$, and $P(L) = 0.03$. The probability of a leakage given the device's signal, is

$$P(L|D) = \frac{P(L) \cdot P(D|L)}{P(L) \cdot P(D|L) + P(\bar{L}) \cdot P(D|\bar{L})} \tag{4.18}$$

$$= \frac{(0.03)(0.98)}{(0.03)(0.98) + (0.97)(0.10)}$$

$$= 0.23.$$

The result states that a leakage exists 23 percent of the time that the device signals a problem situation.

Bayes theorem is frequently used as a mechanism for revising the probability of an event. In the previous illustration, the probability of a leakage was revised after noting the signal from the warning device. The initial probability, called a *prior probability*, of a leakage was 0.03. Following the occurrence of the warning device's signal, the probability of a leakage is revised to 0.23. The revised probability of an event is called a *posterior probability*. The words prior and posterior refer to the probability of an event before and after the *outcome* of some other process, which in this example is the warning device's signal. A tabular form of Bayes theorem, shown in Table 4.5, can be used to show these relationships.

TABLE 4.5: BAYES
THEOREM FOR
RADIOACTIVE LEAKAGE

(1)	(2)	(3)	(4)	(5)
	Prior Probability	*Conditional Probability*	*Joint Probability*	*Posterior Probability*
Event	*P(event)*	*P(outcome|event)*	*P(outcome, event)*	*P(event|outcome)*
L	0.03	0.98	0.0294	0.233
$\bar{L}$	0.97	0.10	0.0970	0.767
	1.00		0.1264	1.000

The tabular format for Bayes theorem uniquely expresses several fundamental concepts of probability. Reference to Equation (4.18) shows that Table 4.5 provides exactly the same computation of the $P(L|D)$. Column (2) gives the prior probabilities $P(L) = 0.03$ and $P(\bar{L}) = 0.97$. The conditional probabilities of the device indicating a problem, given L and $\bar{L}$, are listed in Rows (1) and (2), respectively, of Column (3). The joint probabilities $P(D, L)$ and $P(D, \bar{L})$ are computed in Column (4) as the product of the prior and conditional probabilities. The total of Column (4), the sum of the joint probabilities, is the marginal probability of the device indicating a leakage, $P(D)$. Column (5) contains the posterior probabilities, which are conditional probabilities $P(L|D)$ and $P(L|\bar{D})$. Each posterior probability is computed as the ratio of a joint probability to the marginal probability, which is the basic statement of conditional probability given earlier.

Some important applications of Bayes theorem are made in Chapter 7.

4.11 SUMMARY

Probability is a measure of the chance that an uncertain event will occur. The language of probability is expressed in terms of experiments, events, and outcomes. Probability concepts can be familiarly illustrated, as shown in this chapter, by tossing coins or dice or by dealing a deck of playing cards. In fact, games of chance served to motivate many of the early contributors to probability theory, including Cardano, Galileo, Laplace, Huygens, and Montmort.

Counting the number of ways that events occur is fundamental to evaluating the probability of an event. Permutations describe the number of ways events can occur if both composition and order are important. Combinations describe events where only composition is important.

The mathematical properties of probability are defined by three basic axioms. These axioms state that: (1) the probability that an event will occur is a number between 0 and 1; (2) the sum of the probabilities is 1 for a set of mutually exclusive and collectively exhaustive events; and (3) the probability that either of two mutually exclusive events will occur is the sum of the probabilities that the two events occur separately.

Theories of probability express the different approaches to probability measurement. Under the frequency approach, the probability of an event is viewed as a ratio of the number of occurrences of an event to the number of trials in a long-run experiment. According to the subjective approach, probability represents the degree of belief an individual has about the chances that an uncertain event will occur. The degree of belief can be expressed in terms of a bet or wager.

Probabilities are classified as either joint, marginal, or conditional. Joint probability represents the chance that two or more events occur together, and may be computed as a product of probabilities. If the probability that an event will occur can be expressed as a sum of joint probabilities, then it is called a marginal probability. Conditional probability expresses the chance that one event will occur, given the occurrence of another event, and may be computed as the ratio of a joint probability to a marginal probability. Bayes theorem, an expanded form of conditional probability, allows one conditional probability to be calculated in terms of other conditional probabilities. This is useful for revising the probability that an event will occur, after observing information about a process. The initial and revised probabilities are referred to as prior and posterior, respectively.

4.12 TERMS AND FORMULAS

Axioms of probability	Marginal probability
Complement	Mutually exclusive
Compound event	Odds
Degree of belief	Outcome
Dependent	Personal probability
Elementary event	Posterior probability
Equally likely	Prior probability
Event	Random
Exhaustive	Sample space
Experiment	Statistical regularity
Frequency approach	Subjective probability
Fundamental principle of counting	Trial
Joint event	Venn limit

Bayes theorem

$$P(E_i|A) = \frac{P(E_i) \cdot P(A|E_i)}{\sum_{i=1}^{n} P(E_i) \cdot P(A|E_i)}$$

Combination

$$_nC_x = \frac{n!}{x!(n-x)!}$$

Conditional probability

$$P(E|A) = \frac{P(E \text{ and } A)}{P(A)}$$

Joint probability

$$P(E \text{ and } A)$$

Multiplicative law

$$P(E \text{ and } A) = P(A) \cdot P(E|A)$$

Permutation $\qquad$ $_nP_x = \dfrac{n!}{(n-x)!}$

Probability of event E $\qquad$ $P(E)$

Statistical independence $\qquad$ $P(E) = P(E|A)$

QUESTIONS

A. True or False

1. The possible outcomes for two tosses of a fair coin are zero heads, one head, and two heads. These three outcomes each have a probability of occurrence equal to 1/3.

2. Permutations are groups of items where order, but not composition, is important.

3. A marginal probability expresses a sum of joint probabilities.

4. The terms degree of belief and personal probability are associated with the theory of subjective probability.

5. If five tosses of a fair coin have produced five heads (zero tails), the probability of a head on the sixth toss must be larger than $\frac{1}{2}$.

6. One of the basic axioms of probability states that a numerical probability must always be verifiable by experimental evidence.

7. It is plausible that for two events A and B, $P(A) = 0.7$, $P(B) = 0.5$, and $P(A \text{ or } B) = 1.2$.

8. Bayes theorem is a probability formula for computing marginal probabilities.

9. $P(A|B)$ is always less than $P(B)$.

10. Mutually exclusive events are independent.

B. Fill in

1. If the odds of making a grade of A on the next examination in this course are $1:3$, the probability of an A is _____.

2. The numerical probability assigned to an event that is certain not to occur is _____.

3. Events that cannot occur together are called _____.

4. The _____ specifies that if one event can occur in n_1 ways and another event can occur in n_2 ways, the joint event can occur in $n_1 n_2$ ways.

5. If all possible outcomes in an experiment have the same probability of occurrence, the outcomes are termed _____.

6. If $P(A) = P(A|B)$, then events A and B are _____.

7. Assume that $P(A) = 0.4$, $P(B) = 0.3$, and $P(A|B) = 0.8$. $P(A \text{ and } B) = $ _____.

8. If the probability of event C is 0.4, the probability of the complement of C is _____.

9. The _____ of an experiment is a representation of all possible outcomes.

10. Prior and posterior probabilities are associated with _____ theorem.

EXERCISES

1. The following four events are defined to contain certain numbers.

$$A = (5, 8, 9) \qquad B = (1, 3, 5) \qquad C = (1, 2, 3, 4) \qquad D = (7, 9)$$

Which of the following are mutually exclusive?
a. *A* and *B*.   b. *A* and *C*.   c. *A* and *D*.   d. *B* and *C*.
e. *B* and *D*.   f. *C* and *D*.

2. If a particular event is defined as a passing score on an examination, what is the complement of this event?

3. A standard deck of 52 playing cards contains four suits with 13 cards in each suit. Consider an experiment where the cards are shuffled thoroughly, a single card is dealt, and you wish to know which suit occurs.
   a. Name the four possible events that may occur.
   b. Are the events mutually exclusive? Why?
   c. How many trials are in this experiment?
   d. Identify an event that is the complement of the event "diamond or heart."

4. A class of students was asked what magazines they read regularly in their leisure time. A few, group *A*, reported that they regularly read three magazines; *Time*, *Reader's Digest*, and *Playboy*. Another group, group *B*, regularly read *Time* and *Seventeen*. Group *C* read only *Reader's Digest* while group *D* does not read magazines. Determine whether the following student groups are mutually exclusive.
   a. *A* and *B*.   b. *A* and *C*.   c. *A* and *D*.   d. *B* and *C*.

5. In Exercise 3 above, consider the following partial list of events: an ace, a heart, a jack.
   a. Are an ace and heart mutually exclusive?
   b. Are a heart and jack mutually exclusive?
   c. Are an ace and jack mutually exclusive?
   d. Are these three events exhaustive?

6. The local dealer for NOGASIT, the modern economical people's car, advertises that he has models with all color combinations (a color for the car body combined with a color for the vinyl roof) in his showroom. There are six body colors and eight vinyl roof colors. How many different cars are in the showroom, assuming these are the only cars in the showroom?

7. Five out of eight basketball players are to be selected as the starting team. In how many ways can the starting five be chosen, assuming all eight can play any position?

8. How many permutations of all letters can be made from each of the following words?
   a. Statistics. *b.* Possessive. *c.* Narragansett.
   d. Kankakee.

9. How many different assignments of five problems each could your instructor choose from all those in this set of exercises?

10. Some states allow automobile owners to use personal names or slogans for vehicle license plate identification. License plates with names such as **BIGBOY, MARTHA,** and **DEVIL 2** are common. One Hollywood celebrity is the owner of a car with plates reading **KILLER.** Suppose the name must consist of exactly six characters, with the first five required to be letters of the alphabet while the last character may be either a numeral or letter. Forgetting about pronunciation and that the motor vehicle office may not allow some, how many different license plates are possible?

11. How many elementary events are possible when a pair of dice (six-sided) are tossed? If all possible outcomes are identified as the sums of points on the pair of dice 2, 3, 4, . . . , 12, what is the probability of each? Assume the dice are fair.

12. If the odds are 1:7 that Nikki Gustavas will win the Democratic party nomination in the next U.S. presidential race, what is the probability that he will win the nomination?

13. If the probability of rain tomorrow is 0.10, what are the odds in favor of rain?

14. *The Wall Street Journal* (February 9, 1976) reports that many physicians now have access to a computerized service to determine the odds of a patient having a stroke or heart attack. The service, called Cardio-Dial, is offered free to doctors by Ciba Pharmaceutical Company. A physician dials a toll-free number to report a patient's age, smoking habits, blood pressure, blood cholesterol level, heart size, and results of a blood sugar test. The data is fed into the company's computer, which reports, in seconds, the risk of a heart attack or stroke. Data on a patient might include: Male, age 49, average smoker, blood pressure and cholesterol slightly above average. The computer report reads: Patient has 2.3 times the average risk of his sex and age group of having a stroke or heart attack. Chances of heart attack or stroke in the next eight years are 17 out of 100, compared with 8 out of 100 for the average American male of that age and 7 out of 100 for a 49-year-old male who doesn't smoke and has normal blood pressure and cholesterol levels.
   a. If the chance of a heart attack or stroke is 17 out of 100, what are the odds of a heart attack or stroke?
   b. A patient has odds of a heart attack or stroke of 1:8; what is the probability of a heart attack or stroke?

c. If the average American male at age 49 has a chance of 8 out of 100 of a heart attack or stroke, what are the odds against this?

d. One physician takes special precautions with patients when the odds of a heart attack or stroke are 1:4 or greater. What does this mean in terms of the probability of a heart attack or stroke?

15. Ten well-known sports personalities have been invited to participate in a television quiz program. The ten personalities are classified below by sex and the type of sport. Consider that one of the sports personalities is about to be selected for a question on the quiz program (assume all are equally likely to be selected).

	Sport		
Sex	Football	Baseball	Tennis
Female	—	—	Chris Evert Lloyd Martina Navratilova
Male	Terry Bradshaw Bert Jones Tony Dorsett Earl Campbell John Hannah	Ron Guidry Jim Rice	John McEnroe

a. What is the probability that a baseball personality will be selected for the quiz question?

b. What is the probability that a male sports personality will be selected?

c. What is the probability that a male tennis personality will be selected?

d. If the quiz-master wants to choose a tennis personality, what is the probability of male?

e. If the quiz-master wants to choose one of the ladies, what is the probability of a tennis personality?

f. Are sex and type of sport statistically independent in this classification?

g. Identify the above probabilities as simple, joint, marginal, or conditional.

16. A new type of package of candy tablets contains five different flavors: cherry, lime, peppermint, wintergreen, and orange. Each package consists of a roll of ten pieces of candy with two of each of the five different flavors randomly arranged in the package. If you buy and open a new package, what is the probability that the first two pieces are lime?

17. City residents were surveyed recently to determine readership of newspapers available. Fifty percent of the residents read the morning paper, 60 percent read the evening paper, and 20 percent read both newspapers. Find the probability that a resident selected reads either the morning or evening paper.

18. A salesman makes a sale, on the average, to 40 percent of the people he contacts. If four people are contacted today, what is the probability that

he made sales to exactly two? What assumption is required for your answer?

19. Managers for two new branches of a local bank are to be selected from a group of six candidates; four are male and two are female. The president, who will make the choices, considers all equally competent and decides to choose two at random from the group.

 a. What is the probability that the manager of Branch *A* is male and the manager of Branch *B* is female?
 b. What is the probability that both are female?
 c. What is the probability that one person of each sex is selected for the two branches?
 d. Draw a probability tree for the choices of each branch and use it to verify your results for the previous questions.

20. Questionnaires were mailed to the majority of households during the 1970 Census of Population. Two basic questionnaires were available. A short form asked only the names, ages, and incomes of the members of the household; the long form requested information on the type of housing accommodations in addition to the short-form questions. Assume the same procedure is used in the 1980 Census and that your city is equally and randomly divided into two sectors, I and II. Of the households in sector I, a random choice of 40 percent will receive the long form while the remainder receive the short form. In sector II, 20 percent will receive long forms with the others receiving short forms.

 a. What is the probability of selecting, at random, a person who lives in sector II?
 b. What is the probability that this person receives a long-form questionnaire?
 c. Draw a tree diagram for the possible events and determine the probability of each.

21. The results of a survey of preferences for off-campus housing is given below for a group of graduate and undergraduate students.

	Housing Preference	
	Campus	*Off-Campus*
Undergraduate	7	5
Graduate	2	6

12
8
—
20

9   11   20

 a. What percentage of the students surveyed prefer off-campus housing?
 b. If one of the 20 students is selected at random, what is the probability that he or she is an undergraduate?
 c. What is the probability that the student selected prefers campus housing?

    d. What is the joint probability of selecting a graduate student who prefers off-campus housing?

    e. Prepare a joint probability table and use it to verify your previous answers.

    f. Are housing preferences independent of student status, i.e., undergraduate versus graduate?

22. A local service station operator estimates that 25 percent of the cars coming into the station need an oil change, 40 percent need a tune-up, and 18 percent need both an oil change and a tune-up. Suppose a car arrives at the station. If the car's oil needs changing, what is the probability that a tune-up is needed? If a second car arrives and needs a tune-up, what is the probability that it does not need an oil change?

23. An appliance dealer offers a three-year service contract with each appliance sold. Approximately 40 percent of the customers purchase service contracts. Twenty percent of the sales are for air conditioners. In the past; about 25 percent of those purchasing service contracts were air conditioner purchasers.

    a. What is the probability that a customer purchases an air conditioner and service contract?

    b. If the next customer buys an air conditioner, what are the chances he or she will want the service contract?

24. Today physicians are frequently aided by computer diagnosis of illnesses. Computer diagnosis, like traditional diagnosis, is not infallible. Where both methods are used together, the chance of error should be reduced. Suppose a computer diagnosis is correct for 95 percent of those with a certain disease. Also there is a 0.20 probability that the computer would indicate disease presence for those without the disease. If ten percent of a given population have this disease, what is the probability that a person has the disease given that the computer indicates disease presence?

25. One automobile insurance company insures three types of drivers, *A*, *B*, and *C*. Type *A* drivers have a probability of 0.02 of having an accident in any year while the probability for type *B* is 0.05 and 0.10 for type *C*. According to company records, 60 percent of their insured drivers are type *A*, 30 percent are type *B*, and 10 percent are type *C*. An agent reports that one of the company's insured drivers just had an accident. What is the probability that the accident involved a type *A* driver? Type *B* driver? Type *C* driver?

26. The following problem is reprinted from *The American Statistician*, vol. 29, no. 1 (February 1975).

It is "Let's Make a Deal"—a famous TV show starring Monte Hall.

    Monte Hall: One of the three boxes labeled A, B, and C contains the keys to that new 1975 Lincoln Continental. The other two are empty. If you choose the box containing the keys, you win the car.

    Contestant: Gasp!

Monte Hall: Select one of these boxes.

Contestant: I'll take box B.

Monte Hall: Now box A and box C are on the table and here is box B (contestant grips box B tightly). It is possible the car keys are in that box! I'll give you $100 for the box.

Contestant: No, thank you.

Monte Hall: How about $200?

Contestant: No!

Audience: No!!

Monte Hall: Remember that the probability of your box containing the keys to the car is one-third and the probability of your box being empty is two-thirds. I'll give you $500.

Audience: No!!

Contestant: No, I think I'll keep this box.

Monte Hall: I'll do you a favor and open one of the remaining boxes on the table (he opens box A). It's empty! (**Audience:** applause). Now either box C or your box B contains the car keys. Since there are two boxes left, the probability of your box containing the keys is now one-half. I'll give you $1,000 cash for your box.

WAIT !!!!

Is Monte right? The contestant knows that at least one of the boxes on the table is empty. He now knows it was box A. Does this knowledge change his probability of having the box containing the keys from $\frac{1}{3}$ to $\frac{1}{2}$? One of the boxes on the table has to be empty. Has Monte done the contestant a favor by showing him which of the two boxes was empty? Is the probability of winning the car $\frac{1}{2}$ or $\frac{1}{3}$?

Contestant: I'll trade you my box B for the box C on the table.

Monte Hall: That's weird!!

Do you agree with Monte that the contestant's proposal is "weird," or is the contestant making a logical choice? What is the probability that box C contains the key? (*Hint:* Enumerate all possible outcomes or try a prior/posterior approach.)

REFERENCES

Ewart, Park J., et al. *Probability for Statistical Decision Making.* Englewood Cliffs, N.J.: Prentice-Hall, Inc., 1974.

Hodges, J. L., and Lehmann, E. L. *Basic Concepts of Probability and Statistics*, 2d ed. San Francisco: Holden-Day, Inc., 1970.

Maistrov, L. E. *Probability Theory: A Historical Sketch*, Trans. and ed. by Samuel Kotz. New York and London: Academic Press, 1974.

Mosteller, Frederick; Rouke, R. E. K.; and Thomas, G. B. *Probability with Statistical Applications*, 2d ed. Reading, Mass.: Addison-Wesley Publishing Co., 1970.

Springer, Clifford H., et al. *Probabilistic Models.* Homewood, Ill.: Richard D. Irwin, Inc., 1968.

Zeisel, Hans, and Kalven, Jr., Harry "Parking Tickets and Missing Women: Statistics and the Law." In Judith M. Tanur, et al., *Statistics: A Guide to the Unknown.* San Francisco: Holden-Day, Inc., 1972, pp. 102–11.

Probability Distributions

**Optional material: requires calculus.

5 PROBABILITY DISTRIBUTIONS

The world neither ever saw, nor ever
will see, a perfectly fair lottery . . .
because the undertaker could make
nothing by it.

Adam Smith
Wealth of Nations (1776)

5.1 INTRODUCTION

The process of constructing and using frequency distributions for the purpose of describing data characteristics is given in Chapter 2. Chapter 4 contains the basic concepts and terms used in probability. This chapter examines the use of probability as a description of the distribution of a variable. One concept presented, that of expected value, can be used to demonstrate the accuracy of Adam Smith's statement of over 200 years ago.

5.2 RANDOM VARIABLES

A *variable* is defined in Chapter 2 as quantitative data characterized by different numerical values. The science of statistics centers upon variables whose values are uncertain or unpredictable.

> A *random variable* is a variable whose numerical value is determined by chance.

The selling price of a stock listed on the New York Stock Exchange is a variable whose value is unknown in advance. Selling price for most stocks fluctuates throughout the day so that the value of the variable is unpredictable, at least with certainty. Neither can a retailer predict, with certainty, the number of units of a product demanded at some future date. Yet, selling price and units demanded are random variables whose values must be estimated in daily business operations. Hence the importance of measuring the probability associated with the values of a random variable.

> A *probability distribution* is any representation of the values of a random variable and the associated probabilities.

Symbols such as X and Y are generally used as variable names while x and y are used to indicate values of variables. The probability that variable X has the value x is written as $P(X = x)$, which is often shortened to $P(x)$.

EXAMPLE. A sales representative for a computer company contacts five clients each month in an attempt to sell each a new computer system. Define variable X as the number of clients per month who purchase the computer system; the values of the variable are 0, 1, 2, 3, 4, 5. Relative frequencies for the number of systems sold per month over the past several months are given in Table 5.1 as $P(X = x)$. This is a tabular representation of a probability distribution. The probability of selling five systems may be written as $P(X = 5)$ or as $P(5)$. Figure 5.1 provides a graphic representation of the same probability distribution.

TABLE 5.1:
PROBABILITY
DISTRIBUTION FOR
COMPUTER SYSTEM
SALES

Number Sold Per Month (X)	Probability $P(X = x)$
0	0.05
1	0.12
2	0.25
3	0.30
4	0.20
5	0.08
	1.00

FIGURE 5.1: GRAPHIC
PROBABILITY
DISTRIBUTION FOR
COMPUTER SYSTEM
SALES

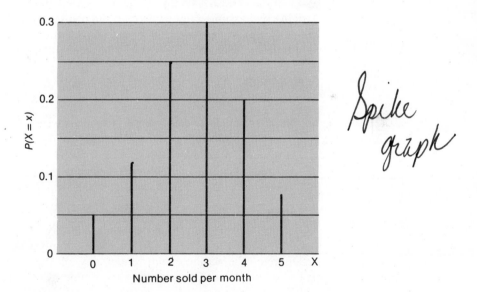

Spike graph

Random variables are classified as either discrete or continuous.

A *discrete* random variable is one whose values are either finite or countably infinite.

The random variable in Table 5.1, number of computer systems sold per month, is discrete because the values of the variable are *finite*, namely 0, 1, 2, 3, 4, and 5. If there is no limit to the number of clients the sales representative may contact in one month, the values of the variable are 0, 1, 2, 3,..., ∞, in which case they are called *countably infinite*. The number of computer systems sold in any month can be counted but, theoretically, an infinite number could be sold in any given month.

Recognizing a discrete random variable involves identifying a counting process. If the values of a random variable can be uniquely associated with the counting numbers (0, 1, 2, etc.), then the random variable is discrete.

A *continuous* random variable is one that can assume any value over the range of the variable.

If the Environmental Protection Agency (EPA) quotes 25 miles per gallon (mpg) as the typical gasoline mileage for highway driving in an automobile, the mileage could in fact be $25\frac{1}{2}$, 25.4, 25.395 mpg, or a similar amount. Depending on the desired degree of accuracy in measurement, the variable "miles per gallon" can assume any value over a certain range (say 0 to 40 mpg); hence it is a *continuous* variable. It is not limited to a finite set of values as in the case of a discrete random variable. We may say that miles per gallon is a random variable that assumes any value over the continuous range between 0 and 40. Weights of packages mailed at a post office and the life (hours) of a tungsten light bulb are examples of other continuous random variables.

EXAMPLE. Determine whether each of the following random variables is discrete or continuous.

   a. Number of defective ball point pens in each carton of twelve.
   b. Number of interruptions (breakdowns) per day at a computer facility.
   c. Distance required for stopping an automobile traveling at 30 miles per hour.
   d. Number of commercial loans processed per day at a bank.
   e. Volume of orange juice in each one-liter container.

Solution.   *a.* Discrete; the variable is limited to the finite values 0, 1, 2, ..., 12. *b.* Discrete; the variable may assume the countably infinite values 0, 1, 2, ..., ∞. *c.* Continuous; the variable (X) may assume any value in the range $X \geq 0$. *d.* Discrete; the variable may equal any of the countably infinite values 0, 1, 2, ..., ∞. *e.* Continuous; the volume (X), in liters, may range over, $0 \leq X \leq 1$.

Probability distributions for discrete random variables are discussed in the following sections. Probability distributions for continuous random variables are discussed in the latter portion of the chapter. In practice, probability distributions are used as models for commonly encountered situations. Each probability model has unique assumptions and characteristics. In choosing a model to represent a given situation, care must be exercised to insure that these assumptions and characteristics are not violated in the applied situation. For each probability distribution, we will summarize the assumptions, state the general form of the model, and give descriptive measures such as the expected value (arithmetic mean), variance, and standard deviation.

THE BERNOULLI FAMILY

Nine members of the Bernoulli family made lasting contributions to mathematics and probability. Some facts about four of them follow. Jacques (1654–1705) and Jean (1667–1748) were sons of a Swiss banking family in Basel. Educated in theology, Jacques refused a church appointment in order to devote himself to mathematics. He was Professor of Mathematics at Basel from 1687 until his death. Many modern probability concepts, such as the binomial distribution and what are now called Bernoulli trials, were introduced in his *Ars Conjectandi*, published posthumously (with the aid of his nephew Nicolaus) in 1713.

Jean received a doctorate in medicine and studied mathematics under the tutelage of his brother Jacques. Both were influenced by Huygens (Chapter 4), who recommended Jean to the post of Professor of Mathematics at Groningen, The Netherlands, an appointment he held until he succeeded Jacques at Basel in 1705. Jean is considered one of the founders of the calculus of variations.

Nicolaus (1695–1726) and Daniel (1700–82), sons of Jean, followed in the footsteps of their father and uncle. At age 18, Nicolaus edited Jacques' *Ars Conjectandi* for publication and defended its contents against critics. He carried on the development of probability theory, corresponding and debating with his peers. Montmort published part of their correspondence in one of his works. Daniel received a doctorate in medicine, held various professorial posts, and made important contributions to astronomy, physics, mathematics, and probability. He is remembered for his work on a classic probability exercise called the St. Petersburg paradox, which was publicized earlier by Nicolaus.

5.3 PROBABILITY MASS FUNCTIONS

Probability distributions for discrete random variables are called probability mass functions.

A *probability mass function* (PMF) is any representation of the values of a discrete random variable and the associated probabilities.

A probability mass function has the following properties. If X is a discrete random variable with values $x_1, x_2, \ldots, x_n$, then:

$$1. \quad P(x_i) \geq 0 \qquad \text{for } i = 1, 2, \ldots, n; \qquad (5.1)$$

and

$$2. \quad \sum_{i=1}^{n} P(x_i) = 1. \qquad (5.2)$$

This states that the probability associated with each value of the variable must be nonnegative, and the sum of these probabilities must equal one. If these properties are met, then the basic axioms of probability in Chapter 4 are satisfied.

Expressions for Probability Mass Functions

Probability mass functions may be expressed in three ways, tabular form, graphically, and in equation form. Table 5.1 is a tabular representation for the number of computer systems sold per month; Figure 5.1 is a graphic portrayal of the same probability mass function. Note that the two properties of a probability mass function are met by both expressions.

A mathematical equation or formula, such as the following, is the most concise way of representing a probability mass function.

$$P(x) = \frac{x}{10} \quad \text{for } x = 0, 1, 2, 3, 4.$$

Does this expression satisfy the two properties for a probability mass function? Substitute values of the variable X to verify that the probabilities are nonnegative and sum to one. (We assume that $P(x) = 0$ for any values of the variable outside the defined range.) Plot a graph of the probability mass function.

A relative frequency distribution (Chapter 2) has all the characteristics of a probability mass function. Refer to Table 2.10 as an illustration. Letting the class midpoints (X) represent the classes, then $P(x) \geq 0$ for all classes, and $\sum P(x) = 1$. Thus our development of probability distributions can be viewed as a refinement of an earlier concept.

Probability mass functions that are used as models for frequently encountered problems are usually expressed in equation form. These are presented after the following measures are defined.

Expected Value and Variance

The principal descriptive measures of a probability distribution are the arithmetic mean and variance. The arithmetic mean is usually referred to as the expected value of the random variable.

Let a discrete random variable X have values $x_1, x_2, \ldots, x_n$. The *expected value* of the discrete random variable is

$$E(X) = \sum_{i=1}^{n} x_i P(x_i). \tag{5.3}$$

The *variance* of the discrete random variable is

$$\sigma^2(X) = \sum_{i=1}^{n} [x_i - E(X)]^2 P(x_i). \tag{5.4}$$

The subscripted notation may be dropped with the understanding that the summation is over all values of X. A more convenient computational formula for the variance is

$$\sigma^2(X) = \sum x^2 P(x) - [E(X)]^2. \tag{5.5}$$

The standard deviation of a random variable, $\sigma(X)$, is the square root of the variance, $\sqrt{\sigma^2(X)}$.

EXAMPLE.   Find the expected value, variance, and standard deviation for the probability mass function of computer system sales, X, given in Table 5.1. Computations required for these measures follow.

x	$P(x)$	$xP(x)$	x^2	$x^2P(x)$
0	0.05	0	0	0
1	0.12	0.12	1	0.12
2	0.25	0.50	4	1.00
3	0.30	0.90	9	2.70
4	0.20	0.80	16	3.20
5	0.08	0.40	25	2.00
Total		2.72		9.02

The expected value of the number sold per month, from Equation (5.3), is

$$E(X) = \sum xP(x) = 2.72.$$

The variance, using Equation (5.5), is

$$\sigma^2(X) = \sum x^2 P(X) - [E(X)]^2$$
$$= 9.02 - (2.72)^2$$
$$= 9.02 - 7.3984 = 1.6216.$$

The standard deviation is

$$\sigma(X) = \sqrt{\sigma^2(X)}$$
$$= \sqrt{1.6216} = 1.27.$$

The expected value of a random variable is the average value over the long run. It should *not* be interpreted as the value of the random variable we "expect" to observe. $E(X)$ may or may not equal an original value of a discrete random variable. It is possible, in the above example, to sell $0, 1, 2, \ldots, 5$ computer systems per month. It is impossible to sell 2.72 systems, the $E(X)$, in a given month; 2.72 is the *average* number sold per month over many months. The most likely number sold per month is 3, the value of X with the highest probability of occurrence.[1]

In the sections that follow, probability mass functions are presented for the following models: uniform, binomial, hypergeometric, and Poisson. For each model, the general equation is given, assumptions and conditions underlying the model are discussed, and illustrations of appropriate applications are presented.

---

[1] The mathematical properties for the mean and variance of linear transformations given in Chapter 2 also hold for random variables. For the linear transformation $Y = a + bX$ on random variable X, the random variable Y has expected value $E(Y) = a + bE(X)$ and variance $\sigma^2(Y) = b^2\sigma^2(X)$.

5.4 UNIFORM PMF

The idea of equally likely elementary events is used in Chapter 4 to introduce some basic probability concepts. The *uniform* probability model is the probability mass function that expresses equally likely events.

> Let X be a discrete random variable that may assume k values with equal probability of occurrence. If the integers $1, 2, \ldots, k$ are used to represent the k values, the variable has the *uniform* probability mass function
>
> $$P(x) = 1/k \qquad \text{for } x = 1, 2, \ldots, k. \qquad (5.6)$$

The expected value and variance of the uniform probability mass function, after substituting $P(x) = 1/k$ into Equations (5.3) and (5.4) and simplifying, are:

$$E(X) = \frac{k + 1}{2} \qquad (5.7)$$

and

$$\sigma^2(X) = \frac{k^2 - 1}{12}. \qquad (5.8)$$

EXAMPLE. Express the possible outcomes of the toss of a fair six-sided die as a uniform probability mass function. Determine the expected value and variance. Let variable X represent the possible number of points on a face of the die; hence, $x = 1, 2, 3, 4, 5, 6$. Variable X is a discrete random variable with equally likely values (fair die). The uniform probability mass function for variable X, the number of points of the face of the die, is

$$P(x) = \frac{1}{6} \qquad \text{for } x = 1, 2, \ldots, 6.$$

A graph of the probability mass function is given in Figure 5.2. Note that any uniform distribution is symmetrical.

FIGURE 5.2: GRAPH OF UNIFORM PROBABILITY MASS FUNCTION

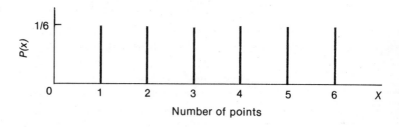

The expected value and variance of the number of points (X) on the face of the die are

$$E(X) = \frac{6 + 1}{2} = 3.5,$$

and

$$\sigma^2(X) = \frac{6^2 - 1}{12} = 2.92.$$

5.5 BINOMIAL PMF   Many processes produce outcomes that are classified into one of two distinct categories. Each item from a production process is frequently labeled either "good" or "defective," classroom attendance is recorded as either "present" or "absent," and a tender offer by one corporation to acquire another is either "accepted" or "rejected." An experiment that results in outcomes of this type is called a *Bernoulli process* (in honor of Jacques Bernoulli).

> A *Bernoulli process* is an experiment that results in one of two mutually exclusive outcomes, usually called success and failure, on each trial. The probability of each outcome remains constant from trial to trial, and repeated trials are independent.

Tossing a fair coin is an experiment that meets the conditions of a Bernoulli process. One of two mutually exclusive outcomes (H = head or T = tail) occurs on each trial. The $P(H)$ and $P(T)$ remain constant for repeated trials, which are independent. The probability that five tosses produce the sequence of outcomes H, H, T, H, T is $(\frac{1}{2})(\frac{1}{2})(\frac{1}{2})(\frac{1}{2})(\frac{1}{2}) = \frac{1}{32}$.

Let the Greek letter π (pi) represent the probability of the outcome "success" and $(1 - \pi)$ represent the probability of the outcome "failure" on each trial of a Bernoulli process (the designation of success and failure is arbitrary). The probability of a *specific* sequence of x successes and $(n - x)$ failures out of n trials is

$$\pi^x(1 - \pi)^{n-x}. \tag{5.9}$$

In the above illustration of five tosses of a fair coin, define success on each trial as the outcome head and failure as the outcome tail. In the present notation, $\pi = P(H) = \frac{1}{2}$ and $(1 - \pi) = P(T) = \frac{1}{2}$. The sequence of outcomes H, H, T, H, T may be referred to as three successes and two failures in a particular order (sequence). Using Equation (5.9), the probability is

$$(\tfrac{1}{2})^3(\tfrac{1}{2})^2 = \tfrac{1}{32}.$$

The experimental result "three successes and two failures" might occur in a different sequence, such as T, T, H, H, H. The probability of this sequence, using Equation (5.9), is also $\frac{1}{32}$. In fact, the probability of three successes and two failures, in any possible sequence, is $\frac{1}{32}$. The number of sequences possible is the number of permutations of three successes and two failures in five trials. From Chapter 4, the number of permutations of x successes and $n - x$ failures out of n trials is

$$_nP_{x, n-x} = \frac{n!}{x!(n - x)!}. \tag{5.10}$$

The number of permutations of three successes out of five trials is

$$_5P_{3,\,2} = \frac{5!}{3!2!} = 10.$$

All possible sequences are enumerated below.

$$
\begin{array}{cc}
H, H, H, T, T & T, H, T, H, H \\
H, H, T, H, T & T, T, H, H, H \\
H, T, H, H, T & H, H, T, T, H \\
T, H, H, H, T & H, T, T, H, H \\
T, H, H, T, H & H, T, H, T, H
\end{array}
$$

Since the probability of each of the ten possible sequences is $\frac{1}{32}$, the probability of three successes ($3H$) and two failures ($2T$), occurring in *any order* (considering all possible sequences) is $(10)(\frac{1}{32}) = \frac{5}{16}$. This result is simply the product of Equations (5.9) and (5.10) for $\pi = \frac{1}{2}$, $n = 5$, and $X = 3$. If X, the number of successes, is the random variable of interest, the *binomial* probability mass function is used to express the probability of X successes as a function of n and π.

The *binomial* probability mass function expresses the probability of x successes out of n trials of a Bernoulli process for fixed values of n and π, the probability of success on a single trial. This is

$$P(X = x \mid n, \pi) = \frac{n!}{x!(n-x)!}\,\pi^x(1 - \pi)^{n-x} \qquad (5.11)$$

for $x = 0, 1, 2, \ldots, n$.

EXAMPLE.   A national advertising agency estimates that only fifty percent of all new products introduced in the U.S. succeed, i.e., become profitable enough to continue being produced. What is the probability that exactly four out of six new products succeed?

Solution.   Define success on each trial of an experiment as a successful product introduction; then $\pi = 0.5$ and $n = 6$. The binomial probability of four successes in six trials is

$$P(X = 4 \mid 6, 0.5) = \frac{6!}{4!2!}\,(0.5)^4(0.5)^2$$

$$= 15(0.015625) = 0.234375.$$

EXAMPLE.   Records for a large airline show that ten percent of its customers request first-class rather than coach accommodations. Of the next four calls for reservations, what is the probability that exactly one requests first class accommodations?

Solution. Let a success on each trial represent a request for first class. For $\pi = 0.1$, $n = 4$, the probability of one success is

$$P(X = 1 | 4, 0.1) = \frac{4!}{1!3!}(0.1)^1(0.9)^3$$

$$= 4(0.0729) = 0.2916.$$

The task of computing binomial probabilities becomes more laborious as n, the number of trials, increases. For this reason, a table of values for the binomial probability mass function is given in Appendix B. Values of the binomial PMF are given to four decimal places, for $\pi = 0.05(0.05)0.50$ (read "π equal to 0.05, by increments of 0.05, to 0.50"), $n = 1(1)20$ and $x = 0(1)n$. Compare the values given in Appendix B with the results given for the previous two examples. Appendix B may also be used to verify that the binomial PMF meets the two properties $P(x) \geq 0$ for all x and $\sum P(x) = 1$. The two probability mass functions for the examples above are illustrated graphically in Figure 5.3.

FIGURE 5.3: BINOMIAL PROBABILITY MASS FUNCTIONS

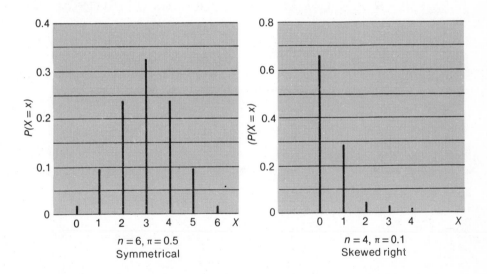

$n = 6$, $\pi = 0.5$
Symmetrical

$n = 4$, $\pi = 0.1$
Skewed right

A probability distribution is defined in terms of its *parameters*, which are numerical values used to specify the distribution. The parameters of the binomial probability mass function are n and π. The parameter values $n = 6$, $\pi = 0.5$ specify one binomial PMF while $n = 4$, $\pi = 0.1$ specify another. The binomial probability mass function is symmetrical if $\pi = 0.5$, skewed right if $\pi < 0.5$, and skewed left if $\pi > 0.5$ (see Figure 5.3).

In each of the above examples, we were interested in evaluating the binomial probability mass function for a single value of the random variable X. Many applications, such as the following example, involve several values of X.

EXAMPLE. Records at a health clinic indicate that 30 percent of the patients visiting the clinic are overweight. What is the probability that, out of the next 20 patients, *a.* between 4 and 6 are overweight, and *b.* at least 5 are overweight?

Solution. Define success as an overweight patient; $\pi = 0.3$ and $n = 20$.
a. The probability that between 4 and 6 are overweight is written as

$$P(4 \leq X \leq 6 | 20, 0.3) = \sum_{x=4}^{6} P(X = x | 20, 0.3)$$

$$= P(X = 4) + P(X = 5) + P(X = 6).$$
$$= 0.1304 + 0.1789 + 0.1916$$
$$= 0.5009 \text{ (using Appendix B).}$$

b. The probability that at least 5 are overweight is

$$P(X \geq 5 | 20, 0.3) = \sum_{x=5}^{20} P(X = x | 20, 0.3)$$

$$= P(X = 5) + P(X = 6) + \cdots + P(X = 20).$$

This is a sum of sixteen values of the binomial PMF where $n = 20$ and $\pi = 0.3$. A more efficient solution recognizes that since

$$\sum_{x=0}^{n} P(X = x | n, \pi) = 1,$$

the complement of "at least 5 successes" may be used to evaluate this. Thus

$$P(X \geq 5 | 20, 0.3) = 1 - P(X \leq 4 | 20, 0.3)$$

$$= 1 - \sum_{x=0}^{4} P(X = x | 20, 0.3)$$

$$= 1 - 0.2374 = 0.7626.$$

The expected value and variance of random variable X, the number of successes, are:[2]

$$E(X) = n\pi \tag{5.12}$$

and

$$\sigma^2(X) = n\pi(1 - \pi). \tag{5.13}$$

---

[2] The expected value and variance of any discrete random variable are defined by Equations (5.3) and (5.4). A simplified form of these measures exists for many probability mass functions. The simplified form for the binomial probability mass function is given by Equations (5.12) and (5.13).

In the previous example, the expected number of overweight patients out of 20 is $(20)(0.3) = 6$; the variance is $(20)(0.3)(0.7) = 4.2$.

BLAISE PASCAL (1623–62)

Pascal, born at Clermont Ferrand, France, was deeply influenced by religious movements. Even though he was in poor health throughout most of his 39 years, he accomplished much in mathematics and physics. He invented an adding machine at age 19.

The mathematical theory of probability and combinatorial analysis evolved from the correspondence between Pascal and Fermat. Pascal was the first to develop a theory of combinations as a triangular arrangement of numbers and to use this for calculating probability. Today, "Pascal's triangle" is recognized as the coefficients in a binomial expansion.

**5.6
HYPERGEOMETRIC
PMF**

Some situations have outcomes that may be classified into one of two categories, such as success and failure, but the probability of each outcome does *not* remain constant from trial to trial, as in the binomial PMF.

EXAMPLE. A repair and service center for household appliances has a carton of ten replacement magnetron tubes for use in certain models of microwave ovens. An oven now being repaired requires one of these replacement tubes and the repairperson randomly chooses a tube from the carton. If two of the ten tubes are defective, the chance of selecting a defective replacement tube is $\frac{2}{10}$. Suppose the repairperson installs one of the tubes into the oven and, without testing the oven, also selects a second tube from the carton and installs it in a second oven also under repair. The probability that the second tube is defective, if the first is defective, is $\frac{1}{9}$. If the first was not defective, the probability that the second is defective is $\frac{2}{9}$. In either case, the probability that the second is defective is unequal to the probability that the first is defective, i.e., the probability of the outcome "defective" does not remain constant in repeated trials.

Let X be a random variable for the number of successes out of a sample of n items selected without replacement from a finite population of N items containing k successes and $N - k$ failures. The probability of x successes, given $n, N,$ and k, follows the *hypergeometric* probability mass function.

$$P(X = x \mid n, N, k) = \frac{(_kC_x)(_{N-k}C_{n-x})}{_NC_n} \tag{5.14}$$

for $x = 0, 1, 2, \ldots, k$.

If four of the ten magnetron tubes (from the above example) are installed in microwave ovens, what is the probability that exactly one is defective? Using

Equation (5.4), $n = 4$, $N = 10$, and $k = 2$,

$$P(X = 1 | 4, 10, 2) = \frac{(_2C_1)(_8C_3)}{_{10}C_4}$$

$$= \frac{(2)(56)}{210} = 0.533.$$

The basis for the hypergeometric PMF lies in the rules of counting given in Chapter 4. In the above application, we evaluated the probability of one defective tube (and three nondefectives) in a sample of four taken from a population consisting of two defective and eight nondefective tubes. The combination formula $_2C_1$ determines the number of ways that one defective can be selected out of two, if order is not important. The number of ways three nondefectives can be obtained out of eight is $_8C_3$. The number of ways that one defective and three nondefectives can occur is the product $(_2C_1)(_8C_3)$. This product relies on the fundamental principle of counting (Chapter 4). The probability of one defective and three nondefectives out of a sample of four is simply the ratio of this product to $_{10}C_4$, the number of ways four items may be selected out of 10.

Use Equation (5.4) to verify that $P(X = 0) = 0.333$ and $P(X = 2) = 0.133$ in the same example. The table of values for the combination formula $_nC_x$ given in Appendix C may be used for this and other applications of the hypergeometric PMF. The hypergeometric probability mass function for X, the number of defective magnetron tubes, is shown graphically in Figure 5.4. The hypergeometric distribution is symmetric only if $k = N - k$.

The expected value and variance of random variable X, the number of successes, are:

$$E(X) = n\left(\frac{k}{N}\right), \tag{5.15}$$

FIGURE 5.4:
PROBABILITY MASS
FUNCTION FOR THE
NUMBER OF DEFECTIVE
MAGNETRON TUBES

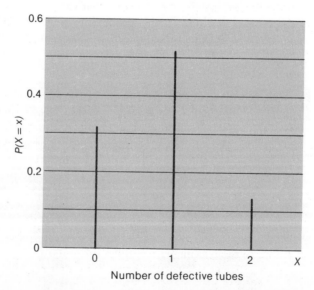

Number of defective tubes

and

$$\sigma^2(X) = \left(\frac{N-n}{N-1}\right)(n)\left(\frac{k}{N}\right)\left(1 - \frac{k}{N}\right). \tag{5.16}$$

The expected number of defective magnetron tubes and variance in the sample of four items are

$$E(X) = 4\left(\frac{2}{10}\right) = 0.8;$$

$$\sigma^2(X) = \left(\frac{10-4}{10-1}\right)(4)\left(\frac{2}{10}\right)\left(1 - \frac{2}{10}\right) = 0.427.$$

EXAMPLE. A merchant's file of 20 accounts contains 6 delinquent and 14 nondelinquent accounts. An auditor randomly selects 5 of these accounts for examination. *a.* What is the probability the auditor finds exactly 2 delinquent accounts? *b.* Find the expected number and variance of delinquent accounts in the sample.

Solution. *a.* Using Equation (5.14) and Appendix C,

$$P(X = 2 | 5, 20, 6) = \frac{(_6C_2)(_{14}C_3)}{_{20}C_5} = \frac{(15)(364)}{15,504} = 0.352.$$

b. Using Equations (5.15) and (5.16),

$$E(X) = 5\left(\frac{6}{20}\right) = 1.5;$$

$$\sigma^2(X) = \left(\frac{20-5}{20-1}\right)(5)\left(\frac{6}{20}\right)\left(1 - \frac{6}{20}\right) = 0.829.$$

5.7 BINOMIAL APPROXIMATION OF HYPERGEOMETRIC PMF

Probability calculations for the hypergeometric PMF defined by Equation (5.14) become very laborious as the population size, N, increases. Appendix C can be used for values of $N \leq 20$. Fortunately, hypergeometric probabilities may be approximated by the binomial probability mass function under fairly general conditions. If the sample size, n, is small relative to the population size, N, then the probability of success differs little from trial to trial. To illustrate, if $N = 1,000$, $k = 200$, and $n = 10$, the probability of a success on each of the first three trials is $\frac{200}{1000}$, $\frac{199}{999}$, and $\frac{198}{998}$. These probabilities are all approximately 0.20. Little error would result here by using the binomial probability mass function with its assumption of a constant probability of success on each trial. Statisticians frequently use the following "rule of thumb" in deciding to approximate the hypergeometric PMF by the binomial PMF.

> The *binomial* probability mass function with parameters n and $\pi = k/N$ provides a good *approximation* of the *hypergeometric* probability mass function if the sample size, n, is no more than five percent of the population size, N, i.e., $n \leq 0.05N$.

EXAMPLE. An Internal Revenue Service (IRS) district office has files on 500 income tax returns that were audited in 1979. After the audit, additional taxes were required on 350 of these. In order to verify that proper audit procedures were followed, a supervisor randomly selects and examines 10 of the 500 returns. What is the probability that additional taxes were required on exactly 6 of the 10 returns sampled? According to the hypergeometric PMF, this would be

$$P(X = 6 \mid 10,500,350) = \frac{_{350}C_6 \, _{150}C_4}{_{500}C_{10}}.$$

Since the sample size ($n = 10$) is 2 percent of the population size ($N = 500$), the binomial PMF with $\pi = \frac{350}{500} = 0.7$ may be used. This is

$$P(X = 6 \mid 10, 0.7) = \frac{10!}{6!4!} (0.7)^6 (0.3)^4 = 0.2001.$$

If a tax return, in the above example, was examined and returned to population before another return is selected, the probability of selecting a return in which additional taxes were required is constant, 0.7, on each trial. This is called *sampling with replacement*. In practice, sampling usually occurs *without replacement* so that the hypergeometric PMF is appropriate when the population is finite. This is often approximated by the binomial PMF, as the above example illustrates.

5.8 POISSON PMF

A *Poisson process* produces a discrete number of occurrences in a continuous interval. The interval may refer to any continuous measure of time, distance, area, etc. The random variable of a Poisson process is the number of occurrences (X) within the interval. Examples of Poisson random variables are: the number of calls occurring at a telephone switchboard in a one-hour interval; the number of defects in a square yard of carpet; and the number of defects in a one-meter length of steel cable. The name Poisson process refers to the French mathematician Simeon D. Poisson (1781–1840).

Any experiment meeting the following conditions is a *Poisson process*.

1. The number of occurrences in one interval is independent of the number of occurrences in another interval.

2. The expected number of occurrences in an interval is proportional to the size of the interval.

3. A very small interval may be identified so that no more than one occurrence is possible in any interval of this size.

Consider the nature of these conditions for calls arriving at a telephone switchboard that averages 10 calls per hour. The number of calls, X, is a random variable; i.e., in some hours only 2 calls are taken, 15 calls in one hour occurs occasionally, etc. Random variable X may assume any of the countably infinite values 0, 1, 2, ... The number of calls in one hour is independent of (has nothing to do with) the number received in another hour (Assumption 1). The expected number of occurrences, λt, is the parameter of the Poisson probability mass function (λ is the Greek letter lambda). λ represents the *intensity* of the Poisson process, or the expected number of occurrences in a specified interval. The value of t is the proportion of this specified interval for the question of interest. If $\lambda = 10$ calls for an interval of one hour, then the expected number of calls in $\frac{1}{2}$ hour is $(10)(\frac{1}{2}) = 5$ (Assumption 2). The expected number of calls in two hours is $(10)(2) = 20$. Assumption 3 implies that no more than one call could occur in some very small time interval, such as one second ($t = \frac{1}{3600}$).

The *Poisson* probability mass function gives, for any Poisson process with parameter λt, the probability of x occurrences in an interval of size t. This is

$$P(X = x | \lambda t) = \frac{(\lambda t)^x e^{-\lambda t}}{x!} \tag{5.17}$$

for $x = 0, 1, 2, \ldots$)

The symbol e in the above formula is the mathematical constant $2.71828\ldots$, the base of the natural logarithms. (Historical note: The symbol is in honor of Leonhard Euler (1707–83), a Swiss mathematician who followed Daniel Bernoulli as Professor of Mathematics at St. Petersburg in 1733.) A table of values for the Poisson probability mass function is given in Appendix D.

EXAMPLE. Patients arrive randomly for treatment at the health clinic of a university hospital. The expected number of patients arriving each hour is four. The probability that exactly two patients arrive in one hour is (using Appendix D)

$$P(X = 2 | 4) = \frac{(4)^2 e^{-4}}{2!} = 0.1465.$$

The Poisson probability mass function has a single parameter, the expected value λt (the expected number of occurrences is the expected value of random variable X). Additionally, the expected value and variance are important descriptive measures for probability distributions. For the Poisson PMF, the expected value and variance are equal. The expected value and variance of the Poisson probability mass function are

$$E(X) = \lambda t \tag{5.18}$$

and

$$\sigma^2(X) = \lambda t. \tag{5.19}$$

In the health clinic example above, the expected number of patients arriving each hour is four; the variance of the number of patients arriving per hour is four.

In general, the Poisson probability mass function is skewed to the right, as shown in Figure 5.5 for three values of the parameter λt. Note that the distribution is less skewed as the value of λt increases.

FIGURE 5.5: POISSON PROBABILITY MASS FUNCTIONS WITH PARAMETER $\lambda t = 0.5, 2.0,$ AND 5.0

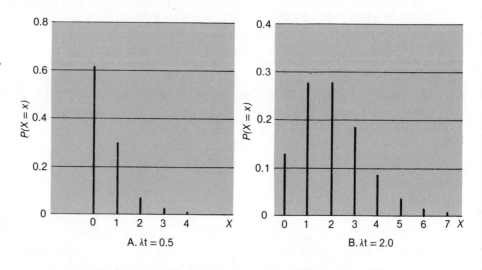

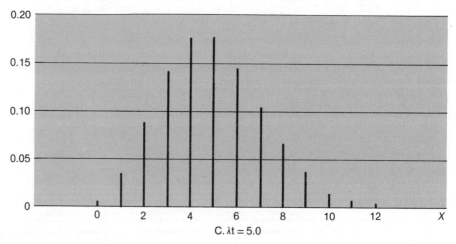

EXAMPLE. Certain types of photographic film are produced and sold in rolls of 100 feet. If the expected number of defects on film rolls is 15 per 100 feet, what is the probability that no more than two defects occur in any ten-foot length?

Solution. For an interval of 100 feet, $\lambda t = 15$. For ten feet, $t = 1/10$ and $\lambda t = (15)(\frac{1}{10}) = 1.5$.

$$P(X \leq 2|1.5) = P(X = 0|1.5) + P(X = 1|1.5) + P(X = 2|1.5)$$
$$= 0.2231 + 0.3347 + 0.2510$$
$$= 0.8088.$$

5.9 POISSON APPROXIMATION OF BINOMIAL PMF

The difficulty of evaluating the binomial probability mass function, without access to extensive tables, increases with sample size. Fortunately, the Poisson probability mass function provides a good approximation for moderately small π and large n.

> The Poisson probability mass function adequately approximates the binomial probability mass function if π, or $(1 - \pi)$, is less than 0.10.

To use the approximation, set the expected successes $n\pi$ equal to expected number of occurrences λt and evaluate $P(X = x|\lambda t)$. If $(1 - \pi) < 0.10$, set $n(1 - \pi)$ equal to the expected number of occurrences and evaluate the probability of $(n - x)$ occurrences. The approximation is satisfactory for n as small as 10 and 20. However, the accuracy of the approximation improves as π decreases and n increases. Since this text has binomial tables for $n \leq 20$, approximation should not be considered unless $n > 20$.

EXAMPLE. A manufacturer of passenger car tires finds that requests for blackwall tires are unpredictable but, in general, amount to 5 percent of orders. If 100 orders are received in one month, what is the probability that fewer than three are for blackwalls?

Solution. The binomial probability of less than three successes, where $n = 100$ and $\pi = 0.05$, is

$$P(X \leq 2|100, 0.05) = \sum_{x=0}^{2} {}_{100}C_x(0.05)^x(0.95)^{n-x}.$$

Since $\pi < 0.10$ (and n is large), the Poisson approximation may be used. The expected number of successes is $n\pi = 100(0.05) = 5$. The Poisson probability of less than three occurrences, given $\lambda t = 5$, is

$$P(X \leq 2|5) = \sum_{x=0}^{2} \frac{(5)^x e^{-5}}{x!}$$
$$= 0.0067 + 0.0337 + 0.0842$$
$$= 0.1246.$$

5.10 PROBABILITY
DENSITY
FUNCTIONS

A probability distribution for a continuous random variable is called a *probability density function* (**PDF**). This is defined as follows.

> A *probability density function* is any function $f(x)$ of a continuous random variable X where:
>
> 1. $f(x) \geq 0$ for all x; and
> 2. The total area under the curve, $f(x)$, equals one.

EXAMPLE. The U.S. Postal Service provides airmail service to foreign countries (except Canada) for packages weighing no more than four pounds. Heavier items may be mailed parcel post. One large post office constructed a probability density function, $f(x)$, for the weight (in pounds, X) of packages airmailed to foreign countries. This is

$$f(x) = \frac{x}{8} \qquad \text{for } 0 \leq x \leq 4.$$

The probability density function is shown graphically in Figure 5.6.

FIGURE 5.6:
PROBABILITY DENSITY
FUNCTION FOR WEIGHT
OF AIRMAIL PACKAGES

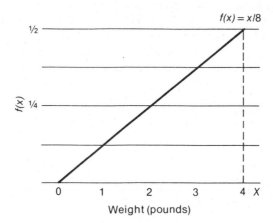

Note that the probability density function for weights of packages meets the two conditions outlined above. For any value x, $f(x) \geq 0$. The total area under curve is one, which may be verified by recognizing that the area between $f(x)$ and the horizontal axis is a right triangle with base 4 and height $\frac{1}{2}$. The area, $\frac{1}{2}$(base)(height), is $\frac{1}{2}(4)(\frac{1}{2}) = 1$.

A probability density function may be expressed graphically or in equation form, as the above example illustrates.

Probability mass functions, probability distributions for discrete random variables, were described in the previous sections of this chapter. The model for any probability mass function determines the probability that the random variable will assume some discrete value, i.e., $P(X = x)$. In a graph of the probability mass function, $P(X = x)$ is represented by the height of the function at the value x. For a continuous random variable, height and probability are not synonymous. Substitution of any value of a continuous random variable into the equation for a probability density function determines only the height or *ordinate* of the function $f(X)$ at the value x. The ordinate of the above probability density function at the value $X = 4$ is $f(X = 4) = \frac{4}{8} = \frac{1}{2}$. Probability measurement for continuous random variables is defined below.

Area as Probability ✳ Probability for a continuous random variable is defined in terms of the *area* ✳ under a curve.

> The probability, $P(a \leq X \leq b)$, that a continuous random variable X takes on values between two limits a and b is the proportion of total area between a and b.

Refer to the probability density function for foreign airmail package weights, $f(x) = x/8$ for $0 \leq X \leq 4$. The probability that a randomly chosen package weighs between 0 and 2 pounds, $P(0 \leq X \leq 2)$, is the proportion of total area between 0 and 2 (shown as the shaded area in Figure 5.7). The area of this small triangle, $\frac{1}{2}$(base)(height), is $\frac{1}{2}(2)(\frac{1}{4}) = \frac{1}{4}$. (The height of the curve at $x = 2$ is found by substituting into the PDF.) Since the total area under the curve is one, the proportion of total area between 0 and 2 is $\frac{1}{4}$, i.e., $P(0 \leq X \leq 2) = \frac{1}{4}$.

FIGURE 5.7:
PROBABILITY THAT A
PACKAGE WEIGHS
BETWEEN 0 AND 2
POUNDS

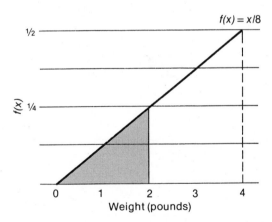

In the case of a continuous variable, there is no area associated with a discrete value. Thus the probability that a continuous random variable takes on a specific discrete value, $P(X = x)$, is 0. For example, the probability that the weight of a foreign airmail package is exactly 2.5 pounds, $P(X = 2.5)$, is zero (see Figure 5.8). *Area* exists only for a *range* of values of a continuous variable. Thus the probability that a foreign airmail package weighs between 2 and 3 pounds, $P(2 \leq X \leq 3)$, is not zero. The area between 2 and 3 is determined by

FIGURE 5.8: NO AREA
ASSOCIATED WITH A
DISCRETE VALUE

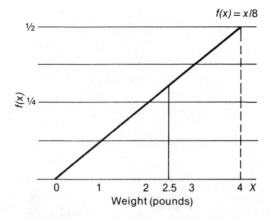

FIGURE 5.8: NO AREA
ASSOCIATED WITH A
DISCRETE VALUE

the difference in area of the two triangles indicated in Figure 5.9. The larger triangle is formed by $f(x)$ and the base 0 to 3. The area of this triangle is $\frac{1}{2}(3)(\frac{3}{8}) = \frac{9}{16}$. The smaller triangle is formed by $f(x)$ and the base 0 to 2. The area, determined earlier, is $\frac{1}{2}(2)(\frac{1}{4}) = \frac{1}{4}$. The area between 2 and 3, $P(2 \leq X \leq 3)$, is the difference in the area of these triangles, $\frac{9}{16} - \frac{1}{4} = \frac{5}{16}$. Also, note that the $P(2 \leq X \leq 3)$ is equivalent to $P(2 < X < 3)$ since $P(X = x) = 0$.

FIGURE 5.9: AREA
BETWEEN 2 AND 3
POUNDS

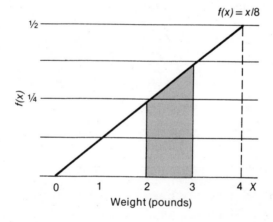

The area under a curve may be determined by a geometric approach, such as the area of a triangle, for only a limited number of mathematical expressions. For most probability density functions, computation of the area under a curve requires the use of integral calculus. Fortunately, computational difficulties in this process may be reduced by utilizing published tables of areas for the more frequently applied probability density functions. Readers familiar with integral calculus will benefit from the discussion in the following section. Others should proceed immediately to the presentation of various probability density functions found in the remainder of the chapter.

**5.11 USE OF INTEGRAL CALCULUS

The definition, characteristics, and application of probability density functions may be formally expressed using integral calculus. The following definition of a probability density function is equivalent to that given in the previous section.

> A *probability density function* is any function $f(x)$ of a continuous random variable X defined over the range R where:
>
> 1. $f(x) \geq 0$ for all x; and
> 2. $\int_R f(x)\,dx = 1$.

parallel to discrete

The second property indicates that the total area over the range of the variable must be one. This is demonstrated as follows for the example in the previous section.

$$\int_R f(x)\,dx = \int_0^4 \left(\frac{x}{8}\right) dx$$

$$= \left(\frac{x^2}{16}\right)\Bigg|_0^4$$

$$= \frac{16}{16} - \frac{0}{16} = 1.$$

The probability that a continuous random variable X assumes any value between two limits a and b is

$$P(a \leq X \leq b) = \int_a^b f(x)\,dx.$$

Using the same example, the probability that a randomly chosen foreign airmail package weighs between two and three pounds is

$$P(2 \leq X \leq 3) = \int_2^3 \left(\frac{x}{8}\right) dx$$

$$= \left(\frac{x^2}{16}\right)\Bigg|_2^3$$

$$= \frac{9}{16} - \frac{4}{16} = \frac{5}{16}$$

Compare the above results with those given in the previous section.

The expected value, $E(X)$, and variance, $\sigma^2(X)$, of a continuous random variable are defined as

$$E(X) = \int_R xf(x)\,dx \tag{5.20}$$

** Asterisks are used to indicate optional material. Those unfamiliar with integral calculus should proceed to the next section.

and

$$\sigma^2(X) = \int_R x^2 f(x)\, dx - [E(x)]^2. \tag{5.21}$$

The expected value and variance of the weight (in pounds) of foreign airmail packages are:

$$E(X) = \int_0^4 x \left(\frac{x}{8}\right) dx$$

$$= \int_0^4 \left(\frac{x^2}{8}\right) dx$$

$$= \left(\frac{x^3}{24}\right)\Big|_0^4 = \frac{8}{3};$$

$$\sigma^2(X) = \int_0^4 x^2 \left(\frac{x}{8}\right) dx - [E(X)]^2$$

$$= \int_0^4 \left(\frac{x^3}{8}\right) dx - \left(\frac{8}{3}\right)^2$$

$$= \left(\frac{x^4}{32}\right)\Big|_0^4 - \frac{64}{9}$$

$$= 8 - \frac{64}{9} = \frac{8}{9}.$$

The standard deviation, defined as $\sigma(X) = \sqrt{\sigma^2(X)}$, is $\sqrt{\frac{8}{9}} = 0.943$.

A final example serves to illustrate these concepts.

EXAMPLE. The time, X (hours), required for scheduled routine maintenance for each airplane in the fleet of Spectre Airlines follows the probability density function

$$f(x) = \frac{3}{500} (10x - x^2) \qquad \text{for } 0 \le X \le 10.$$

a. Verify that $\int_R f(x)\, dx = 1$. b. Find the probability that an airplane, selected at random from the fleet, requires between one and three hours for scheduled routine maintenance (see Figure 5.10). c. Determine the expected value and variance.

Solution.

a. $\quad \int_0^{10} \frac{3}{500} (10x - x^2)\, dx = \frac{3}{500} \left(5x^2 - \frac{x^3}{3}\right)\Big|_0^{10}$

$$= \frac{3}{500} \left(\frac{500}{3}\right) = 1.$$

FIGURE 5.10: TIME
REQUIRED FOR
SCHEDULED ROUTINE
MAINTENANCE AT
SPECTRE AIRLINES

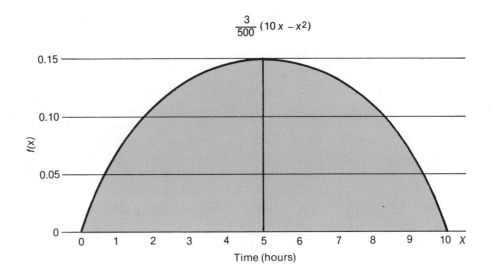

b. $P(1 \leq x \leq 3) = \int_1^3 \dfrac{3}{500}(10x - x^2)\,dx$

$$= \dfrac{3}{500}\left(5x^2 - \dfrac{x^3}{3}\right)\Bigg|_1^3$$

$$= \dfrac{3}{500}\left(\dfrac{94}{3}\right) = 0.188.$$

c. $E(x) = \int_0^{10} xf(x)\,dx$

$$= \int_0^{10} \dfrac{3}{500}(10x^2 - x^3)\,dx$$

$$= \dfrac{3}{500}\left(\dfrac{10x^3}{3} - \dfrac{x^4}{4}\right)\Bigg|_0^{10}$$

$$= \dfrac{3}{500}\left(\dfrac{2500}{3}\right) = 5.$$

$\sigma^2(x) = \int_0^{10} x^2f(x) - [E(x)]^2$

$$= \int_0^{10} \dfrac{3}{500}(10x^3 - x^4)\,dx - (5)^2$$

$$= \dfrac{3}{500}\left(\dfrac{10x^4}{4} - \dfrac{x^5}{5}\right)\Bigg|_0^{10} - 25$$

$$= 30 - 25 = 5.$$

Notice in Figure 5.10 that the distribution of scheduled routine maintenance is symmetrical with an expected time of five hours and variance of five hours.

5.12 UNIFORM PDF   The uniform probability model for a *discrete* random variable was described in Section 5.4. The probability model for a *continuous* uniform random variable is presented here.

> A continuous random variable X that may assume any value x in the interval α to β $(\alpha < \beta)$ follows a *uniform* probability density function if
>
> $$f(x) = \frac{1}{\beta - \alpha}$$
>
> for $\alpha \leq X \leq \beta$.

The expected value and variance of the uniform probability density function are

$$E(X) = \frac{\alpha + \beta}{2} \tag{5.22}$$

and

$$\sigma^2(X) = \frac{(\beta - \alpha)^2}{12}. \tag{5.23}$$

A sketch of the uniform probability density function is given in Figure 5.11. Note that the distribution is symmetrical about the center of the interval α to β, which is the expected value.

FIGURE 5.11: UNIFORM PROBABILITY DENSITY FUNCTION

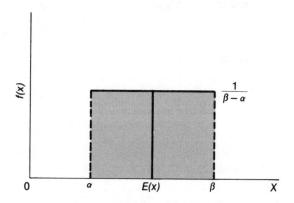

The probability that any continuous random variable assumes any value between limits a and b $(a < b)$ may always be determined by expressing the area between a and b as a proportion to the total area over the interval defined by the probability density function. Since a uniform probability density function defined over the interval α to β has constant height $f(x)$, the probability of any value between limits a and b is simply the distance from a to b expressed as a proportion to the distance from α to β.

Given a uniform probability density function $f(x) = 1/(\beta - \alpha)$ for $\alpha \leq X \leq \beta$, the probability that random variable X assumes values between limits a and b is

$$P(a \leq X \leq b) = \frac{b - a}{\beta - \alpha}.$$

EXAMPLE.   Buses run every 15 minutes along a certain route during rush hour. A person arrives at a bus stop at a random time during rush hour. *a.* What is the probability that this person has to wait between 5 and 10 minutes for the next bus? *b.* What is the probability of a wait of at least 12 minutes? *c.* Determine the expected waiting time and variance of waiting times.

Solution.   Let random variable X be the waiting time (minutes) defined by the uniform probability density function

$$f(x) = \frac{1}{15} \quad \text{for } 0 \leq X \leq 15.$$

a. $P(5 \leq X \leq 10) = \dfrac{10 - 5}{15 - 0} = \dfrac{1}{3}$ (see Figure 5.12, panel A).

b. $P(x \geq 12) = P(12 \leq X \leq 15) = \dfrac{15 - 12}{15 - 0} = \dfrac{1}{5}$ (see Figure 5.12, panel B).

FIGURE 5.12:
DISTRIBUTION OF
WAITING TIME FOR
A BUS

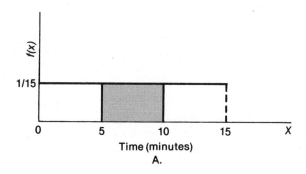

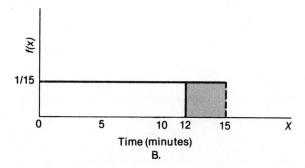

c. $E(x) = \dfrac{0 + 15}{2} = 7.5$ minutes.

$$\sigma^2(x) = \dfrac{(15 - 0)^2}{12} = 18.75 \text{ minutes.}$$

5.13 NORMAL PDF

The *normal* probability density function, often called the *normal distribution*, occupies a place of central importance in statistics. Many phenomena such as student grade-point averages, diameter of machined parts, weight of packages, life of television picture tubes, and breaking strength of steel cables are approximately normally distributed. In addition to its use in modeling such phenomena, the normal distribution is useful for approximating other distributions such as the binomial. Because of its widespread applicability, the normal distribution serves as the basis for many concepts in statistical inference, as shown in the following chapters.

Mathematical Expression

The *normal* probability density function for a continuous random variable X is

$$f(x \mid \mu, \sigma) = \frac{1}{\sigma \sqrt{2\pi}} e^{-\frac{1}{2}\left(\frac{x - \mu}{\sigma}\right)^2}$$

for $-\infty \leq X \leq +\infty$. μ and σ are the parameters of the distribution; π and e are, respectively, the mathematical constants $3.14159\ldots$ and $2.71828\ldots$.

The curve defined by the above expression is a bell-shaped symmetrical distribution for any value of the parameters μ and σ ($-\infty < \mu < +\infty, \sigma > 0$). This is shown graphically in Figure 5.13, where the normal probability density function is defined as $f(x \mid \mu, \sigma)$. Occasionally, the notation $X \sim N(\mu, \sigma)$ is used, which is read "the random variable X is normally distributed with parameters μ and σ." Observe that the distribution is symmetrical about the parameter μ, which is the expected value, $E(X)$, or mean of the distribution. The variance, $\sigma^2(X)$, of a normal distribution is identified as σ^2, while σ is the standard deviation.

The parameters μ and σ, which define a normal distribution, are referred to as the location and shape parameters. The parameter μ determines the

FIGURE 5.13: THE NORMAL PROBABILITY DENSITY FUNCTION

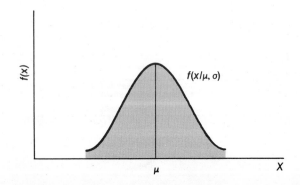

location of the normal distribution along a number line; the parameter σ determines the relative width and height of the distribution. The location parameter μ is illustrated in Figure 5.14 by two distributions where $\mu_1 < \mu_2$ and $\sigma_1 = \sigma_2$. The effect of the shape parameter σ is shown in Figure 5.15 by two distributions where $\mu_1 = \mu_2$ and $\sigma_1 > \sigma_2$. The width of the distribution decreases and the height increases as the standard deviation decreases.

FIGURE 5.14: NORMAL DISTRIBUTIONS WHERE $\mu_1 < \mu_2$ AND $\sigma_1 = \sigma_2$

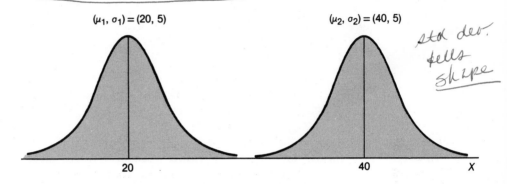

$(\mu_1, \sigma_1) = (20, 5)$ $(\mu_2, \sigma_2) = (40, 5)$

std dev.
tells
shape

20   40   X

FIGURE 5.15: NORMAL DISTRIBUTIONS WHERE $\mu_1 = \mu_2$ AND $\sigma_1 > \sigma_2$

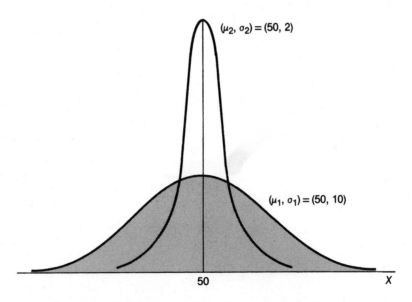

$(\mu_2, \sigma_2) = (50, 2)$

$(\mu_1, \sigma_1) = (50, 10)$

50   X

     The normal distribution is a proper probability density function since (1) $f(x) \geq 0$ for all x, and (2) the total area under the curve $f(x)$ equals one. As in any probability function, the probability that a normally distributed random variable X assumes values between limits a and b, $P(a \leq X \leq b)$, is equal to the proportion of total area under the curve between the limits a and b (see Figure 5.16). For any given value of μ and σ, the area between limits a and b may be determined by integral calculus as follows.

$$P(a \leq X \leq b) = \int_a^b f(x)\,dx = \int_a^b \frac{1}{\sigma\sqrt{2\pi}}\, e^{-\frac{1}{2}\left(\frac{x-\mu}{\sigma}\right)^2}\,dx.$$

FIGURE 5.16: THE AREA
BETWEEN TWO LIMITS IN
A NORMAL DISTRIBUTION

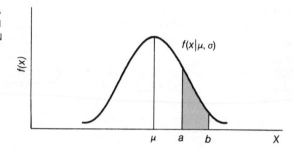

However it is easier to evaluate the area between two limits in a normal distribution by using a table of areas for a special normal distribution called the *standard normal distribution*.

Standard Normal Distribution

A *standardized* random variable is one that has a mean of zero and standard deviation of one. A random variable with any mean and standard deviation can be transformed to a standardized random variable by subtracting the mean and dividing by the standard deviation. For a normal distribution with mean μ and standard deviation σ, the standardized variable Z is obtained as

$$Z = \frac{X - \mu}{\sigma}. \tag{5.24}$$

A value z represents the distance, expressed as a multiple of the standard deviation, that the value x lies away from the mean. The normal distribution has the mathematical property that *any linear function of a normally distributed random variable is also normally distributed*. For any given distribution, μ and σ are constants. And $Z = \dfrac{X - \mu}{\sigma} = -\dfrac{\mu}{\sigma} + \dfrac{1}{\sigma} X = a + bX$ is simply a linear transformation of X. If X is normally distributed, Z is also normal.[3] Thus Z is a normally distributed random variable with mean zero and standard deviation one.

A normal distribution that has a mean of zero and standard deviation of one is called the *standard normal distribution*. If Z is the standardized normally distributed random variable, then Z has the probability density function

$$f(z) = \frac{1}{\sqrt{2\pi}} e^{-\frac{1}{2}z^2}$$

for $-\infty \leq Z \leq +\infty$.

Areas for a normal distribution are tabulated in Appendix E in terms of the standardized variable Z. Since any normally distributed random variable X

---

[3] Recall the linear transformation formulas of Section 5.3. For random variables, if $Y = a + bX$, then (1) $E(Y) = a + bE(X)$ and (2) $\sigma^2(Y) = b^2\sigma^2(X)$. These properties can be used to verify $E(Z) = 0$ and $\sigma^2(Z) = 1$.

with parameters μ and σ can be transformed to the standardized normally distributed random variable Z by use of the standardizing transformation in Equation (5.24), Appendix E may be used to evaluate $P(a \leq X \leq b)$ for any normal distribution.

Before turning to probability calculations, examine Appendix E to learn some valuable characteristics of the normal distribution. Appendix E provides, for any value z, the area between the mean of the distribution (0) and z. Since area is synonymous with probability, denote this as $P(0 \leq Z \leq z)$, the probability that random variable Z assumes values between 0 and z. The $P(0 \leq Z \leq +1) = 0.34134$ is obtained simply by looking for $Z = 1$ in Appendix E. Since the normal distribution is symmetric about the mean, $P(-1 \leq Z \leq 0)$ is the same as $P(0 \leq Z \leq +1)$. The $P(-1 \leq Z \leq +1)$ is the sum of these values, 0.68268, as shown in Figure 5.17. Thus for any normal distribution, approximately 68 percent of all observations are contained within an interval ranging from one standard deviation below the mean ($Z = -1$) to one standard deviation above the mean ($Z = +1$). Likewise, approximately 95 percent of the area is within an interval ranging over 1.96 standard deviations ($Z = 1.96$) on either side of the mean; and about 99 percent is contained within 2.58 standard deviations on either side of the mean. It is precisely this relationship between area and distance away from the mean that is important in the use of the normal distribution. Keep in mind that other distributions may be both bell-shaped and symmetrical without being normal. Only the normal PDF has the properties described above.

FIGURE 5.17:
CHARACTERISTICS OF
AREA FOR A NORMAL
DISTRIBUTION

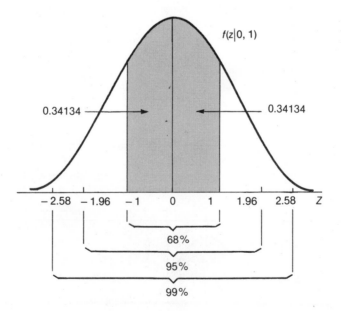

Probability
Calculations

The probability that a normally distributed random variable X assumes values between two limits a and b is determined by transforming these limits into the standardized variable Z and using Appendix E. This is best expressed by means of several illustrations. Keep in mind that Appendix E gives, for any value z, the area between the mean and z units (standard deviations) away from the mean.

EXAMPLE. A manufacturer of lemon-lime drink sells his product in individual cans and six-packs of cans. The volume per can is normally distributed with a mean of 330 milliliters (ml) and standard deviation of 10 milliliters, i.e., $X \sim N(330, 10)$. A can is chosen at random and the volume of its contents measured. What is the probability that the volume is: *a.* between 325 and 340 ml; *b.* less than 345 ml; *c.* greater than 338 ml?

Solution. *a.* The $P(325 \leq X \leq 340)$ is represented by the shaded area shown in upper curve of Figure 5.18. Express the values 325 and 340 in terms of Z, using Equations (5.24), $Z = (X - \mu)/\sigma$. For $X = 325$,

$$Z = \frac{325 - 330}{10} = -0.5.$$

For $X = 340$,

$$Z = \frac{340 - 330}{10} = +1.0.$$

FIGURE 5.18:
TRANSFORMING VALUES
TO THE STANDARD
NORMAL DISTRIBUTION

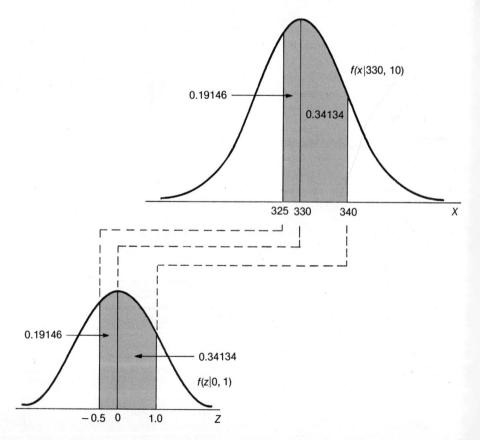

The value $Z = -0.5$ indicates that $X = 325$ is 0.5 standard deviations to the left of the mean in a normal distribution with $(\mu, \sigma) = (330, 10)$. The transformation of the value $X = 325$ in the normal distribution with $(\mu, \sigma) = (330, 10)$ to the value $Z = -0.5$ in the standard normal distribution, where $(\mu, \sigma) = (0, 1)$ is shown by the two curves in Figure 5.18. Similarly, $Z = +1.0$ indicates that $X = 340$ is one standard deviation to the right of the mean in a normal distribution with $(\mu, \sigma) = (330, 10)$. From Appendix E, the area in the standard normal distribution between $Z = 0$ and $Z = -0.5$ is 0.19146 and that between $Z = 0$ and $Z = +1.00$ is 0.34134. The area between $Z = -0.50$ and $Z = +1.00$, $P(-0.50 \leq Z \leq 1.00)$, is the sum $0.19146 + 0.34134 = 0.53280$. Therefore in the normal distribution with $(\mu, \sigma) = (330, 10)$, the $P(325 \leq X \leq 340) = 0.53280$. Approximately 53 percent of the cans have a volume between 325 and 340 ml.

b. The $P(X < 345)$ is represented by the shaded area in Figure 5.19. This area has two components, the area to the left of the mean 330 and the area between the mean and 345. Since a normal distribution is symmetrical, there is 50 percent of the area on either side of the mean (verify from Appendix E). Thus $P(X < 330) = 0.50$. The area between 330 and 345 is obtained by computing

$$Z = \frac{345 - 330}{10} = +1.5,$$

and finding this value in Appendix E, which gives $P(330 \leq X < 345) = 0.43319$. Therefore,

$$P(X < 345) = P(X < 330) + P(330 \leq X < 345)$$
$$= 0.50 + 0.43319$$
$$= 0.93319.$$

Recall that a statement of $P(X < 330)$ is equivalent to a statement of $P(X \leq 330)$ in a continuous distribution.

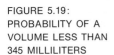

FIGURE 5.19:
PROBABILITY OF A
VOLUME LESS THAN
345 MILLILITERS

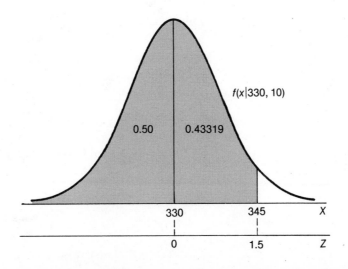

c. The $P(X > 338)$ is represented by the area to the right of 338 in Figure 5.20. Computing

$$Z = \frac{338 - 330}{10} = +0.8$$

and looking in Appendix F, the area between 330 and 338 (the area between $Z = 0$ and $Z = +0.8$) is 0.28814. Since the area to the right of the mean 330 is 0.50, the area to the right of 338, $P(X > 338)$ is $0.50 - 0.28814 = 0.21186$.

FIGURE 5.20:
PROBABILITY OF A
VOLUME GREATER THAN
338 MILLILITERS

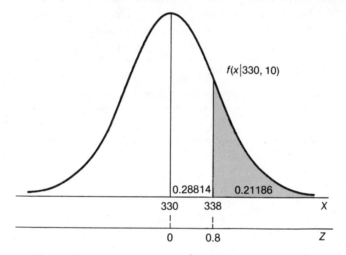

The various parts of the above example illustrate how the information given in Appendix E may be used in probability computations for a normal distribution. In using Appendix E, keep in mind that the table provides the area between the mean $(Z = 0)$ and any value z.

EXAMPLE. A wholesale distributor of fertilizer products finds that annual demand for one type of fertilizer is normally distributed with a mean of 120 tons and standard deviation of 16 tons. If he orders only once a year, what quantity should be ordered to insure that there is only a 5 percent chance of running short?

Solution. Let X be the random variable annual demand (tons) where $X \sim N(120, 16)$. We wish to determine the value x so that $P(X > x) = 0.05$, as shown in Figure 5.21. The area to the right of the desired value x is 0.05. Since the area between the mean and x is 0.45, Appendix E can be used to determine that the desired value x is 1.64 standard deviations to the right of the mean (the area between the mean and $Z = 1.64$ is 0.44950, the closest area to exactly 0.45). Now Equation (5.24) may be solved for the desired value x. Substituting,

$$Z = \frac{X - \mu}{\sigma}$$

$$+1.6 = \frac{X - 120}{16}$$

$$X = 120 + 1.64(16) = 146.24.$$

FIGURE 5.21:
DETERMINING THE 95TH
PERCENTILE OF THE
DISTRIBUTION OF
ANNUAL DEMAND

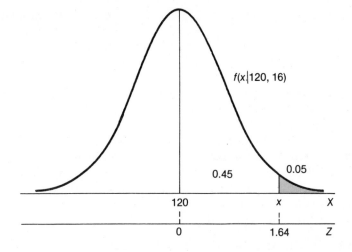

If it is necessary to order in whole units, then the distributor should order 147 tons, in which case the probability of demand exceeding this value is slightly less than 0.05.

The value $X = 146.24$ tons obtained above is the 95th percentile of the distribution of annual demand. Any percentile of a normal distribution can be determined by the same procedure.

5.14 NORMAL APPROXIMATION OF BINOMIAL PMF

The arithmetic difficulties associated with the use of the binomial PMF for large sample size, n, were described earlier. If π, the probability of success on a single trial, is small (less than 0.10) for large n, the binomial PMF may be approximated by the Poisson PMF as described in Section 5.9. Under different conditions, the normal PDF may be used to approximate the binomial PMF. In fact, the normal distribution originated with Abraham DeMoivre's (1667–1754) method of approximating the sum of terms for the binomial.

The normal distribution is bell-shaped and symmetrical with mean μ and standard deviation σ. Recall that the binomial is symmetrical only if $\pi = \frac{1}{2}$. The mean of a binomial PMF, the expected number of successes, is $E(X) = n\pi$; the variance is $\sigma^2(X) = n\pi(1 - \pi)$; and the standard deviation (square root of the variance) is $\sigma(X) = \sqrt{n\pi(1 - \pi)}$. If n is large and π is close to $\frac{1}{2}$, the normal PDF provides a good approximation to the binomial PMF. The following rule is used.

> The normal probability density function with mean $n\pi$ and standard deviation $\sqrt{n\pi(1 - \pi)}$ may be used to approximate the binomial probability mass function if $n \geq 50$ and both $n\pi$ and $n(1 - \pi)$ are greater than 5.

If, for example, $n = 50$ and $\pi = 0.3$, then $n\pi$ $50(0.3) = 15$ and $n(1 - \pi) = 50(0.7) = 35$. Thus random variable X, the number of successes, can be expressed as a normal distribution with mean $n\pi = 15$ and standard deviation

$\sqrt{n\pi(1-\pi)} = \sqrt{50(0.3)(0.7)} = 3.24$. One additional consideration is required in the normal approximation of the binomial. In the binomial distribution, random variable X is discrete while the random variable of a normal distribution is continuous. This difference requires a correction in the approximation. To see why, consider the binomial probability $P(X = 20|50, 0.3)$. The probability of exactly 20 successes is 0.0370 (not given in Appendix B). In a probability density function, such as the normal, the proportion of total area associated with a discrete point is 0 (see Figure 5.8). In order to approximate $P(X = 20)$ we establish a range of values for random variable X by adding and subtracting $\frac{1}{2}$ to the discrete value 20, obtaining 19.5 and 20.5. The proportion of total area between 19.5 and 20.5 in a normal distribution with parameters $(\mu, \sigma) = (15, 3.24)$ approximates the binomial probability $P(X = 20|50, 0.3)$. This approximation is shown by the shaded area in Figure 5.22. To determine this area, express the values 19.5 and 20.5 in terms of the standard normal distribution, $Z = \dfrac{X - \mu}{\sigma}$, and then refer to Appendix E. The Z-values are

$$Z = \frac{19.5 - 15}{3.24} = 1.39$$

and

$$Z = \frac{20.5 - 15}{3.24} = 1.70.$$

Using Appendix E, the proportion of total area between 19.5 and 20.5 is 0.03769. Note that the approximate probability differs little from the exact binomial probability of 0.0370.

FIGURE 5.22: NORMAL APPROXIMATION OF THE BINOMIAL PROBABILITY $P(X = 20|60, 0.3)$

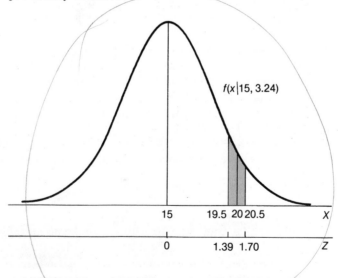

The adjustment of $\pm\frac{1}{2}$ to a discrete value is called a *continuity correction factor*. The correction should be used for any normal approximation of the binomial PMF. The use of this approximation for various binomial probabilities is illustrated in Table 5.2.

TABLE 5.2: USE OF THE CONTINUITY CORRECTION FACTOR FOR APPROXIMATING THE BINOMIAL PMF WITH $(n, \pi) = (50, 0.3)$ BY THE NORMAL PDF WITH $(\mu, \sigma) = (15, 3.24)$

To Approximate the Discrete Values	Use the Continuous Values
$X = 20$	$19.5 \le X \le 20.5$
$12 \le X \le 25$	$11.5 \le X \le 25.5$
$X > 22$	$X \ge 22.5$
$X < 16$	$X \le 15.5$

EXAMPLE. Galaxy Computers, a manufacturer of minicomputers for home use, sells an optional preprogrammed cassette tape for performing the owner's federal income tax computations. Forty percent of the minicomputer buyers purchase the tax cassette. In auditing a random sample of 100 receipts for units sold, what is the probability that tax cassettes were sold with a least 45 of these units?

Solution. Define success as the sale of a tax cassette, $\pi = 0.4$ and $n = 100$. Random variable X, the number of tax cassettes sold, follows a binomial distribution. The required probability is

$$P(X \ge 45 \mid 100, 0.4) = \sum_{x=45}^{100} \frac{100!}{x!(100 - x)!} (0.4)^x (0.6)^{100-x}.$$

The normal approximation may be used since $n = 100$, $n\pi = 100(0.4) = 40$ and $n(1 - \pi) = 100(0.6) = 60$. Represent the number of successes, X, as a normal distribution with mean $n\pi = 100(0.4) = 40$ and standard deviation $\sqrt{n\pi(1 - \pi)} = \sqrt{100(0.4)(0.6)} = 4.90$. The approximate binomial probability is the proportion of area in the normal distribution larger than 44.5 (the area between 44.5 and 45.5 represents the discrete value 45), as shown in Figure 5.23. The value 44.5 is $Z = \dfrac{X - \mu}{\sigma} = \dfrac{44.5 - 40}{4.9} = 0.92$ standard deviations to the right of the mean 40. Using Appendix E, the proportion of area larger than 44.5 is 0.17879. This is the approximate probability of tax cassettes being sold with at least 45 of the 100 units sold.

FIGURE 5.23: NORMAL APPROXIMATION OF BINOMIAL PROBABILITY $P(X \ge 45 \mid 100, 0.4)$

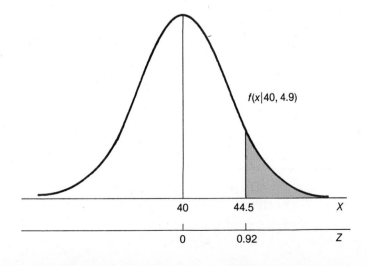

$f(x \mid 40, 4.9)$

5.15 LINEAR
COMBINATIONS OF
RANDOM
VARIABLES

Concepts involving the sum or difference of random variables arise in later chapters. For example, Sections 9.5–9.7 introduce useful inferential procedures based on the difference of two random variables. In this section, important theorems concerning combinations of random variables are given. The proofs are not given, but each theorem's use is demonstrated in an example.

In dealing with sums or differences of random variables, the most general case is termed a linear combination. If $X_1, X_2, \ldots, X_n$ are n random variables, the random variable $Y = a_1 X_1 + a_2 X_2 + \cdots + a_n X_n$ is a *linear combination* of the X_i where the a_i are constants. Three results are of particular interest.

> The expected value $E(Y)$ of a linear combination Y of random variables X_i is given by the linear combination of the expected values:
>
> $$E(Y) = a_1 E(X_1) + a_2 E(X_2) + \cdots + a_n E(X_n). \qquad (5.25)$$

EXAMPLE. In Section 5.4, it was shown that if $X =$ number of points of the face of a die, $E(X) = 3.5$. Suppose a *pair* of dice is thrown and $X_1 =$ outcome of first die and $X_2 =$ outcome from second die, what is the mean of $Y = X_1 + X_2$, the sum of the points on the pair of dice?

The expected value of $Y = X_1 + X_2$ is obtained using Formula (5.25) as:

$$E(Y) = E(X_1) + E(X_2) = 3.5 + 3.5 = 7.0.$$

Another result concerns the variance of a linear combination of random variables. However, this result requires that the n random variables be *independent*. Note that Formula (5.25) does *not* require independent random variables.

> If the X_i are independent random variables, the variance of the linear combination is given by:
>
> $$\sigma^2(Y) = a_1^2 \sigma^2(X_1) + a_2^2 \sigma^2(X_2) + \cdots + a_n^2 \sigma^2(X_n). \qquad (5.26)$$

EXAMPLE. Continuing the previous example, what is the variance of $Y = X_1 + X_2$, the sum of points on a pair of dice?

First, $\sigma^2(X_1) = \sigma^2(X_2) = 2.92$ as illustrated in Section 5.4. Then intuitively, X_1 and X_2 are independent random variables since the outcome of the first die has nothing to do with the outcome of the second. Then, using Equation (5.26):

$$\sigma^2(Y) = (1)^2 \sigma^2(X_1) + (1)\sigma^2(X_2)$$
$$= \sigma^2(X_1) + \sigma^2(X_2)$$
$$= 2.92 + 2.92 = 5.84.$$

EXAMPLE. Suppose X_1 and X_2 are independent random variables. What is the mean and variance for $Y = X_1 - X_2$?

The random variable Y is a linear combination of X_1 and X_2 with $a_1 = 1$ and $a_2 = -1$.

$$Y = X_1 - X_2$$
$$= (1)X_1 + (-1)X_2.$$

Then from Equations (5.25) and (5.26):

$$E(Y) = a_1 E(X_1) + a_2 E(X_2)$$
$$= (1)E(X_1) + (-1)E(X_2)$$
$$= E(X_1) - E(X_2).$$
$$\sigma^2(Y) = a_1^2 \sigma^2(X_1) + a_2^2 \sigma^2(X_2)$$
$$= (1)^2 \sigma^2(X_1) + (-1)^2 \sigma^2(X_2)$$
$$= \sigma^2(X_1) + \sigma^2(X_2).$$

The latter result is somewhat unexpected; in words, the variance of the *difference* is the *sum* of the variances (assuming independence). The utility of this result will become apparent in later chapters.[4] The first two results are especially useful when the X_i are both normally and independently distributed, i.e., the $X_i \sim$ N.I.D.

> If the X_i are normally and independently distributed random variables, their linear combination will be normally distributed. ($E(Y)$ and $\sigma^2(Y)$ are given by Equations (5.25) and (5.26), respectively.)

---

[4] The results of the first three examples of this section may be verified using Formulas (5.3) and (5.4) and the appropriate probability mass functions. If $X_1 =$ outcome on first die and $X_2 =$ outcome on second die, the PMFs in tabular form are given below.

$x_1 + x_2$	$x_1 - x_2$	P
2	-5	1/36
3	-4	2/36
4	-3	3/36
5	-2	4/36
6	-1	5/36
7	0	6/36
8	1	5/36
9	2	4/36
10	3	3/36
11	4	2/36
12	5	1/36
		36/36

The probabilities given are for both $(x_1 + x_2)$ and $(x_1 - x_2)$. Then, for example,

$$E(X_1 - X_2) = \sum (x_1 - x_2) \cdot P(x_1 - x_2) = (-5)(1/36) + \cdots + 5(1/36) = 0.$$

Note that using Formula (5.25) does not require the PMF of $(x_1 - x_2)$ to determine $E(X_1 - X_2)$ if $E(X_1)$ and $E(X_2)$ are known.

EXAMPLE.   The weights of passengers reserving seats on a small commuter airline follow a normal distribution with $\mu = 140$ and $\sigma = 20$ pounds. Assuming independence among the passenger weights, determine the distribution of the *total* weight of ten passengers. If the recommended maximum passenger weight is 1,500 pounds, how often would ten passengers' weight exceed the recommended maximum?

First, the X_i are normally and independently distributed, and the total weight

$$Y = X_1 + X_2 + \cdots + X_{10}$$

is a linear combination of the X_i. Also, $E(X_i) = 140$ and $\sigma(X_i) = 20$ for $i = 1, 2, \ldots, 10$. Then, Y will follow a normal distribution with parameters:

$$E(Y) = E(X_1) + E(X_2) + \cdots + E(X_{10})$$
$$= 140 + 140 + \cdots + 140 = 1400$$
$$\sigma^2(Y) = \sigma^2(X_1) + \sigma^2(X_2) + \cdots + \sigma^2(X_n)$$
$$= (20)^2 + (20)^2 + \cdots + (20)^2 = 4000$$

and

$$\sigma(Y) = \sqrt{4000} = 63.2.$$

The distribution of Y is $N(1400, 63.2)$, as illustrated in Figure 5.24.

FIGURE 5.24:
DISTRIBUTION OF TOTAL
WEIGHT OF TEN
PASSENGERS

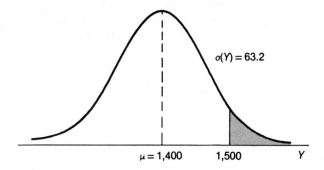

To determine the probability that the total weight exceeds 1,500 pounds, compute:

$$Z = \frac{Y - E(Y)}{\sigma(Y)}$$

$$= \frac{1500 - 1400}{63.2} = 1.58.$$

From Appendix E and $Z = 1.58$, we read 0.44295. Then, $P(Y > 1500) = 0.50 - 0.44295 = 0.05705$.

5.16   SUMMARY

The probability distribution for a random variable models the values of the random variable and provides a mechanism for making probability statements and describing characteristics of the random variable. The choice of a probability model involves a careful comparison of the situation encountered with the unique assumptions and characteristics of the model. One of the probability mass functions is chosen to model a discrete random variable, while a probability density function is used for a continuous random variable.

Each probability mass function or probability density function is defined in terms of its parameters. Different values of parameters describe a different distribution within each general model. In the binomial model, for example, the binomial PMF with parameters $(n, \pi) = (20, 0.3)$ is different from the binomial PMF with $(n, \pi) = (8, 0.5)$. Because of this, it is often said that each probability model, such as the binomial, represents a *family* of distributions. As the parameter values change within a given model, so does the distribution.

Characteristics of the probability distributions discussed in this chapter are summarized in Table 5.3. The table provides, for each model, the general form or equation, values of the variable, parameters, expected value, and variance. The uniform model is used to express equally likely values of a random variable. The uniform PMF is chosen for discrete values while the uniform PDF is appropriate for a continuous random variable.

The binomial probability mass function expresses the probability of X successes out of n trials of a Bernoulli process. A Bernoulli process has two mutually exclusive outcomes, called success and failure, on each trial. Repeated trials are independent and the probability of each outcome remains constant from trial to trial.

A hypergeometric probability mass function determines the probability of X successes out of a sample of n items without replacement from a finite population consisting of k successes and $N - k$ failures. Hypergeometric probabilities may be approximated by the binomial PMF if the sample size, n, is not more than five percent of the population size, N.

If discrete occurrences are observable in a continuous interval where: (*a*) the number of occurrences in one interval is independent of the number in another; (*b*) the expected number of occurrences is proportional to the size of the interval; and (*c*) a large interval can be divided into smaller intervals so that no more than one occurrence is possible in the smallest interval; then the Poisson probability mass function may be used to determine the probability of X occurrences in an interval of size t. The Poisson PMF is also useful for approximating the binomial PMF when π, the probability of success, or $(1 - \pi)$ is less than 0.10.

The normal probability density function is a bell-shaped symmetrical distribution with parameters, the mean and standard deviation, μ and σ. Probability statements are evaluated by expressing values in any normal distribution in terms of the standard normal distribution and using the table of areas in Appendix E. The normal PDF is useful not only as a model for observable phenomena but also as the limiting case for many distributions in statistical inference. It will be encountered frequently in this respect in later chapters. For our present discussion, the normal PDF is used as a limiting case for the binomial as sample size, n, increases. Because of this, the binomial PMF

TABLE 5.3: CHARACTERISTICS OF PROBABILITY DISTRIBUTIONS

Probability Mass Function	*Equation*	*Parameters*	$E(X)$	$\sigma^2(X)$
Uniform	$P(x) = \dfrac{1}{k}$ for $x = 1, 2, \ldots, k$	k	$\dfrac{k+1}{2}$	$\dfrac{k^2-1}{12}$
Binomial	$P(X = x \mid n, \pi) = \dfrac{n!}{x![(n-x)]!}\,\pi^x(1-\pi)^{n-x}$ for $x = 0, 1, \ldots, n$	n, π	$n\pi$	$n\pi(1-\pi)$
Hypergeometric	$P(X = x \mid n, N, k) = \dfrac{{}_kC_x\,{}_{N-k}C_{n-x}}{{}_NC_n}$ for $x = 0, 1, \ldots, k$	n, N, k	$n\left(\dfrac{k}{N}\right)$	$\left(\dfrac{N-n}{N-1}\right)(n)\left(\dfrac{k}{N}\right)\left(1-\dfrac{k}{N}\right)$
Poisson	$P(X = x \mid \lambda t) = \dfrac{(\lambda t)^x e^{-\lambda t}}{x!}$ for $x = 0, 1, \ldots, \infty.$	λt	λt	λt
Probability Density Function				
Uniform	$f(x) = \dfrac{1}{\beta - \alpha}$ for $\alpha \le x \le \beta$	α, β	$\dfrac{\alpha + \beta}{2}$	$\dfrac{(\beta-\alpha)^2}{12}$
Normal	$f(x \mid \mu, \sigma) = \dfrac{1}{\sigma\sqrt{2\pi}}\,e^{-\frac{1}{2}\left(\frac{x-\mu}{\sigma}\right)^2}$ for $-\infty \le x \le +\infty$	μ, σ	μ	σ^2
Standard Normal	$f(z \mid 0, 1) = \dfrac{1}{\sqrt{2\pi}}\,e^{-\frac{1}{2}z^2}$ for $-\infty \le x \le +\infty$	$0, 1$	0	1

may be approximated by the normal PDF if $n \geq 50$ and both $n\pi$ and $n(1 - \pi)$ are greater than 5.

Your understanding of and ability to use these distributions as probability models will be enhanced by exploring their use in applied situations, including the exercises that follow.

5.17 TERMS AND FORMULAS

Bernoulli process	Parameter
Binomial PMF	Poisson process
Continuous random variable	Poisson PMF
Continuity correction factor	Probability distribution
Countably infinite	Probability density function (PDF)
Discrete random variable	Probability mass function (PMF)
Family of distributions	Random variable
Failure	Sampling without replacement
Finite	Sampling with replacement
Hypergeometric PMF	Standardized random variable
Normally and independently	Standard normal PDF
distributed (N.I.D.)	Uniform PDF
Normal PDF	Uniform PMF

Standard deviation units
$$Z = \frac{X - \mu}{\sigma}$$

Expected value of a discrete random variable
$$E(X) = \sum x P(x)$$

Variance of a discrete random variable
$$\sigma^2(X) = \sum x^2 P(x) - [E(X)]^2$$

Expected value of a continuous random variable
$$E(X) = \int_R x f(x)\, dx$$

Variance of a continuous random variable
$$\sigma^2(X) = \int_R x^2 f(x) - [E(X)]^2$$

For the linear combination of random variables,

$$Y = a_1 X_1 + a_2 X_2 + \cdots + a_n X_n,$$
$$E(Y) = a_1 E(X_1) + a_2 E(X_2) + \cdots + a_n E(X_n).$$

If the X_i are independent,

$$\sigma^2(Y) = a_1^2 \sigma^2(X_1) + a_2^2 \sigma^2(X_2) + \cdots + a_n^2 \sigma^2(X_n).$$

Formulas for probability mass functions and probability density functions are summarized in Table 5.3.

QUESTIONS

A. True or False

1. The time required for customer transactions at a bank is an example of a discrete random variable.
2. Any normal distribution is symmetrical.
3. All uniform distributions are not symmetrical.
4. The binomial distribution is symmetrical if $\pi = 1 - \pi$.
5. The number of defects per square foot in a sheet of plate glass is an example of a Poisson process.
6. If $f(x) = 1.5$ for some x, then $f(x)$ cannot be a probability density function.
7. The binomial PMF may be used to approximate the hypergeometric PMF if the sample size is at least 10 percent of the population size.
8. Probability distributions for discrete random variables are called probability density functions.
9. The continuity correction factor is applied when approximating the binomial PMF by the Poisson PMF.

B. Fill in

1. If four cards are dealt (without replacement) from a standard deck of 52 playing cards, the probability of exactly one ace is determined by use of the _____ probability distribution.
2. Probability distributions for continuous random variables are called _____ functions.
3. If a binomial probability mass function has parameters $(n, \pi) = (10, 0.7)$ the expected number of successes is _____.
4. The number of failures in a switching device is Poisson distributed with an expected value of six per 1,000 hours of operation. The expected number of failures for 250 hours of operation is _____.
5. A _____ variable results by subtracting the mean and dividing by the standard deviation.
6. The parameters of the normal distribution are _____.
7. The Poisson distribution may be used to approximate the binomial when _____.
8. The _____ distribution is usually associated with sampling without replacement.
9. If X is a continuous random variable, $P(X = x)$ equals _____.
10. The standard normal distribution has a mean equal to _____ and standard deviation equal to _____.

EXERCISES[5]

1. Given the following PMF:

X	$P(X = x)$
5	0.2
6	0.4
7	0.3
8	0.1

Find the expected value and variance of the random variable X.

2. Historical data has been used to develop the following probability mass function for the number of jobs arriving per minute for processing during peak periods at the input station of a computer facility.

Jobs Per Minute X	$P(X = x)$
0	0.12
1	0.20
2	0.33
3	0.15
4	0.09
5	0.07
6	0.04
	1.00

a. Plot a graph of the probability mass function.
b. Compute the expected number of jobs arriving per minute and locate this on the graph. Also determine $\sigma^2(X)$.
c. Determine $P(1 \leq X < 5)$ and $P(X > 4)$.

3. A probability mass function is described by the following equation.

$$P(x) = \frac{1}{55} x^2$$

for $x = 0, 1, 2, 3, 4, 5$.
a. Prepare a table of the probability mass function showing the possible values of X and $P(x)$.
b. Plot a graph of the probability mass function.
c. Compute $E(X)$, $\sigma^2(X)$, and $\sigma(X)$.

4. Demand for birthday cakes on Fridays at a bakery ranges from 1 through 8 with an equal probability of occurrence. Let demand on Fridays be defined by the discrete random variable X with values $x = 1, 2, \ldots, k$.
a. Write the equation for the uniform probability mass function that describes Friday demand.
b. Find the expected value and variance of Friday demand.
c. What is the probability that fewer than three cakes will be demanded on some Friday?

---

[5] Exercises marked ** require calculus.

5. Ten machines are used in packaging a perishable processed cheese product at a dairy products company. Each package is stamped with a five-digit code to identify the data and machine used. The first two digits indicate a month (01 to 12); the second two digits indicate a day of the month (01 to 31); and the last digit indicates the machine (0 to 9). The package code 09297 represents a product packaged September 29 on machine 7. Packages are examined on a random basis as a quality control measure.

   a. Determine a uniform probability mass function to express the value of the last digit in the package code.

   b. In your solution to (a), what assumption must be made about the number of packages produced by the ten machines?

6. Twenty percent of the guards available from a private security agency are female. One of the agency's clients is a large airport that requires 10 guards. If the agency randomly assigns guards to airport duty, what is the probability the airport is assigned:

   a. four female guards;

   b. fewer than three female guards?

7. A newspaper published in a large metropolitan area estimates that 55 percent of the residents read its morning edition. If telephone calls are made to a random sample of 20 residents, what is the probability that at least 15 read the morning edition? Determine the expected value and variance of the number of residents in the sample who read the morning edition.

8. Perry, a student, must take a ten-question true-false examination today. Since he has not studied for the examination, Perry decides to toss a coin in order to answer each question either true or false. If there must be at least seven correct answers to pass, what is the probability that Perry will pass?

9. A CPA firm is auditing a group of accounts in which they expect that 10 percent of the accounts contain errors. Eight accounts are randomly selected for audit and errors are found in four. What is the probability of this sample result if in fact the proportion of accounts with errors is 10 percent? What does this lead you to believe?

10. A spokesman for an insurance company says that 95 percent of all new policies issued are for term life insurance. If 12 new policies are mailed today, what is the probability that fewer than 2 are for coverage other than term life insurance?

11. A local politician claims that the assessed value of homes (for property tax purposes) is incorrect for 80 percent of the homes in his district. If this is true, what is the probability that a random survey of ten homes in his district finds nine incorrect assessments?

12. A manufacturer of handmade furniture has twenty units of a popular chair available for sale to retailers. Even though three of them have a few

scratches and other blemishes, they can all be sold "as is" because of their scarcity. Riverside Furniture, a retailer, orders ten of these chairs. If ten chairs are selected at random for shipment to Riverside, what is the probability that:

a. none is blemished;

b. at least two are blemished?

13. A box of 20 miniature integrated circuit chips, used in hand-held calculators, includes 4 defectives. If five chips are selected at random from the box, find the probability that a. exactly one is defective, and b. at least three are defective. c. Find the expected value and variance of the number of defective chips in the sample of five.

14. In a statistics class with 18 students enrolled, the instructor decides to choose three names at random from the roll and ask each student to answer a different probability question. If only seven of the students are capable of answering these questions, find the probability that a. all three answer correctly, and b. none answers correctly.

15. An insurance company provides hospitalization coverage at group rates for a firm with 820 employees. Forty-one of the employees are identified as "high-risk" under the terms of the coverage. Hospitalization claims are now being processed for 16 employees.

a. What is the probability that none of these 16 is in the high-risk category?

b. What is the probability that at least one is in the high-risk category?

16. A regional distributor of window-model home air conditioners currently has 600 in stock, of which 360 are "energy efficient". If the distributor's shipping clerk fails to distinguish between these two types of units, what is the probability that the next 20 units shipped will include 10 of each type?

17. In an attempt to be more customer-conscious, automobile manufacturers are encouraging dealers to take additional steps during dealer preparation to insure that the car delivered to the customer is free of minor irritants such as loose bolts, dents and scratches, squeaks and rattles, and improperly operating components. In addition to improving customer satisfaction, the plan also reduces the cost of warranty work for these problems. After following the manufacturer's suggestions for a while, Midwest Autoworks finds an average of three such defects per car inspected. What is the probability that one of the inspected cars had at least six defects?

18. A large manufacturing facility recently initiated a new safety program in an attempt to reduce its workers' injury rate. Since the introduction of the program, the number of injuries requiring medical attention has averaged 1 per 20 working days.

a. What is the probability of no injury that requires medical attention in the next 20 working days?

   b. What is the probability of 2 or more such injuries in the next 10 working days?

   c. What is the probability of no injury of this type in the next working day?

19. An airline reports that its fatality rate over the last decade is 0.005 per million passenger miles. The airline expects to fly about 680 million passenger miles this year.

   a. What is the expected number of fatalities for the airline this year?

   b. What is the probability of fewer than two fatalities this year?

20. During peak periods, customers enter a bank at the rate of 90 per hour.

   a. What is the probability that 15 or more customers enter the bank in a ten-minute interval during a peak period?

   b. What is the most likely number of customers to arrive in a five-minute interval at peak times?*

   c. What is the expected value and variance of the number of customers arriving in a five-minute interval during peak periods?

21. Studies by a telephone company indicate that billing errors occur in only 0.8 percent of the monthly statements mailed to residential customers. A consumer group has asked the state agency governing the telephone company to investigate the accuracy of this claim. The state agency will select a random sample of 500 statements for scrutiny. If the company's claim is correct, *a.* what is the probability that between 2 and 6 statements in the sample have errors; *b.* what is the probability of no statement with errors?

22. About seven percent of the U.S. labor force is between 16 and 24 years old. Assuming your city reflects this age distribution, what is the probability that a random sample of 50 employed persons in your city contains 2 or fewer in the 16–24 age group?

23. A realtor indicates that only 5 percent of all homes sold are sold by the owner; the remainder are sold by agents. If the newspaper contains a listing of 72 homes recently sold, what is the probability that:

   a. exactly three were sold by owners;

   b. at least two were sold by owners?

**24. A machine manufacturer uses the following probability density function to describe the life (in years) of a new machine.

$$f(x) = \tfrac{3}{2}x - \tfrac{3}{4}x^2$$

for $0 \leq X \leq 2$, where X = years of life.

   a. Sketch the probability density function.

   b. Find the expected value and variance of the years of life for the machine.

c. Determine the probability that the machine's life is between 1 and 2 years.

25. The distribution of time (in minutes) between telephone calls at a switchboard can be approximated by the following probability density function.

$$f(x) = \tfrac{1}{8}(4 - x)$$

for $0 \leq X \leq 4$, where $X = $ time (in minutes).
a. Sketch the probability density function.
b. Find the probability that the time between calls is at least 2 minutes.
c. What is the probability that the time between calls is between 1 and 3 minutes, i.e., $P(1 \leq X \leq 3)$.

**26. A probability density function is given by

$$f(x) = kx^3$$

for $0 \leq X \leq 3$.
a. Find k (where $\int_R f(x)\,dx = 1$).
b. Determine $P(X > 2)$.
c. Find $E(X)$, $\sigma^2(X)$, and $\sigma(X)$.

27. The time X (in minutes) for an assembly operation follows a uniform probability density function over the range $2 \leq X \leq 4$.
a. Write the equation for the uniform probability density function.
b. Sketch the probability density function.
c. Find the expected value and variance of assembly times.
d. What is the probability that the assembly time is at least 3 minutes 30 seconds?

28. The demand at any given time on a particular electric utility power plant is approximately normally distributed with a mean of 120,000 kilowatts and standard deviation of 10,000 kilowatts. If the plant cannot generate any more than 145,000 kilowatts at any one time, what is the probability that there will be an overload, i.e., that more than 145,000 kilowatts will be demanded?

29. If Z follows the standard normal distribution, determine the following:
a. $P(0 \leq Z \leq 2)$
b. $P(Z \leq 2)$ $0 \leq .5 \leq .2$
c. $P(-0.75 \leq Z \leq 0)$
d. $P(-0.75 < Z < 0.30)$
e. $P(0.50 \leq Z \leq 1.25)$.

30. Scores on a college entrance exam are normally distributed with $\mu = 500$ and $\sigma = 100$.
a. 75 percent of the scores are below what value?
b. 90 percent of the scores are above what value?
c. What percent of the scores are between 450 and 550?
d. What percent of the scores are above 400?
e. What percent of the scores are less than 500?

31. The number of days' sick leave requested annually by employees of a large insurance company is normally distributed with a mean of 9 days and standard deviation of 2.5 days.
    a. What proportion of the employees request at least 10 days sick leave?
    b. If the company employs 600 people, how many use fewer than 2 days sick leave annually?
    c. Find the 90th percentile of the distribution.

32. The average expenditure per customer in an evening at a local disco is $7.20 with a standard deviation of $2.25. The expenditure per customer is normally distributed.
    a. Find the proportion of customers who spend more than $12.
    b. The middle 50 percent of the expenditures per customer is between what two amounts?
    c. What is the probability that randomly-selected customer spends less than $8 in one evening?

33. A large construction firm estimates that the time required to complete an office complex is normally distributed with a mean of 18 months and standard deviation of 2 months.
    a. What is the probability that it will take at least 20 months to complete the office complex?
    b. What is the probability that it will be completed in less than 21 months?
    c. If the firm wishes to make a bid on this project, quoting a completion time that it has a 90 percent chance of meeting, how many months should it quote?

34. An airline computes capacity and range for its planes using a normal distribution of weights for adult passengers with a mean of 150 and standard deviation of 20 pounds. What proportion of the adult passengers weigh:
    a. more than 185 pounds;
    b. less than 125 pounds;
    c. between 140 and 160 pounds?
    d. Between what weights do the middle 90 percent of the adult passengers fall?

35. Your statistics instructor informs you that on the last examination a score of 95 is 3 standard deviations above the mean while a score of 65 is 1 standard deviation below the mean. If the exam scores are normally distributed, what is the mean and standard deviation?

36. The U.S. banking industry maintains that 30 percent of customers switch banks because of dissatisfaction with the present bank. If a branch of a local bank currently has 200 customer accounts, what is the probability that no more than 50 of these will switch because of dissatisfaction?

37. A domestic airline uses a 120-passenger Boeing 727 on one of its routes where the "seat-load factor" (proportion of seating capacity used) averages 65 percent. The airline estimates that, for this route and equipment, a 60

percent seat-load factor is required to break even. What proportion of the airline's flights on this route lose money?

38. The practice of "overbooking" reserved seats in the airline industry has drawn much attention recently. The industry claims that 10 percent of all reservations are "no shows," and it is generally accepted that the airlines try to offset this revenue loss by selling about 7 percent more reservations than seats on a plane.

    a. If a flight using a 250-seat McDonnell Douglas DC-10 is 100 percent booked, what is the probability of 10 or fewer "no shows"?

    b. If a flight using a 100-seat Concorde is fully booked but there is no overbooking, what is the probability that the flight departs with exactly 95 percent of capacity (called a 95 percent "seat-load factor")?

    c. If a flight using a 300-seat Lockheed L-1011 Tristar is 7 percent overbooked, what is the probability that at least one person holding a reservation will be "bumped" (denied boarding)?

39. A process produces machined parts that are sold in lots of 1,000 units. If 1 percent of the parts are defective and the parts are not inspected before being sold, what is the probability that a lot contains a. fewer than 5 defectives; b. between 8 and 12 defectives?

40. Refer to the last example of Section 5.15. If a DC-9 holds 100 passengers and passenger weights are NID with $E(X_i) = 140$ and $\sigma(X_i) = 20$, what is the distribution of total weight of 100 passengers? How often would the total weight exceed 14,500 pounds?

41. Given two independent random variables with

$$E(X_1) = 10 \qquad E(X_2) = 30$$
$$\sigma^2(X_1) = 10 \qquad \sigma^2(X_2) = 25,$$

determine the mean and variance of the following:

a. $Y = X_1 + X_2$.

b. $R = X_1 - X_2$.

c. $W = 7X_1 + 2X_2$.

42. Suppose that $X_1, X_2, \ldots,$ and X_n are NID random variables and each has the same mean and variance: $E(X_i) = \mu$ and $\sigma^2(X_i) = \sigma^2$ for $i = 1, 2, \ldots, n$. Define:

$$Y = \left(\frac{1}{n}\right)X_1 + \left(\frac{1}{n}\right)X_2 + \cdots + \left(\frac{1}{n}\right)X_n.$$

Use Formulas (5.25) and (5.26) to show that:

$$E(Y) = \mu;$$
$$\sigma^2(Y) = \sigma^2/n.$$

REFERENCES

Ewart, Park J., et al. *Probability for Statistical Decision Making.* Englewood Cliffs, N.J.: Prentice-Hall, Inc., 1974.

Hays, William L. *Statistics for the Social Sciences*, 2d ed. New York: Holt, Rinehart and Winston, Inc., 1973.

Hines, William W., and Montgomery, Douglas C. *Probability and Statistics in Engineering and Management Science*. New York: John Wiley & Sons, Inc., 1972.

Pfaffenberger, Roger C., and Patterson, James H. *Statistical Methods for Business and Economics*. Homewood, Ill.: Richard D. Irwin, Inc., 1977.

Springer, Clifford H., et al. *Probabilistic Models*. Homewood, Ill.: Richard D. Irwin, Inc., 1968.

Sampling and Sampling Distributions

**Optional material.

6 SAMPLING AND SAMPLING DISTRIBUTIONS

*The trick, Fletcher, is that we are trying
to overcome our limitations in order, patiently.
We don't tackle flying through rock until a
little later in the program.*

Richard Bach
Jonathan Livingston Seagull

6.1 INTRODUCTION

The introductory chapter stressed the importance and applicability of inferential statistics to business today. Since Chapter 1, major topics have included descriptive statistics and probability, which, although possessing considerable utility in and of themselves, are also prerequisites to our study of inferential procedures. In this chapter, the concepts of probability distributions, means and variances, and expected value are drawn together to develop the sampling distribution. Sampling distributions represent the last rung of the ladder leading to inferential procedures. Like Fletcher, we are overcoming our limitations in order and, in Chapter 8, we will tackle statistical estimation and hypothesis testing.

> **ABRAHAM DE MOIVRE (1667–1754)**
>
> De Moivre was born in Champagne, France, and studied mathematics at the Sorbonne in Paris. After imprisonment for maintaining his Huguenot beliefs, he moved to England and was employed as a tutor while establishing his reputation. His book, *The Doctrine of Chances*, is considered a landmark in the development of probability. In 1733, De Moivre discovered the normal distribution and proved a version of the central limit theorem for the special case of a Bernoulli (0 or 1 valued) random variable. However, this discovery, first published in Latin as a supplement to another of his works, generated little enthusiasm among his colleagues. In fact, until 1924, the development of the normal curve was erroneously attributed to the German mathematician Gauss.

6.2 SIMPLE RANDOM SAMPLES

The inferential procedures of this text are based upon information obtained via random sampling. The student's intuitive ideas of random selection likely involve such terms as fairness, haphazardness, or chance elements; and visions of "odd man out" and bingo balls are conjured up. To the statistician, random selection has a very specific meaning. In particular we now define:

> A *simple random sample without replacement* is a sample of n elements from a population of N elements selected in such a way that every combination of n elements has an equal probability of selection.

174

Note, *combinations* have an equal probability of selection. Occasionally, students define simple random samples in terms of *elements* having an equal probability of selection. While this is a characteristic of simple random samples, it is not the *distinguishing* characteristic. For instance, suppose a population consists of 100 names on a list and a sample of size 10 is desired. If one of the names in the first ten is randomly chosen (say the seventh) and then every tenth name on the list thereafter (17th, 27th, etc.) one obtains a sample such that every item had an equal chance of selection. However, since a sample chosen in this manner could *not* contain, for example, the 7th and 8th names on the list, every combination of size ten is not equally likely, and the sample is not simple random.[1]

EXAMPLE. Suppose one is given the population below:

Item number:	1	2	3	4	5
Measurement:	12	14	16	18	20.

The item number identifies an element in the population; the elements themselves may be employees, accounts, or departments, while the measurement could be salary, days overdue, or long-distance phone charges. If sampling without replacement with sample size $n = 2$, how many different samples (i.e., combinations) can be selected? If simple random sampling is employed, what is the probability of selecting any particular combination?

Using the combination formula of Chapter 4, the number of possible samples is

$$_NC_n = {_5C_2} = \frac{5!}{2!3!} = 10,$$

and since each sample is equally likely, by definition of simple random sampling, then the probability of selecting any particular sample is

$$\frac{1}{_NC_n} = \frac{1}{10}.$$

6.3 OBTAINING SIMPLE RANDOM SAMPLES

Having defined a simple random sample, let us now turn to the mechanics of obtaining it. While there is more than one procedure (e.g., computer generation of a random selection, or even drawing names from a hat), we shall restrict ourselves to employing tables of random numbers. These tables have been created by means that assure us of their random nature; two of the better known tables are:

1. *Table of 105,000 Random Decimal Digits*, Interstate Commerce Commission, Washington, 1949.

---

[1] This sampling procedure is termed *systematic sampling*. In general, to obtain a systematic sample of n from N elements, let $K = N/n$, then start with a random number between 1 and K, and select every Kth item thereafter.

TABLE 6.1: PAGE FROM A TABLE OF RANDOM NUMBERS[a]

	(1)	(2)	(3)	(4)	(5)	(6)	(7)	(8)	(9)	(10)	(11)	(12)	(13)	(14)
1	10480	15011	01536	02011	81647	91646	69179	14194	62590	36207	20969	99570	91291	90700
2	22368	46573	25595	85393	30995	89198	27982	53402	93965	34095	52666	19174	39615	99505
3	24130	48360	22527	97265	76393	64809	15179	24830	49340	32081	30680	19655	63348	58629
4	42167	93093	06243	61680	07856	16376	39440	53537	71341	57004	00849	74917	97758	16379
5	37570	39975	81837	16656	06121	91782	60468	81305	49684	60672	14110	06927	01263	54613
6	77921	06907	11008	42751	27756	53498	18602	70659	90655	15053	21916	81825	44394	42880
7	99562	72905	56420	69994	98872	31016	71194	18738	44013	48840	63213	21069	10634	12952
8	96301	91977	05463	07972	18876	20922	94595	56869	69014	60045	18425	84903	42508	32307
9	89579	14342	63661	10281	17453	18103	57740	84378	25331	12566	58678	44947	05585	56941
10	85475	36857	53342	53988	53060	59533	38867	62300	08158	17983	16439	11458	18593	64952
11	28918	69578	88231	33276	70997	79936	56865	05859	90106	31595	01547	85590	91610	78188
12	63553	40961	48235	03427	49626	69445	18663	72695	52180	20847	12234	90511	33703	90322
13	09429	93969	52636	92737	88974	33488	36320	17617	30015	08272	84115	27156	30613	74952
14	10365	61129	87529	85689	48237	52267	67689	93394	01511	26358	85104	20285	29975	89868
15	07119	97336	71048	08178	77233	13916	47564	81056	97735	85977	29372	74461	28551	90707
16	51085	12765	51821	51259	77452	16308	60756	92144	49442	53900	70960	63990	75601	40719
17	02368	21382	52404	60268	89368	19885	55322	44819	01188	65255	64835	44919	05944	55157
18	01011	54092	33362	94904	31273	04146	18594	29852	71585	85030	51132	01915	92747	64951
19	52162	53916	46369	58586	23216	14513	83149	98736	23495	64350	94738	17752	35156	35749
20	07056	97628	33787	09998	42698	06691	76988	13602	51851	46104	88916	19509	25625	58104
21	48663	91245	85828	14346	09172	30168	90229	04734	59193	22178	30421	61666	99904	32812
22	54164	58492	22421	74103	47070	25306	76468	26384	58151	06646	21524	15227	96909	44592
23	32639	32363	05597	24200	13363	38005	94342	28728	35806	06912	17012	64161	18296	22851
24	29334	27001	87637	87308	58731	00256	45834	15398	46557	41135	10367	07684	36188	18510
25	02488	33062	28834	07351	19731	92420	60952	61280	50001	67658	32586	86679	50720	94953
26	81525	72295	04839	96423	24878	82651	66566	14778	76797	14780	13300	87074	79666	95725
27	29676	20591	68086	26432	46901	20849	89768	81536	86645	12659	92259	57102	80428	25280
28	00742	57392	39064	66432	84673	40027	32832	61362	98947	96067	64760	64584	96096	98253
29	05366	04213	25669	26122	44407	44048	37937	63904	45766	66134	75470	66520	34693	90449
30	91921	26418	64117	94305	26766	25940	39972	22209	71500	64568	91402	42416	07844	69618
31	00582	04711	87917	77341	42206	35126	74087	99547	81817	42607	43808	76655	62028	76630
32	00725	69884	62797	56170	86324	88072	76222	36086	84637	93161	76038	65855	77919	88006
33	69011	65795	95876	55293	18988	27354	26575	08625	40801	59920	29841	80150	12777	48501
34	25976	57948	29888	88604	67917	48708	18912	82271	65424	69774	33611	54262	85963	03547
35	09763	83473	73577	12908	30883	18317	28290	35797	05998	41688	34952	37888	38917	88050
36	91567	42595	27958	30134	04024	86385	29880	99730	55536	84855	29080	09250	79656	73211
37	17955	56349	90999	49127	20044	59931	06115	20542	18059	02008	73708	83517	36103	42791
38	46503	18584	18845	49618	02304	51038	20655	58727	28168	15475	56942	53389	20562	87338
39	92157	89634	94824	78171	84610	82834	09922	25417	44137	48413	25555	21246	35509	20468
40	14577	62765	35605	81263	39667	47358	56873	56307	61607	49518	89656	20103	77490	18062
41	98427	07523	33362	64270	01638	92477	66969	98420	04880	45585	46565	04102	46880	45709
42	34914	63976	88720	82765	34476	17032	87589	40836	32427	70002	70663	88863	77775	69348
43	70060	28277	39475	46473	23219	53416	94970	25832	69975	94884	19661	72828	00102	66794
44	53976	54914	06990	67245	68350	82948	11398	42878	80287	88267	47363	46634	06541	97809
45	76072	29515	40980	07591	58745	25774	22987	80059	39911	96189	41151	14222	60697	59583

Source: Extracted from "Table of 105,000 Random Decimal Digits," Statement No. 4914 (Washington, D.C.: Interstate Commerce Commission, 1949).

2. *A Million Random Digits*, The RAND Corporation, Free Press, 1955.

Sample pages from one of these are given in Table 6.1.

To obtain a simple random sample of size n using a table of random numbers:

1. Assign every element in the population a number, beginning with 1 and ending with N.
2. Go to the table and "randomly" select a starting point.
3. Record the numbers as they are read from the table; one records as many digits as there are digits in N (e.g., if N is 79, record two-digit numbers), but disregard any numbers chosen that exceed the value of N.
4. When sampling without replacement, if a number is repeated, disregard the duplicate.

When the sample is selected in this manner, every element in the population has an equal chance of selection, and every combination of n elements has an equal chance, i.e., a simple random sample is obtained.

    EXAMPLE. Suppose an auditor desires to draw a simple random sample of size 20 without replacement from the 300 expense account vouchers turned in by employees in May. The auditor first numbers the vouchers 001–300. Second, assume his starting point is Row 1, Column 8 in Table 6–1. Although five-digit numbers appear in the table, he need look at only three of these; for convenience, he reads the last three digits, recording only numbers between 001 and 300 and disregarding duplicates. The sample vouchers are 194, 300, 056, 144, etc., until the $n = 20$ vouchers are selected.[2]

6.4 OTHER PROBABILITY SAMPLES

    Random sampling, sometimes called probability or scientific sampling, is a key element in inferential statistics. Only with random sampling is one guaranteed freedom from bias in the selection of the sample. For example, if one desires a sample of five students from a class, is there a tendency (whether conscious or unconscious) to select the five brightest or most attractive? Or, because of an awareness of such tendencies, does one overcompensate and bias the sample in another direction? With random sampling, selection bias cannot arise. With random sampling, the probabilities of inclusion in the sample are known, and sampling distributions can be determined as in the next section. Finally, *without random sampling, there is no rational basis for the inferential procedures discussed in the following chapters.*

---

[2] The student might observe that modified versions of the above procedure will also yield simple random samples. For example, if a number from the table is 301–600, subtract 300 from it and use the new number. Similarly, subtract 600 from numbers in the table between 601 and 900. This procedure would require examining about two-thirds fewer numbers in the table than that required by the procedure given in the example. Using the same starting point, the sample vouchers selected would now be 194, 102, 230, 237, 005, etc.

The simple random sample is not, however, the only random sampling technique (although in this text random sampling is understood to mean *simple random sampling*). Two other sampling methods are sometimes used in practice, and inferential procedures may be applied to them. *Stratified sampling* divides the population into non-overlapping groups (strata) and then simple random samples are chosen from each stratum. The stratum need not be the same size, and any size sample may be chosen from a stratum. When the variability within strata differs considerably from stratum to stratum, this method will likely have advantages over simple random sampling.

Another random sampling design is *cluster sampling*: the population is divided into non-overlapping groups called clusters, and then a random sample of clusters is chosen. In single-stage cluster sampling, each sample cluster is completely enumerated; in two-stage, a random sample is selected from the sample clusters. This sampling procedure may possess advantages over either simple random or stratified sampling when (1) a complete list of the elements of the population is unavailable or costly to obtain, or (2) the elements of the population are geographically scattered. References given at the end of the chapter deal with these techniques in detail.

6.5 SAMPLING DISTRIBUTION OF THE MEAN (WITHOUT REPLACEMENT)

Having discussed our sampling method, simple random, let us now develop the idea of the sampling distribution. First, recall that a statistic is a measure computed from sample data, and hence the mean of a simple random sample is a statistic. Next recall that a random variable is a variable whose value is dependent upon the outcome of an experiment. If, in this case, the experiment is to select a random sample of size n from a given population and then to compute the sample mean, $\bar{X}$ is a random variable. Why? Because the value of $\bar{X}$ is dependent upon the outcome of the experiment, i.e., dependent upon the particular set of elements that comprise the sample. In other words, from sample to sample $\bar{X}$ may assume different values, just as from toss to toss the number of dots appearing on the face of a die may assume different values. Since every random variable has a probability function (although in some instances the particular function may be unknown or extremely complex), the sample mean has a probability distribution, and we define:

> *The sampling distribution of the mean* is the probability distribution of the sample mean.

Further, the random variable, $\bar{X}$, has an expected value, $E(\bar{X})$, and a standard deviation, which is given a special symbol and name.

> *The standard error of the mean* $(\sigma_{\bar{X}})$ is the standard deviation of the sampling distribution of the mean.

To be more general, we can *define a sampling distribution as the probability distribution of a statistic, and the standard error of the statistic as the standard deviation of the sampling distribution.* An example should clarify these new concepts.

EXAMPLE. In Section 6.2 a population of five items was defined. Develop the sampling distribution of the mean, and determine its expected value and standard deviation when drawing simple random samples of size two without replacement. Our earlier example showed there are 10 possible samples, each having a probability of 0.10 of being selected. These samples and their means are given below.

Sample Members	Sample Values	Sample Mean
1, 2	12, 14	13
1, 3	12, 16	14
1, 4	12, 18	15
1, 5	12, 20	16
2, 3	14, 16	15
2, 4	14, 18	16
2, 5	14, 20	17
3, 4	16, 18	17
3, 5	16, 20	18
4, 5	18, 20	19

Since the values the sample mean can assume are now known, and the probabilities are also known, the construction of the sampling distribution is straightforward. Figure 6.1 illustrates the distribution.

$\bar{X}$	$P(\bar{X})$
13	0.1
14	0.1
15	0.2
16	0.2
17	0.2
18	0.1
19	0.1
	1.0

FIGURE 6.1: SAMPLING DISTRIBUTION OF $\bar{X}$

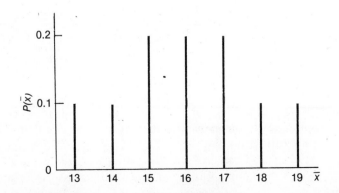

Using Equations (5.3) and (5.4), the expected value and variance are computed as:

$\bar{X}$	$P(\bar{X})$	$\bar{X} \cdot P(\bar{X})$	$\bar{X} - E(\bar{X})$	$[\bar{X} - E(\bar{X})]^2$	$[\bar{X} - E(\bar{X})]^2 \cdot P(\bar{X})$
13	0.1	1.3	-3	9	0.9
14	0.1	1.4	-2	4	0.4
15	0.2	3.0	-1	1	0.2
16	0.2	3.2	0	0	0
17	0.2	3.4	1	1	0.2
18	0.1	1.8	2	4	0.4
19	0.1	1.0	3	9	0.9
	1.0	16.0			3.0

and

$$E(\bar{X}) = \sum \bar{X} \cdot P(\bar{X}) = 16.0$$
$$\sigma_{\bar{X}}^2 = \sum [\bar{X} - E(\bar{X})]^2 \cdot P(\bar{X})$$
$$\sigma_{\bar{X}} = \sqrt{3.0} = 1.73.$$

Instead of being concerned with the sampling distribution in its entirety, let us restrict our attention to the determination of the expected value and standard error of the mean. To obtain these values as shown above, the sampling distribution was first constructed, but was this necessary? In this rather artificial example, determining the sampling distribution was no major problem. But if $N = 30$ and $n = 5$, then $_{30}C_5 = 142,506$ different samples could be selected, and the difficulties of constructing the sampling distribution are apparent. However, if the expected value, $E(X)$, and variance, σ_X^2, of the population are known, the expected value and variance of the sampling distribution of the mean (random sampling without replacement) are given by:

$$E(\bar{X}) = E(X) = \mu_X \tag{6.1}$$

$$\sigma_{\bar{X}}^2 = \frac{\sigma_X^2}{n} \cdot \frac{N - n}{N - 1}. \tag{6.2}$$

In our example, the mean (expected value) and variance of the population are 16 and 8 respectively (the student should confirm these). And using Equations (6.1) and (6.2),

$$E(\bar{X}) = E(X) = 16$$

$$\sigma_{\bar{X}}^2 = \frac{\sigma_X^2}{n} \cdot \frac{N - n}{N - 1} = \frac{8}{2} \cdot \frac{5 - 2}{5 - 1} = \frac{8}{2} \cdot \frac{3}{4} = 3,$$

which agree with our previous computations.

Formulas (6.1) and (6.2) are applicable whenever one is selecting random samples without replacement, regardless of population or sample size. What implications do they pose for us? As concerns Equation (6.1), we can state that the expected value of the sample mean is the mean of the population from which the sample was drawn; we shall discuss this further in Chapter 8.

As concerns Equation (6.2), several observations are in order. First, the variability of the random variable $\bar{X}$, as measured by $\sigma_{\bar{X}}^2$, is influenced by three factors:

1. The variability of the population, as measured by σ_X^2; as σ_X^2 increases, $\sigma_{\bar{X}}^2$ increases.

2. The size of the sample, n; as n increases, $\sigma_{\bar{X}}^2$ decreases.

3. The quantity $(N - n)/(N - 1)$, which is termed the *finite population correction* (fpc); as fpc increases (and holding n constant, this could occur only if N increases), $\sigma_{\bar{X}}^2$ increases.

Note that when $n = N$, the fpc is zero and $\sigma_{\bar{X}}^2$ is zero. Also note that if N is much larger than n, fpc is approximately one. To illustrate, if $N = 10,000$ and $n = 50$,

$$\text{fpc} = \frac{9,950}{9,999} = 0.995 \approx 1.$$

Of course if fpc is one, then $\sigma_{\bar{X}}^2 = \sigma_X^2/n$, which is somewhat easier to remember. As a general rule, *the fpc may be ignored* (i.e., its value is approximately one) *whenever the sample is less than five percent of the population* (i.e., when $n < 0.05\ N$).

EXAMPLE. Given our previous population of 12, 14, 16, 18, 20 with $\mu_X = 16$ and $\sigma_X^2 = 8$, and recall that for $n = 2$, $\sigma_{\bar{X}}^2 = 3$, compute the variance of the sampling distribution when:

a. The population values are 24, 28, 32, 36, and 40 (i.e., the original values multiplied by two). Since from Section 5.3, $\sigma^2(aX) = a^2\sigma^2(X)$, the variance of the new population is now 2^2 times the variance of old population, or $(4)(8) = 32$, and with $n = 2$,

$$\sigma_{\bar{X}}^2 = \frac{32}{2} \cdot \frac{5 - 2}{5 - 1} = \frac{32}{2} \cdot \frac{3}{4} = 12.$$

b. The population is 12, 14, 16, 18, 20 as before, but $n = 3$. Then

$$\sigma_{\bar{X}}^2 = \frac{8}{3} \cdot \frac{5 - 3}{5 - 1} = \frac{4}{3}.$$

c. The population is 12, 12, 14, 14, 16, 16, 18, 18, 20, 20 (i.e., same values as originally but twice as many). Since for this population the mean and variance have not changed, $\mu_X = 16$ and $\sigma_X^2 = 8$, the change in the variance of the sampling distribution will be due to the change in fpc, and

$$\sigma_{\bar{X}}^2 = \frac{8}{2} \cdot \frac{10 - 2}{10 - 1} = \frac{8}{2} \cdot \frac{8}{9} = \frac{32}{9}.$$

**6.6 SAMPLING DISTRIBUTION OF THE MEAN (WITH REPLACEMENT)

The preceding discussion dealt with the sampling distribution of the mean and its expected value and variance, and the square root of the variance was termed the standard error. The sampling method was simple random without replacement. The vast majority of applications in business involve this method, but, on occasion, the sample may be randomly selected *with replacement*. What modifications of our prior work are necessary in such an event? First, define

> *A simple random sample with replacement* is a sample of n elements from a population of N elements such that every sample of size n with elements in a distinct order has an equal chance of selection.

For example, the sample consisting of elements $(1, 2)$ has the same probability as the sample $(2, 1)$ or $(1, 1)$, etc. And the probability of $(1, 2)$, using the multiplication rule for independent events, is

$$P(1, 2) = P(\text{Element 1 is first selected}) \cdot P(\text{Element 2 is second selection})$$
$$= (\tfrac{1}{5})(\tfrac{1}{5}) = \tfrac{1}{25}.$$

And, since the samples are equally likely, the above result implies 25 possible samples. (This can be confirmed using the fundamental principle of counting.) The sampling distribution of the mean for the population 12, 14, 16, 18, 20, when sampling with replacement and $n = 2$, is (see Figure 6.2):

$\bar{X}$	$P(\bar{X})$
12	$\frac{1}{25}$
13	$\frac{2}{25}$
14	$\frac{3}{25}$
15	$\frac{4}{25}$
16	$\frac{5}{25}$
17	$\frac{4}{25}$
18	$\frac{3}{25}$
19	$\frac{2}{25}$
20	$\frac{1}{25}$
	1.0

FIGURE 6.2: SAMPLING DISTRIBUTION OF $\bar{X}$

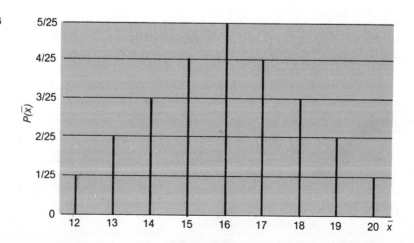

The sampling distribution can be used to show that $E(\bar{X}) = 16$ and $\sigma_{\bar{X}}^2 = 4$, or when sampling with replacement the following can be employed:

$$E(\bar{X}) = E(X) = \mu_X; \tag{6.3}$$

$$\sigma_{\bar{X}}^2 = \frac{\sigma_X^2}{n}. \tag{6.4}$$

Note that the formulas are quite similar to those for sampling without replacement, the only difference being the absence of the fpc in the variance formula. In the remainder of the text, unless specifically noted otherwise, sampling is assumed to be simple random *without* replacement.

6.7 CENTRAL LIMIT THEOREM

Section 6.5 developed the sampling distribution of the mean, and an illustration was given using a small finite population. Formulas (6.1) and (6.2), if the mean and variance of the population are known, provide one with a short-cut method of finding the mean and variance of the sampling distribution. At that time, however, no short cut was provided for determination of the sampling distribution itself. Fortunately, there does exist a most important theorem of statistics that, in many circumstances encountered in the business world, enables one to ascertain the sampling distribution with little effort. For our purposes, the theorem may be stated as:

> *Central limit theorem:* Given any population with mean, μ_X, and variance, σ_X^2, the sampling distribution of the mean will, as n increases, approach a normal distribution with mean, μ_X, and variance, $\dfrac{\sigma_X^2}{n} \cdot \dfrac{N-n}{N-1}$.

Note that the mean and variance of the sampling distribution as given in the theorem are no surprise: they were already known from Equations (6.1) and (6.2), which, reiterating, are completely applicable regardless of the size of population or sample. However, the knowledge that the sampling distribution may be normal is of immense importance, for now, if one knows that the sampling distribution is normal and knows its mean and variance, then, *the sampling distribution is completely specified.*[3]

Again reading the central limit theorem, we see that "... the sampling distribution of the mean will, as n increases, *approach* a normal distribution ..." What does "approach" mean? It means that as n increases the sampling distribution is more and more closely approximated by the normal distribution. And, for some value of n, the approximation is sufficiently accurate for one to say

---

[3] Alan Stuart in *Basic Ideas of Scientific Sampling* remarks: "What happens to the sampling distribution ... when the sample size increases? We know ... that the distribution will become less and less variable ... But even with large samples, we should be optimistic indeed if we expected the sampling distribution generally to approach a regular and predictable form. For once, however, our optimism would be more than vindicated by the event. It is an astounding fact that not only does this sampling distribution approach some regular form as sample size increases, but that it is always the *same* form, the so-called *normal distribution.*"

the sampling distribution is normal; then one may proceed confident that any error introduced by using the approximation (i.e., the normal) will be negligible. Now, for what value of n does the preceding statement hold? As a general rule, if $n \geq 30$, the sampling distribution of the mean may be considered normal.

Figure 6.3 represents graphically our preceding verbal argument. Note that the sampling distribution of the mean is given for samples of size two, five, and thirty from each of four distinctly different populations. The first population is the familiar uniform PDF. The second population is a somewhat unusual bimodal distribution. The third is very heavily skewed to the right, and the last is a normal distribution. Note that even for $n = 5$, the sampling distribution for all cases is beginning to assume the bell-shaped form of the normal PDF. For $n = 30$, regardless of the original population, the sampling distribution of the

FIGURE 6.3: DENSITIES OF $\bar{X}_n$, $n = 2, 5, 30$, FOR VARIOUS POPULATION DENSITIES

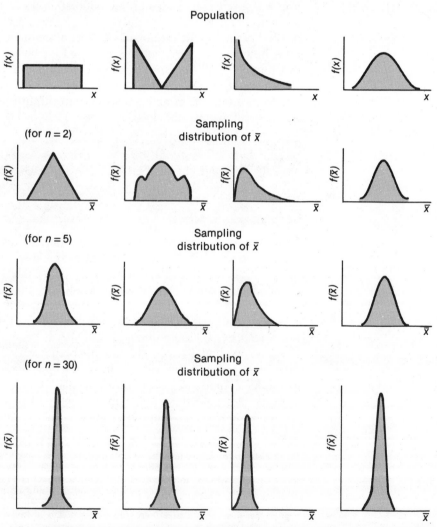

Source: From Ernest Kurnow, Gerald Glaser, and Frederick Ottman, *Statistics for Business Decisions* (Homewood, Ill.: Richard D. Irwin, 1959), pp. 182–83. © 1959 by Richard D. Irwin, Inc.

mean is, for all purposes, that of the normal. (Note also as one moves down a column of the graphs, the mean is unchanging while the variance is constantly decreasing. Why?)

EXAMPLE. Given a sophomore class of 2,000 students with a class mean income for the summer months of $1,200 and standard deviation of $200, determine the sampling distribution of the mean if $n = 64$.

First, since $n \geq 30$, the sampling distribution is assumed to be normal, and

$$E(\bar{X}) = \mu_X = \$1,200$$

and since $n < 0.05N$, ignoring fpc,

$$\sigma_{\bar{X}}^2 = \frac{\sigma_X^2}{n} = \frac{(200)^2}{64} = \frac{40,000}{64} = 625$$

and

$$\sigma_{\bar{X}} = \sqrt{625} = 25.$$

In words, the sampling distribution is normally distributed with mean $1,200 and standard error $25.

If the sampling distribution is normal, the Z transformation of Chapter 5 may be employed to answer probabilistic questions concerning the random variable $\bar{X}$. If the random variable is X, as in Chapter 5,

$$Z = \frac{X - \mu}{\sigma}$$

or

$$Z = \frac{\text{Value of the random variable} - \text{Expected value of the random variable}}{\text{Standard deviation of the random variable}}.$$

Now with $\bar{X}$ as the random variable,

$$Z = \frac{\bar{X} - \mu_X}{\sigma_{\bar{X}}} \qquad (6.5)$$

EXAMPLE. Given the population of the previous example, what is the probability of selecting a sample of size 64 and finding that the sample mean exceeds $1,250?

Graphically the problem is shown in Figure 6.4

FIGURE 6.4

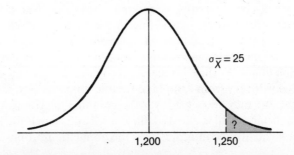

$\sigma_{\bar{X}} = 25$

?

1,200   1,250

Since

$$Z = \frac{\bar{X} - \mu_X}{\sigma_{\bar{X}}} = \frac{1,250 - 1,200}{25} = 2$$

and, from Appendix E,

$$P(0 \leq Z \leq 2) = 0.4772$$

then

$$P(Z > 2) = 0.5 - 0.4772 = 0.0228,$$

which is the $P(\bar{X} > 1,250)$.

For $n \geq 30$, the sampling distribution of the mean may be considered as normal regardless of the population's distribution; the central limit theorem justifies this statement. Under certain circumstances, however, the sampling distribution of the mean is normal for *any n;* the "certain circumstances" being that the population is normally distributed:

> *theorem:* Given a normally distributed population with mean, μ_X, and variance, σ_X^2, the sampling distribution of the mean is normally distributed with mean μ_X and variance σ_X^2/n.

Again, the above theorem holds for any size sample.

EXAMPLE. Given that the distribution of Scholastic Aptitude Test (SAT) scores for 1975 was normally distributed with a mean of 500 and a standard deviation of 100, what is the sampling distribution of the mean for a sample of size 16? By the theorem above, the sampling distribution of the mean is normal, and

$$E(\bar{X}) = \mu_X = 500;$$

$$\sigma_{\bar{X}} = \sqrt{\sigma_{\bar{X}}^2} = \sqrt{\frac{\sigma_X^2}{n}} = \frac{\sigma_X}{\sqrt{n}} = \frac{100}{4} = 25.$$

6.8 SAMPLING DISTRIBUTION OF THE PROPORTION

The sample proportion is another statistic whose sampling distribution will be of considerable use in the next chapter. Define the sample proportion as

$$p = \frac{X}{n}, \tag{6.6}$$

where X is the number of successes (i.e., number of observations possessing a specified characteristic) in a sample size of n. For example if we define the characteristic to be "marketing major" and a sample of size twenty contains five such students, then the sample proportion, p, would be $\frac{5}{20} = 0.25$.

If we consider the sampling procedure to be simple random from a population of size N, then we readily recognize X as a random variable because its value will vary from sample to sample. In particular, if sampling with replacement, we may recognize X as a binomial-distributed random variable with parameters n and π, and as defined in Chapter 5, π is the probability of a success. But π can also be interpreted as the population proportion since:

$$P(\text{success}) = \frac{\text{Number of successes in the population}}{\text{Number of items in the population}}.$$

From Chapter 5, recall that $E(X) = n\pi$ and $\sigma_X^2 = n\pi(1 - \pi)$. Can this be utilized to determine the mean and variance of p? Yes, because the random variable p is merely a linear transformation of the random variable X with a multiplicative constant of $1/n$. Then,

$$E(p) = E\left(\frac{X}{n}\right) = \left(\frac{1}{n}\right)E(X) = \left(\frac{1}{n}\right)n\pi = \pi; \qquad (6.7)$$

$$\sigma_p^2 = \sigma^2\left(\frac{X}{n}\right) = \left(\frac{1}{n}\right)^2 \sigma_X^2 = \left(\frac{1}{n}\right)^2 [n\pi(1 - \pi)] = \frac{\pi(1 - \pi)}{n}. \qquad (6.8)$$

Now, when sampling with replacement, the mean (expected value) and variance of the sample proportion can be determined.)

EXAMPLE. The Normal Company employs 200 sales personnel, 40 of whom will be eligible for retirement by 1985. Determine the mean and variance of the sample proportion for random samples of size five drawn with replacement; a success is defined as eligible for retirement by 1985.

First, π is determined as $\frac{40}{200} = 0.20$.

Using Equations (6.7) and (6.8),

$$E(p) = \pi = 0.20;$$

$$\sigma_p^2 = \frac{\pi(1 - \pi)}{n} = \frac{0.20(0.80)}{5} = \frac{0.16}{5} = 0.032.$$

By analogy to the nomenclature adopted for the sample mean, define the standard deviation of the random variable p as the *standard error of the proportion*, σ_p, and for the example above:

$$\sigma_p = \sqrt{\sigma_p^2} = \sqrt{0.032} = 0.179.$$

What is the mean and variance of the sample proportion when sampling *without* replacement? Just as we saw for the sample mean, the expected value of this statistic is the same whether sampling with or without replacement, and the variances differ only by the finite population correction; i.e., when sampling without replacement,

$$E(p) = \pi; \qquad (6.9)$$

$$\sigma_p^2 = \frac{\pi(1 - \pi)}{n} \cdot \frac{N - n}{N - 1}. \qquad (6.10)$$

As before, the fpc is disregarded when $n < 0.05N$.

EXAMPLE. Suppose we define a population as consisting of the five individuals Jennie, Mary, Andy, Glenn, and Tom. *a.* Develop the sampling distribution of the proportion female for simple random samples of size two drawn without replacement. *b.* Determine the mean and variance of the distribution.

a. As we did earlier with the sample mean (Section 6.5), we could enumerate the $_5C_2 = 10$ possible samples, determine the sample proportion for each, and then construct the distribution. For example, the sample (Jennie, Andy) will be selected with probability of $\frac{1}{10}$, and this sample has a proportion female of 0.50. We leave the enumeration to the student and give the sampling distribution that would be obtained below (see Figure 6.5).

Sample Proportion (p)	Probability
1.0	0.1
0.5	0.6
0.0	0.3

FIGURE 6.5: SAMPLING DISTRIBUTION OF *p*

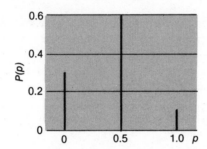

b. One may use the formulas for expected value and variance of a random variable, and use the above distribution to obtain these values for the sample proportion (a suggested exercise for the student). Here we shall use Equations (6.9) and (6.10).

$$E(p) = \pi = \frac{2}{5} = 0.40;$$

$$\sigma_p^2 = \frac{\pi(1 - \pi)}{n} \cdot \frac{N - n}{N - 1}$$

$$= \frac{(0.40)(0.60)}{2} \cdot \frac{5 - 2}{5 - 1} = \frac{0.24}{2} \cdot \frac{3}{4} = 0.09.$$

At this point, we can determine the mean and variance of the sample proportion when sampling with or without replacement. However, if a *complete* specification of the sampling distribution is desired, we could resort to an enumeration of all possible samples (for large N, an almost impossible task). Could we be so fortunate to find something akin to the central limit theorem to save us from this burden? Yes, a similar result holds for the random variable *p*.

The sampling distribution of the proportion will, as n increases, approach a normal distribution with mean π (the population proportion) and variance $\sigma_p^2 = \pi(1 - \pi)/n$.

The above theorem holds when (1) random sampling with replacement or (2) random sampling without replacement but with $n < 0.05N$. If the sampling is without replacement and $n \geq 0.05N$, then the finite population correction must be used and $\sigma_p^2 = \dfrac{\pi(1 - \pi)}{n} \cdot \dfrac{N - n}{N - 1}$. One may naturally ask: For what value of n can the sampling distribution of the proportion be considered normal? Answer: when $n \geq 50$ and *both* $n\pi$ and $n(1 - \pi)$ are greater than five. Under the above conditions, the sampling distribution of the proportion can be *completely* specified; this result is of particular importance in the next chapter.

EXAMPLE. Universal Electronics employs 2,000 persons, 600 of whom are female. *a.* Specify completely the sampling distribution of the proportion (female) for samples of size fifty drawn without replacement. b. Determine the probability of selecting a sample whose proportion female is greater than 0.40.

a. The population proportion (female) is given by $\pi = \frac{600}{2000} = 0.30$. Then, Since $n \geq 50$ and

$$n\pi = 50 \times 0.30 = 15$$

$$n(1 - \pi) = 50 \times 0.70 = 35$$

are both greater than five, the sampling distribution may be considered normal with parameters

$$E(p) = \pi = 0.30;$$

$$\sigma_p^2 = \frac{\pi(1 - \pi)}{n} = \frac{0.30(0.70)}{50} = 0.0042.$$

(Note the fpc was ignored in computing the variance since $n < 0.05N$).

b. With the distribution determined as normal and its parameters known, the Z transformation is used to answer probabilistic questions, and in this instance

$$Z = \frac{p - \pi}{\sigma_p}. \tag{6.11}$$

Substituting, one obtains

$$Z = \frac{0.40 - 0.30}{\sqrt{0.0042}} = \frac{0.10}{0.065} = 1.54$$

and $P(p > 0.40) = P(Z > 1.54) = 0.0618$. The problem is illustrated in Figure 6.6.

FIGURE 6.6: SAMPLING DISTRIBUTION OF p FOR EXAMPLE PROBLEM

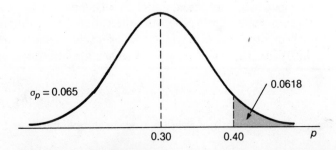

6.9 SAMPLING DISTRIBUTION OF $(\bar{X}_1 - \bar{X}_2)$

The present development of sampling distributions concludes with the distribution of the *difference* between sample means drawn from two populations. This distribution underlies the application of inferential procedures to many practical business problems such as the following.

1. A consumer group is investigating the higher rates that divorced persons pay for car insurance than never-married persons. Is there a difference in claims between the groups that could be used to explain the difference in rates?

2. A breakfast cereal can be treated with either of two approved additives. Do the additives differ in their effects on the shelf-life of the product?

In both of the above, decisions would be made on the basis of sample evidence from each of the two populations, e.g., divorced policy owners would constitute one population and never-married the other.

Suppose we have defined two populations of interest. Each population has a mean and variance, and subscripts shall be used to distinguish between the two. For example, population one has mean μ_1 and variance σ_1^2. We wish to examine the distribution of the difference in sample means, $(\bar{X}_1 - \bar{X}_2)$, where, for example, $\bar{X}_1$ is based on a random sample of size n_1 from the first population. Since $\bar{X}_1$ and $\bar{X}_2$ are both random variables, their difference $(\bar{X}_1 - \bar{X}_2)$ is a random variable. The parameters of the distribution of the difference in sample means are determined as follows.

From Equation (6.1), $E(\bar{X}) = \mu_x$, and from Equation (5.25), the expected value of the difference in two random variables is the difference in their expected values, i.e.,

$$E(X - Y) = E(X) - E(Y).$$

Then,

$$E(\bar{X}_1 - \bar{X}_2) = E(\bar{X}_1) - E(\bar{X}_2)$$

$$= \mu_1 - \mu_2. \tag{6.12}$$

That is, the expected value of the difference of two sample means is the difference of the two population means from which the samples are drawn.

In determining the variance of $(\bar{X}_1 - \bar{X}_2)$, we shall assume that $n < 0.05N$ for both populations. Since the fpc can be ignored, $\sigma_{\bar{x}_1}^2 = \sigma_1^2/n_1$ (Equation 6.2). Now, if $\bar{X}_1$ and $\bar{X}_2$ are based on independently drawn random samples, these two random variables are independent. (Intuitively, will the sample mean obtained from the second population be influenced by the result obtained from the first population? No.) Also, the variance of the difference in two independent random variables equals the *sum* of the variances of the variables (Section 5.15). That is, $\sigma^2(X - Y) = \sigma^2(X) + \sigma^2(Y)$. Therefore,

$$\sigma_{\bar{X}_1 - \bar{X}_2}^2 = \sigma_{\bar{X}_1}^2 + \sigma_{\bar{X}_2}^2 = \frac{\sigma_1^2}{n_1} + \frac{\sigma_2^2}{n_2}. \tag{6.13}$$

Thus the variance of the difference in sample means is the sum of variances of the individual sample means.[4] The standard deviation of the difference in sample means, $\sigma_{\bar{X}_1 - \bar{X}_2}$, is also termed the standard error of the difference in sample means.

As shown by the results of Section 5.15, $(\bar{X}_1 - \bar{X}_2)$ is normally distributed *if* $\bar{X}_1$ and $\bar{X}_2$ are each normally distributed. The latter will occur when sampling from (1) normally distributed populations or (2) non-normal populations when the central limit theorem is applicable (i.e., if n_1 and n_2 are both greater than 30). The distribution is summarized as:

$$N(\mu_1 - \mu_2, \sigma_{\bar{X}_1 - \bar{X}_2}) \tag{6.14}$$

when $n_i < 0.05N_i$ and $n_i > 30$ for $i = 1, 2$.[5] (See Figure 6.7).

FIGURE 6.7: RELATION OF SAMPLING DISTRIBUTION OF $\bar{X}_1 - \bar{X}_2$ TO SAMPLING DISTRIBUTION OF $\bar{X}_1$ AND $\bar{X}_2$

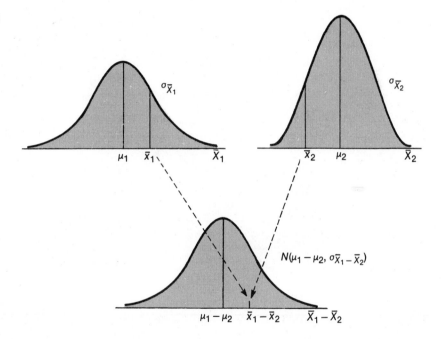

---

[4] The above result holds when (1) sampling with replacement or (2) sampling without replacement and ignoring the fpc's. If more than five percent of a population is sampled without replacement, one must use

$$\sigma_{\bar{X}_1 - \bar{X}_2}^2 = \frac{\sigma_1^2}{n_1} \cdot \frac{N_1 - n_1}{N_1 - 1} + \frac{\sigma_2^2}{n_2} \cdot \frac{N_2 - n_2}{N_2 - 1}.$$

[5] Another distribution of later interest is the sampling distribution of the difference between two proportions, $p_1 - p_2$. The development of this distribution is perfectly analogous to that of the difference in means; the end result is

$$p_1 - p_2 \sim N\left(\pi_1 - \pi_2, \sqrt{\frac{\pi_1(1 - \pi_1)}{n_1} + \frac{\pi_2(1 - \pi_2)}{n_2}}\right)$$

when fpc's are ignored and $n_i\pi_i$ and $n_i(1 - \pi_i)$ are both greater than five for $i = 1, 2$.

EXAMPLE. In the car-insurance example previously mentioned, suppose each of the populations is large (i.e., ignore fpc's) and the distributions of latest annual claims (dollars) for the two populations have the following parameters:

Divorced	Never-Married
$\mu_1 = 200$	$\mu_2 = 190$
$\sigma_1^2 = 100$	$\sigma_2^2 = 150$

Determine the sampling distribution of $(\bar{X}_1 - \bar{X}_2)$ if $n_1 = 100$ and $n_2 = 50$.

First, the expected value is given by Equation (6.12):

$$E(\bar{X}_1 - \bar{X}_2) = \mu_1 - \mu_2 = 200 - 190 = 10.$$

The variance, from Equation (6.13), is:

$$\sigma_{\bar{X}_1 - \bar{X}_2}^2 = \frac{\sigma_1^2}{n_1} + \frac{\sigma_2^2}{n_2} = \frac{100}{100} + \frac{150}{50} = 1 + 3 = 4.$$

Finally, since both $n_i > 30$, $\bar{X}_1$ and $\bar{X}_2$ are both approximately normally distributed via the central limit theorem, then $(\bar{X}_1 - \bar{X}_2)$ is also approximately normally distributed. The distribution, $N(10, 2)$, is shown in Figure 6.8.

FIGURE 6.8:   SAMPLING DISTRIBUTION OF $(\bar{X}_1 - \bar{X}_2)$ IN THE EXAMPLE PROBLEM

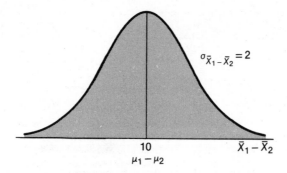

When a sampling distribution is normal, we use the Z transformation to answer probabilistic questions. Since this is always

$$Z = \frac{\text{Value of the random variable} - \text{Expected value of the random variable}}{\text{Standard deviation of the random variable}},$$

the Z-transformation for this sampling distribution is

$$Z = \frac{(\bar{X}_1 - \bar{X}_2) - (\mu_1 - \mu_2)}{\sigma_{\bar{X}_1 - \bar{X}_2}}. \tag{6.15}$$

EXAMPLE. Suppose in the previous example the parameters are changed to:

Divorced	Never-Married
$\mu_1 = 220$	$\mu_2 = 220$
$\sigma_1^2 = 100$	$\sigma_2^2 = 150$

If a sample of $n_1 = 100$ and $n_2 = 50$ is to be drawn, what is the probability that $(\bar{X}_1 - \bar{X}_2) \geq 5$? That is, what is the probability that $\bar{X}_1$ will be at least \$5 greater than $\bar{X}_2$?

Since the n_i and σ_i^2 are the same as the previous example, we know that $(\bar{X}_1 - \bar{X}_2) \sim N(?,2)$. The expected value is then recomputed as:

$$E(\bar{X}_1 - \bar{X}_2) = \mu_1 - \mu_2 = 220 - 220 = 0.$$

Then, using Equation (6.15),

$$Z = \frac{(\bar{X}_1 - \bar{X}_2) - (\mu_1 - \mu_2)}{\sigma_{\bar{X}_1 - \bar{X}_2}}$$

$$= \frac{5 - 0}{2} = 2.5,$$

and

$$P(Z \geq 2.5)$$

$$= 0.0062.$$

Figure 6.9 is an illustration of the solution.

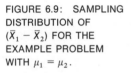

FIGURE 6.9: SAMPLING DISTRIBUTION OF $(\bar{X}_1 - \bar{X}_2)$ FOR THE EXAMPLE PROBLEM WITH $\mu_1 = \mu_2$.

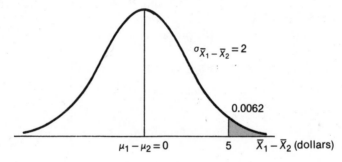

The preceding solution can be stated rather loosely as follows: given the stated sample sizes and population parameters, it would be an extraordinary occurrence for one to obtain sample results such that $\bar{X}_1 - \bar{X}_2 \geq 5$ because the probability of such an event is only 0.0062. In the above problem, the values of μ_1 and μ_2 were stated as *known;* however, if these values had merely been *assumed,* such a sample result would somewhat discredit the assumption. The validation of assumed parameter values is basically what hypotheses tests are all about, and that is the subject of Chapter 8.

6.10 SUMMARY

This chapter introduced the concept of the sampling distribution, which was defined as the probability distribution of a statistic. The sampling distributions of several statistics were investigated. The sampling distributions of other statistics will be developed in subsequent chapters. If the sampling method is random without replacement, the random variable $\bar{X}$ has an expected value equal to the mean of the population, $E(\bar{X}) = \mu_X$, and a variance, whose square root was termed the standard error of the mean, of $\dfrac{\sigma_X^2}{n} \cdot \dfrac{N-n}{N-1}$. The finite population correction, $\dfrac{N-n}{N-1}$, was ignored if $n \le 0.05N$. Next the central limit theorem was presented; this important theorem tells us that the sampling distribution of the mean is approximately normal if the sample size is 30 or more. The sampling distribution of the proportion was also studied, and conditions were set forth under which this distribution may be considered normal with mean π and variance σ_p^2. Finally, the sampling distribution of the difference between two means was developed. The utility of these and other sampling distributions in business applications will become apparent in Chapter 8.

6.11 TERMS AND FORMULAS

Central limit theorem
Difference between two sample means $(\bar{X}_1 - \bar{X}_2)$
Finite population correction (fpc)
Sample proportion (p)
Sampling distribution
Simple random sample
Standard error of the difference between two sample means $(\sigma_{\bar{X}_1 - \bar{X}_2})$
Standard error of the mean $(\sigma_{\bar{X}})$
Standard error of the proportion (σ_p)

Expected value of sample mean $\qquad E(\bar{X}) = \mu$

Variance of $\bar{X}$ (without replacement) $\qquad \sigma_{\bar{X}}^2 = \dfrac{\sigma_X^2}{n} \cdot \dfrac{N-n}{N-1}$

Variance of $\bar{X}$ (with replacement) $\qquad \sigma_{\bar{X}}^2 = \dfrac{\sigma_X^2}{n}$

Finite population correction $\qquad \dfrac{N-n}{N-1}$

Sample proportion $\qquad p = \dfrac{X}{n}$

Expected value of sample proportion $\qquad E(p) = \pi$

Variance of p (without replacement) $\qquad \sigma_p^2 = \dfrac{\pi(1-\pi)}{n} \cdot \dfrac{N-n}{N-1}$

Variance of p (with replacement) $\qquad \sigma_p^2 = \dfrac{\pi(1-\pi)}{n}$

Expected value of difference of two sample means $\qquad E(\bar{X}_1 - \bar{X}_2) = \mu_1 - \mu_2$

Variance of $(\bar{X}_1 - \bar{X}_2)$ assuming
fpc $= 1$

$$\sigma^2_{\bar{X}_1 - \bar{X}_2} = \frac{\sigma^2_1}{n_1} + \frac{\sigma^2_2}{n_2}$$

QUESTIONS

A. True or False

1. The expected value of the sample mean is the population mean.
2. When sampling without replacement and $n = N$, then the fpc is zero.
3. The major advantage of random sampling is that $\bar{X} = \mu$ for every sample.
4. If $n = 5$, the sampling distribution of the mean will be normally distributed if the population is normally distributed.
5. The sample proportion, p, is a random variable.
6. In simple random sampling, every element has an equal chance of selection.
7. The sampling distribution of the proportion will be approximately normal when $n > 30$.
8. The standard error of the mean varies inversely with the standard deviation of the population.
9. The variance of the difference of two sample means is the difference of the variances of $\bar{X}_1$ and $\bar{X}_2$.
10. Given $N = 10$ and $n = 7$, in simple random sampling without replacement, there are seventy possible samples.

B. Fill in

1. The expected value of p is _____ μ _____.
2. The fpc is considered to be approximately one (and hence ignored) when ____ $n < .05N$ ____.
3. The standard deviation of the distribution of the random variable $\bar{X}$ is termed ___ St err of mean ___
4. If sampling without replacement and $n = N$, $\sigma_{\bar{X}}$ equals ___ 0 ___.
5. The central limit theorem is assumed applicable if the sample size is ____ $n \geq 30$ ____.
6. As n increases, $\sigma_{\bar{X}}$ ___ decreases ___.
7. The random variable $(\bar{X}_1 - \bar{X}_2)$ has an expected value of zero when ____ $m_1 - m_2$ ____.
8. One assumes the sampling distribution of the proportion is normal when ____ $n > 50$ ____.
9. The standard error of the proportion is a measure of the ___ variability dispersion ___ of the random variable p.
10. The expected value of $\bar{X}_1 - \bar{X}_2$ is ___ $m_1 - m_2$ ___.

EXERCISES

1. Alan Stuart (*Basic Ideas of Scientific Sampling*) says, "The virtue of a random sampling process is . . . less in its achievement (by which I mean the actual sample it produces on a particular occasion) than in its promise (to be impartial between samples)." Discuss.

2. A large bank has ten vice presidents; a random sample of five vice presidents is to be selected. What is the probability that:
    a. The oldest vice president is in the sample?
    b. The five youngest vice presidents make up the sample?

3. A population consists of the following six families:

Family:	A	B	C	D	E	F
Income ($ Thousands)	12	14	16	18	20	22

    If samples are to be drawn without replacement and $n = 4$:
    a. How many samples are possible?
    b. Specify the sampling distribution of the mean.
    c. Using (*b*), compute the expected value and variance of the distribution.
    d. What are the mean and variance of the population?
    e. Use (*d*) and Formulas (6.1) and (6.2) to confirm your answers in (*c*).

4. Let $n = 5$, and repeat Problem 3, parts (*a*), (*b*), (*c*), and (*e*).

5. Discuss the results of Problem 3 for (*a*) $n = 1$ and (*b*) $n = 6$.

6. A population of $N = 500$ expense account vouchers has a mean of $75 and variance of 100. What is the mean and variance of the sampling distribution of the mean for simple random samples with $n = 10$? What is the standard error of the mean?

7. Littleton has 4,000 population; the distribution of per capita income in Littleton has a mean of 8,000 dollars and standard deviation of 2,000. Metrocity has a population of 800,000; there the distribution of per capita income has mean equal to 11,000 and standard deviation of 2,000.
    a. If samples of $n = 100$ are to be taken from each city, what is the sampling distribution of the mean for Littleton? For Metrocity?
    b. Discuss the relative size of the standard errors in part (*a*).
    c. What is the sampling distribution of $\bar{X}_1 - \bar{X}_2$ (mean income for Littleton − mean income for Metrocity)?

8. Calculate the standard error of the mean for samples with $n = 25$ from a population with standard deviation equal to 50 and:
    a. $N = 50$.
    b. $N = 500$.
    c. $N = 10,000$.
    d. $N = 50,000$.

9. Darrow Co. employs five administrative assistants; three of the five have college degrees.
    a. What is the population proportion of administrative assistants with college degrees?

b. Obtain the sampling distribution of the proportion for $n = 3$ without replacement.

c. Obtain the mean and variance of the sampling distribution of part (b).

d. Use Formulas (6.9) and (6.10) to confirm your answers to part (c).

10. General Rent-A-Car has 4,000 cars available for rental, including 1,000 sub-compacts.

a. Specify the sampling distribution of the proportion (sub-compact) for samples of $n = 100$.

b. Determine the probability of selecting a sample of $n = 100$ with proportion subcompact of 0.20 or less.

11. The weight of a coal truck leaving Strip-Mines Inc. is normally distributed with $\mu = 30$ tons and $\sigma = 1.0$.

a. For $n = 4$, specify the sampling distribution of the mean.

b. For $n = 4$, what is the probability that the sample mean will exceed 31 tons?

c. What is the value of $\bar{X}$ for which 90 percent of the samples will have a larger mean?

12. Taxi fares in Metrocity have $\mu = \$8.00$ and $\sigma = \$1.50$. If a random sample of fifty fares is obtained, determine:

a. The sampling distribution of the mean.

b. The probability that the sample mean will lie in the interval $\$8.00 \pm 0.25$ (i.e., between 7.75 and 8.25).

c. The symmetric interval around $\mu(\$8.00)$ that 95 percent of the sample means will fall.

13. A clothing manufacturer sells factory overruns and clothing with defects to discount retailers. A supervisor reports that, generally, 15 percent of the men's shirts contain defects. In a random sample of 60 shirts from last week's production, 10 shirts with defects are discovered. This is a sample proportion $p = 10/60 = 0.167$.

a. Is a sample proportion of 0.167 or larger unlikely if $\pi = 0.15$?

b. If a second sample of 60 shirts is taken, what is the probability that $p \geq 0.25$, assuming $\pi = 0.15$?

14. State the conditions that must be met in order for the sampling distribution of the mean to be a. exactly normally distributed and b. approximately normally distributed.

15. A random sample of 81 items is drawn from a normal population with parameters $\mu = 125$ and $\sigma = 36$. Find the probability that the sample mean is:

a. Larger than 130.

b. Less than 115.

c. Either less than 121 or greater than 129.

d. Between 123 and 128.

16. A fair coin is tossed 100 times. Find the probability that the proportion of heads is:
    a. Less than 0.43.
    b. Greater than 0.62.
    c. Between 0.45 and 0.55.
    d. Either less than 0.40 or at least 0.60.

17. Two machines are used by a soft drink bottler for filling beverage containers. The amounts dispensed by each machine have the following parameters.

Machine 1	Machine 2
$\mu_1 = 350$ ml	$\mu_2 = 355$ ml
$\sigma_1 = 5$ ml	$\sigma_2 = 7$ ml

    A random sample of $n_1 = 100$ containers is to be chosen from machine 1 and $n_2 = 80$ from machine 2.
    a. Specify the sampling distribution of the difference in means $(\bar{X}_1 - \bar{X}_2)$.
    b. Determine the probability that the difference in sample means is 1.01 ml. or more.

18. The registrars at Playton College and Tech University furnish the following information on undergraduate quality point averages (QPA):

Playton	Tech University
$\mu_1 = 2.20$	$\mu_2 = 1.60$
$\sigma_1 = 0.30$	$\sigma_2 = 0.40$

    a. Specify the sampling distribution of $\bar{X}_1 - \bar{X}_2$ if random samples of size fifty were drawn from each campus.
    b. What is the probability of observing a difference of zero or less?

19. A national health insurer is examining charges for health services in different regions of the U.S. The mean and variance of the fee charged for physician-ordered physical examinations in their New England region are $\mu_1 = \$96$ and $\sigma_1^2 = \$1,024$. In the Northwest Region $\mu_2 = \$84$ and $\sigma_2^2 = \$900$. A random sample of 150 claims filed for physical examinations in the New England region and a random sample of 120 claims filed in the Northwest Region are obtained. What is the probability that the observed difference in sample means $(\bar{X}_1 - \bar{X}_2)$ will be:
    a. At least $16?
    b. Less than $12?
    c. Either less than $6 or greater than $18?

20. An auditor for a small bank uses a random sample of accounts to confirm the account balances. The stated balance for their 40 Christmas Club accounts is listed below. Using the procedure described in Section 6.3 and

Table 6.1, select a random sample of ten accounts. Determine the mean and variance of the stated account balances in the sample.

$86.10	$61.50	$ 6.96	$ 37.56
15.70	14.41	91.47	12.72
47.01	19.36	36.29	17.76
51.12	33.71	48.15	14.92
9.00	4.27	17.50	20.01
5.75	27.06	34.91	19.82
73.00	42.18	98.89	20.25
18.69	13.25	40.56	54.18
29.02	8.97	21.14	138.17
79.66	54.46	8.88	41.23

REFERENCES

American Institute of Certified Public Accountants *An Auditor's Approach to Statistical Sampling*, vols. 1 and 3. New York: American Institute of Certified Public Accountants, Inc., 1968.

Cochran, William G. *Sampling Techniques*. New York: John Wiley & Sons, Inc., 1963.

Mendenhall, William; Ott, Layman; and Schaffer, Richard L. *Elementary Survey Sampling*, 2d ed. Belmont, Cal.: Wadsworth Publishing Co., Inc., 1979.

Stuart, Alan *Basic Ideas of Scientific Sampling*. New York: Hafner Publishing Co. 1968.

Williams, Bill *A Sampler on Sampling*. New York: John Wiley & Sons, 1978.

7

Statistical Decision Theory

7 STATISTICAL DECISION THEORY

One morning (in the early Sixties) we
were faced with a critical business
decision. I was asked to present to the
President and a group of the Company's
executives an analysis of the problem and
the alternatives. I did so and indicated
three possible outcomes, attaching probabilities to each. The President looked
at me and said: "Herb, we don't play
craps with the American Can Company."

Herbert R. Brinberg
The New York Statistician (Vol. 27, No. 2, 1975)

7.1   INTRODUCTION   Statistical decision theory is concerned with the selection of an "optimal" course of action from among several alternatives where the outcome associated with an action is uncertain. Decision making in the face of uncertain outcomes is a common occurrence not only in the business world but in all phases of one's existence. For example, the selection of a senior elective is made with uncertainty about the grade one could obtain in the various courses being considered. In business, a bank must decide whether or not to install 24-hour automated tellers without knowing for certain the customer demand for such a service. An electric utility must decide on the construction of a coal-fired or nuclear generating plant with uncertainty concerning fuel prices, demand for power, and rates as set by the state's Public Service Commission. Major deci-

---

THOMAS BAYES (1702–1761)

Bayes was born in London and educated by private tutors. It is thought that Abraham de Moivre, who first discovered the normal distribution, served for a time as one of Bayes' tutors. Bayes' father was a Presbyterian minister, and Thomas was also ordained and served a congregation in Tunbridge Wells (about 35 miles from London) from 1720 to 1752. Under the pseudonym of John Noon, he published treatises in 1731 and 1736 that led to his election as a Fellow of the Royal Society. Neither of these earlier works dealt with probability. His sole work on probability was published posthumously in 1763, "An Essay Towards Solving a Problem in the Doctrine of Chances." This essay eventually led to the modern formulation of the theorem that bears his name (Chapter 4). Bayes theorem is used extensively in statistical decision theory to combine prior and sample information.

sions by an individual or a corporation are wisely made only if the consequences of the available actions are considered; this chapter provides a useful framework for analysis in a decision problem.

Statistical decision theory has at its root Bayes theorem (Section 4.10), which was published in 1763. However, it was not until the late 1950s that this probability theorem led to the rapid development of today's decision theory. In fact, the author of the chapter's opening quote later added, "Today, probability analysis is very much in use at American Can as well as in many other corporations." This "probability analysis" (statistical decision theory) is also often termed Bayesian statistics or decision theory analysis.

7.2 ELEMENTS OF THE DECISION PROBLEM

A decision problem is usually viewed as having four common elements:

1. *Alternative courses of action.* If alternative courses of action do not exist, no decision is required. For example, this course (statistics) is a degree requirement in many business schools. Assuming the student desires a bachelor's degree in business, there is no alternative to taking the course. However if an elective requirement must be fulfilled, a choice is made from among the available alternatives.

2. *States of nature.* States of nature (sometimes referred to as events or states of the world) are factors that affect the outcome of a decision but are beyond control of the decision maker. For example, rain or clear skies at game-time are states of nature that will affect the outcome of the morning's decision not to carry an umbrella to a football game. Nature, however, is *not* intended to imply restriction to natural phenomena. It does imply a factor not subject to control and possessing neither malice nor goodwill toward the decision maker. In general, states of nature refer to uncertain events that may occur in the decision environment. For example, a decision on the introduction of a new product is made in the face of uncertainty about the demand for the product.

3. *Payoffs.* A payoff is the outcome or consequence associated with a particular action given a specific state of nature. If there are m possible states and n available courses of action, there are mn payoffs (one for each act/state combination).

4. *Uncertainty.* The decision maker is uncertain about what state of nature will occur, e.g., consumer demand for a new product. A major feature of statistical decision theory is the use of probabilities in an assessment of uncertainty and then in the selection of an "optimal" course of action.

In subsequent discussions, courses of actions will be termed simply *acts*, and states of nature simply *states*. When the numbers of acts and states are both finite, the payoffs associated with each act/state combination are displayed in a payoff table. Table 7.1 shows a payoff table with two states and three acts. Note the state is the row variable, and the act is the column variable. In general one may speak of S_i (the ith state), A_j (the jth act), and M_{ij} (the payoff if state S_i occurs when act A_j is the chosen act). For example, M_{23} is the payoff associated with act A_3 when state S_2 occurs.

If the state of nature were *known* before the decision maker selected an act, the decision problem would be greatly simplified. One could look across the

TABLE 7.1: PAYOFF
TABLE

State	Act		
	A_1	A_2	A_3
S_1	M_{11}	M_{12}	M_{13}
S_2	M_{21}	M_{22}	M_{23}

row corresponding to the known state and find the largest payoff; the "best" act would be the one yielding this maximum payoff under the known state. However, since uncertainty about the states is a characteristic of real decision problems, the selection of the best act will require more than a mere examination of the payoff table.

7.3 THE MCDOUGAL'S EXAMPLE

Before formally introducing probabilities (uncertainty) into the decision analysis, a simplified decision problem is introduced and used to illustrate certain concepts. McDougal's, a national chain of fast-food restaurants, has been offering a traditional selection of hamburgers, french fries, soft drinks, etc. Company management has devoted considerable time recently to analyzing ways to increase the revenues and profits generated in their outlets. The hamburger market is becoming quite saturated (especially in view of Windy's rapid expansion), and price-cutting might increase revenues but not profits. Pizza was considered but discarded as an already crowded market. The one strategy that emerged from management's analysis as having real potential was the addition of breakfast items to the menu.

Breakfast items are relatively easy to prepare and would not require a large capital outlay for additional cooking equipment. Most important, such items would be sold in the morning when the demand for the company's traditional products has been very weak. However, because (a) many people are known to skip breakfast and (b) the company does not know how competitors may react, the demand for the new products is uncertain. For our purposes, we shall consider three levels of customer demand: strong, average, and weak.

In this problem, there are two alternative acts available to McDougal's:

A_1: Introduce breakfast items

A_2: Do not introduce breakfast items

and three possible states of nature:

S_1: Strong demand

S_2: Average demand

S_3: Weak demand.

Under the direction of top management, the marketing and financial staffs developed a set of payoffs for each act/state combination. The payoffs considered such items as capital outlay, depreciation policies, training costs, additional advertising expenditures, etc. The outcome of their analysis is given in Table 7.2.

TABLE 7.2: PAYOFF TABLE FOR MCDOUGAL'S (UNITS ARE $ MILLIONS IN PROFIT)

State (demand)	Act	
	A_1 Introduce	A_2 Do Not Introduce
S_1: Strong	30	0
S_2: Average	5	0
S_3: Weak	−15	0

Note act A_2, do not introduce breakfast items, has zero payoffs for all states since there would be no incremental revenue or cost associated with this decision. Also, the number of states could easily be expanded (say to include very strong or below average). If the number of states is quite large, one may need to use a continuous variable for the states rather than a discrete variable. We shall pursue this in a later section.

7.4 DECISION CRITERIA

Given McDougal's payoff table, the optimal act is not apparent. If one were *certain* about the state (demand), choice of an act would be simple. It is not simple because of uncertainty. The decision criteria about to be introduced employ probabilities in the analysis. There are certain nonprobabilistic decision criteria, but they are somewhat unrealistic. For example, the *maximin criterion* says look at the worst (min) that can happen under each act and then choose the act with the highest (max) payoff among those minimums. This is a very, very conservative and pessimistic procedure for making decisions. The decision maker assumes the worst will happen and acts accordingly. (In the McDougal's problem A_2 would be chosen since its minimum of zero is better than A_1's minimum of −15.)

McDougal's faces uncertainty concerning customer demand for breakfast items. Could McDougal's express this uncertainty in probabilistic terms? Yes, subjective and/or objective information could be used to assess the probabilities. Under the subjective approach to probability, probability represents the degree of belief an individual has about an uncertain situation. Further, information about a probability is *not* limited to objective information as gathered in an empirical investigation. Using past experiences in the fast-food business, an assessment of the current market, etc., probabilities can be determined for the three states. What if these probabilities are not accurate? They nevertheless lead the decision maker to a decision *consistent with his beliefs* at the time.

Suppose the decision maker at McDougal's, say the president Ronald McDougal, assigns the following subjective probability distribution to the states in question:

State (demand)	Probability
S_1: Strong	0.2
S_2: Average	0.4
S_3: Weak	0.4

We shall now use these probabilities as inputs to the decision making process.

In Section 5.3, the expected value of a discrete random variable was given by $E(X) = \sum x \cdot P(x)$. The expected value of a random variable was interpreted as the average outcome over a very long series of trials. For example, a fair (not loaded) die has six faces, each of which is equally likely. If X = number of dots observed on a single throw, then $E(X) = \sum x \cdot P(x) = 1(\frac{1}{6}) + 2(\frac{1}{6}) + \cdots + 6(\frac{1}{6}) = \frac{21}{6} = 3.5$. In other words, if one tosses a fair die a very large number of times, the average value of X would be 3.5. (Note that we "expect" X to equal 3.5 even though no single toss of the die can yield this outcome.) Similarly, we can treat the payoffs *under an act* as a random variable and compute the expected payoff of the act.

Consider act A_1, introduce the breakfast items. There are three possible values of the random variable payoff under A_1: 30, 5, and -15 (from Table 7.2). The probabilities of these outcomes are given by the probability distribution for the states of nature. For instance, the probability that the payoff will be 30 under A_1 is the probability that demand is strong: $P(S_1) = 0.2$. Computing the expected payoff of A_1:

TABLE 7.3:
CALCULATION OF THE
EXPECTED PAYOFF OF
ACT A_1

(1) State (demand)	(2) $P(S_i)$	(3) Payoff (M_{i1})	(4) $M_{i1} \cdot P(S_i)$
S_1: Strong	0.2	30	6
S_2: Average	0.4	5	2
S_3: Weak.	0.4	-15	-6
$\qquad \sum M_{i1} \cdot P(S_i)$			2

We could similarly compute the expected payoff under A_2, but since the payoffs are zero for all states, the expected payoff under A_2 is obviously zero. In general, we define the expected monetary payoff of act A_j as:

$$EM(A_j) = \sum_i M_{ij} \cdot P(S_i) \tag{7.1}$$

$$= M_{1j} \cdot P(S_1) + \cdots + M_{mj}P(S_m)$$

$$= \text{expected payoff under } A_j,$$

where m is the number of possible states.

The expected payoffs of the alternatives available to McDougal's are:

$$EM(A_1) = 2$$

$$EM(A_2) = 0$$

(remember units are $ million profit). The decision rule is:

> *Bayesian decision criterion:* Choose the act with the maximum expected payoff.

The act selected using the Bayesian criterion is termed the *optimal* or *Bayes act* and denoted by A^*. For McDougal's, $A^* = A_1$. The expected payoff of the optimal act is also referred to as the *expected payoff under uncertainty* $(EMUU)$:

$$EMUU = EM(A^*). \tag{7.2}$$

For McDougal's,

$$EMUU = EM(A^*) = EM(A_1) = 2.0.$$

Under uncertainty about the state of nature, the decision maker will select the Bayes act (A^*), and his expected payoff under uncertainty is that of the Bayes act. We shall find this terminology useful as we proceed.

7.5 EXPECTED VALUE OF PERFECT INFORMATION

The analysis thus far has utilized the decision maker's current state of knowledge as expressed in the probability distribution for the states of nature (the S_i). The decision maker may, however, obtain additional information that could cause a change (revision) of the states' probability distribution. New information could result from consumer surveys, test marketing, consultation with an expert, etc. In any event, such information is usually *bought* by the decision maker in an effort to make better decisions. In this section we introduce a quantity useful in assessing the need for additional information. In fact, we shall determine an upper bound on the value of new information.

Suppose, in the extreme case, the decision maker could obtain *perfect* information; perfect in that the decision maker would know *for certain* the true state of nature. Selection of the optimal act would then be easy; simply select the act yielding the highest payoff under the "known" state as pointed out in Section 7.2. Define M_i^* as:

$$M_i^* = \text{maximum payoff available under state } S_i.$$

In the McDougal's example, $M_1^* = 30$, $M_2^* = 5$, and $M_3^* = 0$ (Note that $M_3^* = 0$ because 0 is a larger number than -15.) Now define the *expected payoff under certainty* $(EMUC)$ as:

$$EMUC = \sum M_i^* \cdot P(S_i). \tag{7.3}$$

The computation of $EMUC$ for McDougal's is illustrated in Table 7.4.

TABLE 7.4: CALCULATION OF THE EXPECTED PAYOFF UNDER CERTAINTY FOR MCDOUGAL'S

State (S_i)	M_i^*	$P(S_i)$	$M_i^* \cdot P(S_i)$
S_1	30	0.2	6
S_2	5	0.4	2
S_3	0	0.4	0
$EMUC$			8

Note we are computing the *expected* payoff under certainty. This implies that we are able to predict *with certainty* when a given state will occur. The probability distribution indicates that state S_1 occurs 20 percent of the time, state S_2 40 percent, and S_3 40 percent. The average of the payoffs under an assumption of certainty is $EMUC$, or 8.

With perfect information, the decision maker would expect a payoff of $EMUC$. Under uncertainty, the decision maker will elect the Bayes act and expect a payoff of $EM(A^*) = EMUU$. The difference of these two quantities defines the *expected value of perfect information* ($EVPI$):

$$EVPI = EMUC - EMUU. \qquad (7.4)$$

The expected *value* of perfect information tells the decision maker how much *more* he could expect to make with perfect information over his current expectation under uncertainty. For McDougal's,

$$EVPI = EMUC - EMUU$$
$$= 8 - 2 = 6.$$

Under uncertainty, McDougal's will select A_1 using Bayes criterion and the expected payoff ($EPUU$) is 2 ($ million in profit). If McDougal's had access to perfect information before making the decision, the expected payoff would be 8. Hence, the difference of 6 is attributable to the availability of perfect information. Since perfect information removes uncertainty, $EVPI$ is sometimes termed the *cost of uncertainty*.

Since perfect information is seldom available in real business decisions, the decision maker can obtain only imperfect information. If we let the term "sample information" cover any kind of imperfect information, and define $EVSI$ as the *expected value of sample information*, then

$$EVSI \leq EVPI. \qquad (7.5)$$

Hence, the value of $EVPI$ places an upper bound (maximum cost) on what the decision maker would be willing to pay to obtain *any* additional information, whether perfect or imperfect. For example, suppose Marketing Associates, Inc., proposed a massive consumer survey for McDougal's that would yield "high quality" information; and the price for their services is $7 million. This proposal would be declined by McDougal's since the price of this project exceeds the expected value of perfect information.

7.6 OPPORTUNITY LOSS

A concept of considerable use in decision theory is that of an *opportunity loss*. At the moment, this concept offers an alternative but outcome-equivalent procedure for determining the Bayes act and $EVPI$. However, when the analysis proceeds to the continuous case, opportunity losses will prove invaluable. An opportunity loss can be determined for each state/act combination. Define L_{ij} as the opportunity loss under state S_i for act A_j, and

$$L_{ij} = |M_{ij} - M_i^*| \qquad (7.6)$$

where, as previously used, M_i^* is the maximum payoff available under state S_i.

In McDougal's problem, L_{12} would be computed as

$$L_{12} = |M_{12} - M_1^*|$$
$$= |0 - 30| = 30$$

since the payoff under A_2 given S_1 is zero while $M_1^* = 30$. What is the interpretation of this $L_{12} = 30$? If McDougal's selected A_2 and the state were S_1, their payoff would be zero. However, under S_1, had they selected A_1, the payoff would have been 30. Hence, if the state were S_1 and they choose A_2, McDougal's would have missed the chance (suffered an opportunity loss) of earning a payoff of $30 million more. The L_{ij} could be computed for each state/act combination and displayed in an opportunity loss table.[1]

TABLE 7.5:
OPPORTUNITY LOSS
TABLE FOR
MCDOUGAL'S

	Act	
State (demand)	A_1 Introduce	A_2 Do Not Introduce
S_1: Strong	0	30
S_2: Average	0	5
S_3: Weak	15	0

We reiterate: opportunity loss represents a payoff lost by not choosing the best act for a given state. If an L_{ij} is zero, then the act was best for that state. For example, $L_{11} = 0$ since A_1 is the best act given state S_1. Note an opportunity loss is *not* an accounting (or real) loss. For example $L_{22} = 5$, but for accounting purposes there is no effect since under A_2 the firm did nothing.

An equivalent procedure for determining the Bayes act is to choose the act with the minimum expected opportunity loss. Define the expected opportunity loss of act A_j as

$$EOL(A_j) = \sum_i L_{ij} \cdot P(S_i). \qquad (7.7)$$

In our example using Table 7.5 and the probability distribution of the S_i:

$$EOL(A_1) = \sum_i L_{i1} \cdot P(S_i)$$

$$= 0(0.2) + 0(0.4) + 15(0.4) = 6.0;$$
$$EOL(A_2) = \sum L_{i2} \cdot P(S_i)$$
$$= 30(0.2) + 5(0.4) + 0(0.4) = 8.0.$$

---

[1] Two notes on opportunity loss tables. First, a *negative* number cannot appear in the table because L_{ij} is an absolute value: see Formula (7.6). Second, there will be *at least one* zero in every row of the table since for each state there is an optimal act whose L_{ij} would then be zero. If for a a given state, two acts yield the best payoff, they both would have an $L_{ij} = 0$.

Choosing the act with the minimum expected opportunity loss yields $A^* = A_1$ (just as did choosing the act with the maximum expected payoff).

Now examine *EVPI* from an opportunity loss perspective. The expected opportunity loss under uncertainty will be that of the Bayes act, $EOL(A^*)$. But under certainty the decision maker could always select the best act for the given state. And what is the opportunity loss for the best act under a given state? Zero. Given perfect information, the decision maker would *never* suffer an opportunity loss, i.e., the expected opportunity loss with perfect information is zero. As in our discussion concerning payoffs, the difference between the expected losses (1) under uncertainty and (2) with perfect information yields *EVPI*:

$$EVPI = EOL(A^*) - 0 = EOL(A^*). \tag{7.8}$$

For McDougal's,

$$EVPI = EOL(A^*)$$
$$= EOL(A_1) = 6,$$

as determined using the previous method of Formula (7.4). The determination of the Bayes act and *EVPI* yield identical results whether viewed from a payoff or opportunity loss perspective. In later sections, the opportunity loss approach will offer advantages not apparent at this time.

7.7 REVISION OF THE DISCRETE PRIOR

The discussion thus far has utilized what we now term the decision maker's *prior* probability distribution for the state of nature. "Prior" indicates before the acquisition of additional information that causes the decision maker to change (revise) his probabilities concerning the S_i. The prior distribution, whether determined from subjective and/or objective information, can and should be revised if new information is made available. The revision of the states' probability distribution is accomplished using Bayes theorem, Formula (4.17), as discussed in Section 4.10. For convenience, we rewrite Bayes theorem using the notation of this chapter as:

$$P(S_i|N) = \frac{P(S_i) \cdot P(N|S_i)}{\sum_i P(S_i) \cdot P(N|S_i)}, \tag{7.9}$$

where N represents the additional (new) information acquired by the decision maker. As in Chapter 4, the revised distribution will be termed the *posterior probability distribution*.[2]

In the McDougal's example, suppose management decided to acquire additional information before making a decision on the introduction of break-

---

[2] Note prior and posterior are relative terms, relative to the *current* revision. Should further evidence later be acquired, the distribution would be revised again. In the second revision, the first revision's posterior distribution would be the second's prior.

fast items. (The *EVPI* of 6 indicates that additional information might have considerable value.) McDougal's, as well as many other national firms, has test marketed new ideas (products, packaging, etc.) in Seller, a medium-sized midwestern city. Seller is a popular test market because its socio-economic composition is highly representative of the United States as a whole.

McDougal's judges the outcome of the test marketing only as a success or failure. They have tested in Seller a soybean burger (a failure), a banana milk-shake (a success), and other product/price innovations. While Seller's record as a test market is not perfect, it is quite good. For example, when national demand was strong (S_1), 70 percent of the time the test market result had indicated success. Similarly, when national demand was average, the test result had indicated success only 40 percent of the time. However, the test result had indicated success 20 percent of the time when, in fact, national sales were weak. The above indicates the reliability of the test market evidence and can be symbolized as the following conditional probabilities:

$$P(N|S_1) = 0.7$$
$$P(N|S_2) = 0.4$$
$$P(N|S_3) = 0.2$$

where N is the event: test market indicated success.

Suppose that McDougal's introduced breakfast items in the test market and the outcome was a success. This additional information should be used to revise the prior probability distribution concerning the states (demand). Bayes theorem, Formula (7.9), is used to revise the distribution as illustrated in Table 7.6.

TABLE 7.6
CALCULATION OF
POSTERIOR
PROBABILITIES GIVEN
TEST MARKET RESULT
WAS SUCCESS

State S_i	Prior Probability $P(S_i)$	Conditional Probability $P(N\|S_i)$	Joint Probability $P(S_i) \cdot P(N\|S_i)$	Posterior Probability $P(S_i\|N)$
S_1	0.2	0.7	0.14	0.37
S_2	0.4	0.4	0.16	0.42
S_3	0.4	0.2	0.08	0.21
	1.0		0.38	1.00

For convenience we denote the prior probabilities by $P_0(S_i)$ and the posterior probabilities as $P_1(S_i)$. For example, the prior probability of strong demand, $P_0(S_1)$, is 0.2. When the probabilities are revised in light of the test market result, the probability of strong demand, $P_1(S_1)$, has increased to 0.37.

If the decision maker revises his probability distribution, then the posterior probabilities should be used to recompute the expected payoffs of the acts and *EVPI*. With the new probabilities, the expected payoffs will likely change and *perhaps* the optimal act will change under the posterior distribution. In

Table 7.7, the expected payoff of A_1 is recomputed. Note that the payoffs as given in Table 7.2 are unchanged.

TABLE 7.7:
CALCULATION OF $EM(A_1)$
UNDER THE POSTERIOR
DISTRIBUTION

State	$P_i(S_i)$	Payoff (M_{i1})	$M_{i1} \cdot P_1(S_i)$
S_1	0.37	30	11.10
S_2	0.42	5	3.10
S_3	0.21	-15	-3.15
			10.05

Since the payoffs under A_2 are all zeroes, its expected payoff remains zero. Then, under the posterior distribution,

$$EM(A_1) = 10.05$$
$$EM(A_2) = 0,$$

and the Bayes' act remains A_1 (introduce the breakfast items). Calculation of the posterior expected value of perfect information yields $EVPI = 3.15$; the reader may confirm this and also that the posterior $EMUC = 13.20$.

The reader may suspect that the posterior $EVPI$ (or cost of uncertainty) is always less than the prior $EVPI$. The reasoning would be that as more information is obtained, perfect information would become of less value. Maybe. But in fact, the posterior $EVPI$ may actually *exceed* the prior $EVPI$. For example, if the decision maker had little prior uncertainty (say $P(S_1) = 0.90$) and then acquired sample information that was in substantial disagreement with the prior, his uncertainty would be increased. Whether the posterior $EVPI$ increases or decreases, it still places an upper bound on the cost of additional information at that time.

7.8 THE NORMAL PRIOR

In the discussion thus far, the number of possible states of nature, the S_i, has been finite. The McDougal's example had three states: strong, average, and weak demand. The formulas previously given are applicable to any finite number of states. However, in some decision problems the state of nature is a *continuous* (rather than discrete) variable and can assume an infinite number of values. If the state of nature is continuous, payoffs cannot be given in tabular form; rather, the payoffs under a given act must be expressed in functional form. We must develop new ways of computing the $EM(A_j)$, revising the prior, etc.

A continuous state of nature could have any of a variety of probability distributions. Our discussion of continuous distributions is limited to the normal because of two reasons. First, it is frequently a good approximation of the decision maker's prior beliefs. Second, under certain conditions it is very easy to revise a normal prior distribution. In this discussion the state of nature is the mean of some population. The additional evidence that initiates revision of the prior is the result of a sample from the population of interest. We assume that (1) the sampling distribution of the mean ($\bar{X}$) is normal and (2) the population variance, σ^2, is known or well estimated.[3]

Suppose the prior distribution is normal with mean μ_0 and variance σ_0^2. Then a sample of size n is taken; the sample mean is $\bar{X}$ and the variance of the sampling distribution is $\sigma_{\bar{X}}^2 = \sigma^2/n$. We wish to revise the prior distribution in consideration of the sample evidence. Bayes theorem can be used to show that if both the prior and the sampling distribution are normal *the posterior distribution is normal* with mean μ_1 and variance σ_1^2. These parameters are obtained from:

$$\mu_1 = \frac{\dfrac{1}{\sigma_0^2}(\mu_0) + \dfrac{1}{\sigma_{\bar{X}}^2}(\bar{X})}{\dfrac{1}{\sigma_0^2} + \dfrac{1}{\sigma_{\bar{X}}^2}}; \tag{7.10}$$

$$\frac{1}{\sigma_1^2} = \frac{1}{\sigma_0^2} + \frac{1}{\sigma_{\bar{X}}^2}. \tag{7.11}$$

Before illustrating the above in an example, examine the two formulas. Notice first that the posterior mean, μ_1, is a weighted average of the prior mean and the sample mean; each mean being weighted by the *reciprocal* of its variance. This weighting is clearly seen if Formula (7.10) is rewritten as

$$\mu_1 = \frac{\sigma_1^2}{\sigma_0^2}\mu_0 + \frac{\sigma_1^2}{\sigma_{\bar{X}}^2}\bar{X}, \tag{7.12}$$

since $(\sigma_1^2/\sigma_0^2) + (\sigma_1^2/\sigma_{\bar{X}}^2) = 1$, as obtained simply by multiplying both sides of Formula (7.11) by σ_1^2. This weighting leads to two results. One, *the posterior mean will lie between the prior mean and the sample mean.* Second, the mean with the smaller variance receives the larger weight. For example, if n is very large so that $\sigma_{\bar{X}}^2 = \sigma^2/n$ is small relative to σ_0^2, then the sample mean would be weighted more heavily, and μ_1 would be nearer to $\bar{X}$ than to μ_0. As the sample size increases, the weight given to $\bar{X}$ increases. Both results should be intuitively acceptable. As concerns σ_1^2, notice that its reciprocal is the sum of the reciprocals of σ_0^2 and $\sigma_{\bar{X}}^2$. This means the posterior variance, σ_1^2, is always smaller than the prior variance, σ_0^2.

EXAMPLE. Suppose a decision maker has a normal prior distribution with $\mu_0 = 1{,}050$ and $\sigma_0 = 300$. Subsequent to the establishment of the prior, a random sample of size 40 gave the following results: a mean $\bar{X}$ of 1,100 and standard deviation of 400. Assuming the population standard deviation, σ, is 400, revise the prior to take account of the sample information.

---

[3] The first assumption, that the sampling distribution of $\bar{X}$ is normal, is satisfied when the population distribution is normal *or* when the sample size is large because of the central limit theorem of Chapter 6. The population variance is considered well estimated if n is large. Schlaifer and Winkler, cited in the chapter references, both further discuss the problem of an unknown population standard deviation.

Since the prior and sampling distributions are normal, the posterior will be normal. The mean and variance of the posterior distribution are given by Formulas (7.10) and (7.11). We first compute $\sigma_X = \sigma/\sqrt{n} = 400/\sqrt{40} = 63.24$.

$$\mu_1 = \frac{\dfrac{1}{\sigma_0^2}(\mu_0) + \dfrac{1}{\sigma_{\bar{X}}^2}(\bar{X})}{\dfrac{1}{\sigma_0^2} + \dfrac{1}{\sigma_{\bar{X}}^2}}$$

$$= \frac{\dfrac{1}{(300)^2}(1050) + \dfrac{1}{(63.24)^2}(1100)}{\dfrac{1}{(300)^2} + \dfrac{1}{(63.24)^2}}$$

$$= 1{,}098;$$

$$\frac{1}{\sigma_1^2} = \frac{1}{\sigma_0^2} + \frac{1}{\sigma_{\bar{X}}^2} = \frac{1}{(300)^2} + \frac{1}{(63.24)^2} = 0.000261;$$

and

$$\sigma_1^2 = \frac{1}{0.000261} = 3{,}830.$$

The decision maker's revised distribution has a mean of 1,098 and variance of 3,830. As pointed out earlier, note that (1) the posterior mean lies between the prior and sample means and (2) the posterior variance is less than the prior variance, $\sigma_0^2 = (300)^2$.

7.9 THE TWO-ACTION PROBLEM WITH LINEAR PAYOFF FUNCTIONS

We now take up a special category of decision problems: the two-action problem with linear payoff functions. The occurrence of two-action problems in business is frequent, e.g., introduce versus don't introduce product, invest in stocks or invest in bonds, operate or do not operate third shift, etc. When the state of nature is the unknown mean of a population, and the payoff under each act is a linear function of the mean, the determination of the Bayes act is quite simple. Also, when the prior or posterior distribution (depending on the decision maker's current state) is normal, the value of $EVPI$ is easily obtained.

Suppose that the linear payoff functions for acts A_1 and A_2 are

$$\text{Payoff}(A_1) = a_1 + b_1\mu; \tag{7.13}$$
$$\text{Payoff}(A_2) = a_2 + b_2\mu. \tag{7.14}$$

Figure 7.1 shows graphically the two functions. For convenience, we assume that $b_1 > b_2$. (If $b_1 = b_2$, the lines would be parallel, and the act with the larger intercept would always yield the greater payoff.) If in the original formulation of the problem b_1 does not exceed b_2 then simply rename the acts so that the payoff function with the greater slope corresponds to A_1.

Linear payoff functions with unequal slopes must intersect at some point; in Figure 7.1 this intersection takes place at $\mu = \mu_b$. At μ_b, the payoff under each act is the same; hence μ_b is termed the breakeven value of μ (or point of indifference). To the left of μ_b (i.e., $\mu < \mu_b$), A_2 has the higher payoff and would be

FIGURE 7.1: GRAPH OF
THE LINEAR PAYOFF
FUNCTIONS

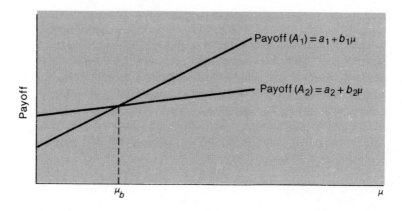

the optimal act. If $\mu > \mu_b$, A_1 is the optimal act. The breakeven value is determined by simply setting the two payoffs equal and solving for μ_b:

$$a_1 + b_1\mu_b = a_2 + b_2\mu_b$$
$$(b_1 - b_2)\mu_b = a_2 - a_1$$
$$\mu_b = \frac{a_2 - a_1}{b_1 - b_2}. \tag{7.15}$$

Corresponding to the payoff functions, each act has an opportunity loss function. We shall develop the loss function for A_1. When $\mu > \mu_b$, A_1 is the optimal act and the opportunity loss is zero. When $\mu < \mu_b$, under A_1 the payoff is $a_1 + b_1\mu$, while under A_2 one could have obtained the higher payoff of of $a_2 + b_2\mu$. The difference in the payoffs is the opportunity loss under A_2: $(a_2 + b_2\mu) - (a_1 + b_1\mu) = (a_2 - a_1) + (b_2 - b_1)\mu$. Summarizing:

$$\text{Loss}(A_1) = \begin{cases} 0 & \text{if } \mu \geq \mu_b \\ (a_2 - a_1) + (b_2 - b_1)\mu & \text{if } \mu < \mu_b. \end{cases} \tag{7.16}$$

The graph of the opportunity loss function for A_1 is shown in Figure 7.2; such a function is called piecewise linear. Similarly,

$$\text{Loss}(A_2) = \begin{cases} (a_1 - a_2) + (b_1 - b_2)\mu & \text{if } \mu > \mu_b \\ 0 & \text{if } \mu \leq \mu_b. \end{cases} \tag{7.17}$$

FIGURE 7.2:
OPPORTUNITY LOSS
FUNCTION FOR ACT A_1

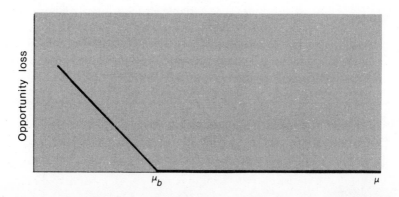

Note that the slope $(b_2 - b_1)$ of the nonzero segment of A_1's loss function has the same *absolute* value as the slope $(b_1 - b_2)$ of the nonzero segment under A_2; symbolize $K = |b_1 - b_2|$. The loss functions are developed because they will later prove useful in determining *EVPI* from the opportunity loss approach.

The Bayesian decision criterion in the continuous case is the same as in the discrete case: choose the act with the maximum *expected* payoff. Looking first at the expected payoff under A_1 by finding the expected value from Formula (7.13):

$$EM(A_1) = E(a_1 + b_1\mu)$$
$$= E(a_1) + E(b_1\mu)$$
$$= E(a_1) + b_1E(\mu)$$
$$= a_1 + b_1E(\mu). \tag{7.18}$$

Similarly for A_2:

$$EM(A_2) = a_2 + b_2E(\mu). \tag{7.19}$$

Therefore, all we need to determine the expected payoffs and then select the optimal act is the expected value of the state variable, μ. We have already seen that if $\mu > \mu_b$, A_1 has the higher payoff. Similarly, looking at Equations (7.18) and (7.19), A_1 will have the higher *expected* payoff if $E(\mu) > \mu_b$. And what is $E(\mu)$? Simply μ_0 if a normal prior is available and μ_1 if a normal posterior is available. The decision rule for the two-action problem with linear payoffs can be stated as:

Select A_1 if $E(\mu) > \mu_b$

Select A_2 if $E(\mu) < \mu_b$. $\tag{7.20}$

Remember that we assume that $b_1 > b_2$. If $E(\mu) = \mu_b$, both acts offer the same payoff and we are indifferent between A_1 and A_2.

EXAMPLE. The management at McDougal's has developed a payoff function in terms of company-wide sales of breakfast items. Their decision is to be made on the basis of a one-year projection. The operations management staff has estimated that fixed costs will be \$4.5 million (primarily the required investment in new equipment) associated with expanding the menu. The staff also estimates that variable costs (primarily labor and the cost of commodities such as eggs) will equal 70 percent of sales. The state variable, μ, is annual sales. Total costs are the sum of fixed costs (\$4.5 million) and variable costs (0.7μ). The annual payoff (profit) to the company is then sales (or revenue) less costs:

$$\text{Payoff} = \mu - (4.5 + 0.7\mu)$$
$$= -4.5 + 0.3\mu.$$

Thus under A_1 (introduce) the payoff function is:

$$\text{Payoff}(A_1) = -4.5 + 0.3\mu.$$

The payoff under A_2 (do not introduce) remains zero:

$$\text{Payoff } (A_2) = 0 + 0\mu = 0.$$

Note $b_1(0.3)$ is greater than $b_2(0)$ as required. See Figure 7.3.

FIGURE 7.3: PAYOFF
FUNCTIONS FOR
MCDOUGAL'S

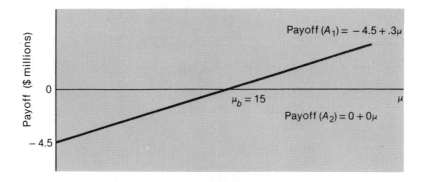

Suppose management has developed a normal prior distribution for μ with parameters $\mu_0 = 17$ and $\sigma_0 = 2$; both units are in \$ millions. Use the payoff functions and the prior distribution to determine the optimal act.

Under the normal prior, $E(\mu) = \mu_0 = 17$ and the expected payoffs under A_1 and A_2 are:

$$EM(A_1) = a_1 + b_1 E(\mu)$$
$$= -4.5 + 0.3(17) = 0.6;$$
$$EM(A_2) = 0.$$

Since A_1 has the greater expected payoff, $A^* = A_1$. The optimal act could also be determined by solving for μ_b:

$$\mu_b = \frac{a_2 - a_1}{b_1 - b_2}$$

$$= \frac{0 - (-4.5)}{0.3 - 0} = \frac{4.5}{0.3} = 15,$$

then using Formula (7.20) to select A_1 since $E(\mu) = 17$ is greater than the breakeven value $\mu_b = 15$.

EVPI

In using management's prior distribution to determine the Bayes act, only μ_0 (the mean of the prior) was needed. Neither normality nor σ_0^2 (the variance of the prior) was required. The assumption of normality and the value of σ_0^2 were required in Section 7.8 in the revision of a normal prior to a normal posterior. The assumptions are required here in order to use a very efficient method in the determination of $EVPI$.

The expected value of perfect information is equal to $EOL(A^*)$. In the discrete case with $A^* = A_1$, $EVPI = EOL(A_1) = \sum L_{i1} \cdot P(S_i)$. In the continuous case, integral calculus is generally required. However, when the prior distribution is normal, the required integration has been tabulated under the name *unit normal loss function*, $L(D)$. In the two-action problem with linear payoffs and a normal prior,

$$EVPI = |b_1 - b_2| \cdot \sigma_0 \cdot L(D), \qquad (7.21)$$

where $L(D)$ is the value of the unit normal loss function as obtained from Appendix F with

$$D = \frac{|\mu_b - \mu_0|}{\sigma_0}. \qquad (7.22)$$

The derivation of this result is beyond the scope of this text, but an examination of Formula (7.21) is in order. Note that three quantities affect the value of *EVPI*:

1. $|b_1 - b_2|$ is the absolute value of the slope of the nonzero segment of the loss function under either A_1 or A_2. As the slope increases, the opportunity losses increase as μ moves away from the breakeven value, μ_b. In Figure 7.4, compare the loss functions of A_1 and A_1''. For any value of μ less than 15, the loss under A_1'' is greater than that under A_1.

FIGURE 7.4:
MCDOUGAL'S NORMAL
PRIOR AND
OPPORTUNITY LOSS
FUNCTIONS FOR A_1, A_1',
AND A_1''

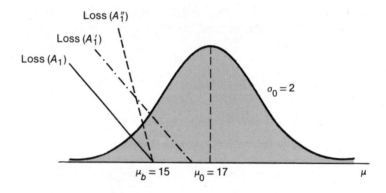

2. Examining Appendix F, we note that $L(D)$ decreases as D increases. Now look at the numerator of D. The greater the difference between μ_b and μ_0, the larger is D and hence the smaller is *EVPI*. Compare the loss functions of A_1 and A_1' in Figure 7.4. For any μ less than 15, the loss is greater under A_1'; some losses will be incurred with μ greater than 15, where A_1 is zero. In our example, A_1' could occur if the *fixed* costs were increased.

3. The prior standard deviation, σ_0, is a measure of uncertainty. The larger σ_0, the more uncertain the decision maker is about possible values of μ. As his uncertainty increases, the cost of uncertainty (or *EVPI*) increases. Graphically this would be shown by a "flatter" or more widely dispersed prior than shown in Figure 7.4. Given a greater σ_0, the probability of $\mu < 15$ is increased.

The unit normal loss function and Formula (7.21) can be used if a normal *posterior* distribution is available. The posterior *EVPI* is obtained by substituting the posterior mean and standard deviation (μ_1 and σ_1) for the prior's μ_0 and σ_0 in Formulas (7.21) and (7.22).

EXAMPLE. In this section, McDougal's payoff functions and breakeven value were given as:

$$\text{Payoff}(A_1) = -4.5 + 0.3\mu$$
$$\text{Payoff}(A_2) = 0$$
$$\mu_b = 15.$$

The normal prior was also stated as having parameters $\mu_0 = 17$ and $\sigma_0 = 2$. Determine *EVPI*.

Using Formula (7.22) to find D,

$$D = \frac{|\mu_b - \mu_0|}{\sigma_0}$$

$$= \frac{|15 - 17|}{2} = 1.$$

From Appendix F, we find $L(1) = 0.08332$. Using Equation (7.21),

$$EVPI = |b_1 - b_2| \cdot \sigma_0 \cdot L(D)$$
$$= |0.3 - 0|(2)(0.08332)$$
$$= 0.05.$$

Thus $EVPI = 0.05$ ($ millions) or $50,000. As before, this places an upper bound on the acquisition costs of any further information.

7.10 SUMMARY

This chapter introduced the student to statistical decision theory. The conceptual framework of decision making, the Bayesian decision criterion, the expected value of perfect information, and opportunity losses are important subjects that were developed and illustrated. We first examined the case of a finite number of state/act combinations where the payoff table was employed. After reviewing the Bayes theorem as applied to normal distributions, a special case with continuous states was analyzed: the two-action problem with linear payoffs. This special case was included for two reasons. First, payoff functions in real decision problems are frequently linear with respect to the state of nature. Second, it is easy to formulate the Bayesian criterion and to revise the normal prior for such problems.

In order to focus clearly on the fundamental principles of the decision-making process, two matters of considerable importance have been neglected. First, the (subjective) prior distribution of the decision maker has merely been given. In practice, the assessment of the prior is frequently a problem that requires expertise beyond that of the decision maker. For example, if you ask a sales manager what will happen if the advertising budget is cut by ten percent, he may respond "Well, I think sales may drop considerably." The conversion of "may" to a probability and "considerably" to a meaningful measure of sales volume is necessary to develop his prior distribution. The quantification of uncertainty is a prerequisite to statistical decision theory; Hogarth and Winkler both offer readings in this area.

The other area glossed over in this chapter concerns payoffs. If you were offered a choice of A_1: \$100,000 for certain *or* A_2: \$210,000 if the toss of a fair coin came up heads and zero if tails, which would you choose? Most people would select A_1 even though $EM(A_1) = \$100,000$ is less than $EM(A_2) = \$105,000$. Would selecting A_1 be irrational or has decision theory failed? Neither: *decisions are based on the utility of the outcomes.* Decisions based on utility require that monetary values be converted to utility measures and the optimal act be chosen by maximizing expected utility. If the utility of \$0, \$100,000, and \$210,000 is 0, 500, and 900 respectively, the expected utility of A_1 is 500 and the expected A_2 is $(0.5)(900) + (0.5)(0) = 450$. Hence A_1 maximizes expected utility. Why are 500 and 900 reasonable measures of the utility of \$100,000 and \$210,000? A student, for example, could take \$100,000 and get a better apartment, new furniture, a super stereo, clothes, a Corvette, a van, etc. After that, would a second \$100,000 bring nearly as much "satisfaction" as the first \$100,000? Probably not, for the economic law of diminishing marginal utility is at work. Throughout the chapter, we have assumed that payoffs expressed in dollars were equivalent to (or a linear function of) those expressed in utils, the measure of utility. For the great majority of business decision problems, this is a reasonable assumption, and expected monetary value is a valid guide to action. However, if the best and/or worst outcomes are extreme (e.g., immense wealth versus bankruptcy), then utilities should be substituted for monetary values. For a further discussion on utility, see Schlaifer or Winkler.

Despite the qualifications above, the framework for decision making should prove useful to the business student in evaluating alternatives. The expected value of perfect information and factors influencing its magnitude also provide reference points in a decision problem. Most important, the effect of the prior in decision making should be understood. For example, we note that providing several individuals with the same objective (sample) information will not always lead them to select the same act if they begin with different priors. However, if σ_0^2 is *very* large relative to $\sigma_{\bar{X}}^2$, the decision maker's posterior distribution depends almost entirely on the information contained in the sample. See Formulas (7.10) and (7.11); if $1/\sigma_0^2$ is negligible compared to $1/\sigma_{\bar{X}}^2$, then $\mu_1 = \bar{X}$ and $\sigma_1^2 = \sigma_{\bar{X}}^2$, and the sample information is said to "overwhelm" the prior information. In such a case, the prior is termed informationless or *diffuse.* In all chapters of this text dealing with statistical inference, estimation and testing are based on sample evidence alone. For Bayesians, this is equivalent to assuming that the prior distribution is diffuse.

7.11 TERMS AND FORMULAS

States of nature (the S_i)

Alternative courses of action (the A_j)

Payoff (M_{ij})

Maximin criteria

Expected payoff of A_j $[EM(A_j)]$

Bayesian decision criterion

Expected payoff under uncertainty ($EMUU$)

Bayes or optimal act (A^*)

Maximum payoff under state S_i (U_i^*)

Expected value with perfect information ($EVPI$)

Opportunity loss (L_{ij})

Normal prior (parameters μ_0 and σ_0^2)

Normal posterior (parameters μ_1 and σ_1^2)

Linear payoff function

Breakeven value of μ (μ_b)

Unit normal loss function $[L(D)]$

Diffuse prior

Expected payoff of act A_j (discrete case)	$EM(A_j) = \sum M_{ij} \cdot P(S_i)$
(continuous case)	$EM(A_j) = a_1 + b_1 E(\mu)$
Expected payoff under uncertainty	$EMUU = EM(A^*)$
Expected payoff under certainty	$EMUC = \sum M_i^* \cdot P(S_i)$
Expected value of perfect information	$EVPI = EMUC - EMUU$
Opportunity loss	$L_{ij} = \left| M_{ij} - M_i^* \right|$
Bayes' theorem	$P(S_i \mid N) = \dfrac{P(S_i) \cdot P(N \mid S_i)}{\sum\limits_{i} P(S_i) \cdot P(N \mid S_i)}$
Posterior mean (normal case)	$\mu_1 = \dfrac{\dfrac{1}{\sigma_0^2}(\mu_0) + \dfrac{1}{\sigma_{\bar{X}}^2}(\bar{X})}{\dfrac{1}{\sigma_0^2} + \dfrac{1}{\sigma_{\bar{X}}^2}}$
Posterior variance (normal case)	$\dfrac{1}{\sigma_1^2} = \dfrac{1}{\sigma_0^2} + \dfrac{1}{\sigma_{\bar{X}}^2}$
Break-even value of μ	$\mu_b = \dfrac{a_2 - a_1}{b_1 - b_2}$
Expected value of perfect information for normal case	$EVPI = \left| b_1 - b_2 \right| \cdot \sigma_0 \cdot L(D)$

QUESTIONS

A. True or False

1. Payoff tables are used when the state of nature is a continuous variable.

2. The expected payoff under uncertainty equals the expected payoff of the Bayes act.

3. With perfect information, the decision maker would expect a payoff of $EVPI$.

4. M_2^* is the maximum payoff available under act A_2.

5. An opportunity loss cannot be negative.

6. In every row of an opportunity loss table, there is at least one zero.

7. The posterior $EVPI$ can be greater than the prior $EVPI$.

8. In the revision of a normal prior, the posterior mean will lie between the prior mean and the sample mean.

9. The discussion of linear payoffs assumed that b_1 was greater than b_2.

10. In the revision of normal priors, the posterior mean moves toward the prior mean as the sample size increases.

B. Fill in

1. _____ are factors that affect the outcome of a decision but are beyond control of the decision maker.

2. _____ says to choose the act with the maximum expected payoff.

3. _____ places an upper bound on what the decision maker would pay to obtain any additional information.

4. If $M_{13} = M_1^*$, then $L_{13} =$ _____.

5. The expected opportunity loss with perfect information is _____.

6. Given a normal prior and a normal sampling distribution of the mean, the posterior distribution will be _____.

7. For linear payoff functions and $\mu =$ _____, the payoff under A_1 and A_2 is the same.

8. As the expected value of μ moves toward μ_b, $EVPI$ will _____.

9. As the slope of the loss function decreases, $EVPI$ will _____.

10. When $1/\sigma_0^2$ is negligible compared to $1/\sigma_{\bar{X}}^2$ (i.e., a diffuse prior), the expected value of μ for the posterior distribution equals_____.

EXERCISES

1. For the following decision makers, identify a problem that each might encounter, and for each problem define at least two states of nature and alternative courses of action:
   a. Loan officer at a commercial bank.
   b. I.R.S. auditor (of federal tax returns).
   c. Advertising manager for a manufacturer of stereo equipment.
   d. Chief purchasing agent for a fast-food chain.
   e. General manager of the New York Yankees.

2. Stix, Inc., is considering submitting a bid for a government contract. Management has determined a bid price that would yield a profit to the firm of $40,000; they feel that Stix has a probability of 0.4 of obtaining

the contract. The cost of preparing the bid (*not* included in the above profit figure) is $5,000.

a. Develop the payoff table.

b. Determine the optimal act.

3. Boyd's of London has been asked to insure a $5 million collection of Chinese porcelain during its shipment from a New York to a Chicago museum. After examining the security arrangement, packaging of the porcelain, etc., Boyd's assesses the probability of a safe shipment (no losses) as 0.999 and the probability of a total loss (theft or catastrophic accident as 0.001.

a. If Boyd's charge for insurance is twice the expected loss, what will they charge?

b. If Boyd's were to charge $4,000 and assuming the shipment could either arrive safe or be a total loss, what probability of safe shipment would give Boyd's an expected payoff of zero for the act: insure shipment?

c. Develop the payoff table for part b.

4. Given the following payoff table and probability distribution for S_i:

	Act			
State	A_1	A_2	A_3	Probability
S_1	30	50	−40	$P(S_1) = 0.40$
S_2	50	0	70	$P(S_2) = 0.30$
S_3	−20	10	20	$P(S_3) = 0.30$

a. Find the expected payoff of each act.

b. What is the optimal act?

c. What is the expected payoff under uncertainty ($EMUU$)?

d. Determine the expected payoff under certainty ($EMUC$).

e. Determine the expected value of perfect information ($EVPI$).

5. Given the following payoff table and probability distribution for S_i:

	Act			
State	A_1	A_2	A_3	Probability
S_1	8	10	5	$P(S_1) = 0.2$
S_2	15	12	10	$P(S_2) = 0.3$
S_3	20	12	5	$P(S_3) = 0.1$
S_4	2	5	20	$P(S_4) = 0.4$

a. Develop the opportunity loss table.
b. Use the loss table to determine the optimal act.
c. Determine *EVPI*.
d. Determine *EMUU*.
e. Determine *EMUC*.
f. Verify that $EOL(A^*) = EMUC - EMUU$.

6. In Problem 4, additional information (N) is obtained, and

S_i	$P(N\|S_i)$
S_1	0.4
S_2	0.6
S_3	0.2

a. Use the above to revise the prior distribution of Problem 4.
b. Using the posterior distribution, determine the optimal act for Problem 4.
c. Determine *EVPI* with the posterior distribution.

7. In Problem 5, additional information (N) is obtained such that $P(N|S_1) = 0.4$, $P(N|S_2) = 0.3$, $P(N|S_3) = 0.2$ and $P(N|S_4) = 0.3$.
a. Revise the prior distribution of Problem 5.
b. Use the revised distribution to determine the optimal act and *EVPI*.

8. A student must travel from Philadelphia to New York; she can travel by plane, bus, or train. Is this a decision problem under undertainty if her sole concern is transportation *cost*? What if transportation *time* were her sole concern?

9. Rocky Coal, Inc., has an option to buy a large tract of land; the option expires in three weeks. The land's value is largely determined by the extent and nature of its coal deposits, e.g., coal can be recovered more cheaply with surface mining than by underground mining. Management has determined the following payoff table (in $ millions):

	Act	
Coal Deposits	Buy	Do Not Buy
Surface coal	5	0
Underground coal	2	0
No coal	−4	0

Management believes the probabilities are: surface coal, 0.4; underground coal, 0.3; and no coal, 0.3.
a. Determine the optimal act.
b. Determine *EVPI*.

10. A local gift shop is about to order candy for Valentine's Day. The candy costs the store \$1.50 per box and is sold for \$4.00 per box through Valentine's Day. After Valentine's Day, the demand for heart-shaped boxes of candy inscribed "Be My Valentine" is somewhat limited, and left-over boxes are donated to a local charity. The owner's estimated sales through Valentine's Day (in *dozens* of boxes) is:

Demand	Probability
5	0.2
6	0.4
7	0.3
8	0.1

    a. Construct the gift shop's payoff table.
    b. What is the optimal act?

11. In Problem 9, one week before the option to buy expired, Rocky Coal received a geologist's report. His opinion, based on experience and test data, was that no significant deposits would be found on the property. If no coal were present, his report would be negative 0.7 of the time. If surface coal were present, he would erroneously issue a negative opinion only 0.1 of the time. However, if the property had underground coal, his report would indicate no coal 0.5 of the time. Use the geologist's report to revise management's prior probability distribution. What is the optimal act under the posterior distribution?

12. Given $\mu_0 = 2{,}500$ and $\sigma_0 = 400$ for normal prior, determine the posterior distribution if:
    a. $\bar{X} = 2{,}000$, $S = 500$, $n = 50$.
    b. $\bar{X} = 2{,}000$, $S = 600$, $n = 30$.
    c. $\bar{X} = 2{,}800$, $S = 300$, $n = 100$.

13. Given a normal prior with $\mu_0 = 400$ and $\sigma_0 = 40$, determine the posterior distribution if $\bar{X} = 300$, $S = 50$, and:
    a. $n = 30$.
    b. $n = 100$.

14. Suppose the payoff functions of two acts are:

$$\text{Payoff}(A_1) = -200 + 4\mu$$
$$\text{Payoff}(A_2) = 800 - 2\mu$$

    a. Graph the payoff functions.
    b. Determine the breakeven value of μ.
    c. Determine and graph the loss functions.
    d. If $\mu = 100$, what is the optimal act?

15. Suppose the payoff functions of two acts are:

$$\text{Payoff}(A_1) = -3000 + 10\mu$$
$$\text{Payoff}(A_2) = 5000$$

and management has a normal prior with $\mu_0 = 750$ and $\sigma_0 = 150$.
a. Determine the optimal act.
b. Determine the breakeven value.
c. Determine $EVPI$.

16. In Problem 14, suppose management has a normal prior with $\mu_0 = 180$ and $\sigma_0 = 38$. Determine the optimal act and $EVPI$.

17. Given the following information and assuming that the prior distribution is normal, determine $EVPI$.
a. $\mu_0 = 200$, $\sigma_0 = 20$, $\mu_b = 205$, $|b_1 - b_2| = 125$.
b. $\mu_0 = 200$, $\sigma_0 = 20$, $\mu_b = 190$, $|b_1 - b_2| = 100$.
c. $\mu_0 = 200$, $\sigma_0 = 30$, $\mu_b = 193$, $|b_1 - b_2| = 200$.

18. A false alarm at a color TV distributor's warehouse triggered the sprinkler system. Water damage is evident from the wet cardboard boxes containing the TV sets. Salvage, Inc., has been offered the lot of 500 sets for $100,000. Salvage knows that a working set can be sold for $250. Based on their experience, Salvage believes that average cost to put a set in working order can be described as a normally distributed random variable with $\mu_0 = \$30$ and $\sigma_0 = 15$.
a. Determine Salvage's payoff functions for the acts: buy and do not buy the lot.
b. At what average repair cost would Salvage be indifferent between the two acts?
c. Based on the prior distribution, should Salvage buy the lot?
d. Determine $EVPI$.

19. Consumer Credit Corporation recently filed for bankruptcy. One of their major assets was a portfolio of 5,000 accounts receivable having a book value of $2 million. The court handling the bankruptcy proceedings has offered the portfolio to JGV Associates for $800,000. JGV knows the value of the portfolio will be determined by the amount spent on collection expenses, and many of the loans are far overdue and have poor collateral. JGV estimates that the average collection expense per loan is described by a normal distribution with mean of $220 and standard deviation of $40.
a. Determine JGV's payoff functions for the acts: buy and do not buy portfolio.
b. What is the optimal act?
c. Determine $EVPI$.

20. Suppose Salvage (see Problem 18) took a random sample of ten sets and the average repair cost was $53 and standard deviation, s, was $17. Assuming that the sampling distribution is normal and that s^2 is a good estimator of the population variance, σ^2:
a. Revise Salvage's prior distribution.
b. Using the posterior distribution, determine the optimal act and $EVPI$.

21. In Problem 19, JGV has been granted the court's permission to sample the portfolio and to collect the amounts outstanding. A random sample

of thirty accounts costs \$140 each in collection expenses; the sample variance, s^2, was \$30.

a. Revise the prior distribution of Problem 19.

b. Using the posterior distribution, determine the optimal act and *EVPI*.

REFERENCES

Dyckman, T. R.; Smidt, S.; and McAdams, A. K. *Management Decision Making Under Uncertainty: An Introduction to Probability and Statistical Decision Theory.* London: The Macmillan Company, 1969.

Hogarth, Robin M. "Cognitive Processes and the Assessment of Subjective Probability Distributions." *Journal of the American Statistical Association* (June 1975), pp. 271–89.

Jones, J. Morgan *Introduction to Decision Theory.* Homewood, Ill.: Richard D. Irwin, Inc., 1977.

Morgan, Bruce W. *An Introduction to Bayesian Statistical Decision Processes.* Englewood Cliffs, N.J.: Prentice-Hall, Inc., 1968.

Schlaifer, Robert *Probability and Statistics for Business Decisions.* New York: McGraw-Hill Book Company, 1959.

Winkler, R. L. *An Introduction to Bayesian Inference and Decision.* New York: Holt, Rinehart and Winston, Inc., 1972.

8

Introduction to Statistical Inference

8 INTRODUCTION TO STATISTICAL INFERENCE

Laws and institutions must go hand in hand with the progress of the human mind. As that becomes more developed, more enlightened, as new discoveries are made, new truths disclosed, institutions must advance and keep pace with the times.

Thomas Jefferson

8.1 INTRODUCTION

Statistical inference is the process of using limited information, a *sample*, for the purpose of reaching conclusions about a large set of data, the *population*. The preceding chapters provide the foundation for discussing the two major areas of statistical inference, *estimation* and *testing*. Examples of situations follow where estimation and testing are desirable.

Estimation of demand. A large food processing company has developed a new breakfast product and now wishes to determine whether it would be profitable to produce and market the product on a national scale. Based on a sample of customer reactions in the test market, the company will estimate demand for the product in all U.S. market areas.

Testing for audit standards. The accountant for a retail merchant must verify that the proportion of accounts with errors does not exceed a certain standard. If a sample of the accounts reveals that the proportion with errors is below a predetermined amount, the accountant will not have to audit the entire population of accounts. Here the accountant uses sample information to test whether the standard set for the population of accounts is met.

Statistical inference enables modern managers to "keep pace with the times" by investigating "new discoveries" and establishing "new truths." Estimation is discussed in the first part of the chapter and is followed by the concepts in testing.

8.2 ESTIMATION

Estimation refers to any procedure where sample information is used to estimate or predict the numerical value of some population measure (called a *parameter*), such as the population mean μ. Some definitions are in order.

An *estimator* is a procedure or function used in estimating a population parameter.
An *estimate* is the numerical value determined from the estimator.

If the sample mean is used as an *estimate* of the population mean, the sample mean is determined from the *estimator*, which is the function (or formula)

$$\bar{X} = \frac{\sum X}{n}.$$

estimate = actual value

When we substitute into the formula and compute a numerical value for $\bar{X}$, the numerical value is the *estimate* of the population mean. Since sample measures are called *statistics*, we may describe estimation as the process of using statistics as estimators of parameters.

Two types of estimators are distinguished. A *point estimator* of a population parameter is a procedure that produces a *single value* as an estimate. The sample mean is a statistic that may be used as a point estimator of the population mean. An *interval estimator* of a population parameter is a procedure that produces a *range of values.* This range of values is useful as a measure of the degree of error that may exist in estimation. Point and interval estimators are discussed at length after the following section.

8.3 CRITERIA FOR POINT ESTIMATORS

In point estimation we seek the sample statistic that is the best estimator of the population parameter. Many criteria have been developed to describe what is "best" for a point estimator. The more general of these are the criteria of *unbiasedness* and *minimum variance*.

> A statistic is an *unbiased estimator* of a parameter if the expected value of the statistic equals the parameter, i.e.,
>
> $$E(\text{statistic}) = \text{parameter}.$$

Any statistic chosen as an estimator is a random variable since the value of that statistic may differ from sample to sample. The expected value of a random variable may be interpreted as the long-run average. Therefore the above definition indicates that a statistic is an unbiased estimator of a parameter if the average value of the statistic is the same as the parameter value. Thus, on the average, the estimator will be correct.

Unbiasedness alone does not guarantee a good estimator. In fact, some parameters may have more than one unbiased estimator. Selection among unbiased estimators is made on the basis of comparing the variances of the estimators.

> If there exist more than one unbiased estimators of a parameter, the estimator with the *minimum variance* is the more accurate.

Even though the average value of an unbiased estimator equals the parameter, an estimator may yield estimates that are not particularly close to the parameter value. The *efficiency* of an estimator is measured by the variance of the estimator. To illustrate, let $\hat{\theta}_1$ and $\hat{\theta}_2$ be two unbiased estimators of a parameter θ.[1] Figure 8.1 represents the distribution of these two estimators where the variance of $\hat{\theta}_2$ is smaller than the variance of $\hat{\theta}_1$. Consequently $\hat{\theta}_2$ is a more efficient estimator of θ. The *minimum-variance unbiased estimator* is the unbiased estimator with the smallest variance.

FIGURE 8.1:
DISTRIBUTION OF TWO
UNBIASED ESTIMATORS
THE PARAMETER θ

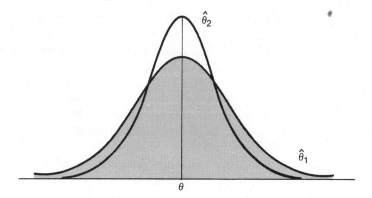

Statisticians often examine other properties of estimators in selecting the best estimator. From the discussion above, we can see that point estimators are not selected haphazardly; rather, they are selected to meet some well-defined criteria. For the point estimators examined below, the concept of a minimum-variance unbiased estimator is adequate. Determination of the statistic that provides a minimum-variance unbiased estimator of a parameter requires a knowledge of the sampling distribution of the statistic. The sampling distribution of the mean $\bar{X}$ and the proportion p are discussed in Chapter 6 and referenced below. The following sections present point estimators for the population mean μ, variance σ^2, standard deviation σ and binomial proportion π.

The estimation and testing procedures discussed in this chapter are generally based on large sample sizes, or what is often described as large sample statistical inference. Recall from Chapter 6 that the sampling distribution of $\bar{X}$ is normal when sampling from a normal population; and it is approximately normal, according to the central limit theorem, for moderately large ($n \geq 30$) samples from non-normal populations. The sampling distribution of the proportion p is approximately normal if the sample size is large ($n \geq 50$) and both $n\pi$ and $n(1 - \pi)$ are greater than five. Inference procedures based on other assumptions are given in later chapters.

Additionally, there is little mention of the finite population correction factor (Section 6.5) in this chapter. The more general applications of statistical inference involve either infinite population conditions or finite populations where the correction factor can be ignored, i.e., where the sample size n is less than 5 percent of the population size N. The procedures presented here apply to both finite and infinite populations. If these methods are used with finite populations where $n \geq 0.05N$, simply incorporate the finite population correction factor as a component of the standard error of the sampling distribution.

---

[1] The symbol "^" is called a "hat" by statisticians and is used to represent an estimator of a parameter. $\hat{\theta}$ is read "theta-hat."

8.4 POINT ESTIMATOR FOR μ

If the population mean μ is unknown and a random sample is taken for the purpose of estimating μ, what sample statistic provides the best estimator? Not only is the sample mean $\bar{X}$ a logical candidate, it is also the minimum-variance unbiased estimator. The unbiasedness criterion implies that

$$E(\bar{X}) = \mu.$$

Refer to Equation (6.1) and the associated example as an illustration of the fact that the expected value of the sample mean equals the population mean. This was demonstrated by taking all possible samples of size two from a population of five items. The findings are

$$\mu = \frac{\sum X}{n} = 16$$

and

$$E(\bar{X}) = \sum \bar{x} \cdot P(\bar{x}) = 16.$$

The variance of the sample means was given by

$$\sigma_{\bar{X}}^2 = \sum [\bar{x} - E(\bar{x})]^2 \cdot P(\bar{x}) = 3.$$

The sample mean $\bar{X}$ and sample median Md are both unbiased estimators of μ. But the variance of $\bar{X}$ is smaller than the variance of Md. Figure 8.1 is illustrative of these estimators of μ if we let $\theta = \mu$, $\hat{\theta}_1 = Md$ and $\hat{\theta}_2 = \bar{X}$. We state without proof or illustration that the sample mean $\bar{X}$ is the minimum-variance unbiased estimator of μ.

EXAMPLE. Quik-Kwik, a convenience food store, wishes to estimate the mean value of dollar sales per customer. Since the sample mean is the best estimator of the population mean, a random sample of 40 customers is taken and the sample mean is determined as

$$\bar{X} = \frac{\sum X}{n} = \frac{\$33.20}{40} = \$0.83.$$

The estimated mean sales per customer is $0.83.

The point estimator of μ is also useful for estimating the total of the raw scores (X) in the population. The mean of a population of N items is defined as

$$\mu = \frac{\sum X}{N}.$$

The total of the scores in the population is

$$\sum X = N\mu.$$

If the population mean is unknown, the point estimator $\bar{X}$ can be substituted for μ to obtain the *estimated population total of raw scores*

$$\text{Estimated } \sum X = N\bar{X}.$$

EXAMPLE. A credit manager wishes to estimate the total dollar value of the accounts receivable in the 31–90 day aging category. Currently there are 120 accounts in this category. The mean dollar value of a random sample of 30 accounts in this category is $42. The estimated dollar value of all accounts receivable in the 31–90 category is $N\bar{X} = (120)(\$42) = \$5,040$.

8.5 POINT ESTIMATOR FOR σ^2 AND σ

What sample statistic provides the best point estimator for the population variance σ^2? Although you might consider the sample variance S^2, the following illustration shows that the sample variance is not an unbiased estimator of σ^2.

EXAMPLE. Consider a population of three items with numerical values 2, 4, and 6. The population mean and variance are

$$\mu = \frac{\sum X}{n} = \frac{12}{3} = 4$$

and

$$\sigma^2 = \frac{\sum (X - \mu)^2}{N} = \frac{8}{3}.$$

To demonstrate that the sample variance is not an unbiased estimator, we find the sample variance for all possible samples of size two *with replacement* and show that $E(S^2) \neq \sigma^2$. Sampling with replacement is used to reflect an *infinite population* and allows us to disregard the finite population correction factor. Later, we also state the results for finite populations.[2]

The possible samples of size two with replacement are given in Column (1) below. The arithmetic mean and variance of each sample are in Columns (2) and (3). To illustrate these computations, the sample mean and variance for the

(1) Sample Values x	(2) Sample Means $\bar{x}$	(3) Sample Variances s^2	(4) Probability $P(s^2)$	(5) $s^2 \cdot P(s^2)$
2, 2	2	0	$\frac{1}{9}$	0
2, 4	3	1	$\frac{1}{9}$	$\frac{1}{9}$
2, 6	4	4	$\frac{1}{9}$	$\frac{4}{9}$
4, 2	3	1	$\frac{1}{9}$	$\frac{1}{9}$
4, 4	4	0	$\frac{1}{9}$	0
4, 6	5	1	$\frac{1}{9}$	$\frac{1}{9}$
6, 2	4	4	$\frac{1}{9}$	$\frac{4}{9}$
6, 4	5	1	$\frac{1}{9}$	$\frac{1}{9}$
6, 6	6	0	$\frac{1}{9}$	0
			1	$\frac{12}{9}$

---

[2] In Chapter 6 the sampling distribution of the mean was illustrated using the method of sampling without replacement.

fourth sample (values 4,2) are

$$\bar{X} = \frac{\sum X}{n} = \frac{6}{2} = 3$$

and

$$S^2 = \frac{\sum(X - \bar{X})^2}{n} = \frac{(4 - 3)^2 + (2 - 3)^2}{2} = \frac{2}{2} = 1.$$

Assuming random sampling, each of the samples in Column (1) has an equal chance of being selected; hence each sample variance in Column (3) has a $\frac{1}{9}$ probability of occurrence, as shown in Column (4). The expected value of the sample variance, from Column (5), is

$$E(S^2) = \sum s^2 \cdot P(s^2) = \frac{12}{9} = \frac{4}{3}.$$

Since the population variance is $\frac{8}{3}$, we know that

$$E(S^2) \neq \sigma^2$$

and conclude that the sample variance S^2 is not an unbiased estimator of the population variance σ^2. Note also that the average value of S^2, $\frac{4}{3}$, is smaller than σ^2, $\frac{8}{3}$. In general, the sample variance is too small as an estimator of the population variance.

The minimum-variance unbiased estimator of the population variance is, for samples from infinite populations,

$$\hat{\sigma}^2 = \frac{\sum(X - \bar{X})^2}{n - 1}. \tag{8.1}$$

This differs from the sample variance only in that a denominator of $n - 1$ is used in place of n. This correction to the computation of the sample variance provides an unbiased estimator, i.e.,

$$E(\hat{\sigma}^2) = \sigma^2.$$

If the population is normal, inferences about σ^2 are made on the basis of the chi-square distribution, a probability distribution discussed in Chapter 10.

An expression for the variance of the estimator $\hat{\sigma}^2$ will not be given here.

EXAMPLE. Use the population of three items above and the possible samples of size two with replacement to show that $\hat{\sigma}^2$ is an unbiased estimator of σ^2. The sample values are given in Column (1) and the values of $\hat{\sigma}^2$ are given in Column (2) on the following page.

(1) Sample Values x	(2) Unbiased Estimator $\hat{\sigma}^2$	(3) Probability $P(\hat{\sigma}^2)$	(4) $\hat{\sigma}^2 \cdot P(\hat{\sigma}^2)$
2, 2	0	$\frac{1}{9}$	0
2, 4	2	$\frac{1}{9}$	$\frac{2}{9}$
2, 6	8	$\frac{1}{9}$	$\frac{8}{9}$
4, 2	2	$\frac{1}{9}$	$\frac{2}{9}$
4, 4	0	$\frac{1}{9}$	0
4, 6	2	$\frac{1}{9}$	$\frac{2}{9}$
6, 2	4	$\frac{1}{9}$	$\frac{4}{9}$
6, 4	2	$\frac{1}{9}$	$\frac{2}{9}$
6, 6	0	$\frac{1}{9}$	0
		1	$\frac{24}{9}$

The expected value of $\hat{\sigma}^2$ is

$$E(\hat{\sigma}^2) = \sum \hat{\sigma}^2 \cdot P(\hat{\sigma}^2) = \frac{24}{9} = \frac{8}{3}.$$

Since $\sigma^2 = \frac{8}{3}$, this illustrates that $\hat{\sigma}^2$ is an unbiased estimator of σ^2. We state without proof or illustration that $\hat{\sigma}^2$ is also the minimum variance unbiased estimator.

Equation 8.1 is useful when the estimate of population variance is computed directly from sample observations. Alternatively, a minimum-variance unbiased estimate of population variance may be determined from the sample variance S^2 as follows:

$$\hat{\sigma}^2 = S^2 \cdot \frac{n}{n-1}. \tag{8.2}$$

The estimator of the population standard deviation is found from the square root of the previous equations:

$$\hat{\sigma} = \sqrt{\frac{\sum (X - \bar{X})^2}{n-1}} \tag{8.3}$$

and

$$\hat{\sigma} = S \sqrt{\frac{n}{n-1}}. \tag{8.4}$$

It should be noted that, while $\hat{\sigma}^2$ (Equation 8.1) is an unbiased estimator of σ^2, $\hat{\sigma}$ (Equation 8.3) is a biased estimator of σ. The amount of bias is generally reduced as sample size increases.

EXAMPLE. A telephone company wants to estimate the variance of the number of directory assistance calls made per month by its customers. A randomly chosen sample of monthly accounts for five customers reveals the following calls per customer for directory assistance: 9, 5, 10, 4, 12. The estimate of the population variance, using Equation (8.1), is

$$\hat{\sigma}^2 = \frac{\sum(X - \bar{X})^2}{n - 1} = \frac{46}{4} = 11.5.$$

Compute the sample variance S^2 and use Equation (8.2) to check the above result. How much is the estimated population standard deviation?

When sampling from a *finite population* the minimum-variance unbiased estimator of σ^2 is

$$\hat{\sigma}^2_f = \frac{\sum(X - \bar{X})^2}{n - 1} \cdot \frac{N - 1}{N}, \tag{8.5}$$

where the sample and population sizes are n and N, respectively. If the sample size is small relative to the population size, the difference in $\hat{\sigma}^2_f$ and $\hat{\sigma}^2$ is negligible. Consequently Equation (8.1) is used in place of Equation (8.5) if the sample size is less than five percent of the population size (i.e., $n < 0.05N$).

8.6 POINT ESTIMATOR FOR π

The minimum-variance unbiased estimator for the binomial proportion π is the sample proportion p. Reference to Equation 6.7 shows that $E(p) = \pi$. Section 6.8 also indicates that the sampling distribution of the proportion p approaches a normal distribution as sample size increases. The variance of the sampling distribution of p, from Equation (6.8), is

$$\sigma^2_p = \frac{\pi(1 - \pi)}{n}.$$

The following example illustrates the use of this point estimator for π.

EXAMPLE. A telephone company randomly sampled the opinions of 80 customers in one calling area to estimate the population proportion who favor a proposed expansion of services at a modest increase in the monthly rate. Thirty-six customers favored the proposal. The estimated population proportion, π, is

$$p = \frac{X}{n} = \frac{36}{80} = 0.45.$$

The sample proportion p may also be used to estimate the total number of items in the population that possess some attribute. Term the attribute a success and let X be the unknown number of successes in a population of size N. From the population proportion of successes

$$\pi = \frac{X}{N}$$

the number of successes is

$$X = N\pi.$$

If π is unknown, the point estimator of π may be substituted to provide the *estimated number of successes in the population:*

$$\text{Estimated } X = Np.$$

EXAMPLE. A manufacturer has an opportunity to buy a lot of 2,000 parts at a reduced price because some of the parts are defective. Neither the number of defectives or proportion defective in the population is known. The manufacturer can rework defective parts at a cost of $0.25 each. He believes the reduced lot price is worthwhile if the total cost for reworking defectives does not exceed $50. Before deciding, he examines a random sample of 100 parts and finds 8 defective. Should the lot be purchased?

Solution. Defining a defective as the success, the sample proportion defectives is $p = X/n = \frac{8}{100} = 0.08$. The estimated number of defective (successes) in the lot is $Np = (2000)(0.08) = 160$. The cost of reworking 160 defective parts at $0.25 each is $(160)(\$0.25) = \$40$. Since this is below $50, the lot should be purchased.

KARL FRIEDRICH GAUSS (1777–1855)

Gauss, the son of a gardener and bricklayer, was born in Brunswick, Germany. He contributed to mathematics, astronomy, and physics. Recognized as a genius in mathematics with few peers, his most lasting contribution was the development of the method of least squares (Chapter 12). He derived, and showed how to solve, the normal equations used to estimate the parameters of a linear function. His proofs of the method of least squares required a development of the probability distribution of errors. Today, because of the Gaussian law of errors, the normal probability distribution is often called the "Gaussian distribution." In fact, the development of the normal distribution was, for a time, erroneously attributed to Gauss rather than De Moivre (Chapter 6), the original (1733) developer.

8.7 INTERVAL ESTIMATION

Point estimators of population parameters, while useful, do not convey as much information as interval estimators. Point estimation produces a single value as an estimate of the unknown population parameter. The estimate may or may not equal the parameter value, i.e., the estimate may be incorrect. Unbiasedness guarantees only that the average value of the estimator determined from repeated samples will equal the parameter value. An interval estimate, on the other hand, is a range of values that conveys the fact that estimation is an uncertain process. The standard error of the point estimator is used in creating a range of values; thus a measure of variability is incorporated into interval estimation. Further, a measure of "confidence" in the interval estimator is provided; consequently, interval estimates are also called *confidence intervals*. For these reasons, interval estimators are considered more desirable than point estimators.

The concept of interval estimation is based on the sampling distribution of point estimators. For the parameters μ and π, the two cases discussed below, the sampling distribution of their point estimators, $\bar{X}$ and p, is approximately normal for moderately large sample sizes. In each case, the interval estimate is established by using the standard error of the sampling distribution to create an interval around the point estimate.

8.8 INTERVAL ESTIMATOR FOR μ

We have seen that the sampling distribution of $\bar{X}$ is approximately normal for moderately large samples from a non-normal population (exactly normal when sampling from a normal population) with mean μ and standard error $\sigma_{\bar{X}}$. The normal distribution of random variable $\bar{X}$ can be transformed to the standard normal distribution using Equation (6.5),

$$Z = \frac{\bar{X} - \mu}{\sigma_{\bar{X}}}.$$

From Chapter 5, we know that

$$P(-1.96 \leq Z \leq +1.96) = 0.95.$$

Since $\bar{X} \sim N(\mu, \sigma_{\bar{X}})$, we may write

$$P\left(-1.96 \leq \frac{\bar{X} - \mu}{\sigma_{\bar{X}}} \leq +1.96\right) = 0.95.$$

Further,

$$P(-1.96\sigma_{\bar{X}} \leq \bar{X} - \mu \leq +1.96\sigma_{\bar{X}}) = 0.95.$$

The above statement indicates that, with 0.95 probability, random variable $\bar{X}$ differs from the population mean μ by at most 1.96 standard errors. Since we are concerned only with the magnitude of difference in $\bar{X}$ and μ, the middle term may be written

$$P(-1.96\sigma_{\bar{X}} \leq \mu - \bar{X} \leq +1.96\sigma_{\bar{X}}) = 0.95.$$

Adding $\bar{X}$ to each term,

$$P(\bar{X} - 1.96\sigma_{\bar{X}} \leq \mu \leq \bar{X} + 1.96\sigma_{\bar{X}}) = 0.95.$$

This is a 95 percent interval estimator for μ. The end points of the interval may be written as

$$\bar{X} - 1.96\sigma_{\bar{X}}$$

and

$$\bar{X} + 1.96\sigma_{\bar{X}}.$$

Now $Z = 1.96$ reflects the middle 95 percent of the area in the standard normal distribution. This may be generalized for any proportion of area in the middle of the distribution by writing the *interval estimator of* μ as

$$\bar{X} \pm Z\sigma_{\bar{X}}. \tag{8.6}$$

Equation (8.6) indicates that the interval estimator of μ is determined by taking the point estimator $\bar{X}$ and creating a range of values by adding and subtracting the product of a factor, Z, and the standard error, $\sigma_{\bar{X}}$. The rationale and interpretation of this is given prior to an illustration.

Recall that in any normal distribution the proportion of the total area contained within the area 1.96 standard deviations on both sides of the mean is 0.95. Consequently, a normal distribution of sample means contains 95 percent of the possible $\bar{X}$-values within $\mu \pm 1.96\sigma_{\bar{x}}$, as illustrated in the upper portion of Figure 8.2. If a random sample of size n is selected and the sample mean $\bar{X}$ is within the limits $\mu \pm 1.96\sigma_{\bar{x}}$, then the interval computed from Equation (8.6) (letting $Z = 1.96$) will contain the population mean μ (see interval A in Figure 8.2). If the sample mean is outside the limits $\mu \pm 1.96\sigma_{\bar{x}}$, then the interval estimate will not contain the population mean within its range of values (such

FIGURE 8.2: 95 PERCENT CONFIDENCE INTERVALS

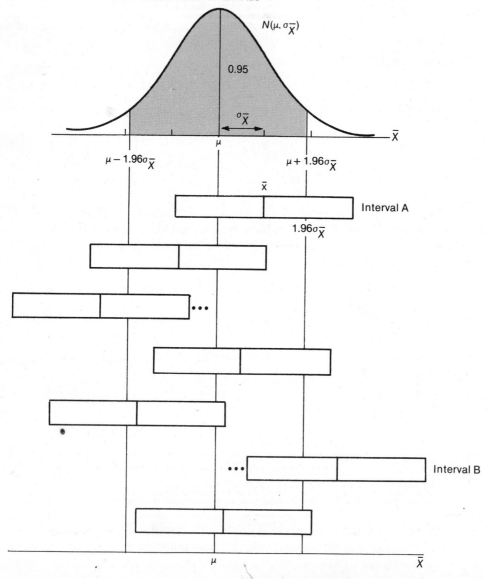

as interval B). Now imagine taking all possible samples of size n and computing interval estimates, $\bar{X} \pm 1.96\sigma_{\bar{X}}$, for each. Since 95 percent of the sample means are within the limits $\mu \pm 1.96\sigma_{\bar{X}}$, 95 percent of the resulting interval estimates would contain μ. The lower portion of Figure 8.2 illustrates several interval estimates that contain μ and two that do not.

Since the use of $Z = 1.96$ in Equation (8.6) is based on the middle 95 percent of the area in a normal distribution, the resulting interval estimate is referred to as a 95 *percent confidence interval*. This simply means that 95 percent of the intervals computed from all possible samples will contain the population mean. In practice a single sample is taken for the purpose of estimating the unknown parameter μ; the resulting interval may or may not include the population mean.

The *confidence level*, expressed as a proportion, is $(1 - \alpha)$. For 95 percent confidence, $1 - \alpha = 0.95$. Confidence levels other than 95 percent are also frequently used. A different confidence level requires a different value of Z. For a 90 percent confidence level, use $Z = 1.64$ in Equation 8.6. As justification, the middle 90 percent of the area in a normal distribution is within a distance of 1.64 standard deviations on either side of the mean. This is determined by locating in Appendix E the Z-value associated with one-half the confidence level. One-half of 90 percent is 0.45; in Appendix E one finds $Z = 1.64$ for an area of 0.45. Verify that $Z = 2.58$ for a 99 percent confidence interval. The following example illustrates the use of confidence intervals.

EXAMPLE. A grocery store manager wishes to estimate the mean sales ($) per customer for today's shoppers. A random sample of 25 customer receipts finds $\bar{X} = \$27.50$. If the population standard deviation is $8.00, determine a 95 percent interval estimate of μ. Assume sales are normally distributed.

Solution. The standard error of the mean is

$$\sigma_{\bar{X}} = \frac{\sigma}{\sqrt{n}} = \frac{\$8}{\sqrt{25}} = \$1.60.$$

This gives

$$\bar{X} \pm Z\sigma_{\bar{X}} = \$27.50 \pm 1.96(\$1.60) = \$27.50 \pm \$3.14.$$

The 95 percent confidence interval is $24.36 to $30.64.

As indicated, an interval estimate is established by using the standard error of the sampling distribution to create an interval around the point estimate.

> If the sampling distribution is normal, the interval estimate is the point estimate $\pm (Z)$(standard error).

The width of a confidence interval is twice the product (Z)(standard error). For the interval estimator of μ, the standard error $\sigma_{\bar{X}}$ is $\sigma/\sqrt{n}$. Therefore, the width of a confidence interval for μ is affected by the quantities Z, σ, and n. How do Z, σ, and n influence width? Increasing the confidence level $(1 - \alpha)$ increases the width of the interval, since Z increases as the confidence level increases. Verify this by computing a 98 percent confidence interval for the example above. The width of the confidence interval is also directly related to the variability of items in the population, as measured by σ. However, the investigator has no control over the value of σ. Finally, the width of the confidence interval is inversely related to the sample size n. Increasing sample size will reduce the width of the confidence interval if other factors remain constant. Sample size is discussed at length in a later section.

The above example assumes that the population standard deviation σ is known. This is rarely the case when estimating μ. If σ is unknown, the standard error of the mean must be estimated in one of the following ways. If a measure of standard deviation will be computed directly from the raw scores in the sample, the best estimator of the population standard deviation, $\hat{\sigma}$, Equation (8.3), should be used. The result is used in place of σ to determine the standard error,

$$\hat{\sigma}_{\bar{X}} = \frac{\hat{\sigma}}{\sqrt{n}}. \tag{8.7}$$

If the sample standard deviation S is computed, the standard error is

$$\hat{\sigma}_{\bar{X}} = \frac{S}{\sqrt{n-1}}. \tag{8.8}$$

Replacing $\sigma_{\bar{X}}$ by $\hat{\sigma}_{\bar{X}}$ does not change the procedure of determining confidence intervals for μ, provided that n is not less than 30.

EXAMPLE. A random sample of 101 customer accounts at a large brokerage firm is selected for the purpose of estimating the mean number of transactions per year for each customer. The sample mean is 43 and sample standard deviation is 12. Determine a 90 percent confidence interval for μ.

Solution. The standard error of the mean is

$$\hat{\sigma}_{\bar{X}} = \frac{S}{\sqrt{n-1}} = \frac{12}{\sqrt{100}} = 1.2.$$

The interval estimate is

$$43 \pm 1.64(1.2) = 43 \pm 1.97, \text{ or } 41.03 \text{ to } 44.97.$$

The result indicates that we have 90 percent confidence that the population mean μ is contained in the interval 41.03 to 44.97. Ninety percent confidence means that if all possible samples of size 101 were taken and the same type interval computed, $\bar{X} \pm (1.64)\hat{\sigma}_{\bar{X}}$, 90 percent of the resulting intervals would include the population mean; 10 percent would not. We do not know whether or not this one interval includes the population mean.

8.9 INTERVAL ESTIMATOR FOR π

The sampling distribution of the proportion p is approximately normally distributed with mean π and standard error σ_p if both the sample size is moderately large ($n \geq 50$) and $n\pi$ and $n(1 - \pi)$ are greater than five. The normal distribution of random variable p can be transformed to a standard normal variable by Equation (6.11):

$$Z = \frac{p - \pi}{\sigma_p}.$$

An interval estimator for π is determined in the same manner as an interval estimator for μ, i.e., an interval is created around the point estimator p by adding and subtracting a multiple of the standard error (σ_p):

$$p \pm Z\sigma_p. \tag{8.9}$$

One further consideration in determining a interval estimate of π concerns the standard error of the sample proportion, σ_p. This is defined as

$$\sigma_p = \sqrt{\frac{\pi(1 - \pi)}{n}}.$$

We wish to use Equation (8.9) to obtain an interval estimate for the unknown parameter π, yet σ_p is a function of π. Thus the standard error of p must be estimated by using the sample proportion p in place of π in the above formula. The estimated standard error is

$$\hat{\sigma}_p = \sqrt{\frac{p(1 - p)}{n}}. \tag{8.10}$$

Replacing the standard error σ_p in Equation (8.9) by its estimator $\hat{\sigma}_p$, the interval estimator of π is

$$p \pm Z\hat{\sigma}_p. \tag{8.11}$$

An interval estimate for π is also called a confidence interval, as in the case for the population mean.

EXAMPLE. In an attempt to control the quality of output for a manufactured part, a sample of parts is chosen randomly and examined in order to estimate the population proportion of parts that are defective. The manufacturing process operates continuously unless it must be stopped for inspection or adjustment. In the latest sample of 60 parts, 9 defectives are found. Determine:

a. A point estimate and

b. A 98 percent interval estimate of π, the population proportion of defective parts.

Solution. a. The point estimate is $p = X/n = \frac{9}{60} = 0.15$. b. For the interval estimate,

$$\hat{\sigma}_p = \sqrt{\frac{p(1 - p)}{n}} = \sqrt{\frac{0.15(0.85)}{60}} = 0.046,$$

and $Z = 2.33$. The confidence interval is

$$p \pm Z\hat{\sigma}_p = 0.15 \pm (2.33)(0.046)$$
$$= 0.15 \pm 0.107$$
$$= 0.043 \text{ to } 0.257.$$

A summary of point and interval estimators is given in Table 8.1.

TABLE 8.1: POINT AND
INTERVAL ESTIMATORS
OF POPULATION
PARAMETERS FOR
SAMPLES FROM INFINITE
POPULATIONS

Parameter	Point Estimator	Interval Estimator
μ	$\bar{X} = \dfrac{\sum X}{n}$	$\bar{X} \pm Z\sigma_{\bar{X}}$
π	$p = \dfrac{X}{n}$	$p \pm Z\hat{\sigma}_p$
σ^2	$\hat{\sigma}^2 = \dfrac{\sum(X - \bar{X})^2}{n - 1}$	*
σ	$\hat{\sigma} = \sqrt{\dfrac{\sum(X - \bar{X})^2}{n - 1}}$	*

* Interval estimation of σ^2 is given in Chapter 10.

8.10
DETERMINATION
OF SAMPLE SIZE

Little attention is given to considerations involved with selecting the sample size in the previous sections. However, collecting valid information through sampling requires careful planning, including determination of an appropriate sample size. Without careful planning, information obtained through sampling may be of little use to a manager. Suppose the manager of a retail store desires an interval estimate of average sales revenue per customer. An investigator reports that average retail sales is between $10 and $50 with 90 percent confidence. The manager replies, "Ten to fifty dollars—I could have guessed that! The interval must be smaller in order to be useful." Of course, the manager and investigator should have agreed on the criteria prior to taking a sample, but this is exactly the point. The manager has specific needs for information. In taking a sample for the purpose of gathering information, the sample design, including sample size, must incorporate these requirements. Procedures are given in this section for determining the sample size needed for estimators of the parameters μ and π that meet certain requirements.

In addition to insisting that estimators have desirable properties such as unbiasedness and minimum variance, it is also useful to specify certain numerical requirements. The numerical requirements placed on interval estimates are the *confidence level* and the *maximum error in estimation*. The confidence level, $1 - \alpha$, is discussed in the previous section.

The *maximum error in estimation* when using the sample statistic $\hat{\theta}$ to estimate the parameter θ is given by the absolute difference in the statistic and parameter, $|\hat{\theta} - \theta|$.

Suppose the retail store manager desires to estimate average sales revenue per customer (μ) with a maximum error in estimation of $5 at the 95.45 percent confidence level. This means that the manager wants 95.45 percent confidence that, when using $\bar{X}$ to estimate μ, the value of $\bar{X}$ will not be more than $5 larger or smaller than the value of μ. The following discussion explains how this is accomplished.

Consider the interval estimator of μ,

$$\bar{X} \pm Z\sigma_{\bar{X}},$$

which is based on the Z-transformation

$$Z = \frac{\bar{X} - \mu}{\sigma_{\bar{X}}}.$$

The standard error, for infinite populations, is

$$\sigma_{\bar{X}} = \frac{\sigma}{\sqrt{n}}.$$

Substituting in the above formula for $\sigma_{\bar{X}}$,

$$Z = \frac{\bar{X} - \mu}{\dfrac{\sigma}{\sqrt{n}}}.$$

Equation (8.12) shows that there are three quantities affecting n, the sample size. First, the value of Z reflects the confidence level, $(1 - \alpha)$. Specifically, this is determined by finding in Appendix E the value of Z associated with the $(1 - \alpha)/2$ proportion of area on either side of the mean. Recall that $Z = 1.96$ for a 95 percent confidence level. Second, the absolute value of the numerator, $|\bar{X} - \mu|$, represents the maximum error in estimation. Third, the sample size is influenced by the variability of items in the population, which is measured by the standard deviation σ. The sample size can be expressed as a function of these three quantities by solving Equation (8.12) for n.

The sample size required to estimate μ with maximum error of estimation $|\bar{X} - \mu|$ and confidence level $(1 - \alpha)$ in a population with standard deviation σ is

$$n = \frac{Z^2\sigma^2}{(\bar{X} - \mu)^2}. \qquad (8.13)$$

If $\sigma = \$15$ and the retail store manager wants an estimate of μ with a maximum error in estimation of $5 at 95.45 percent confidence, the required sample size is

$$n = \frac{(2.00)^2(15)^2}{(5)^2} = 36.$$

The investigator has control over two of the three quantities that influence sample size. The variability of items in a population is inherent; thus the investigator cannot exercise control over the value of σ. But the maximum error in estimation and the confidence level can be selected by the investigator. Suppose the retail store manager wants a smaller maximum error in estimation, deciding that \$2.50 is preferable to \$5. If the confidence level remains 0.9545, the required sample size is

$$n = \frac{(2.00)^2(15)^2}{(2.50)^2} = 144.$$

The reduction in estimation error from \$5 to \$2.50, half the original error, requires a sample size four times the original sample size.

> Reducing the maximum error in estimation of an interval estimate to $1/a$ of the original amount, while holding the confidence level constant, requires a sample size of a^2 times the original sample size.

If, for example, the manager wanted a maximum error in estimation of one-third the original error, while keeping the confidence level at 0.9545, the required sample size is nine times the original sample size. Note that reducing the maximum error in estimation of an interval estimate by a factor $1/a$ is the same as reducing the width of the confidence interval by a factor $1/a$.

The other quantity influencing sample size over which the investigator has some control is the confidence level. While sample size and estimation error are inversely related, the sample size and confidence level are directly related. An increase in the confidence level, while holding estimation error constant, requires a larger sample size, and vice versa. If an estimation error of \$5 is satisfactory, but a confidence level of 0.98 is desired by the store manager, the size must be

$$n = \frac{(2.33)^2(15)^2}{(5)^2} = 48.9, \text{ or } 49.$$

The above discussion indicates that, in order to determine an appropriate sample size, the investigator must have a measure of variability (σ) for the population and specify both the maximum error in estimation and the confidence level. If the population standard deviation is unknown, it must be estimated before using Equation (8.13) to determine a sample size that meets the desired conditions. One method of estimating σ is by taking a pilot sample. A *pilot sample* is any sample of rather arbitrary size obtained for the purpose of estimating the population standard deviation. Equation (8.3) is used to obtain $\hat{\sigma}$, an estimate of σ, from the sample observations. Then $\hat{\sigma}$ is used in place of σ in Equation (8.13) to determine a sample size that will provide an estimate of μ meeting the required conditions. The value of σ may also be estimated by using a measure of variability from a similar process. Also, the range may be estimated subjectively and σ approximated from the relation that the range is four to six times the standard deviation (see Section 2.16).

EXAMPLE. A chain of convenience food stores is considering opening a store in the Westhampton area of the city. They will estimate mean family income (μ) for a specific geographic area by taking a random sample of families. The estimate will be stated as a 90 percent confidence interval with a maximum error in estimation of $1,500. A pilot sample of 15 families finds $\hat{\sigma} = \$6,000$. The required sample size[3] is

$$n = \frac{(1.64)^2(6000)^2}{(1500)^2} = 43.03, \text{ or } 44.$$

The sample size required for interval estimates of π with stated maximum error in estimation and confidence level is determined in the same manner as above. The interval estimator of π is given earlier in Equation (8.10) as

$$\pi = p \pm Z\hat{\sigma}_p,$$

where

$$\hat{\sigma}_p = \sqrt{\frac{p(1-p)}{n}}$$

is the estimated standard error of p. Expressing the statistic p in terms of the standard normal distribution gives

$$Z = \frac{p - \pi}{\sqrt{\dfrac{p(1-p)}{n}}}.$$

Expressing the sample size as a function of the other factors gives

$$n = \frac{Z^2 p(1-p)}{(p - \pi)^2}. \tag{8.14}$$

To determine sample size using Equation (8.14), Appendix E is used to find Z for a $(1 - \alpha)$ confidence level; the absolute value $|p - \pi|$ is the maximum error in estimation; and variability is reflected by p, the proportion of outcomes in a pilot sample, or from prior knowledge, that possess a given attribute.

EXAMPLE. A state income tax auditor wishes to estimate the proportion of returns that contain arithmetic errors in the segment of the return where tax is computed on the basis of taxable income. If a pilot sample of 25 returns has two with arithmetic errors in this segment and an estimate of π with a maximum error in estimation of 0.04 at 99 percent confidence is desired, what size sample should be taken?

Solution. The value of p from the pilot sample is $p = \frac{2}{25} = 0.08$, and $Z = 2.58$ for 99 percent confidence. Sample size, using Equation (8.14), is

$$n = \frac{(2.58)^2(0.08)(0.92)}{(0.04)^2} = 306.2, \text{ or } 307.$$

---

[3] When this procedure produces a noninteger value, sample size is rounded up to the next larger integer.

When a sample size for estimating π is desired and a pilot sample or other method for approximating p in Equation (8.14) is not used, a maximum sample size, for given conditions, can be obtained by setting $p = 0.5$.[4] If $p = 0.5$ in the above example, $n = (2.58)^2(0.5)(0.5)/(0.04)^2 = 1040.06$; a much larger sample size than when $p = 0.08$. If an estimate of p is available, it should be used in place of $p = 0.5$.

TUMBLEWEED

By Tom K. Ryan

© 1970 United Feature Syndicate, Inc.

8.11 TESTING

The process of *testing* involves the investigation of preconceived ideas about the value of population parameters. For example, a chain of retail stores may operate under an assumption that the average (mean) sales revenue per customer is $30. Statistically, this may be substantiated by taking a sample of customers and comparing the sample results with the assumption. If the average sales per customer in the sample does not differ greatly from $30, such as $29 or $31, you might subjectively conclude that average sales of $30 is a reasonable assumption for the population of customers. Since the sample mean is a variable that can differ from the population mean, a small difference, such as $1, would not be surprising. On the other hand, you might be surprised if the average sales per customer in the sample were $38. Questions one might entertain include: Is it reasonable to obtain a sample mean as large as $38 if the population mean is $30? Is this sample result very likely? Is it possible that the population mean is really not $30? Testing is a procedure that attempts to answer questions of this type.

Formally, the preconceived idea about the value of a population parameter is called a *hypothesis*. In the above illustration, we hypothesize that μ, average sales per customer, is $30. Testing attempts to validate the hypothesis; consequently, testing procedures are also referred to as *hypothesis tests*. The hypothesis is tested by determining the likelihood of obtaining certain sample results

---

[4] If the confidence level and the maximum error in Equation (8.14) remain constant, sample size is a function of $p(1 - p)$. The maximum value of $p(1 - p)$, and therefore maximum sample size, occurs when $p = 0.5$, which is determined by setting the first derivative equal to zero and solving for p.

when sampling from a population with the hypothesized parameter value. If the sample results are highly likely (average sales of $29 or $31), then the sample supports the hypothesis. But if the sample results are unlikely, then the sample is in conflict with, or discredits, the hypothesis. The probability of the sample result is computed on the basis of the appropriate sampling distribution. In testing a hypothesis about μ, for example, we compare the sample result $\bar{X}$ with the hypothesized value of μ. The comparison is made using the sampling distribution of $\bar{X}$. The likelihood of certain sample results depends, obviously, on the inherent variability of the items in the population and, in turn, the variability in the sampling distribution. The rationale underlying hypothesis testing is given in the following discussion.

Assume that for the population of customers the distribution of random variable X, sales revenue per customer (in dollars), is normal with mean $\mu = 30$ and standard deviation $\sigma = 10$, i.e., $X \sim N(30, 10)$. If a random sample of 25 customer receipts is taken, the sampling distribution of $\bar{X}$ is normal with mean $E(\bar{X}) = 30$ and standard deviation $\sigma_{\bar{X}} = \sigma/\sqrt{n} = 10/\sqrt{25} = 2$. If the assumption of average sales per customer, $\mu = 30$, is correct, then a sample mean with a value of either 29 or 31 would not be unusual. These values of $\bar{X}$ lie a distance of $1 above and below the assumed population mean $\mu = 30$, i.e., $(\bar{X} - \mu) = +1$ for $\bar{X} = 31$ and $(\bar{X} - \mu) = -1$ for $\bar{X} = 29$ (see Figure 8.3). How unusual is a difference between $\bar{X}$ and μ of at least $1, i.e., what is

$$P[|\bar{X} - \mu| \geq 1 \,|\, N(30, 2)]? \tag{8.15}$$

This probability statement is read as "the probability of an absolute difference in $\bar{X}$ and μ of at least $1, given a sampling distribution of $\bar{X}$ with mean 30 and standard deviation 2." To determine this probability, express the value of $\bar{X}$ in terms of the standard normal distribution,

$$Z = \frac{\bar{X} - \mu}{\sigma_{\bar{X}}},$$

and utilize the table of areas in Appendix E. Referring to Figure 8.3, this is

$$Z = \frac{29 - 30}{2} = -0.5$$

and

$$Z = \frac{31 - 30}{2} = +0.5.$$

Then

$$P[|\bar{X} - \mu| \geq 1 \,|\, N(30, 2)] = P(Z \leq -0.5) + P(Z \geq +0.5) \tag{8.16}$$
$$= 0.30854 + 0.30854$$
$$= 0.61708.$$

The probability of a difference between $\bar{X}$ and μ of at least $1 is almost 0.62, i.e., a sample mean of 29 or less, or 31 or more, occurs about 62 percent of the time in random samples of 25 from a normal population where $\mu = 30$ and $\sigma = 10$.

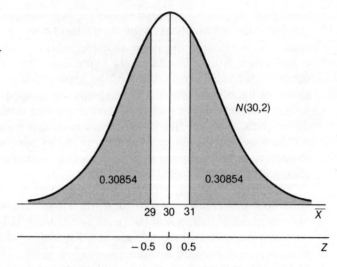

How likely to occur is a sample mean of either 22 or less, or 38 or more, if $\mu = 30$ and $\sigma_{\bar{x}} = 2$? This is a difference in $\bar{X}$ and μ of at least $8. As illustrated in Figure 8.4,

$$Z = \frac{38 - 30}{2} = +4$$

for $\bar{X} = 38$ and

$$Z = \frac{22 - 30}{2} = -4$$

for $Z = 22$. The probability is

$$P[|\bar{X} - \mu| \geq 8 \,|\, N(30, 2)] = P(Z \leq -4) + P(Z \geq +4)$$
$$= 0.00003 + 0.00003$$
$$= 0.00006.$$

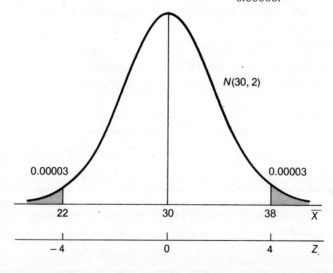

A sample result producing an absolute difference between $\bar{X}$ and μ of at least $8 is extremely unlikely when sampling 25 items from a population where $\mu = 30$ and $\sigma = 10$.

Now what do the two comparisons above imply about the belief that μ, average sales per customer, is $30? If a random sample of 25 items produced a sample mean of $31, we would not at all be surprised. The sample result $\bar{X} = 31$ lends support to a belief that μ equals 30. However, it is extremely unlikely that a sample mean of $38 would occur if μ equals 30. If, in fact, a random sample of 25 produced a sample result $\bar{X} = 38$, we would not be too comfortable with a belief that μ is 30. A sample mean of 38 is more likely to come from a population where the mean μ exceeds 30. Likewise, a sample mean of 22 is more likely to occur from a population where μ is less than 30. Such sample results do not support a belief that μ equals 30.

Now that the rationale underlying hypothesis testing has been presented, we formalize this procedure while introducing additional concepts.

8.12 FORMAL STEPS OF HYPOTHESIS TESTING FOR μ

The concepts outlined in the previous section can be formally expressed in a four-step procedure as follows. The case of testing for the parameter μ is used to introduce the formal procedure.

1. State the null and alternate hypotheses.
2. Specify the sampling distribution of the appropriate statistic, assuming the null hypothesis is true.
3. Compute a probability, called the p-value, based on the sample result.
4. Reach a conclusion regarding the validity of the null hypothesis and state what action, if any, is indicated.

The example in the previous section, investigating average sales revenue per customer in a chain of retail stores, is continued here for the purpose of illustrating these steps.

The *null hypothesis*, identified by the notation H_0, specifies the value of the population parameter that we wish to investigate. In the chain of retail stores, we wish to investigate the assumption that μ, average sales revenue per customer (in dollars), is 30. The *alternate hypothesis*, identified as H_1, indicates what we may conclude about the parameter value if the sample result does not support the assumption. For the retail stores, we would feel that μ has some value other than 30, which is stated as $\mu \neq 30$. This implies that either $\mu < 30$ or $\mu > 30$. Formally, the null and alternate hypotheses are stated as

$$H_0: \quad \mu = 30;$$
$$H_1: \quad \mu \neq 30.$$

To test a hypothesis about μ, a sample of n items is chosen from the population, and the sample mean $\bar{X}$ is compared with the assumed (hypothesized) value of μ. The sampling distribution of the statistic (Step 2) is specified with an assumption that the null hypothesis is correct. Thus the sampling distribution of $\bar{X}$ is normal with mean $E(\bar{X}) = 30$ and standard error $\sigma_{\bar{x}} = \sigma/\sqrt{n} = 10/\sqrt{25} = 2$ (as given in the previous section).

Once a random sample is obtained from the population, a numerical value of the statistic is computed. As described earlier, suppose the sample result $\bar{X} = 29$ occurs in a sample of 25 customer receipts. Using the appropriate sampling distribution, compute the following probability, called a p-value (Step 3):

$$P(\text{a sample result at least as extreme as that observed} | H_0 \text{ true}). \quad (8.17)$$

The p-value represents the probability of a sample result at least as extreme as that observed, given that the null hypothesis is true. How extreme is a sample result such as $\bar{X} = 29$ if the null hypothesis ($\mu = 30$) is true? The term "extreme" is used to describe the probability of a sample result at least as far away from the hypothesized parameter value. The sample mean of 29 lies a distance of 1 (dollar) away from $\mu = 30$. Smaller values of $\bar{X}$ would be even more extreme. Hence Equation (8.15) includes $P(\bar{X} - \mu) < 1$. But according to the alternate hypothesis μ might also be larger than 30, if the null hypothesis is incorrect. Since values of μ above and below 30 are pertinent to the investigation, the extremity of the sample result described in Equation (8.17) can be expressed as the probability of an absolute difference in $\bar{X}$ and μ of at least 1. Thus the general probability statement in Equation (8.17) is determined for the average sales revenue per customer situation by Equation (8.15), which may be written as

$$P(|\bar{X} - \mu| \geq 1 | H_0 \text{ true}) = 0.61708. \quad (8.18)$$

The computed probability or p-value, determined in Equation (8.16), is 0.61708.

In the final step a decision must be reached regarding the validity of the null hypothesis. If the p-value indicates that the sample result is highly likely, then the sample result supports the null hypothesis. This is indicated by stating that the null hypothesis is *accepted*. If the sample result is too unlikely, the sample result is in conflict with (does not support) the null hypothesis. In this case the null hypothesis is *rejected*. Acceptance or rejection of the null hypothesis depends on whether the sample result is likely or unlikely. The obvious question is, "How small must the p-value be in order for the sample result to be considered unlikely?" The cutoff value used to indicate whether a sample result is either likely or unlikely is termed the *level of significance*, which is designated by α (Greek letter alpha).

> The *level of significance*, α, for a statistical test is the minimum probability of sample results that lend support to the null hypothesis.

Considerations in choosing the level of significance are discussed in the following section. For the present discussion, assume $\alpha = 0.05$. This indicates that the null hypothesis should be rejected if the p-value is less than 0.05, otherwise the null hypothesis is accepted. Since the p-value in the current illustration is 0.61708, from Equation (8.18), the null hypothesis is accepted. This indicates that the sample result $\bar{X} = 29$ is not so unusual in this sampling distribution to cause us to disbelieve the claim that average sales revenue per customer (μ) is $30.

In order to verify your understanding of this process, review the following summary of the formal steps of hypothesis testing and the current illustration.

Step 1.   Null and alternate hypotheses:

$$H_0: \quad \mu = 30;$$
$$H_1: \quad \mu \neq 30.$$

Step 2.   Sampling distribution:

$$N(30, 2).$$

Step 3.   The p-value for $\bar{X} = 29$ (see Figure 8.3)

$$P(|\bar{X} - \mu| \geq 1 | H_0 \text{ true}) = 0.61708.$$

Step 4.   Conclusion:

For $\alpha = 0.05$, accept H_0.

Hypothesis testing, as well as other statistical procedures, is a means to an end rather than an end result. Hypothesis testing is a way of investigating a claim about some problem of concern. The original problem is the focus of the decision making effort. A test of hypothesis provides information, the p-value, and a statistical decision, accept or reject H_0. The statistical information is then used in reaching a management decision about the original problem. A management decision may require some action to be taken. Since the statistical conclusion reached in the above illustration is to accept H_0, management may feel that no further investigation of customer sales is needed in order to justify their assumption that average sales revenue per customer is $30. If, on the other hand, a statistical conclusion of reject H_0 occurred, management would probably further investigate customer sales data. Other examples of managerial action based on statistical conclusions include readjustment or no readjustment of a manufacturing process; increase advertising expenditures versus maintaining current level; introduce a new product versus do not introduce; grant credit versus do not grant credit; invest in a business proposition versus do not invest.

EXAMPLE.   Suppose the result $\bar{X} = 22$ occurred in the random sample of 25 customer receipts. Should the null hypothesis $\mu = 30$ be accepted if $\alpha = 0.05$ and $\sigma = 10$ (as before)?

Solution.   The hypothesis test is:

$$H_0: \quad \mu = 30;$$
$$H_1: \quad \mu \neq 30.$$

Sampling distribution is $N(30, 2)$.
The p-value (see Figure 8.4) is

$$P(|\bar{X} - \mu| \geq 8 | H_0 \text{ true}) = 0.00006.$$

Conclusion: For $\alpha = 0.05$, reject H_0.

If the statistical decision in the above example is to reject H_0, what action does this suggest for the management of the retail stores?

EXAMPLE. Beef patties for a chain of fast-food restaurants are prepared and packaged at a processing plant and shipped frozen to the individual restaurants. The best-selling item features a four-ounce beef patty. Because of the preparation process some patties weigh slightly more or less than four ounces. In order to maintain the quantity standard, a sample of prepared patties is taken periodically and weighed to make sure the weight does not deviate too much from four ounces. From past experience, it is known that the weight of individual patties is normally distributed with mean $\mu = 4$ ounces and standard deviation $\sigma = 0.5$ ounces. If a random sample of 25 patties has a mean weight of 4.235 ounces, should management have any reason to feel that patties are not meeting weight specifications? Use a level of significance $\alpha = 0.05$. Based on the statistical decision what action, if any, should management take?

Solution. The hypotheses are:

$$H_0: \quad \mu = 4;$$
$$H_1: \quad \mu \neq 4.$$

The sampling distribution of $\bar{X}$ is normal with mean 4 and standard error $\sigma/\sqrt{n} = 0.5/\sqrt{25} = 0.1$; i.e., $N(4, 0.1)$. The sample result $\bar{X} = 4.235$ lies

$$Z = \frac{\bar{X} - \mu}{\sigma_{\bar{X}}} = \frac{4.235 - 4}{0.1} = 2.35$$

standard errors away from the hypothesized $\mu = 4$. The appropriate p-value is

$$P(|\bar{X} - \mu| \geq 0.235 \,|\, H_0 \text{ true}) = P(|Z| \geq 2.35) = 0.01878,$$

as shown in Figure 8.5. Since the p-value is less than the level of significance, the null hypothesis is rejected. This indicates that an investigation of the preparation

FIGURE 8.5:
PROBABILITY OF AN
ABSOLUTE DIFFERENCE
IN $\bar{X}$ AND μ OF AT LEAST
0.235 FOR $N(4, 0.1)$

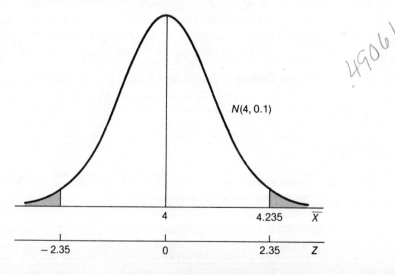

process should be undertaken in order to insure that the mean weight of all patties is four ounces.

8.13 DECISION RULES

A statistical *decision rule* for a hypothesis test is a statement that indicates acceptance or rejection of the null hypothesis for each possible sample result. The decision rule is based on the test procedure used. In the previous example, the null hypothesis was rejected because the p-value, 0.01878, was smaller than the level of significance, $\alpha = 0.05$. A formal decision rule may be stated for comparing computed p-values to the level of significance.

Decision rule for a p-value: Reject H_0 if the p-value $< \alpha$; otherwise accept H_0.

Decision rules may also be stated for measures other than p-values. A decision rule for values of the sample mean is established as follows. In a test about the parameter μ, the level of significance represents the proportion of the total area in the sampling distribution where extreme values of $\bar{X}$ are found. In a test such as those illustrated previously, extreme values of $\bar{X}$ are values of the sample mean that are either too small or too large to support the null hypothesis; i.e., they are unlikely to occur, given that H_0 is true. One-half the value of α is the proportion of extreme sample means in each tail of the sampling distribution, as shown in Figure 8.6. To determine the value of $\bar{X}$ at these points, solve Equation (6.5), $Z = (\bar{X} - \mu)/\sigma_{\bar{X}}$, for $\bar{X}$. The values of $\bar{X}$ are

$$\mu \pm Z\sigma_{\bar{X}}. \tag{8.19}$$

The hypothesized value is used for μ. Values of Z are determined from Appendix E, where $\alpha/2$ is the proportion of area in each tail of the distribution. For $\alpha = 0.05$, $Z = 1.96$. Substituting the appropriate measures for the beef patty example above, the values of $\bar{X}$ are

$$4 \pm (1.96)(0.1) = 4 \pm 0.196$$
$$= 3.804, 4.196.$$

The values of $\bar{X}$ computed from Equation (8.19) are called *decision limits* for $\bar{X}$. In a decision rule for the sample mean, an observed sample mean is compared with the decision limits to determine acceptance or rejection of the null hypothesis.

Decision rule for $\bar{X}$: If an observed sample mean is more extreme than the decision limit(s) of $\bar{X}$, reject H_0; otherwise accept H_0.

In the beef patty example, the observed sample mean, $\bar{X} = 4.235$, is more extreme than the decision limits of $\bar{X}$, 3.804 to 4.196; therefore, H_0 is rejected (see Figure 8.6).

A third decision rule is also used. A decision rule for the standard normal variable Z is determined from the level of significance. In the previous paragraph, $Z = \pm 1.96$ for $\alpha = 0.05$. These are called *critical values* of Z. A computed

FIGURE 8.6: DECISION
LIMITS OF $\bar{X}$ AND
CRITICAL VALUES OF
Z FOR $\alpha = 0.05$

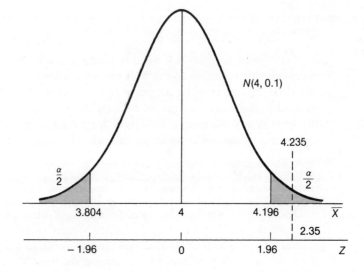

value of Z can be compared with $Z = \pm 1.96$ for a decision about the null hypothesis.

> *Decision rule for Z:* If the computed value of Z is more extreme than the critical value(s) of Z, reject H_0; otherwise accept H_0.

Since the computed Z in the beef patty example is $Z = (4.235 - 4)/0.1 = 2.35$, the null hypothesis is rejected (see Figure 8.6).

All three decision rules are equivalent statements for reaching a decision about the null hypothesis; i.e., they produce the same decision. The choice of a decision rule is largely a matter of personal preference on the part of the investigator. However, stating the results of a test in terms of the p-value provides more flexibility to users of test results. When a p-value is known, each individual may easily reach his or her own conclusion by comparing the p-value with whatever α seems appropriate. For other decision rules, different values of α require additional analysis before reaching a conclusion. Most computer routines for hypothesis testing provide p-values and, usually, the computed Z-value or a similar measure for other theoretical distributions. Since familiarity with all three decision rules is desirable, illustrations of each are given in various sections of this book.

8.14 TYPE I AND TYPE II ERRORS

Decisions made in hypothesis testing may be erroneous. This arises because testing is an uncertain process—using limited (sample) information to reach conclusions about a population parameter. Statisticians seek to design tests that minimize the chances of erroneous decision. It is important to understand the types of errors and how they arise in order fully to appreciate hypothesis testing as an aid to managerial decision making.

TABLE 8.2: CORRECT
AND INCORRECT
DECISIONS IN
HYPOTHESIS TESTING

	Reality: The Null Hypothesis Is	
Statistical Decision for the Null Hypothesis	*True*	*False*
Accept	Correct decision	Incorrect decision (Type II error)
Reject	Incorrect decision (Type I error)	Correct decision

Table 8.2 outlines the possible correct and incorrect decisions in hypothesis testing. The null hypothesis being tested may be either true or false. Statistically, the null hypothesis will be either accepted or rejected. If the null hypothesis is true, the test procedure should indicate that the null hypothesis is accepted a large proportion of the time. Rejection of a true null hypothesis is termed a *Type I error*. The probability of making a Type I error is α, the level of significance. Probabilities of accepting and rejecting a true null hypothesis can be expressed as $(1 - \alpha)$ and α, respectively.

> Type I error: Rejecting a true null hypothesis.
> $P(\text{Type I error}) = P(\text{Rejecting } H_0 \mid H_0 \text{ true}) = \alpha$.
> $P(\text{accepting } H_0 \mid H_0 \text{ true}) = 1 - \alpha$.

Recall the test for average retail sales per customer, H_0: $\mu = 30$ versus H_1: $\mu \neq 30$, in Section 8.12 ($\sigma = 10$, $n = 25$, and $\alpha = 0.05$). The null hypothesis was rejected for a sample result $\bar{X} = 22$ because the p-value of 0.00006 was less than α. Has a Type I error been made? We cannot tell because it is unknown whether the null hypothesis is true, i.e., whether μ equals 30. The p-value indicates that an absolute difference in $(\bar{X} - \mu)$ of at least 8 is extremely unlikely in a sampling distribution defined as $N(30, 2)$. But we do not *know* with certainty, nor have we *proved*, that μ does not equal 30. The chance of making a Type I error in testing based on samples of size $n = 25$ from this population is 0.05, the value of α. The statistical decision to reject H_0 is either correct (if H_0 is really false) or incorrect (if H_0 is really true). The only way to determine the true value of μ is to measure all items in the population.

In the first illustration of a test for average retail sales per customer (Section 8.12) the observed sample mean was $\bar{X} = 29$ (all other information was the same as above). The null hypothesis H_0: $\mu = 30$ was accepted in this case since the p-value, 0.61708, is larger than $\alpha = 0.05$. If the null hypothesis is actually true, then a correct decision has been made. However, it is possible that a sample mean of $\bar{X} = 29$ came from a population where $\mu \neq 30$. If μ is unequal to 30, then a *Type II error*, accepting a false null hypothesis, has been made. The probability of a Type II error is designated β (Greek letter beta). When the null hypothesis is false, the probabilities of the two possible decisions, accept and reject H_0, are respectively β and $(1 - \beta)$.

> Type II error: Accepting a false null hypothesis.
> $P(\text{Type II error}) = P(\text{Accepting } H_0 | H_0 \text{ false}) = \beta$.
> $P(\text{rejecting } H_0 | H_0 \text{ false}) = 1 - \beta$.

In a test of hypothesis there is a single value of α; and the p-value is computed on the basis of an assumption that the null hypothesis is true. In order for a Type II error to occur the null hypothesis must be false. If the hypothesized value of μ is not 30 (in the average retail sales illustration), then what is the true value of μ? This is unknown at the time a test is conducted. But if μ is not 30, the true value of μ could be either larger or smaller than 30. The value of β depends on the true value of the parameter being tested. Consequently, there is not a single value of β. In designing a test, values for β and $(1 - \beta)$ are computed for a given α and several possible values of the parameter being tested.

Let us use the retail sales per customer situation to illustrate the computation of β for different values of μ. Thus we are testing H_0: $\mu = 30$ versus H_1: $\mu \neq 30$; $\sigma = 10$ and $n = 25$ give a sampling distribution that is $N(30, 2)$. Using the level of significance $\alpha = 0.05$, first determine decision limits for the sample mean. The decision limits for the sample mean, determined from Equation (8.19), are

$$\mu \pm Z\sigma_{\bar{X}} = 30 \pm (1.96)(2)$$
$$= 26.08, 33.92.$$

These decision limits for $\bar{X}$ indicate that the null hypothesis is accepted if the mean of a random sample of 25 items is within these values, i.e., accept H_0 if $26.08 \leq \bar{X} \leq 33.92$. Otherwise the null hypothesis is rejected. The decision limits for $\bar{X}$ are shown in Figure 8.7. Suppose now the true population mean is 32 rather than 30. How likely is it that the null hypothesis H_0: $\mu = 30$ would be accepted if the sample actually comes from a population where $\mu = 32$? If μ is actually 32, then the sampling distribution is $N(32, 2)$ as shown in the center of Figure 8.7 (we assume $\sigma = 10$ for both populations; consequently $\sigma_{\bar{X}} = 2$ in both sampling distributions). What is the probability of obtaining a sample mean between 26.08 and 33.92 in the sampling distribution that is $N(32, 2)$? This may be written as

$$P[26.08 \leq \bar{X} \leq 33.92 \,|\, N(32, 2)]$$

and is obtained in the same way the area between two limits in any normal distribution is evaluated. Computing Z-values for the decision limits of $\bar{X}$ where

$$Z = \frac{\bar{X} - \mu}{\sigma_{\bar{X}}},$$

$$Z = \frac{26.08 - 32}{2} = -2.96$$

and

$$Z = \frac{33.92 - 32}{2} = +0.96.$$

FIGURE 8.7:
DETERMINING THE
PROBABILITY OF A
TYPE II ERROR

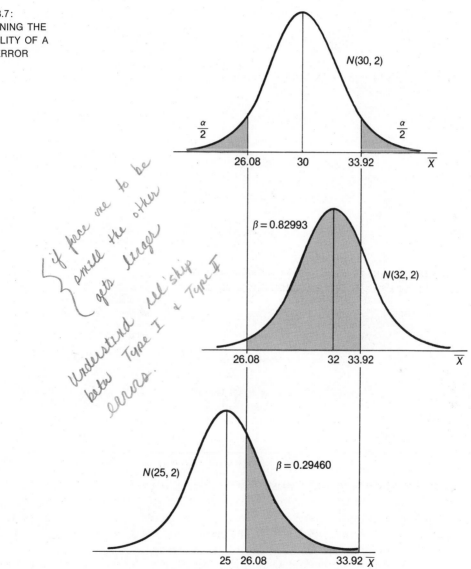

Using Appendix E, the probability is

$$P[26.08 \leq \bar{X} \leq 33.92 \,|\, N(32,2)] = P(-2.96 \leq Z \leq 0) + P(0 \leq Z \leq +0.96)$$
$$= 0.49846 + 0.33147$$
$$= 0.82993.$$

Thus there is a 0.82993 probability of accepting the null hypothesis H_0: $\mu = 30$ if the true μ is 32. Accepting H_0: $\mu = 30$ if μ is really 32 is a Type II error. This is summarized as

$$P(\text{Type II error}|\mu = 32) = 0.82993.$$

Obviously the probability of a Type II error depends on the true value of μ. Since μ could have many possible values, if $\mu \neq 30$, it is necessary to examine the behavior of β for several different values of μ. The curve at the lower portion of Figure 8.7 represents a sampling distribution that is $N(25, 2)$. The value of β here is

$$P(\text{Type II error}|\mu = 25) = P[26.08 \leq \bar{X} \leq 33.92 | N(25, 2)]$$
$$= 0.29460.$$

TABLE 8.3: VALUES OF β AND $(1 - \beta)$ FOR SELECTED VALUES OF μ

μ	β	$1 - \beta$
22	0.02068	0.97932
24	0.14917	0.85083
26	0.48401	0.51599
28	0.82993	0.17007
(30)*	(0.95000)	(0.05000)
32	0.82993	0.17007
34	0.48401	0.51599
36	0.14917	0.85083
38	0.02068	0.97932

* A type II error cannot occur when $\mu = 30$. The values in parentheses indicate that, if $\mu = 30$, there is a $(1 - \alpha) = 0.95$ probability of correctly accepting the null hypothesis and an $\alpha = 0.05$ probability of incorrectly rejecting the null hypothesis.

The probability of a Type II error is summarized for several values of μ in Table 8.3. Note that β gets smaller as the distance between the true μ and the hypothesized μ increases. Values of $(1 - \beta)$ are given in the third column of Table 8.3. Recall that

$$P(\text{rejecting } H_0 | H_0 \text{ false}) = 1 - \beta.$$

The value of $(1 - \beta)$ is called the *power of a test*. Power represents the probability of correctly rejecting the null hypothesis. A curve of the values of $(1 - \beta)$, as given in Figure 8.8, is called a *power function*. The values of β and $(1 - \beta)$ for different values of μ indicate that the ability of the test procedure to discriminate between a true and false null hypothesis increases as the distance between the true μ and hypothesized μ increases. The power function has a value equal to the level of significance α at the hypothesized $\mu = 30$. (In Table 8.3 the values in parentheses at $\mu = 30$ indicate that 0.95 is the probability, $(1 - \alpha)$, of correctly accepting null hypothesis and 0.05 is the probability, α, of incorrectly rejecting the null hypothesis.) The power function approaches a value of one as the distance between the hypothesized μ and true μ increases. Statisticians examine power functions for different decision rules in order to select the more powerful test.

FIGURE 8.8: POWER CURVE FOR THE TEST OF $H_0: \mu = 30$

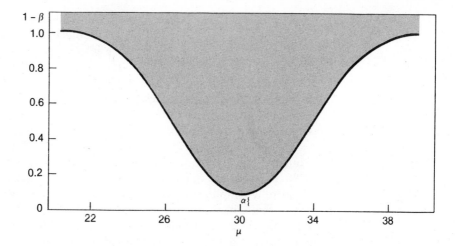

Ideally, α and β should be as small as possible. However, *for a fixed sample size, any reduction in one occurs at the expense of an increase in the other.* In practice, a value of α such as 0.01, 0.02, 0.05, or 0.10 is usually selected and the corresponding power function examined. Some knowledge of the consequences or costs of a wrong decision can help in selecting appropriate values of α and β that, in turn, can be used to determine a sample size meeting these specifications. Such procedures are beyond the scope of the present discussion.

Some final comments about reporting the findings of a statistical test are in order. We have mentioned earlier that by comparison of a p-value with α, a decision to accept or reject H_0 can be made. Investigators find it helpful to state the p-value along with the statistical decision. This allows different investigators to examine the results and use whatever level of significance they might consider appropriate. Stating only that H_0 is rejected at the 0.05 level of significance does not provide this added flexibility.

Finally, it is important to keep in mind the implication of "accepting H_0" and "rejecting H_0." If the sample result is highly unlikely (p-value is less than α), H_0 is rejected. When H_0 is rejected, we say the sample result is significantly different from the parameter value specified by the null hypothesis. This indicates that the sample result is more supportive of the alternate hypothesis than the null. When H_0 is accepted, the sample result is not significantly different from the parameter value specified by the null hypothesis. This does not say that H_0 is true; a sample result cannot "prove" anything about a population. *A statement of "accept H_0" simply means that the sample result is insufficient evidence to reject H_0.* Because of this, some statisticians prefer the statement "fail to reject H_0" rather than "accept H_0." This philosophical argument is based largely on semantics, however; we prefer to state simply that H_0 is either "accepted" or "rejected" and imply the above interpretations.

8.15 ONE-TAILED VERSUS TWO-TAILED TESTS

The central concept in hypothesis testing is evaluating the probability of obtaining sample results at least as extreme as observed, when sampling from a population with the parameter value specified by the null hypothesis (Equation

8.14). This is what we have termed the p-value. In specifying the alternate hypothesis and computing the p-value, we must determine exactly what sample results should be considered extreme in relation to the problem at hand. If the investigator considers that both extremely small and large values of the sample result do not support the null hypothesis, a *two-tailed test* is appropriate. If only extremely small values of the sample result or only extremely large values, but not both, do not support the null hypothesis, then a *one-tailed test* is appropriate. One- and two-tailed tests are contrasted in this section.

The tests of average retail sales per customer illustrated in Section 8.11 are two-tailed tests. Management assumes that average retail sales per customer is \$30 (μ) and uses a null hypothesis H_0: $\mu = 30$. A check on this assumption is obtained by sampling customer receipts. To determine whether a test should be one- or two-tailed, we must answer the following. If the null hypothesis is not true, what values of μ would the management of the retail store chain consider important? Would they be interested only in the possibility that $\mu > 30$, or would they also want to consider that possibly $\mu < 30$? If μ is not 30, it is important for management to recognize both directions away from the value 30, i.e., $\mu < 30$ and $\mu > 30$. The alternate hypothesis incorporates both of these inequalities into the single statement H_1: $\mu \neq 30$. When it is important to recognize that, if the null hypothesis is incorrect, the parameter being tested can have both larger and smaller values than hypothesized, a two-tailed test is appropriate.

The form of the p-value for two-tailed tests is illustrated by Equation (8.18). What sample results, values of $\bar{X}$, would not support H_0 if we consider that μ may either be larger or smaller than 30? Either an extremely large or extremely small value of $\bar{X}$. Therefore a p-value based on the absolute difference $|\bar{X} - \mu|$ is used. The computed p-value for the observed sample result $\bar{X} = 29$, as given in Equation (8.18), is

$$P(|\bar{X} - \mu| \geq 1 \,|\, H_0 \text{ true}) = 0.61708.$$

This type of p-value evaluates some area in both tails of the sampling distribution, which reflects the term *two-tailed test* (see Figure 8.9).

A *two-tailed test* of hypothesis is conducted when it is important to consider both directions of difference away from the parameter value specified in the null hypothesis. The p-value for a *two-tailed test* evaluates the probability of an *absolute difference* between the sample result and hypothesized parameter value at least as large as observed.

To perform a two-tailed test that the parameter μ equals some amount a, the null and alternate hypotheses are of the form

$$H_0: \quad \mu = a;$$
$$H_1: \quad \mu \neq a.$$

The p-value evaluates the probability that the *absolute difference* in $\bar{X}$ and μ is at least as extreme as that observed, given that H_0 is true. Letting d represent

FIGURE 8.9:
$P(|\bar{X} - \mu| \geq 1 | H_0 \text{ true})$

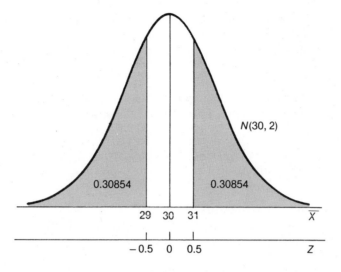

the observed difference in $(\bar{X} - \mu)$, this is

$$P(|\bar{X} - \mu| \geq d | H_0 \text{ true}). \tag{8.20}$$

In some situations an investigator may not be concerned with both possible directions of difference away from a parameter value. A *one-tailed test* is appropriate for these cases.

A *one-tailed test* of hypothesis is conducted when it is important to consider only a single direction of difference away from the parameter specified in the null hypothesis. The p-value for a *one-tailed test* evaluates the probability of a *directional difference* between the sample result and hypothesized parameter value at least as extreme as observed.

Consider a manufacturer who purchases steel cable for industrial use that requires a mean breaking strength (μ) of at least 200 lb. As long as this specification is met, the buyer is satisfied with the product. The purchaser periodically examines a random sample of lengths of the cable for breaking strength in an attempt to insure that the cable meets specifications. What values of the sample result $\bar{X}$ would cause the purchaser to suspect that the specification might not be met? Values of $\bar{X} > 200$, such as 210, would not imply this. Only values of $\bar{X} < 200$ would give the puchaser cause to suspect that this sample came from a population that does not have a mean breaking strength of at least 200 lb. Thus the purchaser is only concerned that μ might be less than 200, i.e., he is concerned about only one direction of difference away from the value of $\mu = 200$. Consequently a one-tailed test is appropriate. Assume the supplier specifies that the breaking strength of the steel cable is normally distributed with a standard deviation of 21 lb. If the purchaser finds a mean breaking strength of 194.6 lb. in a random sample of 49 lengths, should he conclude at the 0.05 level of significance that specifications are not met?

The null hypothesis for any test must specify a parameter value that we can use as a standard to compare against sample results. The value of 200 lb. is the standard here, which produces a null hypothesis

$$H_0: \quad \mu = 200.$$

The steel cable fails to meet specificiations only if $\mu < 200$. Thus the alternate hypothesis states a *directional difference* away from the standard of 200 lb. This is

$$H_1: \quad \mu < 200.$$

The sampling distribution is normal with a mean

$$E(\bar{X}) = 200$$

and standard error

$$\sigma_{\bar{X}} = \frac{\sigma}{\sqrt{n}} = \frac{21}{\sqrt{49}} = 3.$$

The observed difference in the sample result and hypothesized value is

$$(\bar{X} - \mu) = (194.6 - 200) = -5.4.$$

Notice that the difference is directional. The p-value in a one-tailed test determines the probability of *directional difference in* $(\bar{X} - \mu)$ at least as extreme as observed. Values of $\bar{X}$ smaller than that observed would be considered more extreme; these are differences of $(\bar{X} - \mu) < -5.4$. Thus the p-value is

$$P[(\bar{X} - \mu) \leq -5.4 | H_0 \text{ true}].$$

Using the Z-transformation the observed difference in $(\bar{X} - \mu)$ is

$$Z = \frac{\bar{X} - \mu}{\sigma_{\bar{X}}} = \frac{194.6 - 200}{3}$$

$$= \frac{-5.4}{3} = -1.80$$

standard errors, as shown in Figure 8.10. In the standard normal distribution, the p-value is (from Appendix E)

$$P(Z \leq -1.80) = 0.03593.$$

Since the p-value is less than $\alpha = 0.05$, the null hypothesis is rejected. Thus the sample result in this one-tailed test causes the purchaser to suspect that $\mu < 200$. Further action should be taken to investigate the situation. Note that a decision rule based on the critical value of $Z = -1.64$ indicates the same conclusion.

In conclusion, for a one-tailed test that the parameter μ equals some amount a, the null hypothesis is the same as for a two-tailed test, $H_0: \mu = a$. The alternate hypothesis for a one-tailed test, depending on the *directional difference* that is important, is either $H_1: \mu < a$ or $H_1: \mu > a$. The p-value determines the probability that the *directional difference* in $\bar{X}$ and μ is at least as extreme as

FIGURE 8.10:
DETERMINING p-VALUE
FOR A ONE-TAILED TEST

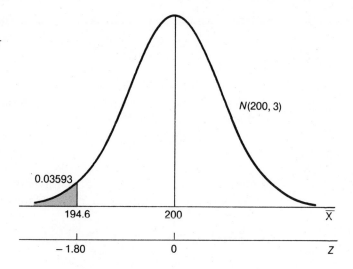

that observed. Table 8.4 summarizes the p-value for one- and two-tailed tests about the parameter μ.

TABLE 8.4: p-VALUES
FOR ONE- AND
TWO-TAILED TESTS
ABOUT THE
PARAMETER μ

Alternate Hypothesis H_1:	*Null Hypothesis H_0: $\mu = a$*
$\mu \neq a$	$P(|\bar{X} - \mu| \geq d \mid H_0 \text{ true})$
$\mu < a$	$P[(\bar{X} - \mu) \leq d \mid H_0 \text{ true}]$
$\mu > a$	$P[(\bar{X} - \mu) \geq d \mid H_0 \text{ true}]$

Note: d is the observed difference $(\bar{X} - \mu)$

In determining whether a one- or two-tailed test is appropriate, it is helpful to express the problem in terms of some phrases that indicate whether a single direction or both directions of difference away from a parameter value is important. If the basic question of interest can be expressed as "Has there been an increase?" "Is the new better than the old?", "Is there a decrease?" or "Has there been a decline?", then a one-tailed test is appropriate. If the question can be expressed as "Is there any change?" or "Is there a difference?", then a two-tailed test is appropriate.

Hypothesis testing for parameters other than the population mean μ follows the general procedure outlined in Sections 8.11–8.15. This chapter concludes with a test for the binomial proportion π. Tests for other parameters are given in the following chapters.

8.16 HYPOTHESIS TEST FOR π

In a hypothesis about the value of the population proportion π, the statistic of interest is the sample proportion p. The sampling distribution is normally distributed with mean π and standard error (for infinite populations)

$$\sigma_p = \sqrt{\frac{\pi(1-\pi)}{n}}$$

if the sample size is moderately large ($n \geq 50$) and both $n\pi$ and $n(1-\pi)$ are greater than five. The p-value for either a one- or two-tailed test of hypothesis about π can be determined from the use of the Z-transformation

$$Z = \frac{p-\pi}{\sigma_p}$$

and Appendix E. A summary of p-values and hypothesis statements for one- and two-tailed tests that the parameter π equals some amount a is given in Table 8.5. As in Section 8.13, decision rules for π can also be established as (1) decision limits for p and (2) critical values of Z.

TABLE 8.5: *p*-VALUES FOR ONE- AND TWO-TAILED TESTS ABOUT THE PARAMETER π

Alternate Hypothesis H_1:	Null Hypothesis H_0: $\pi = a$
$\pi \neq a$	$P(|p-\pi| \geq d \mid H_0 \text{ true})$
$\pi < a$	$P[(p-\pi) \leq d \mid H_0 \text{ true}]$
$\pi > a$	$P[(p-\pi) \geq d \mid H_0 \text{ true}]$

Note: d is the observed difference $(p-\pi)$

EXAMPLE. The plant manager of a manufacturing company is investigating the possibility of replacing a machine because the proportion of defectives is 0.15, a figure he considers too large. In the hope of selling a machine to this company, Dynamics Inc. lets the plant manager use a demonstrator model of their new machine on a trial basis. A random sample of 50 items from the new machine contains 4 defectives. Does the new machine produce fewer defectives than the old machine? Use a 0.05 level of significance for the test of hypothesis.

Solution. The standard for which we wish to compare the new machine is $\pi = 0.15$. The new machine is really better than the old if the proportion of defectives from the new machine is less than 0.15 (a one-tailed test). The null and alternate hypotheses are

$$H_0: \quad \pi = 0.15;$$

$$H_1: \quad \pi < 0.15.$$

Since $n = 50$ and both $n\pi$ and $n(1 - \pi)$ are greater than 5, we may assume that the test statistic p is normally distributed with mean $\pi = 0.15$ and standard error

$$\sigma_p = \sqrt{\frac{(0.15)(0.85)}{50}} = 0.0505.$$

The observed difference in the sample result and the hypothesized value is $(p - \pi) = (0.08 - 0.15) = -0.07$, which is

$$Z = \frac{0.08 - 0.15}{0.0505} = -1.39$$

standard errors below the hypothesized value $\pi = 0.15$. The p-value for the observed difference $(p - \pi)$, as shown in Figure 8.11, is

$$P[(p - \pi) \le -0.07 \,|\, H_0 \text{ true}] = P(Z \le -1.39)$$
$$= 0.08226.$$

Since the p-value is larger than $\alpha = 0.05$, the conclusion is to accept the null hypothesis. The sample result $p = 0.08$ is not significantly less than $\pi = 0.15$. The only action the plant manager should contemplate, based on these results, is not to buy the new machine.

FIGURE 8.11:
PROBABILITY OF A
DIFFERENCE
$(p - \pi) \le -0.07$

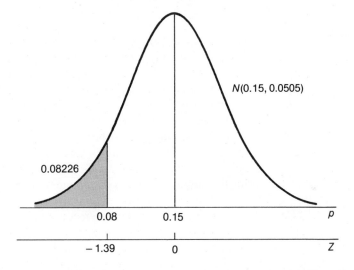

A decision rule for Z reaches the same conclusion as the p-value. The critical value is $Z = -1.64$ for a one-tailed test at the 0.05 level of significance. Since the observed value of $Z = -1.39$ is larger than the critical value, the null hypothesis is accepted.

8.17 SUMMARY

Statistical inference is the process of using sample information to reach conclusions about the population. The inference procedures discussed in this chapter are based on the assumption that the sampling distribution of the appropriate statistic is normal. This condition is met, for the parameters examined in this chapter, if samples are drawn from normally distributed populations or if moderately large samples are drawn from non-normal populations. Generally, the discussion concentrates on infinite populations so that the finite population correction factor is ignored.

The two major areas of statistical inference are estimation and testing. Two types of estimators are distinguished—point and interval estimators. A point estimator of a population parameter is a procedure that produces a single value as an estimate. Unbiasedness and minimum variance are the criteria used for selecting among estimators. A statistic is an unbiased estimator of a parameter if the expected value of the statistic equals the parameter value. The unbiased estimator that has the smallest variance is the minimum-variance unbiased estimator. Point estimators, which are minimum-variance unbiased estimates, of μ, π, and σ^2 are $\bar{X} = \sum X/n$, $p = X/n$ and $\hat{\sigma}^2 = \sum(X - \bar{X})^2/(n - 1)$, respectively. Further, $\hat{\sigma}$ is the best estimator of σ, even though it is a biased estimator.

Interval estimators provide a range of values that incorporate the inherent variability in the population and the confidence level desired. Interval estimators of the parameters μ and π are determined as

$$\text{point estimator} \pm (Z)(\text{standard error}).$$

The width of the confidence interval is influenced by the confidence level $(1 - \alpha)$; variability in the population, measured by σ; and the sample size. The investigator may exercise some control over the width of the interval by varying the confidence level and sample size. For a given measure of variability in the population, an investigator determines sample size on the basis of the maximum error in estimation and confidence level desired.

Hypothesis tests are procedures for investigating preconceived ideas about the value of population parameters. Sample information is compared with hypothesized parameter values to determine whether a given sample result is unusual. Testing is desirable because various actions are associated with different parameter values. Examples of actions based on parameter values include readjusting a machine, marketing a new product, increasing advertising budget, and not investing in a business venture.

Formally, testing begins with the statement of the null and alternate hypotheses. If it is critical to consider that the true parameter value might be either larger or smaller than that specified by the null hypothesis, a two-tailed test is appropriate. Otherwise a one-tailed test is conducted. The sampling distribution of the statistic must be specified assuming that the null hypothesis is true. A p-value is computed, which gives the probability of a sample result at least as extreme as that observed, given that the null hypothesis is true. The p-value is compared to a level of significance in order to reach a statistical decision. The level of significance α represents the minimum probability of sample results that supports the null hypothesis. The null hypothesis is rejected if the p-value is less than α, otherwise the null hypothesis is accepted. Decision rules may also be stated in terms of critical values of Z and decision limits for $\bar{X}$ and p.

Choosing the level of significance for a test involves a consideration of the probability and consequences of making either a Type I or Type II error. A Type I error is made if a true null hypothesis is rejected. The probability of a Type I error is the chosen level of significance α. A Type II error results if a false null hypothesis is accepted. The probability of a Type II error, β, depends on the difference between the hypothesized and true value of a parameter. Statisticians examine the power of a test, $(1 - \beta)$, for alternative parameter values in selecting an appropriate test. Hypothesis tests for the parameters μ and π are described in this chapter.

A relationship exists between confidence intervals and hypothesis tests that allows confidence intervals to be used in deciding between acceptance or rejection of the null hypothesis. For a test of $H_0: \mu = a$ versus $H_1: \mu \neq a$ at the α level of significance, compute the $(1 - \alpha)$ confidence interval. If the hypothesized value a is inside the $(1 - \alpha)$ confidence limits, accept H_0; otherwise H_0 is rejected. This procedure yields the same conclusion as any of the decision rules outlined in the chapter.

Additional concepts in statistical inference are given in the following chapters.

8.18 TERMS AND FORMULAS

Absolute difference	Null hypothesis
Alternate hypothesis	One-tailed test
Confidence interval	Parameter
Confidence level	Pilot sample
Critical value	Point estimator
Decision limits	Population
Decision rule	Power function
Directional difference	Power of a test
Estimate	Sample
Estimation	Sample size
Estimator	Sampling distribution
Finite population	Sampling with replacement
Hypothesis	Statistic
Hypothesis test	Statistical inference
Infinite population	Testing
Interval estimator	Two-tailed test
Level of significance	Type I error
Maximum error in estimation	Type II error
Minimum-variance unbiased estimator	Unbiased estimator

Estimated population total of raw scores	Estimated $\sum X = N\bar{X}$
Estimated number of successes in the population	Estimated $X = Np$
Interval estimator	Point estimator $\pm (Z)$(standard error)
Sample size for estimating μ	$n = \dfrac{Z^2 \sigma^2}{(\bar{X} - \mu)^2}$

Sample size for estimating π

$$n = \frac{Z^2 p(1-p)}{(p-\pi)^2}$$

p-value

P(a sample result at least as extreme as that observed $| H_0$ true)

Point and interval estimation formulas are summarized in Table 8.1. Formulas for the standardized variables used in testing follow.

Parameter	Standardized Value
μ	$Z = \dfrac{\bar{X} - \mu}{\sigma_{\bar{X}}}$
	$\sigma_{\bar{X}} = \dfrac{\sigma}{\sqrt{n}}; \ \hat{\sigma}_{\bar{X}} = \dfrac{S}{\sqrt{n-1}}$
π	$Z = \dfrac{p - \pi}{\sigma_P}$

QUESTIONS

A. True or False

T 1. Sample size is one factor the investigator may control in order to meet numerical requirements for estimating a population parameter.

F 2. The minimum-variance unbiased estimator of the population variance is the sample variance $\sum(X - \bar{X})^2/n$.

F 3. The power of a test measures the probability of a Type II error.

T 4. The null hypothesis is rejected if the p-value is less than the level of significance.

F 5. The width of a confidence interval decreases as the sample size decreases.

T 6. A Type I error cannot be made if the null hypothesis is accepted.

T 7. In a hypothesis test, the sampling distribution of the test statistic is based on the assumption that the null hypothesis is true.

B. Fill In

1. The minimum-variance unbiased estimator of the population mean is the _Sample mean_

2. _Statistical inference_ is the process of using sample information to reach conclusions about the population.

3. The _level of significance_ for a statistical test is the minimum probability of sample results that lend support to the null hypothesis.

4. The average diameter of a machined part should be 0.125 inches. After observing the thickness of parts in a random sample, the machine was readjusted to produce thicker parts. But now virtually all parts are too thick. A _Type I_ error has been made.

5. A _____one_____ -tailed test is appropriate when it is important to consider the possibility of a directional difference away from the hypothesized value.

6. Rejecting a true null hypothesis is a _Type II_ error.

7. A statistic is an _unbiased_ estimator of a parameter if E(statistic) = parameter.

8. The width of an interval estimator is _decreased_ if the confidence level $(1 - \alpha)$ is increased.

9. The critical value of Z for a one-tailed test when $\alpha = 0.01$ is (ignore sign) _2.33_ .

EXERCISES

1. Describe the meaning of the following.
    a. Unbiased estimator.
    b. Point estimator.
    c. Interval estimator.
    d. Type I error.
    e. Type II error.
    f. Significant difference in sample result and hypothesized value.
    g. Accept the null hypothesis.
    h. p-value.

2. Consider a population made up of the four numerical values (2, 4, 6, 8) where each numerical value occurs with equal frequency.
    a. Make a list of all different possible samples of size $n = 2$ with replacement and compute the mean of each.
    b. Sketch the distribution of sample means.
    c. Show that $E(\bar{X}) = \mu$.

3. Compute the unbiased estimator of σ^2 for each of the possible samples of size $n = 2$ in Exercise 2 and show that $E(\hat{\sigma}^2) = \sigma^2$.

4. Six of the eight members of a board of directors are in favor of the latest proposal, i.e., $\pi = \frac{6}{8} = 0.75$. Express the possible outcomes, number in favor of proposal out of a sample of four, as values of the sample proportion $p = X/n$. For example, if 1 out of 4 favors the proposal, the sample proportion is $p = \frac{1}{4} = 0.25$. Use the binomial formula to find the probability of each possible outcome, e.g., $P(X = 1|4, 0.75) = 0.0469$. Then show that $E(p) = \pi$.

5. A state agency wants to determine the average amount spent on gasoline per year by auto owners. A sample of 75 auto owners is taken at random. The average expenditure of this sample is $618 with a standard deviation of $86.
    a. What single value would you use to estimate the average expenditure for gasoline for all auto owners in the United States?
    b. Compute a 92 percent interval estimate of μ. Interpret the result.

6. The pervasiveness of television in American family life was indicated in a recent nationwide survey. In a random sample of 2,500 TV-owning

households, children watched TV an average of 23 hours per week with a standard deviation of 8 hours.

   a. Determine a 98 percent interval estimate of the average time per week children spend watching TV.

   b. Estimate the number of hours per year that children watch TV.

7. A factory is conducting an efficiency study. One measure desired is the average time for production workers to perform an assembly operation. A pilot study provides an estimate of 15 minutes for the standard deviation of assembly time. What size sample should be taken to estimate average assembly time with a maximum error in estimation of 5 minutes and 98 percent confidence?

8. The manager of a popular restaurant examines a random sample of sales receipts for 40 customers to find mean sales per customer of $4.86 and standard deviation $2.10.

   a. Determine a 95 percent confidence interval for the population mean sales per customer.

   b. What size sample is required if the manager wants the width of a 95 percent confidence interval estimate of μ to be $1.00?

9. An audit of a random sample of 50 sales reports (one for each customer contacted) for salesman Joe Max shows average sales revenue per contact of $430.

   a. If Joe contacts 200 customers in a year, estimate annual sales revenue.

   b. Management wants a 98 percent interval estimate of average sales revenue per contact with a maximum error in estimation of $25. How many sales reports must be audited, if the estimated population standard deviation is $120?

10. Forty of the 400 lung cancer victims monitored in a recent study were cured of the disease.

   a. What is the estimated cure rate, i.e., the estimated population proportion cured.

   b. Determine a 90 percent interval estimate for the proportion cured.

11. A leading airline made a survey of 100 Concorde passengers on the London-to-New York route, finding that 43 percent had flown the Concorde more than once.

   a. Determine a 96 percent interval estimate of the proportion flying the Concorde more than once.

   b. If the airline wants an interval estimate of the proportion of passengers flying Concorde more than once, an indication of repeat business, what size sample is required for a 5 percent maximum error in estimation and 99 percent confidence?

12. Past studies show that 3 out of every 10 bank customers switch banks because of human relations or personal service reasons, such as rude or unhelpful employees and bank errors in accounts.

a. Stanley National Bank wishes to estimate this proportion for its own customers with a maximum error in estimation of 0.04 and confidence level of 0.95. What size sample is required?

b. If the maximum error is 0.08 with 95 percent confidence, what is the sample size?

c. How many of the bank's 15,000 customers would be expected to switch banks because of these reasons?

13. A tennis player observes that he double-faulted 6 times in 80 serves today. How many serves must be observed, at random, in order to estimate the proportion of double-faults at 95 percent confidence with a maximum error in estimation of 0.03?

14. An electric utility has filed a request with the state regulatory agency for permission to construct a nuclear power generating station. The utility says that the new station is needed to meet power needs in the future, claiming that the average annual residential use is currently 13,000 kilowatt hours with a standard deviation of 4,000 KWH. The regulatory agency doubts that average annual usage is that high. They survey 400 residential customer records, finding an average of 12,500 KWH used annually. Is the sample result significantly less than the 13,000 KWH claim at the 0.05 level?

15. The telephone company claims that the average length of local calls is 4.5 minutes. A random sample of 50 calls are observed; the mean call length is 3.9 minutes with a standard deviation of 2.8 minutes. Is the sample result significantly different from the claim if $\alpha = 0.08$?

16. Automobiles manufactured in the United States must average 27 miles per gallon (mpg) by 1984, according to the Department of Transportation. This standard is based on the average fuel consumption of all passenger automobiles in a manufacturer's fleet for a particular model year. One manufacturer claims that its cars meet the standard this year. A consumer group tests a random sample of 37 cars in the manufacturer's fleet, finding an average of 25.8 mpg and standard deviation of 4.5 mpg. Does this evidence support a claim that the cars average at least 27 mpg? Let $\alpha = 0.05$.

17. The insurer for hospitalization and health care programs for state employees indicates that, because of the increased cost of providing medical care, a rate increase will go into effect at the time of the annual renewal. One of the costs cited in support of the rate increase is a state average daily charge for hospital care of $210 per patient. Personnel in the insurance commissioner's office suspect this figure is inflated and obtain a statewide random sample of 145 daily charges for hospitalization. The sample average daily charge is $198 with standard deviation $60. Does the sample result conflict with the insurer's claim at the 0.05 level of significance?

18. An airline computes capacity and range of their planes based on average weights of 170 pounds for men and 115 pounds for women. Recently the airline selected 100 male and 100 female passengers at random and, with their permission, weighed each. The men had an average weight of 185 pounds and standard deviation of 35 pounds. The women's average weight was 121 pounds with standard deviation 28 pounds. Is the airline underestimating weights, in which case the capacity and range computations should be revised upward? Conduct a test of average weight for women and a test for men; use $\alpha = 0.05$ for both.

19. The average diameter (μ) of a machined part is 0.250 inches with standard deviation of 0.060. A sample of 36 parts is taken periodically and measured. If the sample mean diameter is less than 0.235 inches or more than 0.265 inches, the machine is readjusted; otherwise the machine continues operating without adjustment.

    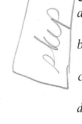

    a. What is the probability that the machine will continue operating without adjustment if μ remains 0.250 inches?
    b. What is the probability that the machine will be readjusted unnecessarily, i.e., when μ is 0.250 inches?
    c. What is the probability the machine will not be adjusted if μ shifts to 0.260 inches?
    d. What is the probability that the machine will be readjusted if $\mu = 0.235$?
    e. Indicate whether the above situations describe no error in hypothesis testing, a Type I error, or a Type II error. Identify the probabilities for each question as α, $1 - \alpha$, β, or $1 - \beta$.

20. A filling machine is designed to fill barrels with a chemical so that the mean volume per barrel is 200 liters with a standard deviation of 10 liters. The volume per barrel is normally distributed. Samples of the volume in 25 barrels are examined periodically to see if the filling machine is operating properly. A 0.05 level of significance is used to test the null hypothesis that the machine is filling barrels so that the mean volume is 200 liters. If the hypothesis is rejected, a costly inspection of the machine is undertaken to see what adjustment is required.

    a. What are the decision limits of the sample mean for accepting and rejecting the null hypothesis in this test?
    b. What is the probability of a Type II error if the machine actually fills barrels with a mean volume (μ) of 194 liters? If $\mu = 198$? If $\mu = 208$?
    c. Sketch the power function for this test.
    d. What is the probability that the null hypothesis is rejected if the mean volume per barrel is 200 liters?

21. The campaign manager of a candidate for the U.S. Senate believes that the proportion of potential voters currently in favor of the candidate is 0.60. A random sample of 200 potential voters is taken; 139 favor the candidate. Does the sample result support the campaign manager's belief? Use $\alpha = 0.02$.

22. The local telephone company has agreed to extend the non-toll calling area in one exchange for a small increase in the monthly rate if more than half of the customers are in favor of the expansion. In a preliminary study, 150 out of a random sample of 250 customers favor expansion. Does the sample evidence support an expansion of service, using $\alpha = 0.01$?

23. Economists for an airline say their new supersonic airliner can be profitable on transatlantic routes if, based on the present schedule, the seat-load-factor averages 0.60 (60 percent seat utilization). An analyst determines that the seat-load-factor is 0.51 for a random sample of 50 recent flights. Does the analyst have sufficient data to report that the 0.60 average is not being met, if a 0.02 level of significance is used?

REFERENCES

Bhattacharyya, Gouri K., and Johnson, Richard A. *Statistical Methods and Concepts.* New York: John Wiley & Sons, Inc., 1977.

Gibbons, Jean D., and Pratt, John W. "P-values: Interpretation and Methodology." *The American Statistician*, vol. 29, no.1 (February 1975), pp. 20–25.

Hays, William L. *Statistics for the Social Sciences*, 2d ed. New York: Holt, Rinehart and Winston, Inc., 1973.

Kirk, Roger E. *Statistical Issues: A Reader for the Behavioral Sciences.* Belmont, Cal.: Wadsworth Publishing Co., Inc., 1972.

McClave, James T., and Benson, George *Statistics for Business and Economics.* San Francisco, Cal.: Dellen Publishing Co., 1978.

Pfaffenberger, Roger C., and Patterson, James H. *Statistical Methods for Business and Economics.* Homewood, Ill.: Richard D. Irwin, Inc., 1977.

Springer, Clifford H., et al. *Statistical Inference.* Homewood, Ill.: Richard D. Irwin, Inc., 1966.

9

Additional Topics in Statistical Inference

9 ADDITIONAL TOPICS IN STATISTICAL INFERENCE

The universe, so far as known to us, is
so constituted, that whatever is true in
one case, is true in all cases of a certain description; the only difficulty is,
to find what description.

John Stuart Mill
A System of Logic

9.1 INTRODUCTION

The concepts of estimation and testing under general conditions are presented in Chapter 8. In the present chapter, we first examine the conditions under which the sampling distribution of $\bar{X}$ is not normal. A new distribution, called Student's t distribution, is introduced for use in inference procedures for μ under certain conditions. Then tests for the difference in two population means, $\mu_1 - \mu_2$, and the difference in two population proportions, $\pi_1 - \pi_2$, are provided. Statistical inference, as discussed in these chapters, is an inductive procedure that helps to determine ". . . whatever is true."

9.2 STUDENT'S t DISTRIBUTION

The inference procedures for the mean presented in Chapter 8 are based on an assumption that the sampling distribution of $\bar{X}$ is normal. If samples are drawn from a normally distributed population, the sample means are normally distributed. Consequently the statistic

$$Z = \frac{\bar{X} - \mu}{\sigma_{\bar{X}}}$$

has the standard normal distribution. Since $\sigma_{\bar{X}} = \sigma/\sqrt{n}$, the above statistic requires that the population standard deviation be known. If σ is unknown, we saw in the last chapter that $\sigma_{\bar{X}}$ can be estimated as either $\hat{\sigma}_{\bar{X}} = \hat{\sigma}/\sqrt{n}$ or $\hat{\sigma}_{\bar{X}} = S/\sqrt{n} - 1$. When σ is unknown, the statistic

$$\frac{\bar{X} - \mu}{\hat{\sigma}_{\bar{X}}}$$

approximates the standard normal distribution if $n \geq 30$. If the sample size is less than 30, the approximation is inadequate. This requires a different distribution.

This chapter provides inference procedures for means when samples are drawn from normal populations where the population standard deviation is unknown. For this case the statistic

$$\frac{\bar{X} - \mu}{\hat{\sigma}_{\bar{X}}}$$

follows what is called *Student's t distribution*. Thus the t statistic is

$$t = \frac{\bar{X} - \mu}{\hat{\sigma}_{\bar{X}}}. \tag{9.1}$$

Substituting the two methods of determining $\hat{\sigma}_X$, the t statistic may be expressed as

$$t = \frac{\bar{X} - \mu}{\dfrac{S}{\sqrt{n-1}}} \tag{9.2}$$

and

$$t = \frac{\bar{X} - \mu}{\dfrac{\hat{\sigma}}{\sqrt{n}}}. \tag{9.3}$$

A close examination of these forms for $\hat{\sigma}_{\bar{X}}$ reveals an important characteristic of Student's t distribution. The estimated population standard deviation $\hat{\sigma}$ in Equation (9.3) is described in Chapter 8 as the square root of the unbiased estimator of σ^2,

$$\hat{\sigma}^2 = \frac{\sum(X - \bar{X})^2}{n - 1}. \tag{9.4}$$

The denominator of this unbiased estimator, $n - 1$, is called the *degrees of freedom*, the parameter of the t distribution. While a technical explanation of degrees of freedom is beyond the scope of this text, some understanding of its use in the present discussion follows. The number of degrees of freedom (df) is a function of sample size, df $= n - k$. Writing Equation (9.4) in terms of the general expression for df,

$$\hat{\sigma}^2 = \frac{\sum(X - \bar{X})^2}{n - k},$$

we may ask what value of k makes $E(\hat{\sigma}^2)$ equal to σ^2. As illustrated in Chapter 8, $k = 1$. Thus the parameter, degrees of freedom, is a function of sample size, df $= n - k$, where k is the appropriate constant required to produce an unbiased estimator of population variance. The equivalence of $S/\sqrt{n-1}$ in Equation (9.2) and $\hat{\sigma}/\sqrt{n}$ in Equation (9.3) and their use of the quantity df $= n - 1$ is shown by the squared standard error

$$\hat{\sigma}_{\bar{X}}^2 = \frac{\hat{\sigma}^2}{n} = \frac{\dfrac{\sum(X - \bar{X})^2}{n-1}}{n} = \frac{\sum(X - \bar{X})^2}{n} \cdot \frac{1}{n-1} = \frac{S^2}{n-1}.$$

The probability density function for the t distribution with parameter v (Greek letter "nu"), the degrees of freedom, is

$$f(t|v) = \frac{[(v-1)/2]!}{\sqrt{v\pi}[(v-2)/2]!} \cdot \left[1 + \frac{t^2}{v}\right]^{-\frac{1}{2}(v+1)} \qquad \text{for } -\infty < t < +\infty. \tag{9.5}$$

The range of t is exactly the same as in the standard normal distribution. The mean and variance are

$$E(t) = 0$$

and

$$V(t) = \frac{v}{v-2} \quad \text{for } v > 2.$$

Note that the population variance σ^2 is not a factor of Equation (9.5) as in the case of a normal distribution, i.e., the t distribution does not depend on σ^2. Thus the t distribution provides a way to make inferences about the population mean of a normal distribution without knowledge of the population variance.

> If $\bar{X}$ and S are the mean and standard deviation of a random sample from a normal population with mean μ, then the statistic
>
> $$\frac{\bar{X} - \mu}{\left(\dfrac{S}{\sqrt{n-1}}\right)}$$
>
> has a t distribution with $n - 1$ degrees of freedom.

The t distribution, like the normal, is bell-shaped and symmetrical about the mean. But there is not a "standard" t distribution, as there is a standard normal distribution. The exact shape of a t distribution depends on its parameter, the number of degrees of freedom. The amount of dispersion in the t distribution is generally more than in the normal distribution, but the dispersion decreases as df increase. A comparison of the standard normal distribution with two t distributions is provided in Figure 9.1. The diagram shows, for each distribution, the number of standard deviations away from the mean that includes the middle 95 percent of the area; thus a total of 5 percent of the area remains in the tails. This is 1.96 standard deviations for the standard normal, i.e.,

$$P(|Z| > 1.96) = 0.05.$$

For the t distribution with df $= 3$,

$$P(|t| > 3.182) = 0.05.$$

With df $= 9$, the t distribution has

$$P(|t| > 2.262) = 0.05.$$

These values of t are given in Appendix G, a table of values of t for various values of degrees of freedom and tail probabilities (areas in the tails of the distribution). Both Figure 9.1 and Appendix G show that the distance away from the mean required for a certain area (probability) is generally larger for t than for the standard normal. But the distance (dispersion) decreases as the

FIGURE 9.1:
COMPARISON OF
STANDARD NORMAL
AND TWO *t*
DISTRIBUTIONS

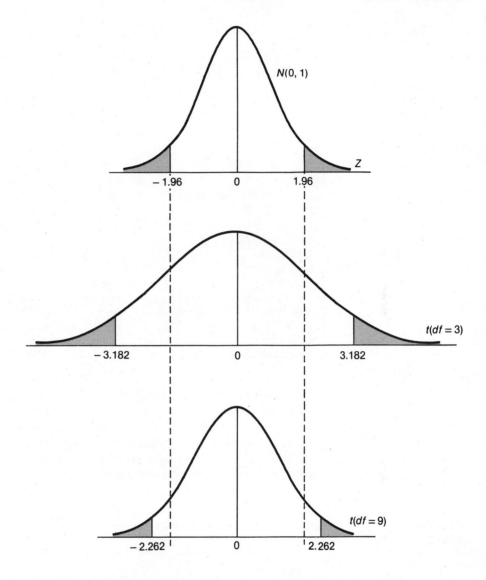

degrees of freedom increase. Reading down the column of Appendix G for a two-tailed 0.05 probability, the number of standard errors approaches 1.96. Hence *the t distribution approaches the standard normal as n, and therefore df, increases.*

Since the *t* distribution approaches the standard normal as degrees of freedom increase, there is little difference in the use of *t* and the standard normal for $n \geq 30$. Thus in practice, we approximate the *t* distribution with the standard normal distribution when sampling from normal populations with unknown standard deviation σ and $n \geq 30$. The practical use of the standard normal and *t* distributions for statistical inference on population means is summarized in Table 9.1.

TABLE 9.1:
DISTRIBUTION OF THE
STATISTIC $\dfrac{\bar{X} - \mu}{S/\sqrt{n-1}}$
FOR RANDOM SAMPLES
FROM A NORMAL
POPULATION

n	σ Unknown
< 30	Student's t
≥ 30	Approximate t by standard normal

WILLIAM SEALY GOSSET (1876–1937)

Unlike most of the other contributors to statistics highlighted in this book, Gosset was not an academician, but practiced statistics in an industrial setting. Born in Canterbury and educated in chemistry and mathematics, Gosset entered the employment of the Guinness Brewery in Ireland in 1899. (The Guinness brand is still widely known.) One of many science graduates employed by the brewery, Gosset devoted his attention to analysis of small samples of experimental data as required in brewery work. In 1908 he published a paper entitled "On the Probable Error of the Mean" under the pseudonym "Student." Ignored for a while, the work evolved into what today is known as Student's t distribution. Recognition of this major contribution came in the 1920s from perhaps the most prolific contributor to statistics, R. A. Fisher. The dates cited here are indicative of the recency of development of some of the major concepts in statistics.

The use of the t distribution in statistical inference is presented in the following sections. These are inference procedures regarding population means when the t distribution is appropriate, i.e., when sampling from normal populations where σ is unknown and $n < 30$. Since the practical application of the t distribution involves, among other things, small sample sizes, its use is frequently referred to as *small sample inference*. Since the use of the normal distribution usually involves larger sample sizes, its use is referred to as *large sample inference*.

9.3 INTERVAL ESTIMATOR FOR μ: SMALL SAMPLES

It was pointed out in Chapter 8 that an interval estimator based on the normal distribution is given by the expression

Interval estimator = Point estimator $\pm$ (Z)(standard error).

The form of the estimator using t is analogous to that using Z.

If a random sample of size n is taken from a normal population with unknown standard deviation, the $1 - \alpha$ confidence limits for μ are

$$\bar{X} \pm t\hat{\sigma}_{\bar{X}},$$

where t is based on a two-tailed probability of α and df $= n - 1$.

EXAMPLE. A consumer testing agency is examining a new automobile for gasoline mileage performance. Ten readings of miles per gallon (mpg) are selected at random, resulting in $\bar{X} = 30.6$ and $S = 1.5$. Assuming that mpg is normally distributed, determine a 95 percent interval estimate of μ, the mean mpg for the population.

Solution. Since a random sample of $n = 10$ is taken from a normal population where σ is unknown, the interval estimate is based on the t distribution with df $= n - 1 = 9$. The estimated standard error is $\hat{\sigma}_{\bar{x}} = S/\sqrt{n-1} = 1.5/\sqrt{9} = 0.5$ and the t distribution with 9 df has a value of $t = 2.262$ for a two-tailed probability of $\alpha = 0.05$, i.e., $P(|t| > 2.262) = 0.05$. The confidence interval is

$$\bar{X} \pm t\hat{\sigma}_{\bar{x}} = 30.6 \pm (2.262)(0.5)$$
$$= 29.469; 31.731.$$

The interpretation of this interval is the same as in Chapter 8. There is 95 percent confidence that the mean performance for this automobile is between 29.469 and 31.731 miles per gallon.

If 95 percent confidence intervals for μ are computed for repeated samples of size n from a normal distribution *where σ is known*, the width of these intervals will be the same. The distance between $\bar{X} - 1.96\sigma_{\bar{x}}$ and $\bar{X} + 1.96\sigma_{\bar{x}}$ is the same for different values of $\bar{X}$ because $\sigma_{\bar{x}} = \sigma/\sqrt{n}$ is constant for known σ and fixed n. This was illustrated in Figure 8.2. This is not the case for confidence intervals based on the t distribution since the standard error must be estimated from the sample by either $\hat{\sigma}_{\bar{x}} = S/\sqrt{n-1}$ or $\hat{\sigma}_{\bar{x}} = \hat{\sigma}/\sqrt{n}$. The measure of variability may differ from sample to sample so that $\hat{\sigma}_{\bar{x}}$ does not remain constant. Consequently, the width of $1 - \alpha$ confidence intervals for μ from repeated samples of fixed size n from a normal distribution with unknown σ is not constant.

EXAMPLE. Another investigator for the consumer testing agency randomly selects ten mileage readings independently of the first investigator and finds $\bar{X} = 29.8$ mpg and $S = 2.1$ mpg. Construct a 95 percent interval estimate for μ and compare the width of this interval with the interval in the previous example.

Solution. Here $\hat{\sigma}_{\bar{x}} = 2.1/\sqrt{9} = 0.7$, and t remains 2.262 for 95 percent confidence and df $= 9$. The interval estimate is $29.8 \pm (2.262)(0.7) = 28.2166$; 31.3834. The width of the interval, determined as either the upper limit minus the lower limit or $2t\hat{\sigma}_{\bar{x}}$, is 3.1668. The width of the confidence interval in the previous example is 2.262.

9.4 HYPOTHESIS TEST FOR μ: SMALL SAMPLES

A hypothesis test for the parameter μ is based on the t distribution if the sample result $\bar{X}$ is determined by a random sample from a normal population where the standard deviation σ is unknown and n is small. The p-value for the test is determined from the t-statistic

$$t = \frac{\bar{X} - \mu}{\hat{\sigma}_{\bar{x}}}$$

and the table of values given in Appendix G. As indicated earlier, the t-statistic and its accompanying p-value would, in practice, be estimated by the standard normal distribution if the sample size is at least 30. To illustrate the use of the t distribution in testing, consider the following.

EXAMPLE. A nationwide survey indicates that children spend an average of 23 hours per week watching television. A city councilwoman wishes to determine whether the time that children in her district spend watching television is significantly different from 23 hours. She obtains a random sample of time spent watching television in a week for 26 children in her district. A summary of the sample results on hours per week (X) is $\bar{X} = 20$ and $S = 8.9$. Assuming that random variable X is normally distributed, conduct the appropriate test at $\alpha = 0.05$.

Solution. For a normally distributed population with σ unknown and $n < 30$, the t distribution must be used. A two-tailed test is appropriate with hypotheses

$$H_0: \quad \mu = 23$$
$$H_1: \quad \mu \neq 23.$$

With $\hat{\sigma}_{\bar{X}} = S/\sqrt{n-1} = 8.9/\sqrt{25} = 1.78$, the t-statistic is

$$t = \frac{\bar{X} - \mu}{\hat{\sigma}_{\bar{X}}} = \frac{20 - 23}{1.78} = -1.685.$$

The exact p-value may be unobtainable for tests utilizing the t distribution since values of t are given in Appendix G only for selected probability levels. At least an approximate p-value can be determined as follows. Locate the row of Appendix G for the particular t distribution used for the test, which is identified by the number of degrees of freedom. Read across that row until the values of t in the table closest to the absolute value of the computed t are located. The p-value is written as an inequality using the one- or two-tailed probability levels given as column headings. In this particular test, df $= 25$. Reading across this row of Appendix G, we find that $|t| = 1.685$ is between the values 1.316 and 1.708. Using the two-tailed probability levels for these columns, we know that $0.10 <$ p-value < 0.20. Writing the p-value in its appropriate form,

$$0.10 < P(|\bar{X} - \mu| \geq 3 \,|\, H_0 \text{ true}) < 0.20.$$

The p-value is illustrated in Figure 9.2. Since $\alpha = 0.05$ and we know that the exact p-value exceeds 0.10, the null hypothesis is accepted. Thus the sample evidence does not indicate that the time children in this district spend watching TV is significantly different from that determined in the nationwide survey.

Decision rules based on measures other than p-values, as discussed in Chapter 8, may also be used here. A *decision rule for the critical value(s) of* t is analogous to that given for Z in Chapter 8. Critical values of t are found in Appendix G. For the previous example, $t = 2.060$ is given in the row-and-column intersection representing 25 df and a two-tailed $\alpha = 0.05$. This indicates that the null hypothesis is accepted if the computed t is between -2.060 and $+2.060$; otherwise H_0 is rejected. Since the computed $t = -1.685$, H_0 is accepted as before.

FIGURE 9.2: AREA
REPRESENTED
BY THE p-VALUE
$P(|\bar{X} - \mu| \geq 3|H_0 \text{ TRUE})$

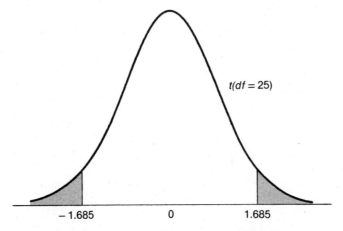

$t(df = 25)$

−1.685   0   1.685

A *decision rule for the decision limits of* $\bar{X}$ follows the same procedure given in Chapter 8. The decision limits of $\bar{X}$ for a two-tailed test are $\mu \pm t\hat{\sigma}_{\bar{X}}$, where μ is the hypothesized value. The null hypothesis is accepted if the observed sample result $\bar{X}$ is within the decision limits; otherwise H_0 is rejected. The decision limits of $\bar{X}$ for the previous example are $23 \pm (2.060)(1.78)$, or 19.33 to 26.67. Since the observed $\bar{X} = 20$, the null hypothesis is accepted. As indicated in Chapter 8, all three decision rules lead to exactly the same conclusion. The choice of a decision rule is a matter of the investigator's preference.

The following example illustrates a one-tailed test using the t distribution.

EXAMPLE. An accountant uses a sampling procedure in auditing a client's statements of accounts payable for possible monetary errors in the balance payable. A random sample of 17 accounts is selected, the balance payable on each is confirmed, and the sample results are used to test the null hypothesis that the average monetary error for the population of accounts (μ) does not exceed $50. The accountant uses $\alpha = 0.01$ and assumes that monetary errors in accounts ($\bar{X}$) are normally distributed. For the sample of 17 accounts, $\bar{X} = \$56$ and $S = \$8.24$. Does this indicate that μ does not exceed $50?

Solution. The hypotheses are

$$H_0: \quad \mu = 50;$$
$$H_1: \quad \mu > 50.$$

With $\hat{\sigma}_{\bar{X}} = 8.24/\sqrt{16} = 2.06$, the t-statistic is

$$t = \frac{56 - 50}{2.06} = 2.91.$$

The p-value, shown in Figure 9.3, is estimated from the one-tailed probabilities in Appendix G where df = 16. This is

$$0.005 < P[(\bar{X} - \mu) \geq 6|H_0 \text{ true}] < 0.01.$$

Since the p-value is less than α, the null hypothesis is rejected. Since it appears unlikely that the standard is met, the accountant will examine more of the

FIGURE 9.3: AREA
REPRESENTED
BY THE p-VALUE
$P[(\bar{X} - \mu) \geq 6 | H_0 \text{ TRUE}]$

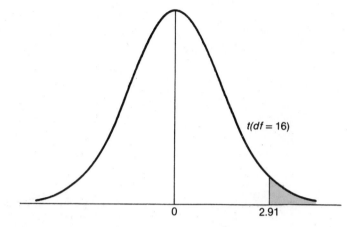

$t(df = 16)$

0   2.91

accounts, perhaps all of them, to determine the average monetary error. Note that a critical value of $t = 2.583$ also indicates rejection of H_0 for the computed $t = 2.91$.

9.5 HYPOTHESIS TEST FOR $\mu_1 - \mu_2$: LARGE SAMPLES

Often we wish to compare characteristics of two different populations. We might wish to know whether two populations have the same mean or whether two population proportions are equal. Comparison of two population means is accomplished by a hypothesis test of the *difference in two population means.*

Let μ_1 and μ_2 represent the means of two populations. The parameter under investigation in a test for the difference in population means is $\mu_1 - \mu_2$. The null hypothesis specifies that there is no difference in the population means, which is written $\mu_1 - \mu_2 = 0$. The alternate hypothesis is $\mu_1 - \mu_2 \neq 0$ for a two-tailed test. For a one-tailed test, H_1 is either $\mu_1 - \mu_2 < 0$ or $\mu_1 - \mu_2 > 0$. The statistic used here is the difference in two sample means, $\bar{X}_1 - \bar{X}_2$, obtained from two independently selected random samples. The sampling distribution of this statistic, called the sampling distribution of differences in sample means, $\bar{X}_1 - \bar{X}_2$, was presented in Section 6.9.

If random samples of large size are chosen independently from two populations, the distributions of both $\bar{X}_1$ and $\bar{X}_2$ are, according to the central limit theorem, approximately normal. The difference $(\bar{X}_1 - \bar{X}_2)$ is normally distributed because it is a linear combination of normally and independently distributed random variables (see Section 5.15). Independence specifically requires that the sample observations from one population be chosen independent of the sample observations from the other population. Under the assumption that the null hypothesis is true, the mean of the sampling distribution is zero. That is,

$$E(\bar{X}_1 - \bar{X}_2) = \mu_1 - \mu_2 = 0.$$

The standard error of this distribution, for infinite populations, is

$$\sigma_{\bar{X}_1 - \bar{X}_2} = \sqrt{\frac{\sigma_1^2}{n_1} + \frac{\sigma_2^2}{n_2}}.$$

Thus the sampling distribution of $\bar{X}_1 - \bar{X}_2$ is identified as $N(\mu_1 - \mu_2 = 0, \sigma_{X_1 - \bar{X}_2})$. If the variances of the populations are unknown, which is often the case, the standard error of the difference in sample means, $\sigma_{\bar{X}_1 - \bar{X}_2}$, is estimated (for large samples) from the sample standard deviations as

$$\hat{\sigma}_{\bar{X}_1 - \bar{X}_2} = \sqrt{\frac{S_1^2}{n_1} + \frac{S_2^2}{n_2}}.$$

The p-value for the test is calculated, as before, by expressing an observed difference in sample means $\bar{X}_1 - \bar{X}_2$ in terms of the standard normal distribution. As given in Section 6.9, the Z-transformation for this sampling distribution is

$$Z = \frac{(\bar{X}_1 - \bar{X}_2) - (\mu_1 - \mu_2)}{\sigma_{\bar{X}_1 - \bar{X}_2}}.$$

The p-value for a two-tailed test, the probability of an absolute difference in the sample result $(\bar{X}_1 - \bar{X}_2)$ and the hypothesized value $(\mu_1 - \mu_2)$ at least as extreme as observed, given that H_0 is true, is

$$P\left[\left|(\bar{X}_1 - \bar{X}_2) - (\mu_1 - \mu_2)\right| \geq d \,\middle|\, H_0 \text{ true}\right]$$

for an observed difference d. When the parameter $\mu_1 - \mu_2$ has a hypothesized value of zero, the observed difference d in the sample result and hypothesized value reduces to the sample result $(\bar{X}_1 - \bar{X}_2)$. Consequently, the p-value may be shortened to

$$P(|\bar{X}_1 - \bar{X}_2| \geq d \,|\, H_0 \text{ true}).$$

If a directional difference in $\mu_1 - \mu_2$ is important, then a one-tailed test is appropriate. Table 9.2 summarizes the p-values and hypotheses for one- and two-tailed tests.

TABLE 9.2: p-VALUES
FOR ONE AND
TWO-TAILED TESTS
ABOUT THE PARAMETER
$\mu_1 - \mu_2$

Alternate Hypothesis H_1:	*Null Hypothesis H_0: $\mu_1 - \mu_2 = 0$*
$\mu_1 - \mu_2 \neq 0$	$P(|\bar{X}_1 - \bar{X}_2| \geq d \,|\, H_0 \text{ true})$
$\mu_1 - \mu_2 < 0$	$P[(\bar{X}_1 - \bar{X}_2) \leq d \,|\, H_0 \text{ true}]$
$\mu_1 - \mu_2 > 0$	$P[(\bar{X}_1 - \bar{X}_2) \geq d \,|\, H_0 \text{ true}]$

Note: d is the observed difference $(\bar{X}_1 - \bar{X}_2)$.

EXAMPLE. Two different methods of instruction are used in a management training program for a large group of supervisors at Steel City Metals. Supervisors without any training in the subject matter were randomly assigned to either the Personalized Instruction Method (PIM) or the more traditional Lecture Method (LM). In the PIM, supervisors used programmed materials on their own and proceeded at their own pace during scheduled periods. The LM used a training leader in a classroom setting. In order to determine whether the method of instruction makes any difference in learning, a standardized test was given to all participants. For a random selection of 40 PIM participants, the mean score is 72 and standard deviation is 15. For 50 LM participants, the mean is 81 with standard deviation 20. Is the observed difference in mean scores significant at the 0.05 level?

Solution.   Let $\bar{X}_1$ and $\bar{X}_2$ represent the mean scores for PIM and LM, respectively. A two-tailed test is appropriate since a specific directional difference is not implied. The hypotheses are

$$H_0: \quad \mu_1 - \mu_2 = 0;$$
$$H_1: \quad \mu_1 - \mu_2 \neq 0.$$

The statistic $\bar{X}_1 - \bar{X}_2$ is approximately normally distributed with mean zero and standard error

$$\hat{\sigma}_{\bar{X}_1 - \bar{X}_2} = \sqrt{\frac{(15)^2}{40} + \frac{(20)^2}{50}} = 3.69.$$

To determine the p-value, compute

$$Z = \frac{(\bar{X}_1 - \bar{X}_2) - (\mu_1 - \mu_2)}{\hat{\sigma}_{\bar{X}_1 - \bar{X}_2}}$$

$$= \frac{(72 - 81) - 0}{3.69}$$

$$= \frac{-9 - 0}{3.69} = -2.44.$$

The p-value, as shown in Figure 9.4, is

$$P(|\bar{X}_1 - \bar{X}_2| \geq 9 \,|\, H_0 \text{ true}) = 0.01468.$$

FIGURE 9.4: p-VALUE
FOR OBSERVED
DIFFERENCE IN MEAN
SCORES

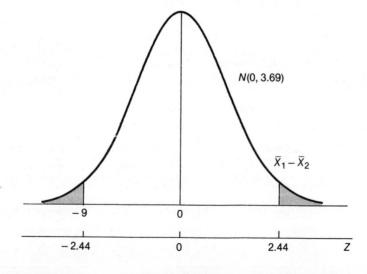

$N(0, 3.69)$

$\bar{X}_1 - \bar{X}_2$

Since this is less than $\alpha = 0.05$, the null hypothesis is rejected. The sample result indicates that the method of instruction may make a difference in the mean scores.

9.6 HYPOTHESIS TEST FOR $\mu_1 - \mu_2$: SMALL SAMPLES

The t distribution is also useful for testing the difference in means of two normally distributed populations when the population standard deviations are unknown and the sample sizes are small. A hypothesis test of $\mu_1 - \mu_2$ for large samples is given in the previous section. In that case, the statistic

$$\frac{(\bar{X}_1 - \bar{X}_2) - (\mu_1 - \mu_2)}{\hat{\sigma}_{\bar{X}_1 - \bar{X}_2}}$$

approximates the standard normal distribution if independent random samples of at least 30 observations are taken from each population. If either sample size is below 30, the statistic may not have the standard normal distribution. Provided certain conditions are met, the t distribution may be used.

Let a random sample of n_1 items be selected from a normally distributed population, identified as $N(\mu_1, \sigma_1)$ where the standard deviation is unknown. A random sample of n_2 items is selected, independent by of the first sample, from the normal population $N(\mu_2, \sigma_2)$ where the standard deviation is unknown. If the unknown population standard deviations are equal, i.e., $\sigma_1 = \sigma_2$, the statistic

$$\frac{(\bar{X}_1 - \bar{X}_2) - (\mu_1 - \mu_2)}{S_{\bar{X}_1 - \bar{X}_2}} \tag{9.6}$$

has a t distribution with $n_1 + n_2 - 2$ degrees of freedom, where $S_{\bar{X}_1 - \bar{X}_2}$ is the estimated standard error of the difference in sample means. Since the population standard deviation, which is the same for both populations, is unknown, it must be estimated from the two samples. Technically, a test for equality of population variances should be performed in order to justify the equality assumption. This is rarely done in practice unless there is reason to suspect that the assumption is unjustified. Such a test is described in Chapter 11.

Let us examine the basis for estimating the population standard deviation and using it to obtain the estimated standard error $S_{\bar{X}_1 - \bar{X}_2}$. To simplify the explanation, we concentrate on variance rather than standard deviation. The variance of a sample of n items is

$$S^2 = \frac{\sum(X - \bar{X})^2}{n}. \tag{9.7}$$

If two independent samples are taken from populations with equal, but unknown, variances, the unbiased estimator of the population variance is

$$S_{po}^2 = \frac{\sum(X_1 - \bar{X}_1)^2 + \sum(X_2 - \bar{X}_2)^2}{(n_1 - 1) + (n_2 - 1)} \tag{9.8}$$

where $n - 1$ is the number of degrees of freedom associated with each sample and the subscripts refer to results for samples 1 and 2. This is often called a *pooled estimator of population variance*, as the symbol S_{po}^2 reflects. We are pooling the results of two samples to estimate a common value. Since from Equation (9.7),

$$\sum(X - \bar{X})^2 = ns^2,$$

the pooled estimator may be written as

$$S_{po}^2 = \frac{n_1 S_1^2 + n_2 S_2^2}{n_1 + n_2 - 2}. \tag{9.9}$$

Equations (9.8) and (9.9) are equivalent methods for computing S_{po}^2. The former is more efficient for computing S_{po}^2 directly from the raw scores in the two samples, while the latter is appropriate if the sample results are summarized by means and variances.

Next the estimated variance of the difference in sample means, $S_{\bar{X}_1 - \bar{X}_2}^2$, is determined from the pooled estimator of population variance, S_{po}^2. To do this, recall that the variance of a sample mean is the population variance divided by sample size. Using our pooled estimate, the estimated variance of a sample mean is

$$S_{\bar{X}}^2 = \frac{S_{po}^2}{n}. \tag{9.10}$$

Since the variance of a difference in sample means equals the sum of the variances from Equation (6.13), the estimated variance of the difference in sample means is

$$S_{\bar{X}_1 - \bar{X}_2}^2 = S_{\bar{X}_1}^2 + S_{\bar{X}_2}^2$$

$$= \frac{S_{po}^2}{n_1} + \frac{S_{po}^2}{n_2}. \tag{9.11}$$

The estimated standard error of the difference in sample means, the square root of Equation (9.11), is

$$S_{\bar{X}_1 - \bar{X}_2} = \sqrt{\frac{S_{po}^2}{n_1} + \frac{S_{po}^2}{n_2}}. \tag{9.12}$$

In summary, the estimated standard error $S_{\bar{X}_1 - \bar{X}_2}$, the denominator of the t-statistic in a test for $\mu_1 - \mu_2$, is essentially a two-step computation. First, determine the pooled estimate of population variance S_{po}^2, Equation (9.9). Then use this result to determine the estimated standard error of difference in sample means $S_{\bar{X}_1 - \bar{X}_2}$, Equation (9.12). An alternate form for Equation (9.12) is

$$S_{\bar{X}_1 - \bar{X}_2} = S_{po} \sqrt{\frac{1}{n_1} + \frac{1}{n_2}}. \tag{9.13}$$

Likewise, the two-step procedure for the standard error may be combined into the single formula

$$S_{\bar{X}_1 - \bar{X}_2} = \sqrt{\left(\frac{n_1 S_1^2 + n_2 S_2^2}{n_1 + n_2 - 2}\right)\left(\frac{1}{n_1} + \frac{1}{n_2}\right)}. \tag{9.14}$$

Note that the symbols for standard errors in this section differ from those in Section 9.4. This is done so that small and large sample methods may be easily distinguished.

EXAMPLE. A discrimination suit has been filed against Technical Mills, Inc., charging that female employees are paid less than male employees for the same type of work. Attorneys for the company immediately obtain a random sample of hourly wages for 16 female employees and an independent random sample of the wages of 10 male employees with the same job classification. Letting X_1 = hourly wage (\$) for females and X_2 = hourly wage (\$) for males, the sample results are:

$$n_1 = 16 \qquad n_2 = 10$$
$$\bar{X}_1 = 5.45 \qquad \bar{X}_2 = 6.25$$
$$S_1 = 0.90 \qquad S_2 = 0.75.$$

If wages for both females and males are normally distributed, test at $\alpha = 0.05$ to determine whether the claim that females are paid less might be justified.

Solution. The null and alternate hypotheses are

$$H_0: \quad \mu_1 - \mu_2 = 0;$$
$$H_1: \quad \mu_1 - \mu_2 < 0.$$

The pooled estimate of population variance, from Equation (9.8), is

$$S_{po}^2 = \frac{n_1 S_1^2 + n_2 S_2^2}{n_1 + n_2 - 2}$$

$$= \frac{16(0.90)^2 + 10(0.75)^2}{16 + 10 - 2}$$

$$= 0.774.$$

The estimated standard error is

$$S_{\bar{X}_1 - \bar{X}_2} = \sqrt{\frac{S_{po}^2}{n_1} + \frac{S_{po}^2}{n_2}}$$

$$= \sqrt{\frac{0.774}{16} + \frac{0.774}{10}}$$

$$= 0.355.$$

The t-statistic, for the observed difference $\bar{X}_1 - \bar{X}_2 = 5.45 - 6.25 = -0.80$, is

$$t = \frac{(\bar{X}_1 - \bar{X}_2) - (\mu_1 - \mu_2)}{S_{\bar{X}_1 - \bar{X}_2}}$$

$$= \frac{-0.80 - 0}{0.355}$$

$$= -2.25.$$

The p-value statement has the same format as those in the last section. The directional p-value for a one-tailed t-test with 24 df is, from Appendix G,

$$0.01 < P[(\bar{X}_1 - \bar{X}_2) \leqq -0.80 | H_0 \text{ true}] < 0.025.$$

This is illustrated in Figure 9.5. Since the p-value is less than $\alpha = 0.05$, the null hypothesis is rejected. It does appear that, on a statistical basis, the suit may have some validity.

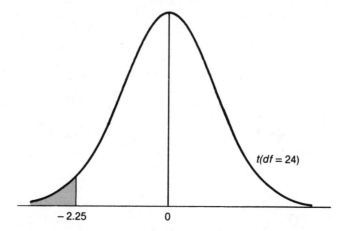

Note that other decision rules may be used for testing differences in means. For example, the critical value of t for the above example is -1.711 (one-tailed test with $\alpha = 0.05$ and 24 df). Since the computed $t = -2.25$ is less than the critical value, H_0 must be rejected.

9.7   HYPOTHESIS TEST FOR $\pi_1 - \pi_2$: LARGE SAMPLES

The test for a difference in two population proportions parallels the test for a difference in two population means given in Section 9.5. For two population proportions π_1 and π_2, the null hypothesis specifies that the parameter $\pi_1 - \pi_2$ has a value of zero. Either a one- or two-tailed test may be performed. The appropriate statistic is the observed difference in two sample proportions. The characteristics of the sampling distribution follow.

Let a sample of n_1 items from one population have X_1 successes, or a sample proportion $p_1 = X_1/n_1$. A second sample of n_2 items, obtained independently of the first sample, has X_2 successes, which is a sample proportion $p_2 = X_2/n_2$. If each sample size is large ($n \geqq 50$), the sampling distribtions of p_1 and p_2 are approximately normal, according to the central limit theorem. Then the distribution of differences $(p_1 - p_2)$ is also normal, since it is a linear combination of normally and independently distributed random variables (see Section 5.15). If the null hypothesis is true, the mean of the sampling distribution is zero;

$$E(p_1 - p_2) = \pi_1 - \pi_2 = 0. \tag{9.15}$$

The standard error of the sampling distribution is, for infinite populations,

$$\sigma_{p_1-p_2} = \sqrt{\frac{\pi_1(1 - \pi_1)}{n_1} + \frac{\pi_2(1 - \pi_2)}{n_2}} \tag{9.16}$$

where π is the common population proportion specified by the null hypothesis. Since π is unknown, we obtain an estimate from the sample information. The estimator of π is

$$\hat{\pi} = \frac{X_1 + X_2}{n_1 + n_2}. \tag{9.17}$$

The estimator $\hat{\pi}$ is used in place of π in the above formula to provide an estimated standard error of the sampling distribution

$$\hat{\sigma}_{p_1 - p_2} = \sqrt{\frac{\hat{\pi}(1 - \hat{\pi})}{n_1} + \frac{\hat{\pi}(1 - \hat{\pi})}{n_2}}. \tag{9.18}$$

An equivalent form of Equation (9.18) is

$$\hat{\sigma}_{p_1 - p_2} = \sqrt{\hat{\pi}(1 - \hat{\pi})\left(\frac{1}{n_1} + \frac{1}{n_2}\right)}. \tag{9.19}$$

Since the sampling distribution is specified as $N(\pi_1 - \pi_2 = 0, \hat{\sigma}_{p_1 - p_2})$, the p-value is determined by expressing the sample result $(p_1 - p_2)$ in terms of the standard normal distribution:

$$Z = \frac{(p_1 - p_2) - (\pi_1 - \pi_2)}{\hat{\sigma}_{p_1 - p_2}}. \tag{9.20}$$

For a two-tailed test the alternate hypothesis is $H_1: \pi_1 - \pi_2 \neq 0$. The p-value gives the probability of an absolute difference in the sample result $(p_1 - p_2)$ and the hypothesized value $(\pi_1 - \pi_2)$ at least as extreme as the observed difference d, assuming that the null hypothesis is true. This is

$$P[|(p_1 - p_2) - (\pi_1 - \pi_2)| \geqq d | H_0 \text{ true}]. \tag{9.21}$$

Again, since $\pi_1 - \pi_2$ is hypothesized as zero, the observed difference d is simply the sample result $(p_1 - p_2)$ and the p-value may be shortened to

$$P(|p_1 - p_2| \geqq d | H_0 \text{ true}). \tag{9.22}$$

Hypotheses and p-values for one and two-tailed tests are given in Table 9.3.

TABLE 9.3: p-VALUES FOR ONE- AND TWO-TAILED TESTS ABOUT THE PARAMETER $\pi_1 - \pi_2$

Alternate Hypothesis H_1:	Null hypothesis $H_0: \pi_1 - \pi_2 = 0$
$\pi_1 - \pi_2 \neq 0$	$P(|p_1 - p_2| \geqq d | H_0 \text{ true})$
$\pi_1 - \pi_2 < 0$	$P[(p_1 - p_2) \leqq d | H_0 \text{ true}]$
$\pi_1 - \pi_2 > 0$	$P[(p_1 - p_2) \geqq d | H_0 \text{ true}]$

Note: d is the observed difference $(p_1 - p_2)$.

EXAMPLE. A dairy products company is considering two different television advertisements for promotion of a new product. Management believes that advertisement A is more effective than advertisement B. Two test market areas with virtually identical consumer characteristics are selected; ad A is used in one area and ad B in the other area. In a random sample of 60 customers who saw ad A, 18 tried the product. In a random sample of 100 customers who saw ad B, 22 tried the product. Does this indicate that ad A is more effective than ad B, if a 0.05 level of significance is used?

Solution. Let π_1 and π_2 represent the proportion of customers who would buy the product after being exposed to ad A and ad B, respectively. A one-tailed test with the following hypotheses is appropriate:

$$H_0: \quad \pi_1 - \pi_2 = 0;$$
$$H_1: \quad \pi_1 - \pi_2 > 0.$$

The sample proportion of customers buying the product after viewing ad A is $p_1 = x_1/n_1 = \frac{18}{60} = 0.30$. For those viewing ad B, the proportion is $p_2 = X_2/n_2 = \frac{22}{100} = 0.22$. This gives an observed sample result $(p_1 - p_2) = (0.30 - 0.22) = 0.08$. The estimated population proportion is

$$\hat{\pi} = \frac{X_1 + X_2}{n_1 + n_2} = \frac{18 + 22}{60 + 100} = 0.25.$$

The statistic $(p_1 - p_2)$ is normally distributed with mean zero and standard error

$$\hat{\sigma}_{p_1 - p_2} = \sqrt{\frac{\hat{\pi}(1 - \hat{\pi})}{n_1} + \frac{\hat{\pi}(1 - \hat{\pi})}{n_2}}$$

$$= \sqrt{\frac{(0.25)(0.75)}{60} + \frac{(0.25)(0.75)}{100}}$$

$$= \sqrt{0.005} = 0.07.$$

This distribution is portrayed in Figure 9.6. The p-value gives the probability

FIGURE 9.6: p-VALUE FOR COMPARING TWO ADVERTISEMENTS

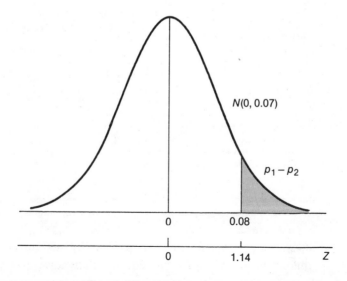

$N(0, 0.07)$

$p_1 - p_2$

0   0.08

0   1.14   Z

that the sample result $(p_1 - p_2) = 0.08$ differs from the hypothesized value of $(\pi_1 - \pi_2) = 0$ by an amount of at least 0.08, assuming H_0 is true. The Z-value is

$$Z = \frac{(p_1 - p_2) - (\pi_1 - \pi_2)}{\hat{\sigma}_{p_1 - p_2}}$$

$$= \frac{0.08 - 0}{0.07} = 1.14.$$

The p-value, using Appendix E, is

$$P[(p_1 - p_2) \geq 0.08 \,|\, H_0 \text{ true}] = P(Z \geq 1.14) = 0.12714.$$

Since the p-value is larger than $\alpha = 0.05$, the null hypothesis is accepted. The sample result is insufficient evidence for management to conclude that advertisement A is more effective than advertisement B.

As before, other decision rules may be used. In the above example, comparing the critical value of $Z = 1.64$ with the computed value $Z = 1.14$ indicates that H_0 is accepted.

Tests for proportions based on small samples ($n < 50$) are made with the binomial distribution. These tests are not discussed in this book. The interested reader may consult the reference by Mendenhall and Reinmuth.

9.8 SUMMARY

Interval estimators and tests for μ based on the t distribution are covered in this chapter. The use of the t distribution for testing the difference in two population means is also presented. These are inference procedures for means based on small samples. Large-sample tests for the difference in two population means and the difference in two population proportions are also included in this chapter.

This chapter suggests that caution is required in specifying the sampling distribution of means when small samples are taken from populations with unknown standard deviation. If, in those situations, the population is normally distributed, the test is based on the t-distribution. Where two populations are involved, as in the test for a difference in population means, an additional assumption of equal population variances is required.

How restrictive are these assumptions? First, the assumption of a normal population is met for a wide variety of applications. Though not illustrated, a test for normality is available. However, we know that mild departures from normality affect the results very little. Consequently the procedures outlined in this chapter are widely used for many populations with unimodal distributions that are not highly skewed. An obvious word of caution is appropriate. The larger the departure from normality, or from any other assumption, the less reliable are the results. The same caution applies to the assumption of equal population variances in the test for $\mu_1 - \mu_2$. As stated in the chapter, it is possible to test for equality of variances (see Chapter 11). When the assumptions underlying these tests are unacceptable, the methods of Chapter 15 are often applicable.

When should the t distribution be used in place of the normal in the process of statistical inference for population means? If the population is normal and the standard deviation is known, inference procedures based on the standard normal distribution apply, regardless of sample size. If samples are taken from normal populations where the standard deviation is unknown, the exact distribution for the appropriate statistic is Student's t-distribution. However, t is generally used only in those cases where the sample size is small ($n < 30$). If $n \geq 30$, the t distribution is approximated by the standard normal distribution.

9.9 TERMS AND FORMULAS

Degrees of freedom

Estimated standard error

Large sample

Level of significance

Pooled estimator of population
 variance

p-value

Sampling distribution

Small sample

Standard error of sampling
 distribution

Standard normal distribution

Student's t distribution

t statistic

Variance of sampling distribution

Interval estimator for μ with
 $df = n - 1$

$$\bar{X} \pm t\hat{\sigma}_{\bar{X}}$$

$$\text{where } \hat{\sigma}_{\bar{X}} = \frac{S}{\sqrt{n-1}}$$

$$\text{or } \frac{\hat{\sigma}}{\sqrt{n}}$$

t statistic in test for μ with
 $df = n - 1$

$$t = \frac{\bar{X} - \mu}{\hat{\sigma}_{\bar{X}}}$$

t statistic in test for $\mu_1 - \mu_2$ with
 $df = n_1 + n_2 - 2$

$$t = \frac{(\bar{X}_1 - \bar{X}_2) - (\mu_1 - \mu_2)}{S_{\bar{X}_1 - \bar{X}_2}}$$

$$\text{where } S_{\bar{X}_1 - \bar{X}_2} = \sqrt{\frac{S_{po}^2}{n_1} + \frac{S_{po}^2}{n_2}}$$

$$S_{po}^2 = \frac{n_1 S_1^2 + n_2 S_2^2}{n_1 + n_2 - 2}$$

Z statistic in test for $\mu_1 - \mu_2$

$$Z = \frac{(\bar{X}_1 - \bar{X}_2) - (\mu_1 - \mu_2)}{\sigma_{\bar{X}_1 - \bar{X}_2}}$$

$$\sigma_{\bar{X}_1 - \bar{X}_2} = \sqrt{\frac{\sigma_1^2}{n_1} + \frac{\sigma_2^2}{n_2}}$$

$$\hat{\sigma}_{\bar{X}_1 - \bar{X}_2} = \sqrt{\frac{S_1^2}{n_1} + \frac{S_2^2}{n_2}}$$

Z statistic in test for $\pi_1 - \pi_2$

$$Z = \frac{(p_1 - p_2) - (\pi_1 - \pi_2)}{\hat{\sigma}_{p_1 - p_2}}$$

$$\hat{\sigma}_{p_1 - p_2} = \sqrt{\frac{\hat{\pi}(1 - \hat{\pi})}{n_1} + \frac{\hat{\pi}(1 - \hat{\pi})}{n_2}}$$

$$\hat{\pi} = \frac{X_1 + X_2}{n_1 + n_2}$$

QUESTIONS

A. True or False

~~F~~ 1. In a hypothesis test for μ, the t distribution is used if a random sample of fewer than 30 items is drawn from a normal population where σ is known.

~~T~~ 2. The width of confidence intervals for μ based on the t distribution is not constant for fixed values of n and $1 - \alpha$.

~~F~~ 3. The standard normal distribution has larger dispersion than Student's t distribution for small sample sizes.

~~F~~ 4. The assumption of a normal population is not critical in the use of the t distribution.

~~T~~ 5. A t test of $H_0: \mu_1 - \mu_2 = 0$ assumes that the population variances are equal.

~~T~~ 6. The expected value of the difference of two random variables equals the difference in their expected values.

~~T~~ 7. The pooled estimator S_{po}^2 obtained from independent random samples is an unbiased estimator of the population variance.

~~F~~ 8. The samples need not be independent in the t test for $H_0: \mu_1 - \mu_2 = 0$.

B. Fill in

1. The t distribution approaches the ___normal___ distribution as sample size increases.

2. A random sample of 15 items is chosen from a normal population with unknown standard deviation for the purpose of obtaining an interval estimate of μ. The number of degrees of freedom for the appropriate t distribution is ___14___.

3. In the t distribution with 8 df, the value of c such that $P(|t| > c) = 0.05$ is ___2.306___.

4. The only parameter of the t distribution is the ___degree of freedom___.

5. In a hypothesis test, if $\alpha = 0.05$ and $0.05 < \text{p-value} < 0.10$, the null hypothesis should be ___accepted___.

6. The mean of a t distribution with 15 degrees of freedom equals ___0___.

7. In testing $H_0: \mu_1 - \mu_2 = 0$ with $n_1 = n_2 = 15$, the appropriate t distribution has ___26___ degrees of freedom.

8. In the t test $H_0: \mu = 140$ where $\sigma = 18$, $n = 10$, and $\alpha = 0.02$, the decision rule for the critical value is reject H_0 if $|t| >$ ___2.821___.

9. The decision limits of $\bar{X}$ for accepting H_0 in Question 8 are ___123.074___ and ___156.926___.

EXERCISES

1. A random sample of 10 credit account balances at a local department store shows an average balance of $82.70 with a standard deviation of $20.15. Assuming account balances are normally distributed, determine a 99 percent interval estimate for the average balance in all accounts.

2. Salaries, in thousands of dollars, for a random sample of chemists at the Dull Chemical Company are: 16, 13, 15, 17, 12. Assume that chemists' salaries are normally distributed.
   a. Determine a 0.95 interval estimate of the average salary for all chemists at the Dull Chemical Company.
   b. Test the hypothesis that average salary (μ) is $17,000 at the 0.10 level of significance.

3. The telephone company finds that the average number of directory assistance calls this last month for a random sample of 28 customers is 8 with a standard deviation of 2.75. Find a 98 percent confidence interval for μ, the average number of directory assistance calls for all customers this last month. The number of directory assistance calls per customer is normally distributed.

4. The state personnel department believes that the average number of days' sick leave requested annually (X) by state employees is 8. A random sample of 15 employees records is selected; the sample results are $\bar{X} = 5$ and $S = 2.1$ days. Assume that X is normally distributed. Does the sample result differ significantly from the 8-day belief, if $\alpha = 0.05$?

5. A builder tells a prospective purchaser of an all-electric ranch-style home that the average number of kilowatt-hours (KWH) used per home in the coldest month this year, January, does not exceed 4,500. The prospective purchaser interviews a random sample of eight owners of the same model of home in this subdivision and finds sample results of $\bar{X} = 6,260$ KWH and $S = 1,500$ KWH. Should the prospective buyer believe the builder? Base your answer on a test of hypothesis at $\alpha = 0.05$. Assume that KWH used is normally distributed.

6. A California wine producer states that the volume of wine in its standard size bottles averages 750 milliliters (ml). A state Alcoholic Beverage Control Board examines a random sample of 17 of these bottles, finding an average volume of 721 ml and standard deviation of 48 ml. Does the ABC Board have any reason to suspect that the average volume in all these bottles is less than 750 ml? Volume is normally distributed; let $\alpha = 0.01$.

7. The manager of a large high-rise condominium development expresses to his lender that the average family income of his tenants is $42,000. Since

the lender also holds mortgages on a large number of these units, a sample of reported family income can be easily obtained. A random sample of 20 files finds average family income $\bar{X} = \$36,000$ and standard deviation $S = \$16,000$. Assume that family income (X) is normally distributed. Has the manager overstated average family income? Use $\alpha = 0.01$.

8. In a study to compare surgical charges among U.S. cities, the average surgical cost for an appendectomy is $510 for a random sample of 50 patients in New York City; the standard deviation is $70. A random sample of 50 San Francisco patients incurred average surgical charges of $465 for an appendectomy with standard deviation of $50. Does this indicate that the average surgical cost for appendectomies is different in these two cities at the 0.01 level of significance?

9. The admissions officer at a large university wishes to determine whether scores on the Scholastic Aptitude Test (SAT) are different for female and male applicants. Random samples of applicants' files are taken and summarized below for both the verbal and mathematics sections of the SAT.

		Applicants	
		Female	Male
Section		(n = 399)	(n = 204)
Verbal	$\bar{X}$	502.1	510.5
	S	86.2	90.4
Mathematics	$\bar{X}$	512.2	551.1
	S	81.9	84.1

Using the above sample data and $\alpha = 0.05$, test the null hypothesis that:
a. The average verbal score is the same for the population of male and female students.
b. The average mathematics score is the same for the population of male and female students.

10. Public school teachers in one county (A) claim that their average salary is lower than in an adjacent county (B). Their claim is based on random samples of salaries in the two counties. The average annual salary of 50 teachers in county A is $11,586 with a standard deviation of $1,200. In county B, the average annual salary of 80 teachers is $12,450 with standard deviation $1,500. Does this evidence support the claim of county A teachers? Use $\alpha = 0.05$.

11. Salaries, in thousands of dollars, for a random sample of chemists at the Boring Chemical Company are: 14, 19, 20, 21, 11. Assume that salaries are normally distributed for the population of chemists at Boring. Refer to Exercise 2 and test the hypothesis that there is no difference in the average salaries of chemists at the Dull and Boring chemical companies. Let $\alpha = 0.10$.

12. In an effort to promote energy conservation through car-pooling by employees, a company is considering the institution of a rule at all plants requiring at least three passengers in each car that is allowed free parking. Parking attendants at the Southside plant have provided the results of a random sample of 15 cars. A random sample of 12 cars is obtained at the West End plant. Letting X_1 and X_2 represent the passengers per car at Southside and West End plants, respectively, the results are:

$$\bar{X}_1 = 1.8 \qquad \bar{X}_2 = 2.9$$
$$S_1 = 1.5 \qquad S_2 = 1.6.$$

Is there a difference in the average passengers per car for all cars parking at these two plants? Assume passengers per car is normally distributed; let $\alpha = 0.10$.

13. The research department of a historical society has developed a chemical that they claim will lengthen the life of paper treated with the chemical. Before agreeing to allow some very old and valuable manuscripts to be treated with the chemical, the society's governing board requests statistical evidence that the paper life is lengthened by the chemical treatment. Some identical papers are selected for comparison. Twelve sheets are randomly selected and treated with the chemical. Nine sheets are left untreated. Then all papers are aged artificially by an oven process. After aging the papers are tested for tear resistance by a machine that precisely measures the force required to tear the papers. The force required to tear the treated papers averaged 0.052 grams with a standard deviation of 0.015 grams. For the untreated papers, average tear force was 0.036 grams with a standard deviation of 0.010 grams. Does the sample evidence indicate that treatment with the chemical actually improves paper life as measured by the tear force? Let $\alpha = 0.01$; assume tear force is normally distributed.

14. A government researcher investigating automobile safety suspects that female drivers are more conscientious about fastening seat belts than males. Data for this characteristic, as well as other information, have been obtained by observing drivers at intersections throughout the U.S. In a random sample of 500 male drivers, 85 wore seat belts. Eighty in a random sample of 400 female drivers wore seat belts. Use these data to test the researcher's claim at the 0.05 level of significance.

15. A national food chain is test-marketing unbranded food products, which are priced about a third below major brands, in its retail outlets. In order to decide whether they should continue to offer unbranded products, they are surveying customer opinion. One concern is whether customer preference for unbranded products differs between east-coast and west-coast stores. Seventy out of a sample of 200 customers in west-coast stores prefer to purchase unbranded items. In a random sample of 300 customers in east-coast stores, 75 prefer unbranded products. Should the chain conclude, based on the sample information, that there is no difference in

the proportion of customers in east- and west-coast stores who prefer unbranded products? Let $\alpha = 0.05$.

16. A CPA firm wishes to know if there is any difference in the ability of accounting graduates from Canon University and those from Kenmore University. Using the pass rate on the CPA examination as an indicator, they obtain data for a random sample of 60 students from each school who recently took the examination. Twelve of the Canon graduates and 18 of the Kenmore graduates passed the examination. Test the hypothesis that there is no difference in the proportion of accounting graduates from the two schools who pass the CPA examination. Use $\alpha = 0.10$.

17. Use the following sample results for a two-tailed test that $H_0: \pi_1 - \pi_2 = 0$ at the 0.01 level of significance.

$$p_1 = 0.40 \qquad p_2 = 0.46$$
$$n_1 = 100 \qquad n_2 = 200.$$

18. Decision rules for p-values, critical values of t and decision limits for $\bar{X}$ produce the same conclusion in a test of hypothesis. Establish all three decision rules for Exercise 4 and verify that their application produces the same conclusion.

19. Interval estimation of the difference in two random variables follows the same procedure as for other parameters by using the general form

$$\text{Point estimator} \pm (Z)(\text{standard error}).$$

Use this concept and the data in Exercise 8 to establish a 95 percent interval estimate of the difference in the mean surgical cost for appendectomies in New York City and San Francisco. Interpret the result.

20. Use the general form of an interval estimator

$$\text{Point estimator} \pm (Z)(\text{standard error})$$

and the data in Exercise 15 to establish a 90 percent interval estimate of the difference in the proportion of customers in east- and west-coast stores who prefer unbranded products.

21. Use the general form of an interval estimator

$$\text{Point estimator} \pm (t)(\text{standard error})$$

and the data in Exercise 13 to establish a 95 percent interval estimate of the difference in the mean force required to tear the treated and untreated papers. What interpretation can you give to the result?

REFERENCES

Bhattacharyya, Gouri K., and Johnson, Richard A. *Statistical Methods and Concepts.* New York: John Wiley & Sons, Inc., 1977.

Eisenhart, Churchill "On the Transition from 'Student's'z to 'Student's't." *The American Statistician*, vol. 33, no. 1 (February 1979), pp. 6–10.

Kuebler, Roy R., and Smith, Harry Jr., *Statistics, A Beginning.* New York: John Wiley & Sons, Inc., 1976.

McClave, James T., and Benson, George *Statistics for Business and Economics.* San Francisco: Dellen Publishing Company, 1978.

Mendenhall, William, and Reinmuth, James E. *Statistics for Management and Economics,* 3d ed. North Scituate, Mass.: Duxbury Press, 1978.

Pfaffenberger, Roger C., and Patterson, James H. *Statistical Methods for Business and Economics.* Homewood, Ill.: Richard D. Irwin, Inc., 1977.

The Chi-Square Distribution

10

**Optional material.

303

10 THE CHI-SQUARE DISTRIBUTION

To throw in a fair game at Hazards only three spots . . .
is a natural occurrence and deserves to be so deemed;
and even when they come up the same way for a second
time, if the throw be repeated. If the third and
fourth plays are the same, surely there is occasion
for suspicion on the part of a prudent man.

Girolamo Cardano
De Vita Propria Liber (1574)

**10.1
INTRODUCTION**

The interest of Cardano, as cited above, lay in a game of chance that involved the tossing of dice. We have noted before that the analysis of games of chance was a major attraction of the early probabilists, and Cardano was no exception. In this chapter, one illustration of the inferential procedures to be introduced is a test of the hypothesis that a die is fair.

The introduction of a new probability distribution must precede the further study of inferential procedures. Several probability distributions have thus far been presented and used in various applications, for example, hypothesis tests of proportions and means. After the initial portion of this chapter, the student will have added another distribution to his repertoire: the chi-square (χ^2). The definition of a random variable with a chi-square distribution is given, as well as its parameters and procedures for determining probabilities associated with the distribution.

The chapter then reviews three important applications of the chi-square distribution. First, it is used in testing hypotheses about the variance of a normally distributed population. Second, it is used in the goodness-of-fit test developed by Karl Pearson. And lastly, we employ the distribution to test for

> ### KARL PEARSON (1857–1936)
>
> Pearson was born in London, the son of a Queen's Counselor (lawyer). His degree in mathematics was granted by Cambridge in 1879. His early interests were quite varied, including poetry, philosophy, and religion, but in 1890 he met W. F. R. Weldon. Both were faculty members at University College, London, and Weldon channeled Pearson's considerable talents toward statistics. His numerous contributions include the development of multiple and partial correlation and the chi-square goodness-of-fit test. He was the originator of the terms coefficient of variation, standard deviation, mode, array, and others. He was co-founder and editor from 1901 to 1936 of the statistical journal *Biometrika*. Overall, his contributions are such that many consider Pearson "the founder of the science of statistics."

independence between two variables of classification. Beyond the utility of the distribution in the above applications, the *F*-distribution of the following chapter is defined in terms of chi-square.

10.2 THE CHI-SQUARE DISTRIBUTION

Suppose we are given a random variable X *known* to be normally distributed with a mean of μ and variance of σ^2; i.e., X is $N(\mu, \sigma)$. Then standardize X, using the equation

$$Z = \frac{(X - \mu)}{\sigma},$$

to obtain the random variable Z; Z is $N(0, 1)$. Now square and obtain

$$Z^2 = \frac{(X - \mu)^2}{\sigma^2}.$$

The random variable Z^2 has a chi-square distribution, and it shall be denoted symbolically by χ_1^2.

The variable χ_1^2 will be of interest to us, so let us describe its distribution. First, since χ_1^2 is a squared value, it can never be negative. Second, the value of Z, the standard normal variable, will lie between -1 and 1 approximately 68 percent of the time (recall the normal distribution); hence $\chi_1^2(Z^2)$ will be between 0 and 1 approximately 68 percent of the time. Similarly, χ_1^2 will lie between 1 and 4 approximately 27 percent of the time, etc. The distribution of χ_1^2 can be presented graphically as in Figure 10.1. The notation χ_1^2 is read chi-square with one degree of freedom.

Now suppose we take two independent, normally distributed random variables X_1 and X_2, standardize each, square, and sum:

$$\chi_2^2 = \sum_{i=1}^{2} \left[\frac{X_i - \mu_i}{\sigma_i} \right]^2.$$

FIGURE 10.1: THE CHI-SQUARE DISTRIBUTION WITH ONE DEGREE OF FREEDOM.

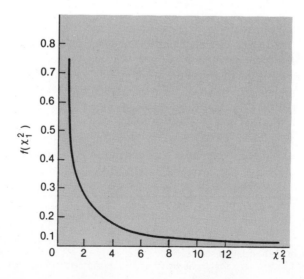

This is a value of the χ^2 variable for two degrees of freedom. In general for n independent, normal random variables

$$\chi_n^2 = \sum_{i=1}^{n} \left[\frac{X_i - \mu_i}{\sigma_i} \right]^2 = \sum_{i=1}^{n} Z_i^2. \qquad (10.1)$$

In words, the sum of the standardized, squared values of n independent, normal random variables is chi-square distributed with n degrees of freedom.

The χ^2 distributions with four and ten degrees of freedom, v (read nu), are shown in Figure 10.2. Several observations can be made by examining the graphs. First, as noted in Figure 10.1, the height of the density function decreases as the value of the random variable increases; this is also true for $v = 2$. However, for small values of $v > 2$, this is no longer true, but the distribution remains highly skewed to the right. As v becomes large, say greater than 10, the distribution becomes more symmetrical. In fact, as v becomes infinite, its limit is the normal distribution.

FIGURE 10.2: GRAPH OF CHI-SQUARE DISTRIBUTION WITH FOUR AND TEN DEGREES OF FREEDOM

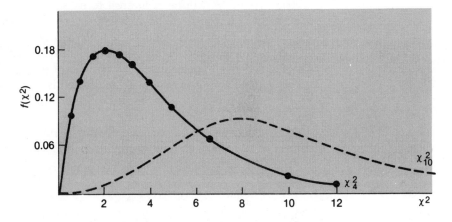

As a probability density function, the total area under a χ^2 distribution is equal to one, and area under the curve is interpreted as a probability. For example, the probability of a χ_1^2 variable exceeding 1.0 is approximately 0.32. A table of the distribution is found in Appendix H. Given a probability, α, and a number of degrees of freedom, v, the table gives a value such that the probability of a χ_v^2 distributed random variable *exceeding* the tabular value is α. Our notation will be $\chi_{v,\alpha}^2$.[1]

EXAMPLE. Suppose we have random variable known to be χ^2 distributed with $v = 12$. The probability is 0.10 that a randomly chosen value of χ^2 will exceed 18.55, i.e. $\chi_{12,0.10}^2 = 18.55$. The probability is 0.95 that the value of the random variable will exceed 5.23 (i.e., $\chi_{12,0.95}^2 = 5.23$); conversely, the prob-

---

[1] Our investigation of the chi-square distribution thus far has been limited to a description based on the definition of the random variable, $\chi_n^2 = \sum Z_i^2$, and, using our knowledge of the standard normal random variable, an inference of the characteristics of the chi-square distribution. For

ability is 0.05 that the random value is *less than* 5.23. (Recall that in continuous distributions the probability that the random variable will assume a *particular* value is zero.)

The sole parameter of the chi-square distribution is v, *the number of degrees of freedom*. Given v, the chi-square distribution is completely specified. It can be mathematically proven that the expected value (mean) of χ_v^2 is v and that the variance is $2v$.

$$E(\chi_v^2) = v. \qquad (10.2)$$
$$\text{Var}(\chi_v^2) = 2v. \qquad (10.3)$$

It can also be shown that the mode of χ_v^2 equals $v - 2$ *if* $v > 2$. If v equals 1 or 2, the mode is zero.

$$Mo(\chi_v^2) = v - 2 \quad \text{if } v > 2$$
$$Mo(\chi_v^2) = 0 \qquad \text{if } v = 1, 2. \qquad (10.4)$$

Note that $Mo(\chi_v^2) < E(\chi_v^2)$. This is because of the positively skewed nature of the χ^2 distribution.

EXAMPLE.   Sketch the chi-square distribution for $v = 7$. Determine the mean, mode, and variance of this distribution. Determine the value of $\chi_{7,\,0.25}^2$.

From the preceding discussion, we recall that for small v, the distribution is positively skewed. The mode will be $v - 2 = 7 - 2 = 5$. With this knowledge we can sketch the distribution as in Figure 10.3. The mean is $E(\chi_7^2) = 7$, and the variance is $\text{Var}(\chi_7^2) = 2(7) = 14$. From Appendix H, $\chi_{7,\,0.25}^2 = 9.04$.

We conclude this section with the additive property of the chi-square distribution.

Additive property of χ^2: If X and Y are two *independent* random variables possessing χ^2 distributions of v_1 and v_2 degrees of freedom respectively; then if $W = X + Y$, W will be χ^2 distributed with $v_1 + v_2$ degrees of freedom.

instance, it has been observed that chi-square is a continuous distribution; this, of course, implies the existence of some probability density function which is, for $X > 0$,

$$f(x) = \frac{1}{(2)^{v/2}\Gamma(v/2)} x^{(v-2)/2} e^{-x/2},$$

where $\Gamma(v/2)$ is a gamma function and involves mathematics beyond that required for this course. However, in the special case that b is a positive integer, $\Gamma(b) = (b - 1)!$ Thus, in certain instances, we can simplify $f(x)$. To illustrate, if $v = 4$, $\Gamma(v/2) = \Gamma(2) = 1! = 1$, and

$$f(x) = \frac{1}{(2)^{4/2}(1)} x^{(4-2)/2} e^{-x/2}$$

$$= \frac{1}{4} x e^{-x/2}.$$

Selected values of x and $f(x)$ can be plotted, enabling one to sketch the graph in Figure 10.2.

FIGURE 10.3

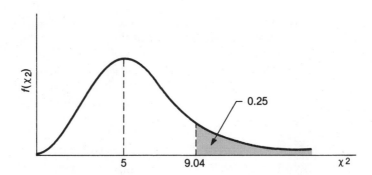

EXAMPLE. Suppose X is chi-square distributed with six degrees of freedom, and Y possesses a chi-square distribution with nine degrees of freedom. If X and Y are independent, what is the distribution of $W = X + Y$?

By the additive property of χ^2, W is chi-square distributed with 15 degrees of freedom.

**10.3 TESTS ON A VARIANCE

The earlier discussion of statistical inference was concerned with hypothesis tests of means, proportions, and differences. In this section, we shall develop a procedure for testing hypotheses concerning population variances.

Suppose we are given a sample of a size n from a population known to be $N(\mu, \sigma)$. The deviations about the mean could be written as

$$(X_i - \mu) = (X_i - \bar{X}) + (\bar{X} - \mu),$$

and the squared deviations would then be

$$(X_i - \mu)^2 = [(X_i - \bar{X}) + (\bar{X} - \mu)]^2$$
$$= (X_i - \bar{X})^2 + 2(\bar{X} - \mu)(X_i - \bar{X}) + (\bar{X} - \mu)^2.$$

The sum of the squared deviations over the sample is

$$\sum_{i=1}^{n} (X_i - \mu)^2 = \sum_{i=1}^{n} [(X_i - \bar{X})^2 + 2(\bar{X} - \mu)(X_i - \bar{X}) + (\bar{X} - \mu)^2]$$
$$= \sum(X_i - \bar{X})^2 + 2\sum(\bar{X} - \mu)(X_i - \bar{X}) + \sum(\bar{X} - \mu)^2.$$

Since $(\bar{X} - \mu)$ is a constant, and $\sum(X_i - \bar{X}) = 0$ (recall the mathematical properties of the mean of Section 2.9), the above expression simplifies to

$$\sum(X_i - \mu)^2 = \sum(X_i - \bar{X})^2 + n(\bar{X} - \mu)^2.$$

Dividing both sides by the variance σ^2 and rearranging the last term, we obtain

$$\frac{\sum(X_i - \mu)^2}{\sigma^2} = \frac{\sum(X_i - \bar{X})^2}{\sigma^2} + \frac{(\bar{X} - \mu)^2}{\sigma^2/n} \tag{10.5}$$

The last term now represents the standardized, squared value of a sample of size 1 from the sampling distribution of $\bar{X}$, which has a mean of μ and variance of $\sigma_{\bar{X}}^2 = \sigma^2/n$. This distribution is normal for *any* n because the population distribution was specified as normal. Therefore, the statistic defined by this term is chi-square distributed with one degree of freedom (see Section 10.2). The statistic defined by the first term is distributed according to χ_n^2; utilizing the additive property of the χ^2 distribution we determine that the second term of Equation (10.5) is χ^2 distributed with $v = n - 1$.[2]

In Section 2.4 the sample variance was defined to be

$$S^2 = \frac{\sum(X_i - \bar{X})^2}{n}.$$

The second term of Equation (10.5) is now rewritten as

$$\frac{\sum(X_i - \bar{X})^2}{\sigma^2} = \frac{nS^2}{\sigma^2}. \tag{10.6}$$

A most important conclusion has been obtained: Given a population known to be $N(\mu, \sigma)$, the sampling distribution of nS^2/σ^2 is χ_{n-1}^2. In other words, if we have a normally distributed population with a variance of σ^2 and take a random sample of size n from this population and compute nS^2/σ^2, then the distribution of this statistic is chi-square with $v = n - 1$. (See Figure 10.4).

FIGURE 10.4

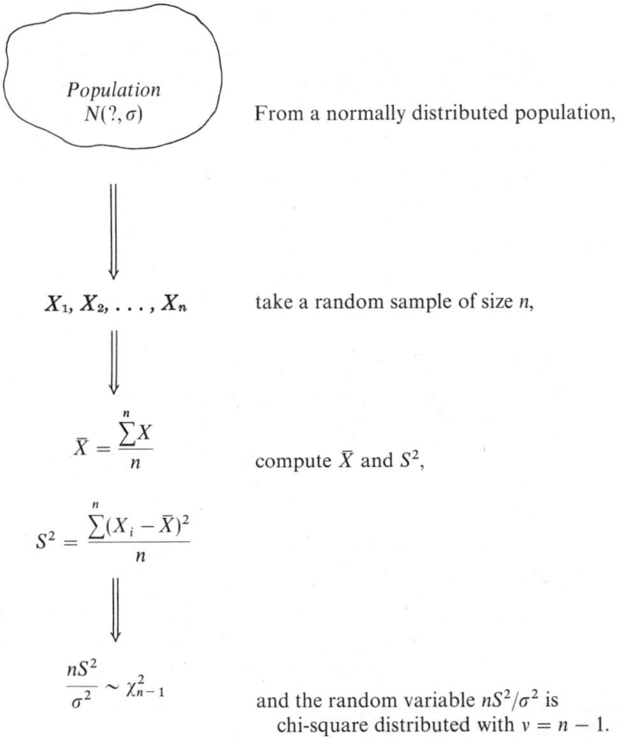

Population
$N(?, \sigma)$

From a normally distributed population,

$X_1, X_2, \ldots, X_n$ take a random sample of size n,

$$\bar{X} = \frac{\sum^n X}{n}$$

compute $\bar{X}$ and S^2,

$$S^2 = \frac{\sum^n (X_i - \bar{X})^2}{n}$$

$$\frac{nS^2}{\sigma^2} \sim \chi_{n-1}^2$$

and the random variable nS^2/σ^2 is chi-square distributed with $v = n - 1$.

---

[2] To utilize the additive property of χ^2, the first and last terms of Equation (10.5) must be independent. This proves to be true, but such a proof is beyond the scope of this book.

The student should anticipate the procedure for testing hypotheses concerning variances because it is precisely the procedure used in Chapter 8. We shall take a random sample of size n from a population assumed to be normally distributed and hypothesize that:

$$H_0: \quad \sigma^2 = \sigma_0^2$$
$$H_1: \quad \sigma^2 \neq \sigma_0^2,$$

where σ_0^2 is some hypothesized value. If the null hypothesis, H_0, is true, the distribution of nS^2/σ_0^2 is χ_{n-1}^2 (and the expected value of nS^2/σ_0^2 is $n-1$). A *test statistic* is an observed or computed value used to test the hypothesis. And, if the test statistic, nS^2/σ_0^2, is a "rare" event, i.e., its probability of occurrence (p-value) is less than α, the null hypothesis is rejected. Otherwise accept H_0. As before, α is the probability of rejecting a true hypothesis. A decision rule based on the critical value of χ^2 specifies that H_0 is rejected in a two-tailed test if the computed value is less than $\chi_{1-\alpha/2}^2$ or greater than $\chi_{\alpha/2}^2$. The critical values of χ^2 are obtained from Appendix H.

EXAMPLE. Suppose that national medical studies have established the variance of weights of adult women to be 120. A sociologist studying a small religious sect incidentally noticed relatively few obese or lean women (the sect practiced severe dietary restrictions). Assuming the weights of the women are normally distributed and that a random sample of 30 women had a variance, S^2, of 60, test the hypothesis that the sect's variance is equal to that for adult women nationally, i.e., $H_0: \sigma^2 = 120$. Use a significance level of 0.05.

The pertinent information is summarized symbolically as:

$$H_0: \quad \sigma^2 = 120 \qquad n = 30$$
$$H_1: \quad \sigma^2 \neq 120 \qquad S^2 = 60$$
$$\alpha = 0.05.$$

If H_0 is true, nS^2/σ^2 is chi-square distributed with $30 - 1 = 29$ degrees of freedom. The computed value of the random variable in this instance is:

$$\chi^2 = \frac{nS^2}{\sigma^2} = \frac{30(60)}{120} = 15.$$

The hypothesized distribution with the appropriate critical values is sketched in Figure 10.5. The decision rule for the critical value of χ^2 indicates that H_0 should be rejected if the computed value is less than 16.05 or greater than 45.7. Since the computed value of chi-square is less than 16.05, H_0 is rejected.

From our study of Chapter 8, we should be able to make one-tailed tests without further explanation. For example, if the previous problem's hypotheses had read:

$$H_0: \quad \sigma^2 = 120$$
$$H_1: \quad \sigma^2 < 120$$

FIGURE 10.5

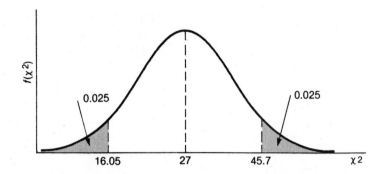

the rejection region would now lie entirely in the left tail and the critical value of $\chi^2 = 17.71$. The null hypothesis would still be rejected.

Finally, the χ^2 distribution can be used to compute confidence intervals for the population variance. The rationale is exactly that of Section 8.8, where the concept of confidence intervals for means was developed. We first note that

$$P\left[\chi^2_{n-1,\,1-\alpha/2} \leq \frac{nS^2}{\sigma^2} \leq \chi^2_{n-1,\,\alpha/2}\right] = 1 - \alpha$$

and therefore

$$P\left[\frac{nS^2}{\chi^2_{n-1,\,\alpha/2}} \leq \sigma^2 \leq \frac{nS^2}{\chi^2_{n-1,\,1-\alpha/2}}\right] = 1 - \alpha. \tag{10.7}$$

EXAMPLE. Using the information of the previous example, construct the 95 percent confidence interval for σ^2.

Using Equation (10.7),

$$P\left[\frac{nS^2}{\chi^2_{n-1,\,\alpha/2}} \leq \sigma^2 \leq \frac{nS^2}{\chi^2_{n-1,\,1-\alpha/2}}\right] = 1 - \alpha,$$

then the lower confidence limit is given by:

$$\frac{nS^2}{\chi^2_{29,\,0.025}} = \frac{30(60)}{45.7} = 39.39.$$

Similarly, the upper confidence limit is:

$$\frac{nS^2}{\chi^2_{29,\,0.975}} = \frac{30(60)}{16.05} = 112.15.$$

This indicates that, with 95 percent confidence, σ^2 is between 39.39 and 112.15.

10.4

GOODNESS-OF-FIT TESTS

This section is concerned with another important application of the chi-square distribution: *goodness-of-fit tests*. This is used to test the hypothesis that a random variable follows a specified (i.e., hypothesized) distribution. The procedure employs a criterion proposed by Karl Pearson in 1899.

Suppose we are given a set of J categories and, from sample data, the observed number of occurrences of the random variable in each category: O_j

where $j = 1, 2, \ldots, J$. Under the null hypothesis, assume that a set of expected frequencies for the J categories can be computed: E_j where $j = 1, 2, \ldots, J$. The data could be presented as in Table 10.1.

TABLE 10.1

Category	1	2	3 . . . . J
Observed frequency	O_1	O_2	O_3 . . . . O_J
Expected frequency	E_1	E_2	E_3 . . . . E_J

Utilizing Pearson's criterion we compute a statistic known as Pearson's χ^2 (although no distinction is usually made between this and the previous χ^2):

$$\chi^2 = \sum_{j=1}^{J} \frac{(O_j - E_j)^2}{E_j}, \tag{10.8}$$

which is an index of the agreement between the observed and expected frequencies.[3] The numerator in Equation (10.8) reflects the absolute differences between the observed and expected frequencies; the denominator inversely weights the squared differences by the expected frequencies for each category. For example, if $|O_j - E_j| = 10$, this is a substantial difference if $E_j = 10$ but not if $E_j = 200$. Subject to conditions to be subsequently given,

> If the sample was taken from the distribution specified under the null hypothesis, Pearson's χ^2 will be approximately chi-square distributed with $v = J - 1$.[4]

The approximation is sufficiently accurate if: (1) when there are two categories, each $E_j \geq 10$; (2) when more than two categories, each $E_j \geq 5$. When using Pearson's χ^2 in goodness-of-fit tests, note that only *large* values of χ^2 discredit the null hypothesis. For example, in the case where $E_j = O_j$ for all J categories, the computed χ^2 equals zero, and this indicates perfect agreement between the observed and expected frequencies. Similarly, other values of χ^2 in the left tail of the distribution indicate that the null hypothesis is credible. Hence, *goodness-of-fit tests are one-tail tests with the critical region lying in the right tail.* Thus the decision rule for the critical value of χ^2 is: Reject H_0 if the computed χ^2

---

[3] Helen M. Walker relates the following in "The Contributions of Karl Pearson" [*Journal of the American Statistical Association*, vol. 53, no. 281 (March 1958), pp. 11–22]: "Quetelet and others who wanted to demonstrate the closeness of agreement between the frequencies . . . observed . . . and the frequencies calculated . . . merely printed the two series side by side and said in essence, 'Behold!' They had no measure of discrepancy and so far as I know they were not made uncomfortable by the lack of such a measure. Pearson not only devised the measure but he worked out its distribution . . ."

[4] For the goodness-of-fit tests illustrated in this section, the number of degrees of freedom (v) is always $J - 1$. One degree of freedom is lost because the sum of the expected frequencies is forced to equal the sum of the observed frequencies. In other goodness-of-fit tests, further restrictions are imposed and $v \neq J - 1$. For example, when testing the fit of a normal curve to observed data, the expected and observed frequencies are also forced to have an equal mean and variance. In such a problem, $v = J - 3$.

exceeds the critical value, otherwise accept. The decision rule for a p-value is: Reject H_0 if the p-value is less than α. As in the case of the t distribution, an exact p-value may not be obtainable from Appendix H. The form of the p-value is $P(\chi_v^2 \geq \text{Computed value} | H_0 \text{ true})$. An example is in order.

EXAMPLE. Suppose we are given a die that we hypothesize to be fair, i.e., all faces equally likely to appear. Then, if p_j equals the probability that the jth face appears:

$$H_0: \quad p_j = \tfrac{1}{6}$$
$$H_1: \quad p_j \neq \tfrac{1}{6}, \qquad j = 1, 2, \ldots 6$$

or equivalently:

$$H_0: \quad \text{the die is fair}$$
$$H_1: \quad \text{the die is not fair.}$$

Suppose that the die is thrown 48 times, and the one-through-six faces are observed 4, 7, 8, 13, 11, and 5 times, respectively. If the die is fair we would expect to see each face 8 times. The data are presented in Table 10.2.

TABLE 10.2

	Face					
	1	*2*	*3*	*4*	*5*	*6*
O_j	4	7	8	13	11	5
E_j	8	8	8	8	8	8

Then,

$$\chi^2 = \sum_{j=1}^{6} \frac{(O_j - E_j)^2}{E_j}$$

$$= \frac{(4-8)^2}{8} + \frac{(7-8)^2}{8} + \frac{(8-8)^2}{8} + \frac{(13-8)^2}{8} + \frac{(11-8)^2}{8} + \frac{(5-8)^2}{8}$$

$$= 7.5.$$

If the null hypothesis is true, the above statistic follows a chi-square distribution with $v = J - 1 = 5$. Setting $\alpha = 0.01$, the critical value is $\chi_{5,\,0.01}^2 = 15.09$. Since our computed value is 7.5, we accept H_0. We could reject H_0 only if the computed χ^2 were to *exceed* the critical value of χ^2.

Equation (10.8) can be simplified for computation as follows:

$$\sum_{j=1}^{J} \frac{(O_j - E_j)^2}{E_j} = \sum_{j=1}^{J} \frac{O_j^2 - 2O_j E_j + E_j^2}{E_j}$$

$$= \sum \frac{O_j^2}{E_j} - 2\sum \frac{O_j E_j}{E_j} + \sum \frac{E_j^2}{E_j}$$

$$= \sum \frac{O_j^2}{E_j} - 2\sum O_j + \sum E_j.$$

Since the total of the expected frequencies and the total of the observed frequencies are equal, $\sum O_j = \sum E_j = N$. Then,

$$\sum \frac{(O_j - E_j)^2}{E_j} = \sum \frac{O_j^2}{E_j} - 2N + N$$

$$= \sum \frac{O_j^2}{E_j} - N. \tag{10.9}$$

Equation (10.9) is computationally easier since it involves $(J - 1)$ fewer subtractions than Equation (10.8).

As indicated earlier, the Pearson χ^2 is closely approximated by the chi-square distribution if: (1) when there are two categories, each $E_j \geq 10$; (2) when more than two categories, each $E_j \geq 5$. However, if the appropriate condition is not *initially* satisfied, one may combine categories in order that the expected frequencies in all categories will satisfy the requirement. Theoretically, one may combine categories in any manner, but generally adjacent categories are pooled. When combining categories, v is based upon the number of categories in the *revised* table.

EXAMPLE. The Normal Company routinely purchases a certain type of bolt in lots of 1,000. The purchasing agent has been instructed to spread the orders among suppliers A, B, C, and D in the ratio of $2:2:1:1$. As a spot check, 24 purchase orders are randomly selected from the last six months' orders. Suppliers A, B, C, and D have received 13, 4, 4, and 3 orders, respectively. Does this indicate the instructions are being followed? Test at the 0.10 level.

The null hypothesis is:

H_0: purchases are distributed in the ratio of $2:2:1:1$;

H_1: purchases are not distributed in the ratio $2:2:1:1$.

Under H_0, supplier A is expected to receive $2/(2 + 2 + 1 + 1) = \frac{2}{6} = \frac{1}{3}$ of the purchase orders. Thus, of twenty-four sample orders, we expect A to receive eight. Similarly, suppliers B, C, and D are expected to receive $\frac{1}{3}, \frac{1}{6}$, and $\frac{1}{6}$, respectively, of the total number of orders. The data are presented as:

	Supplier			
	A	B	C	D
O_j	13	4	4	3
E_j	8	8	4	4

Note that in categories C and D, the expected frequencies are less than five. Combining these categories yields:

	Supplier		
	A	B	C and D
O_j	13	4	7
E_j	8	8	8

Counting C and D as a single supplier (category), the requirement that $E_j \geq 5$ is met, and H_0 becomes:

H_0: purchases are distributed according to 2:2:2;

H_1: purchases are not distributed according to 2:2:2.

Then using Equation (10.9),

$$\chi^2 = \sum_{j=1}^{3} \frac{O_j^2}{E_j} - N$$

$$= \frac{(13)^2}{8} + \frac{(4)^2}{8} + \frac{(7)^2}{8} - 24 = 5.25.$$

Since $\chi^2_{2,\,0.10} = 4.61$, H_0 is rejected. On the basis of the sample evidence, we conclude the agent is not following his instructions. Similarly, the p-value is $0.05 < P(\chi^2_2 \geq 5.25 \,|\, H_0 \text{ true}) < 0.10$, which also indicates rejection of H_0.

10.5 TESTS FOR INDEPENDENCE

This chapter concludes with a most useful application of the chi-square distribution: a test of the statistical independence of two variables of classification. This will be seen to be a natural extension of the goodness-of-fit tests.

This particular application is best introduced by an example. Suppose that a certain elective is offered to freshmen and sophomores on a pass-fail basis only. An advisor is interested in determining whether there is a relationship between the students' grades and class standings. The null hypothesis (H_0) that there is *no* relationship, i.e., that the two variables of classification are statistically independent, can be tested by the methods of this section; it follows that H_1 states that there is a relationship.

Sufficient data for the test were obtained from last semester's classes. The information is presented in Table 10.3. Table 10.3 is called a *contingency table*. In particular it is termed a 2 × 2 contingency table; there are two categories of

TABLE 10.3

	Class Standing		
Grade	Freshman	Sophomore	Total
Pass	150	250	400
Fail.	50	50	100
Total	200	300	500

the row variable (grade) and two categories of the column variable (class standing). One observes that there are four cells in the table, i.e., a student who took the course must belong in one of the four cells. (That the possibilities number four could also have been deduced from our knowledge of the fundamental principle of counting).

The numbers in the cells are the observed frequencies. O_{ij} will refer to the observed frequency of the cell in the ith row and jth column. For example $O_{12} = 250$. It remains for us to determine the expected frequency, E_{ij}, for each cell and the number of degrees of freedom.

If the null hypothesis is true, the two variables of classification are independent. Recall the multiplication rule for independent events in Section 4.7: $P(E \text{ and } A) = P(A) \cdot P(E)$. For example, if H_0 is true, then $P(\text{pass and freshman}) = P(\text{pass}) \cdot P(\text{freshman})$, and estimates of the marginal probabilities can be obtained from the table. Then,

$$P(\text{pass and freshman}) = P(\text{pass}) \cdot P(\text{freshman})$$

$$\frac{E_{11}}{500} = \frac{400}{500} \cdot \frac{200}{500},$$

$$E_{11} = 200 \cdot \frac{400}{500}$$

$$= 160.$$

In other words, if H_0 is true, in predicting the number of passes from a group of 200 students, knowing the class standing of the group is of no help whatsoever since $P(\text{pass}|\text{freshman}) = P(\text{pass})$. So that in a group of 200 we would expect $400/500 = \frac{4}{5}$, or 160, to pass regardless of their class standing. Similarly,

$$E_{22} = 300 \cdot \tfrac{100}{500} = 60.$$

In a 2×2 table, note that when one has determined any *one* of the four cells' expected frequency, the remaining three are then uniquely determined. The row and column totals of the expected frequencies must equal the totals shown in Table 10.3. Hence, after determining $E_{11} = 160$, E_{12} must equal 240 (although it could still be calculated using probabilities). Let $r =$ number of categories of the row variable and $c =$ number of categories of the column variable. *Then for an $r \times c$ contingency table, the number of degrees of freedom equals $(r - 1)(c - 1)$.*[5]

---

[5] For an $r \times c$ contingency table, the sum of the expected frequencies for a row *or* column must equal the sum of the observed frequencies for that row or column. Because of this constraint the expected frequencies in the last row and last column can be obtained by subtraction. Hence, of the rc expected frequencies, only $(r - 1)(c - 1)$ must be computed; the difference, $r + c - 1$, represents the degrees of freedom "lost" because of the constraints placed on totals.

Given the observed and expected frequencies and the number of degrees of freedom, we now modify Equation (10.8) to compute χ^2:

$$\chi^2 = \sum_{j=1}^{2} \sum_{i=1}^{2} \frac{(O_{ij} - E_{ij})^2}{E_{ij}}$$

$$= \frac{(150 - 160)^2}{160} + \frac{(250 - 240)^2}{240} + \frac{(50 - 40)^2}{40} + \frac{(50 - 60)^2}{60}$$

$$= 0.625 + 0.417 + 2.50 + 1.67$$

$$= 5.21.$$

Testing at the 0.05 level, $\chi^2_{1,0.05} = 3.84$. Since the observed value exceeds the critical value, the null hypothesis is rejected. For any $r \times c$ table, the E_{ij} can be computed from:

$$E_{ij} = \sum_{i} O_{ij} \cdot \frac{\sum_{j} O_{ij}}{N} \qquad (10.10)$$

Chi-square is then computed by:

$$\chi^2 = \sum_{j=1}^{c} \sum_{i=1}^{r} \frac{(O_{ij} - E_{ij})^2}{E_{ij}}. \qquad (10.11)$$

As a general rule, this test should not be employed unless all E_{ij} are at least five. The null hypothesis is that there is no relationship between the two variables of classification, i.e., they are statistically independent. An example shows how this may be extended to larger values of r and c.

EXAMPLE. Suppose we are interested in determining whether there is a relationship between one's education and one's voting in the 1976 presidential election. A random sample of 150 persons of voting age is taken and presented in Table 10.4.

TABLE 10.4

Highest Education	Voting		
	Yes	No	Total
Grade school	10 (14)	20 (16)	30
High school	30 (33)	40 (37)	70
College.	30 (23)	20 (27)	50
Total	70	80	150

H_0: There is no relationship between education and voting.

H_1: There is a relationship between education and voting.

$$\alpha = 0.01$$

The E_{ij} given in parentheses were determined using Equation (10.10). For example,[6]

$$E_{31} = \sum O_{i1} \cdot \frac{\sum O_{3j}}{N} = 70 \cdot \frac{50}{150}$$

$$= 23\tfrac{1}{3} \approx 23.$$

Using Equation (10.11),

$$\chi^2 = \sum_{j=1}^{2} \sum_{i=1}^{3} \frac{(O_{ij} - E_{ij})^2}{E_{ij}}$$

$$= \frac{(10 - 14)^2}{14} + \frac{(20 - 16)^2}{16} + \frac{(30 - 33)^2}{33} + \frac{(40 - 37)^2}{37}$$

$$= + \frac{(30 - 23)^2}{23} + \frac{(20 - 27)^2}{27}$$

$$= 6.59.$$

The number of degrees of freedom is given by

$$v = (r - 1)(c - 1) = (3 - 1)(2 - 1) = (2)(1) = 2.$$

The critical value of chi-square, $\chi^2_{2,\,0.01}$, equals 9.21. Since the computed value is less than the critical value, H_0 is accepted at the 0.01 level.

10.6 SUMMARY

This chapter served as an introduction to the continuous probability distribution, chi-square. The distribution has been defined, and we have observed that its sole parameter is its number of degrees of freedom, v.

The chi-square distribution was employed in three useful applications. First, it was used to test the hypothesis: $H_0: \sigma^2 = \sigma_0^2$ on the basis of a random sample of size n from a normally distributed population. Second, the chi-square distribution was used in goodness-of-fit tests where the hypothesis that a random variable follows some specified distribution is tested. Finally, the distribution was used with contingency tables to test for the statistical independence of the two variables of classification.

The latter two tests are referred to as nonparametric tests. The rationale of the term "nonparametric" and other tests of this nature are the subject matter of Chapter 15.

10.7 TERMS AND FORMULAS

Chi-square (χ^2)

Contingency table

Degrees of freedom (v)

Expected frequency (E_j)

---

[6] The E_{ij} in this example are rounded for convenience. In general, this is not a recommended practice.

Observed frequency (O_j)

Pearson's chi-square ($\sum(O_j - E_j)^2/E_j$)

Test statistic for $H_0: \sigma^2 = \sigma_0^2$

$$\chi^2 = \frac{nS^2}{\sigma_0^2}$$

Confidence interval for σ^2

    Lower limit

$$\frac{nS^2}{\chi_{n-1,\,\alpha/2}^2}$$

    Upper limit

$$\frac{nS^2}{\chi_{n-1,\,1-\alpha/2}^2}$$

Pearson's χ^2 (test statistic for goodness-of-fit)

$$\chi^2 = \sum_j \frac{(O_j - E_j)^2}{E_j}$$

Expected frequency in test for independence

$$E_{ij} = \sum_i O_{ij} \cdot \frac{\sum_j O_{ij}}{N}$$

Degrees of freedom in test for independence

$$v = (r-1)(c-1)$$

Pearson's χ^2 (test statistic in tests for independence)

$$\chi^2 = \sum_j \sum_i \frac{(O_{ij} - E_{ij})^2}{E_{ij}}$$

QUESTIONS

A. True or False

T 1. A random variable with a chi-square distribution is always nonnegative.

T 2. The sole parameter of a chi-square distribution is its number of degrees of freedom.

F 3. In goodness-of-fit tests, the rejection region lies in both tails of the distribution.

F 4. In the test of independence, H_0 states that the two variables are not independent.

T 5. Use of the chi-square distribution to test an hypothesis about a population variance assumes that the population is normally distributed.

F 6. The symbol E_{32} refers to the expected frequency of the cell of the third column and second row.

F 7. Large values of chi-square indicate agreement between the observed and expected frequencies.

F 8. For $v = 7$, the chi-square distribution would be negatively skewed.

B. Fill in

1. The mode of a chi-square distribution with v degrees of freedom is, for $v > 2$, _____$v - 2$_____.

2. The additivity property of chi-square distributed random variables holds only if they are _____independent_____.

3. For a goodness-of-fit test with two categories, each expected frequency must be greater than or equal to ___*10*___ .

4. The goodness-of-fit test was developed by ___*Karl Pearson*___ .

5. The data for a chi-square test of independence is usually presented in a ___*Contingency*___ table.

6. For small values of v, the chi-square distribution is ___*Positively*___ skewed.

7. In testing hypotheses about population variances, the number of degrees of freedom is ___*n − 1*___ .

8. For an $r \times c$ contingency table, the number of degrees of freedom is ___*(r − 1)(c − 1)*___ .

EXERCISES

1. Given a chi-square distributed random variable with ten degrees of freedom:
   a. Determine its mean, mode, and variance.
   b. Sketch the distribution.

2. For a chi-square distribution with $v = 20$:
   14.34 a. Determine the median (i.e., $\chi^2_{20,\,0.5}$).
   10.85 b. Determine $\chi^2_{20,\,0.95}$.
   28.4 c. Determine $\chi^2_{20,\,0.10}$.

3. Determine the following:
   a. $\chi^2_{14,\,0.05}$.
   b. $\chi^2_{3,\,0.01}$.
   c. $\chi^2_{5,\,0.90}$.
   d. The area to the right of 18.31 for χ^2_{10}.

4. If $X \sim \chi^2_5$ and $Y \sim \chi^2_{10}$, and $Z = X + Y$:
   a. Determine the distribution of Z if X and Y are independent.
   0.01 b. Determine the probability of Z exceeding 30.578.

5. A sample of size 20 from a normally distributed population shows a variance $S^2 = 40$.
   a. Test the hypothesis that the population variance is equal to 60; let $\alpha = 0.05$.
   b. Construct a 90 percent confidence interval for the population variance.

6. Last year the variability of patients' ages at University Hospital was determined to be $\sigma^2 = 64$. This year a random sample of sixty patients showed the variability of age as $S^2 = 90$. Assuming patients' age follow a normal distribution, is this year's variability significantly higher than last year's? Test at the 0.10 level.

$$\chi^2 = \frac{ns^2}{\sigma^2}$$ *84.375*   *Reject*

7. Salaries for professors in a university's economics department were determined to have a standard deviation, σ, of $4,000. A sample of ten English professors' salaries had a standard deviation of $2,800. Assume that

English salaries follow a normal distribution and test to see if there is a significant difference in the variability of the English salaries from that of economics. Use $\alpha = 0.05$.

8. An advertising manager believes that in Normal City, 40 percent of the residents receive only the morning paper, 30 percent receive only the evening paper, while 20 percent receive both and 10 percent neither. A random sample of fifty residents show 12 morning only, 18 evening only, 12 both, and the rest receive neither. Do these data support the manager's belief? Use a 0.10 level of significance.

9. In a World Series pool, each participant antes up the same amount (e.g., $2) and then draws a number from a hat. A participant's number, which will be a digit from zero to nine, is the winning number if it is the last digit in the sum of both teams' score for that game. For example, the number 5 would be a winner if the score were 3–2, 4–1, 9–6, 8–7, etc. Bryan Wilkinson, in the October 1968 issue of *The American Statistician*, furnished the winning digit for 300 World Series games.

Digit	Winner
0	15
1	27
2	23
3	49
4	31
5	41
6	27
7	38
8	23
9	26

Use his data to test the hypothesis that all digits have an equal chance to be the winner. Let $\alpha = 0.01$.

10. For School of Business undergraduate transfer students, the dean said he believed that $\frac{1}{2}$ came from Jeff State, $\frac{1}{4}$ from Adams College, and $\frac{1}{4}$ from others. The registrar checked this by selecting a random sample of 28 transfer students, and 9 were from Jeff State, 9 from Adams, and 10 from elsewhere. Do these data support the dean? (Use 0.05 level of significance.)

11. Commonwealth Insurance Company prices its policies for commercial business at either a discount rate, standard rate, or a premium rate by estimating the degree of risk involved. A company executive believes that 20 percent of commercial business is priced at a discount, 60 percent is standard, and 20 percent is premium. To check this, a company underwriter examines 50 current accounts, finding 12 discount, 33 standard, and 5 premium. Does this sample evidence support the executive's belief? Use a 0.10 level of significance.

12. A random sample of tax returns for 1976 has the characteristics given in the table below. Use these data to determine if there is a relationship between returns selected for audit and preparer. Let $\alpha = 0.05$.

	Audit	
Preparer	Yes	No
Self	32	118
Others	8	42

13. A random sample of current students at United University yields the following data:

	Sex	
Major	Male	Female
Business	20	10
Engineering	15	5
Liberal arts	30	20
Science	30	20

Use the data to determine if major is independent of sex. Let $\alpha = 0.01$.

14. A bookie handicapping college football games wants to determine if the home team winning is independent of the conference. He compiles the following data from a random sample of last fall's games.

	Home Team Wins	
Conference	Yes	No
Big 10	15	11
Pac 8	10	5
Big 8	13	20
SWC	20	14
SEC	12	10

Use the above to determine if home team winning is independent of the conference at the 0.10 level.

15. The Standard Co. is considering a change in its medical insurance package, a fringe benefit covering all employees. A recent sample of employees' opinions concerning the change is summarized below.

	Opinion		
Employee Classification	Favorable	Unfavorable	No Opinion
Administrative	19	5	6
Sales	10	5	5
Clerical	15	4	9
Production	16	16	20

Is employee opinion independent of job classification? Test at the 0.10 level of significance.

16. An electronic manufacturer has completed a market survey studying consumer perceptions of quality for the firm's various products. Part of the data shows:

	Rating		
Product	Superior	Average	Inferior
Radio	30	40	5
Stereo	10	20	5
BW TV	30	20	10
Color TV	10	40	10

Does the above indicate that consumer rating and product are independent?

17. Vacancies on the U.S. Supreme Court during a 96-year period are given below as reported by W. A. Wallis in the *Journal of the American Statistical Association*, vol. 31 (1936).

Number of Vacancies/Year	Frequency
0	59
1	27
2	9
3	1
4 or more	0

Note that the mean number of vacancies per year is $\bar{X} = \sum f_i X_i / \sum f_i = \frac{48}{96} = 0.5$. Use the above data to test the hypothesis that vacancies on the court follow a Poisson distribution with λ (mean) of 0.5. (Hint: The probability of zero vacancies in a year, $P(X = 0)$, is

$$e^{-0.5}(0.5)^{0}/0! = 0.6065;$$

in a 96-year period one would then expect (96×0.6065) years with no vacancy.) Note *two* degrees of freedom are lost because not only are the sums of E_j and O_j forced to be equal, but their means as well.

REFERENCES

Gibbons, Jean Dickinson *Nonparametric Methods for Quantitative Analysis*. New York: Holt, Rinehart and Winston, 1976.

Noether, Gottfried E. *Introduction to Statistics: A Nonparametric Approach*, 2nd ed. Boston: Houghton Mifflin Company, 1976.

The F Distribution and Analysis of Variance

11 THE *F* DISTRIBUTION AND ANALYSIS OF VARIANCE

In one sense, everything we discover is already there: a sculptured figure and the law of nature are both concealed in the raw material.

J. Bronowski
The Ascent of Man

**11.1
INTRODUCTION**

This chapter introduces another continuous probability distribution: the *F* distribution. This distribution was developed in 1924 by the great English statistician R. A. Fisher. It was later named in his honor by G. W. Snedecor.

The *F* distribution shall be employed in two important applications. First, the *F* distribution will enable one to test the equality of two population variances, $H_0: \sigma_1^2 = \sigma_2^2$. You will recall that it was necessary to make the assumption of equal variances to use Student's *t* distribution to test $H_0: \mu_1 - \mu_2 = 0$ or $\mu_1 = \mu_2$. The *F* distribution provides a check on the validity of this assumption.

The *F* distribution further serves as the basis of analysis of variance techniques. These techniques are extremely useful tools and possess applicability to a variety of fields. Analysis of variance as employed in experimental design by trained statisticians is indispensable in many areas of research.

Two experimental designs are presented in the chapter. The first, the completely randomized design, is used to test the hypothesis $H_0: \mu_1 = \mu_2 = \cdots = \mu_c$. In testing for the equality of means, our earlier procedures (using the normal and *t* distributions) were limited to two populations. The *F* distribution can test for the equality of two *or more* population means. The randomized block design is a more sophisticated procedure used in testing for the equality of means. With this tool, the statistician attempts to "refine" the data so that the

---

SIR RONALD A. FISHER (1890–1962)

Fisher, the youngest of eight children, was born in a London suburb. He studied at Cambridge, graduating with a B. A. in astronomy in 1912. His early employment included serving as chief statistician at the Rothamsted Agricultural Station, where much of his work was concerned with the design and analysis of experiments. From 1943 until 1957 he held the Balfour Chair of Genetics at Cambridge. Fisher's contributions are numerous in genetics and both mathematical and applied statistics, but his advances in experimental design represent, perhaps, his greatest achievements. His book, *Statistical Methods for Research Workers*, first published in 1925, has been translated in seven languages. The term "statistic," as used today to represent a measure calculated from a sample, is attributed to Fisher, who was knighted in 1952.

basic hypotheses being tested in a given experimental situation may be more precisely analyzed. In this sense, the statistician acts as a sculptor attempting to free a figure (to test a hypothesis precisely) from the raw material (experimental data). Experimental design is a well developed field of statistics. More advanced designs, widely used in industry, are quite complex. Interested students should consult the chapter references.

We shall initially undertake a description of the *F* distribution. Its parameters and the relationship to the chi-square distribution will be examined. The use of the *F* distribution in testing the equality of two variances will be shown. The chapter concludes with an introduction to analysis of variance and experimental design.

11.2 THE *F* DISTRIBUTION

The random variable defined by the ratio of two *independent*, chi-square distributed random variables, each divided by its respective degrees of freedom,

$$\frac{\chi_1^2/v_1}{\chi_2^2/v_2},\qquad(11.1)$$

follows the *F* distribution. *Its parameters are the number of degrees of freedom for the numerator, v_1, and for the denominator, v_2.* Given v_1 and v_2, an *F* distribution is completely specified.

The values of *F* range from zero to positive infinity; this should be obvious when one recalls that chi-square random variables and degrees of freedom are both nonnegative. Generally, the *F* distribution is positively skewed. However, for large values of v_1 and v_2 it approaches normality. Let F_{v_1, v_2} symbolize an *F* distributed random variable with v_1 and v_2 degrees of freedom. Expressions for the mean and variance are given by:

$$E(F_{v_1, v_2}) = \frac{v_2}{v_2 - 2} \quad \text{for } v_2 > 2 \qquad (11.2)$$

$$\text{Var}(F_{v_1, v_2}) = \frac{2v_2^2(v_1 + v_2 - 2)}{v_1(v_2 - 2)^2(v_2 - 4)} \quad \text{for } v_2 > 4. \qquad (11.3)$$

Figure 11.1 illustrates the *F* distribution.

FIGURE 11.1: THE *F* DISTRIBUTION

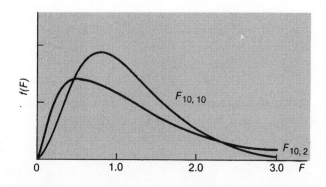

A series of tables for the F distribution is found in Appendix I. The tables have varying levels of α, where α is the *right* tail area. Given α, v_1, and v_2, the appropriate table gives the value of F such that the probability of the value of the random variable F *exceeding* the tabular value is α. This value is symbolized by $F_{v_1, v_2, \alpha}$.

EXAMPLE. Suppose we are given a random variable known to follow an F distribution with $v_1 = 6$ and $v_2 = 8$. The probability is 0.10 that the value of the random variable will exceed 2.67, i.e., $F_{6, 8, 0.10} = 2.67$. Similarly one can determine that $F_{10, 5, 0.05} = 4.74$.

Critical values for the *left* tail of the standard normal, t, and chi-square distributions were obtained directly from their tables. However, a different procedure will be required for the F distribution. The value of F_{v_1, v_2} such that α of the distribution lies to the *left* of this value is equivalent to stating the value of F such that $1 - \alpha$ of the distribution lies to the *right* of this value, i.e., $F_{v_1, v_2, 1-\alpha}$ (recall the rule of complementation). Values of $F_{v_1, v_2, 1-\alpha}$ where $(1 - \alpha) > 0.10$ are not given in Appendix I. However, they may be found by taking the reciprocal of the F value for the same area in the right tail with the degrees of freedom for numerator and denominator reversed:

$$F_{v_1, v_2, 1-\alpha} = \frac{1}{F_{v_2, v_1, \alpha}}. \tag{11.4}$$

EXAMPLE. Given an F distribution with $v_1 = 10$ and $v_2 = 12$, find the critical values of F such that 0.01 of the distribution lies in each tail.

From Appendix I,

$$F_{10, 12, 0.01} = 4.30$$

and using Equation (11.4),

$$F_{10, 12, 0.99} = \frac{1}{F_{12, 10, 0.01}} = \frac{1}{4.71} = 0.21.$$

See Figure 11.2.

FIGURE 11.2: CRITICAL VALUES FOR THE F DISTRIBUTION WITH 10 AND 12 DEGREES OF FREEDOM

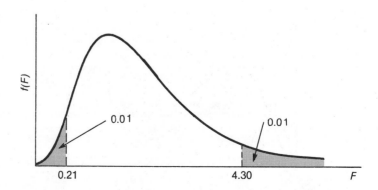

To summarize, in using the F distribution to test hypotheses such that the critical region lies in both tails, the right tail critical value, $F_{v_1, v_2, \alpha/2}$, is found in Appendix I. The left tail critical value, $F_{v_1, v_2, 1-\alpha/2}$, is found from $1/F_{v_2, v_1, \alpha/2}$, where the denominator is obtained in Appendix I.

**11.3 TESTING FOR THE EQUALITY OF VARIANCES

Section 10.3 showed that the chi-square distribution could be used to test the hypothesis: $H_0: \sigma^2 = \sigma_0^2$. We shall now see that the F distribution will enable us to test for the equality of two population variances: $H_0: \sigma_1^2 = \sigma_2^2$.

Recall that the sampling distribution of nS^2/σ^2 was chi-square with $n-1$ degrees of freedom. Also in Chapter 2 we saw that $\hat{\sigma}^2 = nS^2/(n-1)$. This allows Equation (11.1) to be re-written:

$$F = \frac{\chi_1^2/v_1}{\chi_2^2/v_2}$$

$$= \frac{n_1 S_1^2/\sigma_1^2 \cdot \dfrac{1}{(n_1 - 1)}}{n_2 S_2^2/\sigma_2^2 \cdot \dfrac{1}{(n_2 - 1)}}$$

$$= \frac{\hat{\sigma}_1^2/\sigma_1^2}{\hat{\sigma}_2^2/\sigma_2^2}.$$

If the null hypothesis is true, $\sigma_1^2 = \sigma_2^2$, then

$$\frac{\hat{\sigma}_1^2}{\hat{\sigma}_2^2} \sim F_{v_1, v_2}. \tag{11.5}$$

FIGURE 11.3

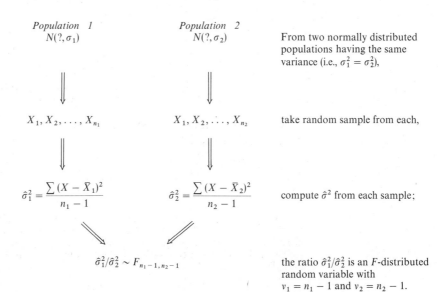

Population 1
$N(?, \sigma_1)$

Population 2
$N(?, \sigma_2)$

From two normally distributed populations having the same variance (i.e., $\sigma_1^2 = \sigma_2^2$),

$X_1, X_2, \ldots, X_{n_1}$

$X_1, X_2, \ldots, X_{n_2}$

take random sample from each,

$$\hat{\sigma}_1^2 = \frac{\sum (X - \bar{X}_1)^2}{n_1 - 1}$$

$$\hat{\sigma}_2^2 = \frac{\sum (X - \bar{X}_2)^2}{n_2 - 1}$$

compute $\hat{\sigma}^2$ from each sample;

$$\hat{\sigma}_1^2/\hat{\sigma}_2^2 \sim F_{n_1 - 1, n_2 - 1}$$

the ratio $\hat{\sigma}_1^2/\hat{\sigma}_2^2$ is an F-distributed random variable with $v_1 = n_1 - 1$ and $v_2 = n_2 - 1$.

In other words, if $\hat{\sigma}_1^2$ and $\hat{\sigma}_2^2$ (unbiased estimates of the population variances σ_1^2 and σ_2^2) are computed from random samples of size n_1 and n_2 taken from two normally distributed populations having a common variance (i.e., $\sigma_1^2 = \sigma_2^2$), then the ratio $\hat{\sigma}_1^2/\hat{\sigma}_2^2$ will follow an F distribution with $v_1 = (n_1 - 1)$ and $v_2 = (n_2 - 1)$. See Figure 11.3. Then, to test the hypothesis $\sigma_1^2 = \sigma_2^2$, compare the observed value $(\hat{\sigma}_1^2/\hat{\sigma}_2^2)$ to the critical values of the distribution with the appropriate degrees of freedom.[1]

EXAMPLE. In Section 9.6 the hypothesis $H_0\colon \mu_1 - \mu_2 = 0$ (or $\mu_1 = \mu_2$) was tested using Student's t distribution. At that time we assumed $\sigma_1^2 = \sigma_2^2$. Test the validity of this assumption at the 0.02 level via the F distribution.

The data are:

$$n_1 = 16 \qquad n_2 = 10$$
$$S_1^2 = 0.81 \qquad S_2^2 = 0.56$$

and

$$H_0\colon \quad \sigma_1^2 = \sigma_2^2$$
$$H_1\colon \quad \sigma_1^2 \neq \sigma_2^2$$
$$\alpha = 0.02.$$

Then,

$$\hat{\sigma}_1^2 = n_1 S_1^2/(n_1 - 1) = \tfrac{16}{15} \cdot 0.81 = 0.864.$$

Similarly,

$$\hat{\sigma}_2^2 = 0.622.$$

Using Equation (11.5),

$$F = \frac{\hat{\sigma}_1^2}{\hat{\sigma}_2^2} = \frac{0.864}{0.622} = 1.39.$$

The critical values of F are:

$$F_{15,\,9,\,0.01} = 4.96;$$

$$F_{15,\,9,\,0.99} = \frac{1}{F_{9,\,15,\,0.01}} = \frac{1}{3.89} = 0.257.$$

Since the computed value of F lies between the critical values, the null hypothesis is accepted.

---

[1] A confidence interval for σ_1^2/σ_2^2 can also be computed, but the procedure will not be given in this text. We should also note that, in testing $H_0\colon \sigma_1^2 = \sigma_2^2$, some tests always place the larger estimated variance in the *numerator* when computing the F value. When this is done, one must also modify the selection of critical values; there would be a *single* right-tail critical value with $\alpha/2$ to the right.

B. C.   by Johnny hart

B.C. by permission of Johnny Hart and Field Enterprises, Inc.

11.4 THE COMPLETELY RANDOMIZED DESIGN (ONE-WAY ANOVA)

Experimental design is an integral part of scientific research. In general, experimental design refers to the planning of experiments for the purpose of testing one or more hypotheses so that the necessary data can be collected and analyzed in an efficient manner. A basic tool in most experimental designs is the analysis of variance.

Analysis of variance (ANOVA) was primarily developed as an aid in agricultural research. It will be seen that the present terminology still reflects its agrarian origin. However, experimental design is today a basic tool in such areas as marketing research, personnel training, drug testing, and industrial experimentation. The development of this area of statistics was primarily the work of R. A. Fisher.

This chapter studies two of the more elementary models of experimental designs. References containing a detailed treatment of experimental design and its more complex models are found at the end of the chapter.

The Completely Randomized Design

The normal and Student's t distributions have, under their respective requirements, been used to test the hypothesis $H_0: \mu_1 = \mu_2$. The completely randomized design (*CRD*) is used to test for the equality of two *or more* population means, i.e., $H_0: \mu_1 = \mu_2 = \cdots = \mu_c$. The completely randomized design is so termed because the experimental units are randomly assigned to the various treatments. The procedure used to evaluate the experimental results is sometimes referred to as *one-way analysis of variance*.

For example, suppose that United University offers three hours' credit in economics for sophomore students who pass a standardized examination. As an aid to the students in preparing for this exam, the economics department is about to recommend a programmed-learning text. The department has agreed that three of the available texts seem superior to the others, but no consensus has been reached on which one of these three to adopt. In order to determine whether there is a significant difference in effectiveness, the three texts are randomly distributed to the sophomore students who have registered for the exam. The null hypothesis is that there is no significant difference among the mean scores of the three groups, i.e., $H_0: \mu_1 = \mu_2 = \mu_3$. This test is an application of one-way analysis of variance.

TABLE 11.1

	Treatment	
1	*2*	*3*
x_{11}	x_{12}	x_{13}
x_{21}	x_{22}	x_{23}
$\vdots$	$\vdots$	$\vdots$
$x_{n_1 1}$	$x_{n_2 2}$	$x_{n_3 3}$

The data from such an experiment are usually presented in the form of Table 11.1. Initially, the problem will be dealt with symbolically; numbers and computations will come later. The notation x_{21} refers to the score of the second student who received the first text. The subscript n_3 gives the total number of students in the third group. One should note that the groups need not be of equal size. In ANOVA nomenclature, the textbooks are referred to as treatments. In our example, the treatments (textbooks) are applied to the experimental units (students) and the x_{ij} is the measurement (test score) obtained from the ith unit receiving the jth treatment. Other examples are given in Table 11.2.

TABLE 11.2

Experimental Units	*Treatments*	*Measurements* (X_{ij})
Hybrid seed	Fertilizers	Yield
Patients	Hospitals	Length of stay
Tires	Brands	Mileage
Machine operators	Shift	Productivity
Retail stores	Product display	Sales

One-way ANOVA and more advanced methods of analysis of variance are often expressed in the form of a mathematical model. To be general, we shall consider c treatments. Then, define:

μ = A constant common to all observations.

τ_j = Effect of treatment j (τ is the Greek letter "tau").

$\mu_j = \mu + \tau_j$ = Mean of jth treatment population.

$\varepsilon_{ij} = (X_{ij} - \mu_j) = (X_{ij} - \tau_j - \mu)$ = Random error associated with the ith observation in the jth treatment (ε is the Greek letter "epsilon").

The model for one-way ANOVA is:

$$X_{ij} = \mu + \tau_j + \varepsilon_{ij} \qquad (11.6)$$

where the ε_{ij} are NID $(0, \sigma)$

Hence, each X_{ij} is the sum of three components: (1) the common constant μ, (2) a treatment effect, and (3) a random or experimental error. Note that since the ε_{ij} are independent $N(0, \sigma)$, it follows that the observations in the jth treatment group represent a random sample of size n_j from a population that is $N(\mu + \tau_j, \sigma)$. Figure 11.4 illustrates these relationships when $c = 3$.

FIGURE 11.4: GRAPHICAL PRESENTATION OF THE ONE-WAY ĀNOVA MODEL: $X_{ij} = \mu + \tau + \varepsilon_{ij}$ (NOTE THAT THE THREE TREATMENT POPULATIONS ARE NORMAL AND POSSESS IDENTICAL VARIANCES)

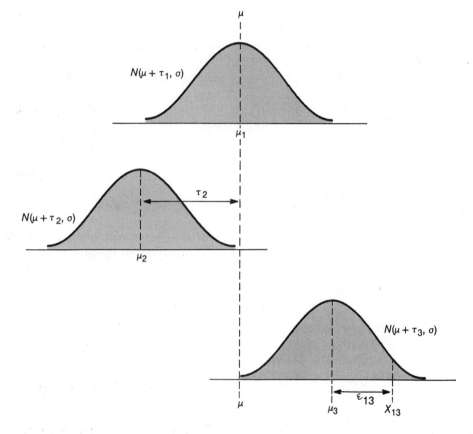

The error term is required because of variability within treatment groups. For example, one would not expect every student that received text three to make the same score. Within the third treatment population the scores will vary. This variation is due to the myriad of factors (other than text) that influence one's score, e.g., I.Q., hours studied, luck, etc. Hence, the error term is included in order to complete the model of the individual scores. The variation of the individual scores within a treatment population is measured by σ^2. (The so-called error variance, σ^2, is not subscripted because it is assumed to be same for all treatment populations.)

The null and alternate hypotheses for one-way ANOVA have been given by:

H_0: All μ_j are equal.

H_1: Not all μ_j are equal.

If the μ_j are all equal, then the treatment effects must all have been zero. Hence an equivalent statement is:

$$H_0: \quad \text{All } \tau_j = 0$$
$$H_1: \quad \text{Not all } \tau_j = 0.$$

Although the null hypothesis is concerned with the equality of means (i.e., the absence of treatment effects), the test of this hypothesis is performed via the F distribution, which was previously used to test for the equality of two population variances. We shall see how the two are related.

11.5 TESTING FOR THE EQUALITY OF MEANS

The preceding section introduced the student to the completely randomized design, its model, and one-way ANOVA. As we have indicated, ANOVA and its use of the F distribution is *not* for the *ultimate* purpose of testing for the equality of two population variances (although such a test will be performed in the course of the analysis). Our purpose is the testing for the equality of three or more population means (or, equivalently, for the absence of treatment effects). The fact that the F distribution can be employed in the latter test may seem rather remarkable to the student; the necessary concepts are now explained.

A customary starting point in ANOVA is appropriately termed the partitioning of the sum of squares. In one-way ANOVA, the total sum of squares (SST) is partitioned into (1) the treatment sum of squares ($SSTR$) and (2) the error sum of squares (SSE) (see Figure 11.5). We define:

$$T_j = \sum_{i=1}^{n_j} X_{ij} = \text{Sum of the observations in the } j\text{th treatment group;}$$

$$T = \sum_{j=1}^{c} T_j = \text{Sum of } all \text{ observations;}$$

$$\bar{X}_j = \frac{T_j}{n_j} = \text{Sample mean of the } j\text{th treatment group;}$$

$$N = \sum_{j=1}^{c} n_j = \text{Total number of observations in all groups;}$$

$$\bar{\bar{X}} = \frac{T}{N} = \text{Sample mean of } all \text{ observations.}$$

The sums of squares can then be defined as:

$$SST = \sum_{j=1}^{c} \sum_{i=1}^{n_j} (X_{ij} - \bar{\bar{X}})^2 \tag{11.7}$$

$$SSTR = \sum_{j=1}^{c} n_j (\bar{X}_j - \bar{\bar{X}})^2 \tag{11.8}$$

$$SSE = \sum_{j=1}^{c} \sum_{i=1}^{n_j} (X_{ij} - \bar{X}_j)^2. \tag{11.9}$$

FIGURE 11.5: SUM-OF-SQUARES PARTITIONING FOR THE COMPLETELY RANDOMIZED DESIGN

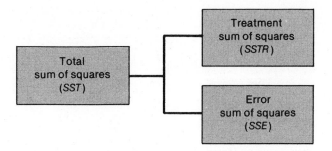

SST can be seen to measure the variation among all observations (treated as a single group). $SSTR$ measures variation among treatment means. And SSE measures the variation within treatment groups (remember, according to the CRD model, variation within a group is attributed to random error). See Table 11.3. A fundamental relation in one-way ANOVA is:[2]

$$SST = SSTR + SSE. \tag{11.10}$$

Now define the error mean square (MSE) and the treatment mean square ($MSTR$) as:

$$MSE = SSE/(N - c); \tag{11.11}$$

$$MSTR = SSTR/(c - 1). \tag{11.12}$$

The denominators of Equations (11.11) and (11.12) are seen to be their respective degrees of freedom. For example, given $\bar{X}_j$, one can freely assign only $(n_j - 1)$ values in the jth group; over all groups, one can then freely assign $(N - c)$ values.

---

[2] The identity can be proved as follows:

$$SST = \sum_{j=1}^{c} \sum_{i=1}^{n_j} (X_{ij} - \bar{\bar{X}})^2 = \sum_{j=1}^{c} \sum_{i=1}^{n_j} [(X_{ij} - \bar{X}_j) + (\bar{X}_j - \bar{\bar{X}})]^2$$

$$= \sum_{j=1}^{c} \sum_{i=1}^{n_j} (X_{ij} - \bar{X}_j)^2 + 2 \sum_{j=1}^{c} \sum_{i=1}^{n_j} (X_{ij} - \bar{X}_j)(\bar{X}_j - \bar{\bar{X}}) + \sum_{j=1}^{c} \sum_{i=1}^{n_j} (\bar{X}_j - \bar{\bar{X}})^2.$$

Since $(\bar{X}_j - \bar{\bar{X}})$ and $(\bar{X}_j - \bar{\bar{X}})^2$ in the second and third terms respectively involve only the outer index of summation, they can be factored outside the inner summation sign (i.e., with respect to the inner summation, these terms are constants). Then,

$$SST = \sum_{j=1}^{c} \sum_{i=1}^{n_j} (X_{ij} - \bar{X}_j)^2 + 2 \sum_{j=1}^{c} (\bar{X}_j - \bar{\bar{X}}) \sum_{i=1}^{n_j} (X_{ij} - \bar{X}_j) + \sum_{j=1}^{c} n_j(\bar{X}_j - \bar{\bar{X}})^2.$$

Since a property of the arithmetic mean is that the sum of the deviations about it is zero,

$$\sum_{i=1}^{n_j} (X_{ij} - \bar{X}_j) = 0 \text{ for all } j,$$

and the middle term drops out. Then,

$$SST = \sum_{j=1}^{c} \sum_{i=1}^{n_j} (X_{ij} - \bar{X}_j)^2 + \sum_{j=1}^{c} n_j(\bar{X}_j - \bar{\bar{X}})^2$$

$$= SSE + SSTR.$$

TABLE 11.3: THE
VARIATION BEING
MEASURED BY *SSTR*
AND *SSE*

	Treatment	
1	*2*	*3*

$$\left.\begin{matrix} x_{11} \\ x_{21} \\ \vdots \\ x_{n_1 1} \end{matrix}\right\} SSE_1 \qquad \left.\begin{matrix} x_{12} \\ x_{22} \\ \vdots \\ x_{n_2 2} \end{matrix}\right\} SSE_2 \qquad \left.\begin{matrix} x_{13} \\ x_{23} \\ \vdots \\ x_{n_3 3} \end{matrix}\right\} SSE_3 = \sum(x_{i3} - \bar{x}_3)^2$$

$$T_1 = \sum x_{i1} \qquad T_2 = \sum x_{i2} \qquad T_3 = \sum x_{i3}$$

$$\bar{x}_1 \qquad\qquad \bar{x}_2 \qquad\qquad \bar{x}_3$$

SSTR measures variation among the treatment means

SSE measures variation within treatment groups ($SSE_1 + SSE_2 + SSE_3$)

Under the assumptions of the model given in Section 11.4, it can be shown that:

$$E(MSE) = \sigma^2; \tag{11.13}$$

$$E(MSTR) = \sigma^2 + \frac{\sum\limits_{j=1}^{c} n_j \tau_j^2}{c - 1}. \tag{11.14}$$

Now if the null hypothesis is true, then there are no treatment effects, and all $\tau_j = 0$. Hence, the second term in Equation (11.14) is zero. Then if H_0 is true, *MSE* and *MSTR* are two unbiased, independent estimators of a common population variance, and the ratio *MSE/MSTR* will follow an *F* distribution with $v_1 = (c - 1)$ and $v_2 = (N - c)$. Further, if H_0 is true, one would expect the computed *F* value to be "close" to one. Now suppose H_0 is false and some $\tau_j \neq 0$, then the second term of Equation (11.14) is some *positive* value, and the expected value of *MSTR* is some value larger than σ^2. (Note that in either case, the $E(MSE)$ remains σ^2.) Hence, only large values of *F* indicate treatment effects. Therefore, *the critical region lies entirely in the right tail.*

The conceptual foundation of one-way analysis of variance is complete. The approach, however, has neglected the computational burden, which the student may rightfully view as quite laborious. However, one can simplify the computations of the sums of squares defined in Equations (11.7) and (11.8) through the use of more efficient formulas.[3] They are given without proof as:

$$SST = \sum_{j=1}^{c} \sum_{i=1}^{n_j} X_{ij}^2 - \frac{T^2}{N} \tag{11.15}$$

---

[3] Two other observations are also in order. (1) It was previously shown that if $Y = a + bX$ then $\sigma_Y^2 = b^2 \sigma_X^2$. By coding the observed values in ANOVA problems, one may greatly simplify the hand computations; the resulting ANOVA table will be *exactly* the same with or without coding. A most instructive exercise for the student would be to code the values of the following example by an additive constant (e.g., let $a = -75$) and recompute *SST* and *SSTR*. (The first group values would now be -6, $+1$, -4, and $+9$.) (2) Computer programs for many ANOVA models are readily accessible at most colleges, and hand computations are then not required. However, one must still have a conceptual understanding of the technique in order properly to interpret the results of the computer output.

$$SSTR = \sum_{j=1}^{c} \frac{T_j^2}{n_j} - \frac{T^2}{N}. \tag{11.16}$$

We shall present an alternate formula for SSE; however, it is routinely obtained by subtraction: $SSE = SST - SSTR$.

$$SSE = \sum_{j=1}^{c} \sum_{i=1}^{n_j} X_{ij}^2 - \sum_{j=1}^{c} \frac{T^2}{n_j}. \tag{11.17}$$

Finally, the computational results of an ANOVA problem are usually summarized in a tabular format. For one-way ANOVA, Table 11.4 is used.

TABLE 11.4: ONE-WAY ANOVA TABLE

Source of Variation	SS	v	MS	F_{v_1, v_2}
Treatments	SSTR	$v_1 = c - 1$	$MSTR = SSTR/v_1$	$MSTR/MSE$
Error	SSE	$v_2 = N - c$	$MSE = SSE/v_2$	—
Total	SST	$N - 1$	—	—

EXAMPLE. Assume the Economics Department of United University has randomly distributed the three texts to fourteen students and, following the exam, recorded their grades as shown in Table 11.5.

TABLE 11.5: EXPERIMENTAL DATA AND PRELIMINARY COMPUTATIONS

Treatment (text)		
1	*2*	*3*
69	74	71
76	79	92
71	63	85
84	74	68
	70	84

$n_1 = 4$	$n_2 = 5$	$n_3 = 5$	$N = 14$
$T_1 = 300$	$T_2 = 360$	$T_3 = 400$	$T = 1{,}060$
$T_1^2 = 90{,}000$	$T_2^2 = 129{,}600$	$T_3^2 = 160{,}000$	$T^2 = 1{,}123{,}600$

The hypotheses and level of significance are:

$$H_0: \text{ all } \tau_j = 0$$
$$H_1: \text{ not all } \tau_j = 0 \qquad j = 1, 2, 3$$
$$\alpha = 0.05.$$

One should first compute the sums of squares and then construct the ANOVA table.

$$SST = \sum\sum X_{ij}^2 - \frac{T^2}{N}$$

$$= (69^2 + 76^2 + \cdots + 84^2) - \frac{1{,}123{,}600}{14}$$

$$= 81{,}106 - 80{,}257 = 849.$$

$$SSTR = \sum \frac{T_j^2}{n_j} - \frac{T^2}{N}$$

$$= \left(\frac{90{,}000}{4} + \frac{129{,}600}{5} + \frac{160{,}000}{5} \right) - 80{,}257$$

$$= 80{,}420 - 80{,}257 = 163.$$

$$SSE = SST - SSTR = 849 - 163 = 686.$$

The remaining calculations ($MSTR$, MSE, and F) are straightforward, and the results are given in Table 11.6.

Source of Variation	SS	v	MS	$F_{2,11}$
Treatments	163	2	81.50	1.31
Error.	686	11	62.36	
Total	849	13		

Since the critical value of $F_{2,11,0.05} = 3.98$, the null hypothesis is not rejected. On the basis of the experimental evidence, we conclude that there is no treatment effect (that is, the population means do not differ significantly). The decision on what text to adopt can then be made on other criteria, e.g., price. Note that the p-value can also be used for a decision rule. Since $P(F > 1.31 | H_0$ true$) > 0.10$, H_0 is accepted.

**11.6

**CONFIDENCE
INTERVALS**

The data used to test the null hypothesis of no treatment effects (i.e., equality of treatment means) may also be used to construct confidence intervals for a treatment (e.g., μ_1) or for a difference in means (e.g., $\mu_1 - \mu_2$). The procedures used are essentially those discussed in Chapter 9.

The observations in the completely randomized design represent independent, random samples drawn from each of the treatment populations. Hence, a sample mean, $\bar{X}_j$, is an unbiased estimator of the population mean, μ_j; and any two sample means, $\bar{X}_i$ and $\bar{X}_j$, are independent. Further, the treatment populations are assumed to be *normal* with *identical variances;* this common variance, σ^2, is estimated by MSE since we have stated that $E(MSE) = \sigma^2$. The estimated variance necessitates using the t distribution; this is appropriate because of the normality assumption.

The confidence interval for a single mean, μ_j, is given by

$$\bar{X}_j \pm t(\sqrt{MSE/n_j}). \tag{11.8}$$

Note that $\sqrt{MSE/n_j}$ is simply an estimated standard error of the mean, $\hat{\sigma}_{\bar{X}} = S/\sqrt{n}$. The t value will be chosen from the distribution with $N - c$ degrees of freedom (i.e., the degrees of freedom associated with MSE) and area $\alpha/2$ in the tail for level of confidence $(1 - \alpha)$. This interval is constructed as deemed useful by the analyst. However, a confidence interval for difference in means is ordinarily constructed only *after rejection* of the null hypothesis of equality of means, i.e., one would not usually proceed to estimate a difference when a difference has been deemed not to exist. This interval is of the form:

$$\bar{X}_i - \bar{X}_j \pm t\hat{\sigma}_D \tag{11.19}$$

where

$$\hat{\sigma}_D^2 = MSE\left(\frac{1}{n_i} + \frac{1}{n_j}\right). \tag{11.20}$$

Note that $\hat{\sigma}_D^2$ is simply a version of $S_{\bar{X}_1 - \bar{X}_2}^2$ in Chapter 9. The number of degrees of freedom for t in this instance is $N - c$.

EXAMPLE. The Horton Business School took random samples from each of three groups of last year's graduates: those accepting jobs in (1) sales, (2) banking, and (3) government. The starting salary of each individual was ascertained, and then one-way ANOVA was used to test $H_0: \mu_1 = \mu_2 = \mu_3$. The null hypothesis was rejected. Selected statistics are given below (data are in $ thousands; $MSE = 1.44$).

Group	Mean ($\bar{x}_j$)	Sample Size (n_j)
1. Sales.	13.0	8
2. Banking.	14.0	10
3. Government	10.5	6

A 95 percent confidence interval is desired for

a. The mean salary of those accepting jobs in banking, μ_2, and

b. The difference in the mean salaries of those employed in sales and government, $\mu_1 - \mu_3$.

A confidence interval for μ_2 uses Formula (11.18). Since $N = n_1 + n_2 + n_3 = 8 + 10 + 6 = 24$, and c (number of treatments) $= 3$, the appropriate t distribution has degrees of freedom:

$$df = N - c = 24 - 3 = 21.$$

The t value for 95 percent confidence is then 2.08. Substituting in Formula 11.18,

$$\bar{X}_2 \pm t(\sqrt{MSE/n_2}) = 14.0 \pm 2.08(\sqrt{1.44/10})$$
$$= 14.0 \pm 2.08(0.379)$$
$$= 14.0 \pm 0.788 = 13.212 \text{ to } 14.788.$$

Thus, with 95 percent confidence, the mean starting salary of Horton's graduates in banking is between $13,212 and $14,788.

A point estimate of $\mu_1 - \mu_3$ is given by $\bar{X}_1 - \bar{X}_3 = 2.5$. The estimated variance, Formula (11.20), is:

$$\hat{\sigma}_D^2 = MSE\left(\frac{1}{n_1} + \frac{1}{n_2}\right)$$

$$= 1.44\left(\frac{1}{8} + \frac{1}{6}\right)$$

$$= 0.42.$$

Then, $\hat{\sigma}_D = \sqrt{0.42} = 0.648$. The t value is the same as previously used since the confidence level and degrees of freedom are unchanged. Using Formula (11.19):

$$\bar{X}_1 - \bar{X}_3 \pm t\hat{\sigma}_D = 2.5 \pm 2.08(0.648)$$

$$= 2.5 \pm 1.348 = 1.152 \text{ to } 3.848.$$

With 95 percent confidence, the mean starting salary for graduates in sales is between \$1,152 and \$3,848 higher than that for government.

11.7 THE RANDOMIZED BLOCK DESIGN (TWO-WAY ANOVA)

Section 11.5 presented the most elementary experimental design, the completely randomized design. This section introduces a somewhat more sophisticated technique: a two-way analysis of variance that is appropriately termed a randomized block design. We shall continue with the previous example.

In one-way ANOVA, the denominator of the F test is MSE, an estimator of the variance of the error term. The variability of the individual observations within a treatment group could be due to a variety of factors, e.g., intelligence, previous instruction in economics, math background, or flu on exam day. If one can isolate a factor that *significantly* contributes to the variability of the scores, then the variance of the error term is decreased, and the F test for treatment effects is made more sensitive. One attempts to accomplish this through what is called *blocking*.

The randomized block design, *RBD*, groups the experimental units into homogeneous blocks, and then treatments are randomly assigned to the experimental units within each block. In our example, suppose that a significant amount of the variability of the error term is due to individual differences in scholastic ability; an effective blocking might then be obtained by grouping individual students according to their quality point average (*QPA*). However, if one blocked indiscriminately, e.g., by hair color, the randomized block design will offer little or no improvement over the less complex *CRD*.[4]

---

[4] Blocking, as described above, is usually accomplished by explicit recognition of a variable associated with the experimental unit. However, blocking is sometimes accomplished by isolating a variable associated with the experimental setting. For example, if the exam given to our economics students were of an essay nature (as opposed to multiple-choice), and if the papers were divided among three faculty members for grading, then the variable "grader" might be considered as a blocking variable. As another example, consider an investigation of the quality of coal (e.g., percent ash) being purchased by an electric-power company. The treatments would be the firm's coal suppliers. Blocking associated with the experimental unit might be method of delivery: truck, rail, or barge. Blocking by experimental setting might be the laboratory employee that analyzed the sample. In any case, effective blocking reduces the variability of the error term. (The student may

In this section, x_{ij} refers to the observed value of the experimental unit with treatment j in block i. This design has *one* observation per treatment-block combination.[5] The model for the *RBD* is given as:

$$X_{ij} = \mu + \beta_i + \tau_j + \varepsilon_{ij}, \qquad (11.21)$$

where the ε_{ij} are NID $(0, \sigma)$

As before τ_j is a treatment effect. And,

$$\beta_i = \text{effect of block } i$$
$$\varepsilon_{ij} = (X_{ij} - \beta_i - \tau_j - \mu) = \text{random error.}$$

In general, there will be c treatments and r blocks. Since each treatment is applied once within each block, there are $N = rc$ observations. In words, the model states that (1) the observations constitute random samples of size one from each of the N populations, each population being $N(\mu_{ij}, \sigma)$, and (2) the block and treatment effects are additive, i.e., a treatment effect is the same *from block to block*. For example, the latter statement says that if text one raises the score of a high-*QPA* group by five points, it also raises the score of a low-*QPA* group by five points. When this is not the case, block and treatment effects are said to *interact* (and a more sophisticated model is required).

As before, the analysis is based on a partitioning of the sum of squares. See Figure 11.6. First, define:

$$T_{i.} = \sum_{j=1}^{c} X_{ij} = \text{sum of observations in the } i\text{th block;}$$

$$\bar{X}_{i.} = \frac{T_{i.}}{c} = \text{sample mean of block } i;$$

$$T_{.j} = \sum_{i=1}^{r} X_{ij} = \text{sum of observation in } j\text{th treatment group;}$$

$$\bar{X}_{.j} = \frac{T_{.j}}{r} = \text{sample mean of } j\text{th treatment group;}$$

$$N = rc = \text{total number of observations;}$$

$$T = \sum_{i=1}^{r} T_{i.} = \sum_{j=1}^{c} T_{.j} = \text{sum of } all \text{ observations;}$$

$$\bar{\bar{X}} = \frac{T}{N} = \text{sample mean of all observations.}$$

---

suspect, and correctly so, that more advanced analyses may employ more than one blocking variable.)

[5] More advanced models include *replications* within each treatment-block combination; this necessitates a triple subscript notation (e.g., X_{314} would refer to the fourth observation with treatment one in block three). Such models are beyond the scope of this text.

FIGURE 11.6: SUM-OF-
SQUARES PARTITIONING
FOR THE RANDOMIZED
BLOCK DESIGN

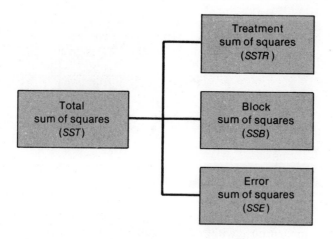

The definitional formulas for sum of squares are given by:

$$SST = \sum\sum(X_{ij} - \bar{\bar{X}})^2 = r\sum_{j=1}^{c}(\bar{X}_{.j} - \bar{\bar{X}})^2 + c\sum_{i=1}^{r}(\bar{X}_{i.} - \bar{\bar{X}})^2$$

$$+ \sum\sum(\bar{X}_{ij} - \bar{X}_{i.} - \bar{X}_{.j} + \bar{\bar{X}})^2$$

$$= SSTR + SSB + SSE. \qquad (11.22)$$

The *SSB* in Equation (11.22) is the block sum of squares. The notation and variation being measured by *SSTR* and *SSB* are shown in Table 11.7. Computational formulas, degrees of freedom, and expected mean squares are given in Table 11.8.[6]

TABLE 11.7: THE
VARIATION BEING
MEASURED BY *SSTR*
AND *SSB*

Block	*Treatment* 1	2	$\cdots$	c	$\sum$	$\bar{x}_{i.}$	
1	x_{11}	x_{12}	$\cdots$	x_{1c}	$T_{1.}$	$\bar{x}_{1.}$	
2	x_{21}	x_{22}	$\cdots$	x_{2c}	$T_{2.}$	$\bar{x}_{2.}$	*SSB* measures variation
$\vdots$	$\vdots$	$\vdots$	$\vdots$	$\vdots$	$\vdots$	$\vdots$	among block means
r	x_{r1}	x_{r2}	$\cdots$	x_{rc}	$T_{r.}$	$\bar{x}_{r.}$	
$\sum$	$T_{.1}$	$T_{.2}$	$\cdots$	$T_{.c}$	T		
$\bar{x}_{.j}$	$\bar{x}_{.1}$	$\bar{x}_{.2}$	$\cdots$	$\bar{x}_{.c}$	—	$\bar{\bar{x}}$	

SSTR measures
variation among
treatment means

---

[6] As before, the data may be coded for simplifying the computations.

TABLE 11.8: ANOVA TABLE FOR THE *RBD*

Source of Variation	*SS*	*df*	*MS*	*E(MS)*	*F*
Treatments........	$SSTR = \dfrac{\sum T_{.j}^2}{r} - \dfrac{T^2}{N}$	$c-1$	$MSTR = \dfrac{SSTR}{c-1}$	$\sigma^2 + \dfrac{r}{c-1}\sum \tau_j^2$	$MSTR/MSE$
Blocks	$SSB = \dfrac{\sum T_{i.}^2}{c} - \dfrac{T^2}{N}$	$r-1$	$MSB = \dfrac{SSB}{r-1}$	$\sigma^2 + \dfrac{c}{r-1}\sum \beta_i^2$	
Error............	$SSE = SST - SSTR - SSB$	$(c-1)(r-1)$	$MSE = \dfrac{SSE}{(c-1)(r-1)}$	σ^2	
Total	$SST = \sum\sum X_{ij}^2 - \dfrac{T^2}{N}$	$N-1$			

The ratio $MSTR/MSE$ is the statistic used to test the null hypothesis of no treatment effects. The student may question whether a test of an hypothesis of no block effects is possible. The answer is yes, by using the test statistic MSB/MSE and comparing it to an appropriate critical value: $F_{r-1,\,(c-1)(r-1),\,\alpha}$. (Note that $E(MSB) = \sigma^2$ if all $\beta_i = 0$.) However, such a test is usually of little interest. Blocking is typically employed for the purpose of increasing the likelihood of detecting the treatment effects.

EXAMPLE. Suppose the Economics Department has blocked the students into three groups by QPA: Block 1 (2.00–3.00), Block 2, (1.50–2.00), and Block 3 (1.00–1.50). A student with an average less than 1.00 may not take the exam. Each treatment is applied once within each block; the exam results are recorded in Table 11.9.

TABLE 11.9:
EXPERIMENTAL DATA
AND PRELIMINARY
CALCULATIONS

Block	*Treatment (text)*			
(QPA)	*1*	*2*	*3*	
1 81		78	88	$T_{1.} = 247 \ T_{1.}^2 = 61{,}009$
2 78		73	81	$T_{2.} = 232 \ T_{2.}^2 = 53{,}824$
3 66		65	71	$T_{3.} = 202 \ T_{3.}^2 = 40{,}804$
	$T_{.1} = 225$	$T_{.2} = 216$	$T_{.3} = 240$	$T = 681$
	$T_{.1}^2 = 50{,}625$	$T_{.2}^2 = 46{,}656$	$T_{.3}^2 = 57{,}600$	$T^2 = 463{,}761$

The interest lies in testing for treatment effects; i.e.,

$$H_0: \quad \text{all } \tau_j = 0$$
$$H_1: \quad \text{not all } \tau_j = 0 \qquad j = 1, 2, 3$$
$$\alpha = 0.05.$$

Using the computational formulas of Table 11.8,

$$SST = \sum\sum X_{ij}^2 - \frac{T^2}{N}$$

$$= (81^2 + 78^2 + \cdots + 71^2) - \frac{463{,}761}{9}$$

$$= 51{,}985 - 51{,}529 = 456.$$

$$SSTR = \frac{\sum T_{.j}^2}{r} - \frac{T^2}{N}$$

$$= \frac{(50{,}625 + 46{,}656 + 57{,}600)}{3} - 51{,}529$$

$$= 51{,}627 - 51{,}529 = 98.$$

$$SSB = \frac{\sum T_{i.}^2}{c} - \frac{T^2}{N}$$

$$= \frac{(61{,}009 + 53{,}824 + 40{,}804)}{3} - 51{,}529$$

$$= 51{,}879 - 51{,}529 = 350.$$

$$SSE = SST - SSTR - SSB$$

$$= 456 - 98 - 350 = 8.$$

The remaining computations are given in Table 11.10.

TABLE 11.10: ANOVA TABLE FOR *RBD* EXAMPLE

Source of Variation	SS	v	MS	$F_{2,4}$
Treatments	98	2	49	24.5
Blocks	350	2	175	
Error	8	4	2	
Total	456	8		

The critical value of $F_{2,4,0.05} = 6.94$, and H_0 is rejected. The student might observe that the treatment means of this example are the same as in the *CRD* example (75, 72, 80). The difference in these means is now deemed significant because effective blocking was used to reduce *MSE*.

11.8 SUMMARY

The subject matter of this chapter has been the *F* distribution and its applications. This continuous distribution has parameters v_1 and v_2. Its relation to the chi-square distribution of the previous chapter was observed. The initial application was the test of the hypothesis $H_0: \sigma_1^2 = \sigma_2^2$ where the two populations were assumed to be normally distributed. The remainder (and major portion) of the chapter was devoted to techniques for the analysis of variance.

Analysis of variance was discussed in the context of experimental design. Two of the more elementary models were presented: the completely randomized design and the randomized block design. Assumptions associated with each were explicitly noted. Although not previously pointed out, the two models of this chapter are referred to as *fixed-effects models*. This means that our concern was limited to that of the populations that were actually sampled. When one's interest lies in some larger set of populations and the populations sampled represent a random sample from this larger set, one employs a *random-effects model*. Detailed discussions on such topics as random-effects models, interaction, violation of assumptions, and more sophisticated experimental designs are found in the references. The book by Neter and Wasserman is especially recommended.

11.9 TERMS AND FORMULAS

F distribution

Analysis of variance (ANOVA)

Completely randomized design (CRD)

Treatment

Experimental unit

Effect of treatment $j(\tau_j)$

Random error (ε_{ij})

Error variance (σ^2)

Randomized block design (RBD)

Effect of block $i(\beta_i)$

Left-tail values of F	$F_{v_2, v_1, 1-\alpha} = \dfrac{1}{F_{v_1, v_2, \alpha}}$
Test statistic for $H_0: \sigma_1^2 = \sigma_2^2$	$F = \dfrac{\hat{\sigma}_1^2}{\hat{\sigma}_2^2}$

One-way ANOVA

Partition of the sum of squares

$$SST = SSTR + SSE$$

Total sum of squares

$$SST = \sum\sum (X_{ij} - \bar{\bar{X}})^2$$

$$= \sum\sum X_{ij}^2 - \frac{T^2}{N}$$

Treatment sum of squares

$$SSTR = \sum n_j (\bar{X}_j - \bar{\bar{X}})^2$$

$$= \frac{\sum T_j^2}{n_j} - \frac{T^2}{N}$$

Error sum of squares

$$SSE = \sum\sum (X_{ij} - \bar{X}_j)^2$$

$$= \sum\sum X_{ij}^2 - \frac{\sum T_j^2}{n_j}$$

Degrees of freedom, means squares and test statistic

See Table 11.4

Confidence interval for a mean

$$\bar{X}_j \pm t\sqrt{\frac{MSE}{n_j}}$$

Confidence interval for difference in two means

$$(\bar{X}_i - \bar{X}_j) \pm t\sqrt{MSE\left(\frac{1}{n_i} + \frac{1}{n_j}\right)}$$

Two-way ANOVA

Partition of the sum of squares

$$SST = SSTR + SSB + SSE$$

Sums of squares, mean squares, degrees of freedom, and test statistics

See Table 11.7

QUESTIONS

A. True or False

1. The test of $H_0: \sigma_1^2 = \sigma_2^2$ assumes the two populations are normally distributed.

2. The alternate hypothesis in the completely randomized design is that not all μ_j are equal.

3. The completely randomized design (CRD) assumes that each treatment population is F distributed.

4. In one-way ANOVA, $SSE \le SST.$

5. SST measures the variation within a treatment group.

6. In analysis of variance, the critical region lies in the left tail.

7. Interval estimation of differences in means is usually done only if H_0: all μ_j are equal has been rejected.

8. In the CRD, $E(MSTR) = E(MSE) = \sigma^2$ if the null hypothesis is false.

9. An F distributed random variable can never be negative.

10. Blocking is employed in order to increase the total sum of squares.

B. Fill in

1. The parameters of the F distribution are _____.

2. One-way analysis of variance is associated with the _____ design.

3. In the CRD, H_0: All μ_j are equal can be stated in terms of treatment effects as _____.

4. In the CRD, $SST - SSTR =$ _____.

5. In the randomized block design (RBD), SST is partitioned into three parts: _____.

6. In both experimental designs, $E(MSE) =$ _____.

7. In the CRD, one assumes the variances of the treatment populations are _____.

8. The "F" in F distribution stands for _____.

9. When v_1 and v_2 are small, the F distribution is _____ skewed.

10. The denominator for testing hypotheses with ANOVA is _____ for all tests in the chapter.

1. Given $v_1 = 5$ and $v_2 = 8$ for the F distribution,
    a. Determine the values such that:
        (1) 0.01 of the distribution lies to the right.
        (2) 0.10 of the distribution lies to the right.
        (3) 0.05 of the distribution lies to the left.
        (4) 0.05 of the distribution lies to the right.

  b. Sketch the distribution. Hint: The mean equals 1.33 using Formula (11.2).

2. Find the critical values of F when the critical region is in the right tail and

  a. $v_1 = 2, v_2 = 4, \alpha = 0.05$.

  b. $v_1 = 8, v_2 = 10, \alpha = 0.01$.

  c. $v_1 = 10, v_2 = 6, \alpha = 0.10$.

3. For the following individuals, describe a situation where the completely randomized design might be used; specify the treatments and two or more causes of variation within the treatment populations:

  a. Customer services manager for a hotel chain.

  b. Training director for a computer services firm.

  c. Admissions chairman of a prestigious law school.

  d. Chief purchasing agent for an automobile rental company.

4. For Problem 3, describe a randomized block design that each individual might use instead of the completely randomized design.

5. Given two random samples of size 10 each with $\hat{\sigma}_1^2 = 90$ and $\hat{\sigma}_2^2 = 78$, test $H_0: \sigma_1^2 = \sigma_2^2$ at the 0.10 level of significance.

6. Given the following:

$$n_1 = 25 \qquad n_2 = 21$$
$$\hat{\sigma}_1^2 = 160 \qquad \hat{\sigma}_2^2 = 138,$$

test $H_0: \sigma_1^2 = \sigma_2^2$ versus $H_1: \sigma_1^2 < \sigma_2^2$ at the 0.05 level.

7. There are two operators of front-end loaders (a piece of heavy equipment) used in loading trucks at a coal mine. All trucks have the same capacity and weight when empty and are weighed loaded before leaving the mine. A previous test did not show a significant difference in the mean loads of the two operators. The data were $n_1 = 16$, $\bar{X}_1 = 20.1$ tons, $S_1 = 0.63$ ton and $n_2 = 16$, $\bar{X}_2 = 19.9$ tons, $S_2 = 0.71$ ton. Use the data to test for a significance difference in variability of loads for the two operators. (Use the 0.05 level of significance.)

8. Given the following results:

$$n_1 = 3 \qquad n_2 = 3 \qquad n_3 = 3 \qquad n_4 = 3$$
$$\bar{X}_1 = 46.33 \qquad \bar{X}_2 = 49.67 \qquad \bar{X}_3 = 45 \qquad \bar{X}_4 = 51$$
$$SST = 110.7;$$
$$SSTR = 70.7;$$
$$SSE = 40.0.$$

  a. Test $H_0: \mu_1 = \mu_2 = \mu_3 = \mu_4$ at the 0.05 level.

  b. Construct a 95 percent confidence interval for (1) $\mu_1 - \mu_2$ and (2) $\mu_3 - \mu_4$.

9. Given the following:

Source of Variation	SS	v
Treatments	15	3
Error	24	8
Total	39	11

    a. What was the total number of observations?
    b. How many treatments were used?
    c. Test the appropriate hypothesis.

10. Given the following calculations from a completely randomized design with three treatments and a total of twelve observations: $SSTR = 276$; $SSE = 302$.
   Complete the ANOVA table and test $H_0: \mu_1 = \mu_2 = \mu_3$ at the 0.05 level.

11. Given the following data from a completely randomized design:

	Treatment	
1	*2*	*3*
5	9	8
6	3	10
4	5	9
6	4	7
4	4	6

   Develop the ANOVA table and test the appropriate hypothesis at the 0.05 level.

12. A management instructor recently performed an experiment on type of exam. He made up three exams (true-false, fill in, and essay) of the same level of difficulty and randomly distributed them among his class. The following grades were determined.

	Exam	
True-False	*Fill in*	*Essay*
75	72	64
90	78	78
70	94	70
80	82	90
85	78	96
60	—	60

Use the above data to test $H_0: \mu_1 = \mu_2 = \mu_3$ at the 0.05 level of significance.

13. A national mail-order firm bought a number of thirty-second advertisements at different radio stations across the country. The stations were randomly selected from those whose music programming was popular, classical, or country/western. Each different station gave a different address so the number of buyers from each could be determined; the results are given below.

Programming		
Popular	*Classical*	*Country/Western*
32	17	22
30	19	24
31	16	25
28	15	20
26	13	16
24	17	19

a. Test $H_0: \mu_1 = \mu_2 = \mu_3$ at the 0.05 level.
b. Construct a 95 percent confidence interval for the mean number of buyers from a popular music station.
c. Construct a 95 percent confidence interval for the difference in mean numbers of buyers for classical and country/western stations.

14. Three manufacturers of stereo equipment offer receiver kits for home assembly. Three receiver models, one from each manufacturer, are selected that are comparable in price and design specifications. Consumers' United has randomly assigned the kits to a group of 20–30-year-old males for assembly. The following data is the assembly time in hours.

Stereo Receiver		
A	*B*	*C*
40	30	23
52	41	36
47	38	30
43	34	—

a. Are the mean assembly times for the three receivers different? Test at the 0.05 level.
b. Construct a 95 percent confidence interval for the mean assembly time for receiver *B*.
c. Construct a 95 percent confidence interval for the difference in mean assembly times for receivers *A* and *C*.

15. Given the following partial ANOVA table for a randomized block design:

Source of Variation	SS	v
Treatments	160	2
Blocks		4
Error	90	
Total	410	

    a. Complete the table and compute *MSTR*, *MSB*, and *MSE*.

    b. Test the appropriate hypothesis for treatments at the 0.05 level of significance.

16. Boctor and Gamble, Inc., are about to introduce a new laundry detergent. A major element in their marketing strategy is the packaging design. The firm has developed three designs, and test results are desired before making a final choice. The firm believes that consumers in different geographic regions have different characteristics and have used this as a blocking factor in their test. The three displays were randomly allocated to suburban supermarkets in each region. The data represents a week's sales.

Region	Design A	B	C
North	64	67	72
East	79	78	89
South	61	76	79
West	53	61	65

Does it seem to matter which design Boctor and Gamble uses?

17. Given the following results from a randomized block design: $SST = 350$; $SSTR = 150$; $SSB = 100$; $MSTR = 50$; $MSB = 25$.

    a. How many treatments were used?

    b. How many blocks were used?

    c. Was the blocking effective? (I.e., test H_0: All β_i equal zero at the 0.05 level of significance.)

    d. Test H_0: All τ_j equal zero with $\alpha = 0.05$.

18. A manufacturer of 100 percent cotton sport shirts is concerned with the fading of colors in his product after repeated washings. A major determinant in fading is the dye used in the production process. There are four types of dye, and an experiment is performed to see how many washings take place before fading becomes noticeable. Since the detergent used is also a factor, the three leading detergents are used as blocking factors.

		Dye		
Detergent	1	2	3	4
1	16	9	25	20
2	44	38	54	53
3	40	32	43	45

a. Use ANOVA to compare the four dyes.

b. Was blocking effective in this experiment?

19. The data in the following table resulted from a particular manufacturing operation that can be performed by one of four different machines. Although each machine performs the same function (e.g., drill press), each has unique characteristics. In collecting the data, the firm used operators as a blocking factor. Each observation represents the units processed in one day.

		Machine		
Operator	A	B	C	D
1	93	108	123	133
2	98	153	143	163
3	80	123	150	168
4	88	158	165	145
5	60	143	140	130

At the 0.01 level of significance, test the hypothesis of no treatment effects. (Hint: The calculations are simplified if a constant, e.g., 100, is subtracted from all observed values.)

REFERENCES

Daniel, Cuthbert *Applications of Statistics to Industrial Experimentation.* New York: John Wiley & Sons, 1976.

Denenberg, Victor H. *Statistics and Experimental Design for Behavioral and Biological Workers.* Washington, D.C.: Hemisphere Publishing Corp., 1976.

Guenther, William C. *Analysis of Variance.* Englewood Cliffs, N.J.: Prentice-Hall, Inc., 1964.

Kirk, Roger E. *Experimental Design: Procedures for or the Behavioral Sciences.* Belmont, Cal.: Brooks/Cole Publishing Company, 1968.

Neter, John, and Wasserman, William *Applied Linear Statistical Models.* Homewood, Ill.: Richard D. Irwin, Inc., 1974.

12

Simple Linear Regression

12 SIMPLE LINEAR REGRESSION

It was Galton who first freed me from the prejudice
that sound mathematics could only be applied
to natural phenomena under the category of causation.
Here for the first time was a possibility, I will not
say a certainty, of reaching knowledge—as valid as
physical knowledge was then thought to be—in the
field of living forms and above all in the field of
human conduct.

Karl Pearson (1934)

**12.1
INTRODUCTION**

The reader is likely familiar with expressions such as $F = ma$ (Newton's second law) or $a^2 = b^2 + c^2$ (Pythagorean theorem) as examples of relationships that have been discovered in the natural sciences and mathematics. Relationships such as these are termed deterministic, for when the values of the independent variables (mass and acceleration) are specified the value of the dependent variable (force) is completely determined. In the behavioral sciences, such relationships are rarely encountered. Instead, we find that while the value of the independent variable may exert an influence on the dependent variable, the dependent variable's value is not completely determined, and we are concerned with what is termed a statistical relationship. For example, economic theory tells us that consumption is related to income. Nevertheless, knowledge of an individual's income does not completely determine his consumption expenditure. Why? Because there is a myriad of other factors also influencing consumption such as psychological expectations, family size, thrift habits, etc. Hence, an expression of this relationship would be:

$$C = f(I) + \varepsilon$$

i.e., consumption is determined by a function of income, $f(I)$, *plus* another term ε, which represents the total effect of all other factors affecting consumption.

The motive of regression analysis lies in ascertaining the nature of the statistical relationships, if any, between variables. This chapter is restricted to simple linear regression, the adjective simple denoting that only one independent variable will be considered. The adjective linear denotes that the functional portion of the expression is of the form $\beta_0 + \beta_1 X$, where X is the independent variable. The discovery that two variables are related may be the single goal of a regression study, but, more often, the analyst seeks to exploit the relationship in "sharpening" predictions concerning the independent variable. Such applications will be illustrated later in the chapter.

The concept of regression was developed by Sir Francis Galton, in the late 19th century, with later contributions by Pearson, Yule, Edgeworth, and others. Applications of the technique in psychology, marketing, finance, etc., became more numerous with the widespread availability of the electronic computer. And today, regression analysis is one of the most important tools available to the business statistician.

> **SIR FRANCIS GALTON (1822–1911)**
>
> Galton was born near Birmingham, England, into a prosperous Quaker family. At age sixteen he was studying medicine in Birmingham and later in London, but he abandoned this career after his father's death in 1844. Financially independent, he traveled for some years in Egypt and Africa, often publishing accounts of these adventures. Around 1865, sparked by his half-cousin Charles Darwin's book *Origin of Species,* Galton's interest turned to heredity. From subsequent studies, including an analysis of characteristics of parent and offspring, he was led to develop the concept of regression, which was published in his book *Natural Inheritance* in 1889. Refined and extended by Karl Pearson and others, this discovery evolved to the widely used analytical technique of today. A man of diverse interests, he was the first to chart weather patterns systematically, was instrumental in developing the fingerprint identification system, and spent some months in an attempt to concoct the perfect cup of tea.

12.2 THE SCATTER DIAGRAM

As an initial example, consider investigating the nature of the statistical relationship, if any, between students' exam scores in some junior class, say Quantitative II, and their quality point average (QPA); the latter's maximum value is 3.00. We have observed that most students carrying a high QPA have done well in this course, but not all. We also note that some of the course's better students are not considered scholars (if scholarship is evidenced by QPA). To pursue the investigation, data obtained from a current class of nineteen students are given in Table 12.1.

While an examination of the data in tabular form is helpful, an easier and and more informative look at the data is accomplished after plotting. In regression analysis, a plot of the data is termed a *scatter diagram* (See Figure 12.1). The horizontal axis is the independent variable (QPA), and the vertical axis is the dependent variable (score). Each point in the scatter diagram represents a sample observation (student); the sample size, n, is the total number of points or paired (X, Y) observations.

What sort of information can be gleaned from an examination of the data? First we note that, overall, higher scores tend to be associated with higher QPAs. (This is true despite the fact that while the student with the highest score (96) has a QPA of 2.9, the next highest score (95) is that of a student with a QPA of 1.8.) Also note that more than one student possesses a particular QPA, e.g., two students have a QPA of 1.7, yet their scores were 86 and 63. This clearly

TABLE 12.1: EXAM
SCORES AND QPA FOR
STUDENTS IN
QUANTITATIVE II

Observation Number (i)	Exam Score (Y_i)	QPA (X_i)
1	89	2.0
2	80	1.6
3	86	1.7
4	74	1.2
5	63	1.7
6	67	2.1
7	87	1.5
8	83	1.9
9	79	1.8
10	72	1.0
11	95	1.8
12	83	2.1
13	78	2.4
14	85	2.7
15	67	1.3
16	65	1.5
17	85	1.8
18	96	2.9
19	93	2.5

FIGURE 12.1: SCATTER
DIAGRAM FOR DATA OF
TABLE 12.1

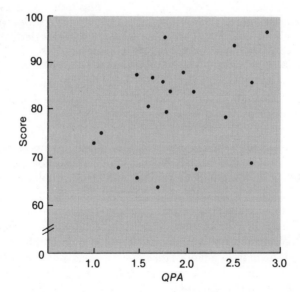

reveals, as we already knew, that score and QPA are not *deterministically* related; if they were, a given value of the independent variable would always be
associated with a unique value of the dependent variable. In other words, in a
deterministic relationship students with the same QPA would make identical
scores. To determine whether there is a statistical relationship must await the
development of the simple linear regression model.

12.3 THE SIMPLE LINEAR REGRESSION MODEL

The simple linear regression model to be used throughout this chapter is

$$Y_i = \beta_0 + \beta_1 X_i + \varepsilon_i, \tag{12.1}$$

where:

Y_i = Value of the dependent variable for the ith observation;

X_i = Value of the independent variable for the ith observation;

β_0, β_1 = Parameters;

ε_i = Value of the error term for the ith observation of Y.

We shall assume that ε_i is a normally distributed random variable with mean of zero and variance σ_ε^2 for all $i = 1, 2, \ldots, n$, i.e., the $\varepsilon_i \sim N(0, \sigma_\varepsilon)$. Further, the error term for some observation, say ε_i, is independent of the error term for any other observation, say ε_j.

What effect do the above assumptions concerning the error term have on our conceptualization of the relationship between X and Y? Suppose the parameters β_0 and β_1 were known, and some value of X, X_h, were specified. Then,

$$Y_h = \beta_0 + \beta_1 X_h + \varepsilon_h$$
$$= \text{Constant} + \text{Random variable.}$$

Now, a constant plus a random variable yields a new random variable; hence, Y_h is a random variable. What is the probability distribution of Y_h? Answer: Except for the mean, exactly that of ε_h. Recall that (1) the linear transformation of a normally distributed random variable is itself normally distributed and (2) if $Z = a + bX$, then $E(Z) = a + bE(X)$ and $\sigma_Z^2 = b^2 \sigma_X^2$. Now $\beta_0 + \beta_1 X_h$ is the additive constant, and the multiplicative constant is equal to 1.0. Hence, Y_h is normal since ε_h is normal; variance of Y_h equals variance of ε_h; and $E(Y_h) = \beta_0 + \beta_1 X_h + E(\varepsilon_h) = \beta_0 + \beta_1 X_h$.

Figure 12.2 illustrates this relationship.

FIGURE 12.2: PROBABILITY DISTRIBUTIONS OF THE RANDOM VARIABLES ε_h AND Y_h

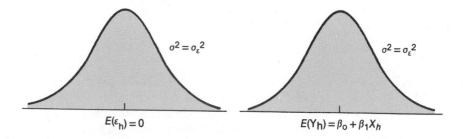

$\sigma^2 = \sigma_\varepsilon^2$ $\sigma^2 = \sigma_\varepsilon^2$

$E(\varepsilon_h) = 0$ $E(Y_h) = \beta_0 + \beta_1 X_h$

To summarize, for any value of the independent variable X, there exists a probability distribution for Y (and an observed value of Y represents a sample of size one from this distribution). Further, the *means* of these probability distributions are linearly related to X, and each probability distribution is normal and has an identical variance.

A review of the model in terms of our initial example would perhaps be helpful. First, given a particular QPA (X), there is a probability distribution of scores (Y); this explicitly recognizes, as we have seen, that for any given QPA one may observe different scores. Second, the distribution of scores, given any QPA, is normal and the variation of scores is the same. For example, if QPA is 1.3 or 2.5, the scores' distributions for each are normal with identical variances. Lastly, the means of these distributions are linearly related to QPA. Figure 12.3 illustrates the model.

FIGURE 12.3:
ILLUSTRATION OF THE
SIMPLE LINEAR
REGRESSION MODEL

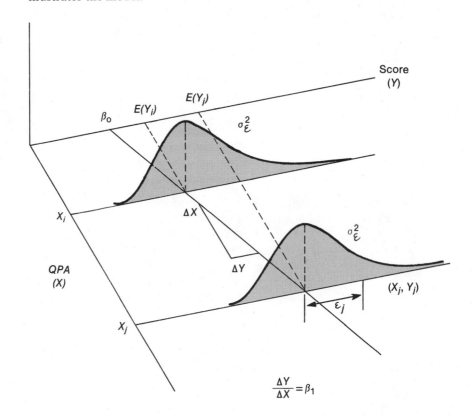

12.4 FITTING THE MODEL

In Table 12.1 and Figure 12.1 are given a set of observed data. Equation 12.1 and Figure 12.3 express the simple linear regression model. The problem is: How does one proceed from a scatter diagram to the construction, i.e., estimation, of a regression model? Estimating the model parameters β_0 and β_1 is naturally our next topic.

Recall that the means, $E(Y_i)$, of the Y distribution are, according to the model, linearly related to X:

$$E(Y_i) = \beta_0 + \beta_1 X_i.$$

Now define:

$$\hat{Y}_i = b_0 + b_1 X_i$$

as the *regression equation* (sometimes termed the *estimating equation*) where

b_0 and b_1 are estimators of β_0 and β_1, respectively, and $\hat{Y}_i$ is a point estimator of $E(Y_i)$. Further, define the *residual* for the ith observation, e_i, as:

$$e_i = Y_i - \hat{Y}_i$$
$$= \text{(Observed value of } Y \text{ for the } i\text{th observation)}$$
$$- \text{(Estimated value of } Y \text{ for } i\text{th observation)}.$$

(A witty acquaintance once defined the residual as the difference between "what you see and what you get.") The procedure for estimating β_0 and β_1 is the method of *least squares*. Consider:

Population error (handwritten)

$$\sum_{i=1}^{n} e_i^2 = \sum (Y_i - \hat{Y}_i)^2$$
$$= \sum [Y_i - (b_0 + b_1 X_i)]^2.$$

The quantity $\sum e_i^2$ is termed the *error sum of squares* (also called residual sum of squares) and abbreviated as *SSE*. For a given set of data, X_i and Y_i are numerical values for $i = 1, \ldots n$. Hence, for a given set of data, *SSE* is a function of b_0 and b_1. When b_0 and b_1 are chosen in such a way so as to minimize *SSE*, they are termed *least squares estimators*, i.e., b_0 and b_1 define a line for which the error sum of squares is the *least* (minimum). The formulas for these estimators are given as:[1]

$$b_1 = \frac{\sum XY - n\bar{X}\bar{Y}}{\sum X^2 - n\bar{X}^2}; \tag{12.2}$$

$$b_0 = \frac{\sum Y - b_1 \sum X}{n}. \tag{12.3}$$

In order to minimize the computational drudgery, a small data set is introduced in Columns (1) and (2) of Table 12.2; Columns (3) and (4) show additional computations necessary for calculating b_0 and b_1, while Column (5) is for later use.

---

[1] The deviation of the least squares estimators is accomplished by taking the partial derivatives of *SSE* with respect to b_0 and b_1:

$$\frac{\partial SSE}{\partial b_0} = -2\sum [Y_i - (b_0 + b_1 X_i)];$$

$$\frac{\partial SSE}{\partial b_1} = -2\sum X_i [Y_i - b_0 + b_1 X_i].$$

Setting the above equal to zero and rewriting yields the so called *normal* equations:

$$nb_0 + b_1 \sum X = \sum Y;$$
$$b_0 \sum X + b_1 \sum X^2 = \sum XY.$$

These two linear equations when solved simultaneously for b_0 and b_1 yield Formulas (12.2) and (12.3). For those interested, there is an algebraic (noncalculus) derivation of the least squares estimators by Stanley and Glass in the February 1969 issue of *The American Statistician*.

TABLE 12.2: DATA AND
PRELIMINARY
CALCULATIONS FOR
EXAMPLE PROBLEM

(1) X	(2) Y	(3) XY	(4) X^2	(5) Y^2
0	3	0	0	9
1	5	5	1	25
2	10	20	4	100
3	12	36	9	144
4	15	60	16	225
$\sum X = 10$	$\sum Y = 45$	$\sum XY = 121$	$\sum X^2 = 30$	$\sum Y^2 = 503$

Then,

$$b_1 = \frac{\sum XY - n\bar{X}\bar{Y}}{\sum X^2 - n\bar{X}^2}$$

$$= \frac{121 - (5)(2)(9)}{30 - (5)(2)^2}$$

$$= \frac{121 - 90}{30 - 20} = \frac{31}{10} = 3.1;$$

$$b_0 = \frac{\sum Y - b_1 \sum X}{n}$$

$$= \frac{45 - (3.1)(10)}{5}$$

$$= \frac{14}{5} = 2.8.$$

And the equation is then:

$$\hat{Y} = b_0 + b_1 X$$
$$= 2.8 + 3.1X.$$

The scatter diagram and the regression equation are shown in Figure 12.4.

FIGURE 12.4: SCATTER
DIAGRAM AND
REGRESSION EQUATION
FOR EXAMPLE PROBLEM

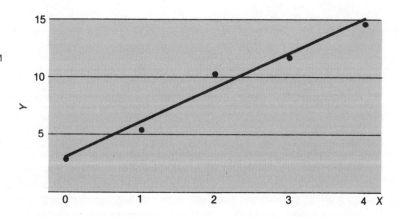

The regression equation estimates the relationship between X and Y. The $b_0 = 2.8$ is simply the Y-axis intercept of the regression equation. The slope of the regression equation is given by b_1, and estimates the change in Y per unit change in X. Since $b_1 = 3.1$, this indicates that Y increases by 3.1 units when X increases by 1.0.

12.5 THE SAMPLING DISTRIBUTIONS OF b_0 AND b_1

The calculation of the least squares line does not necessitate the postulation of a regression model with its accompanying assumptions. The model becomes important when we wish to go beyond the mathematics of least squares to statistical inference and estimation. First, recognize that in general b_0 and b_1 can be regarded as random variables. For example, suppose our population of interest is the 4,000-member sophomore class at United University, and a model of ACT Score (X) versus QPA (Y) is desired. If two individuals independently selected random samples from the sophomore class and, using their observations, each computed a regression equation, be assured that the two equations would be different. Why? Because b_0 and b_1 are variables whose values are dependent upon the outcome of an experiment (the experiment here being: Take a random sample of observations from the population and compute a regression equation).

Before turning to the sampling distributions of b_0 and b_1, one further development need be made, that of estimating the variance of the error term, σ_ε^2. This term measures the variability of the Y's given a value of X, i.e., a measure of the variability about any $E(Y_i)$. Since the $E(Y_i)$ lie on the population regression line, $\beta_0 + \beta_1 X$, σ_ε^2 may be viewed as a measure of the observed values about the population regression line. The regression equation, $b_0 + b_1 X_1$, is an estimate of $\beta_0 + \beta_1 X$; hence a measure of the variability of the observed values about $b_0 + b_1 X$ will serve as an estimate of σ_ε^2.
Define

$$\hat{\sigma}_\varepsilon^2 = \frac{\sum(Y_i - \hat{Y}_i)^2}{n - 2}$$

$$= \frac{SSE}{n - 2};$$

(12.4)

we state without proof that $\hat{\sigma}_\varepsilon^2$ is an unbiased estimator of σ_ε^2; i.e., $E(\hat{\sigma}_\varepsilon^2) = \sigma_\varepsilon^2$. Formula (12.4) is the definitional formula for the estimated error variance or, as it is frequently termed, the *error (or residual) mean square* (abbreviated *MSE*).[2] A computationally more efficient formula for $\hat{\sigma}_\varepsilon^2$ (*MSE*) is

$$\hat{\sigma}_\varepsilon^2 = MSE = \frac{\sum Y^2 - b_0 \sum Y - b_1 \sum XY}{n - 2}.$$

(12.5)

---

[2] Compare Formula (12.4) to the estimated variance of the random variable Y_i: $\hat{\sigma}_Y^2 = \sum(Y_i - \bar{Y})^2/(n - 1)$. Clearly, Formula (12.4) measures variability of observed values about a line ($\hat{Y}_i = b_0 + b_1 X_i$) while $\hat{\sigma}_Y^2$ measures variability about a single value, the sample mean.

EXAMPLE. Estimate the error variance for the data given in Table 12.2,

$$S^2 \; \hat{\sigma}_\varepsilon^2 = \frac{\sum Y^2 - b_0 \sum Y - b_1 \sum XY}{n-2}$$

$$= \frac{503 - (2.8)(45) - (3.1)121}{3}$$

$$= \frac{503 - 126 - 375.1}{3}$$

$$= \frac{1.9}{3} = 0.633.$$

Of course, utilizing Equation (12.4) yields an identical result:

X_i	Y_i	$\hat{Y}_i = 2.8 + 3.1X_i$	$(Y_i - \hat{Y}_i)$	$(Y_i - \hat{Y}_i)^2$
0	3	2.8	0.2	0.04
1	5	5.9	−0.9	0.81
2	10	9.0	1.0	1.00
3	12	12.1	−0.1	0.01
4	15	15.2	−0.2	0.04
			0	1.90

$$S^2 \quad \hat{\sigma}_\varepsilon^2 = \frac{\sum (Y_i - \hat{Y}_i)^2}{n-2} = \frac{1.9}{3} = 0.633.$$

Comparison of the two methods shows that the computational effort using Equation (12.5) is less than using Equation (12.4). (Note that $\sum (Y_i - \hat{Y}_i) = 0$; this is a property of any least squares line.) The square root of the estimate of the variance of the error term would naturally be referred to as the estimate of the standard deviation of the error term; however, it is often called the *standard error of estimate*. For the above problem, $\hat{\sigma}_\varepsilon = \sqrt{0.633} = 0.796$.

Now the probability distribution of b_1 can be given; for the model and assumptions of Section 12.3.

$$b_1 \sim N(\beta_1, \sigma_{b_1}) \tag{12.6}$$

where

$$\sigma_{b_1}^2 = \frac{\sigma_\varepsilon^2}{\sum X^2 - n\bar{X}^2}. \tag{12.7}$$

In words, b_1 is a normally distributed random variable whose expected value is β_1 and whose variance, symbolized by $\sigma_{b_1}^2$, is dependent upon (1) the variance of the error term and (2) the dispersion of the X values.[3]

---

[3] Note the denominator of Equation (12.7) is the same as for Equation (12.2). This quantity appears frequently in regression computations and may be written in other sources as $\sum X^2 - \bar{X}\sum X$ or $\sum (X - \bar{X})^2$.

In order to compute $\sigma_{b_1}^2$ one would need to know σ_ε^2; since the latter is unknown

$$\hat{\sigma}_{b_1}^2 = \frac{\hat{\sigma}_\varepsilon^2}{\sum X^2 - n\bar{X}^2} \tag{12.8}$$

and Equation (12.8) is an unbiased estimator of $\sigma_{b_1}^2$.

EXAMPLE. Compute the estimated variance of b_1 for the data of Table 12.2.

$$\hat{\sigma}_{b_1}^2 = \frac{\hat{\sigma}_\varepsilon^2}{\sum X^2 - n\bar{X}^2}$$

$$= \frac{0.633}{10} = 0.0633.$$

If $b_1 \sim N(\beta_1, \sigma_{b_1})$, then we know that

$$\frac{b_1 - \beta_1}{\sigma_{b_1}} \sim N(0, 1),$$

which is the familiar standardizing transformation. But since σ_{b_1} must be estimated, it should not be unexpected that

$$\frac{b_1 - \beta_1}{\hat{\sigma}_{b_1}} \sim t_{n-2}, \tag{12.9}$$

and it is this result that enables hypothesis tests and confidence intervals about the unknown parameter β_1.

A similar result holds for b_0:

$$b_0 \sim N(\beta_0, \sigma_{b_0})$$

where

$$\sigma_{b_0}^2 = \sigma_\varepsilon^2 \left[\frac{1}{n} + \frac{X^2}{\sum X^2 - n\bar{X}^2} \right]$$

and

$$\hat{\sigma}_{b_0}^2 = \hat{\sigma}_\varepsilon^2 \left[\frac{1}{n} + \frac{X^2}{\sum X^2 - n\bar{X}^2} \right]. \tag{12.10}$$

However, inferences concerning β_0 are usually of little interest to the analyst because the range of the model rarely extends to the Y-axis. For example, the score of a person with a QPA of 0 has no meaning for such an individual would be on academic suspension long before he reached a junior level course; hence, such an observation cannot occur. Therefore, inferential procedures for β_0 are not discussed further.

12.6 TESTING AND ESTIMATION

Inferences on β_1

The hypothesis of initial interest in regression analysis is:

$$H_0: \quad \beta_1 = 0;$$
$$H_1: \quad \beta_1 \neq 0.$$

Recall that the postulated model was of the form

$$Y_i = \beta_0 + \beta_1 X_i + \varepsilon_i.$$

Now if one tests and accepts the hypothesis that β_1 equals zero, then the model reduces to

$$Y_i = \beta_0 + \varepsilon_i$$

for the X term would be dropped. Therefore a test of $\beta_1 = 0$ is a test of no linear relationship between X and Y. If this hypothesis is true, then X is of no use in predicting Y. Suppose one obtained data on score (Y) versus height (X) for the nineteen students of Quantitative II. If a regression equation were obtained for this data and a test of $\beta_1 = 0$ performed, the hypothesis would be accepted. Of course this is not surprising. We do not expect height and score to be related, and the knowledge of someone's height would be of no value in predicting his or her exam score.

The test of H_0: $\beta_1 = 0$ can be accomplished by the knowledge that $(b_1 - \beta_1)/\hat{\sigma}_{b_1}$ is t-distributed with $(n - 2)$ degrees of freedom. Hence, the appropriate test statistic is:

$$t = \frac{b_1 - 0}{\hat{\sigma}_{b_1}} = \frac{b_1}{\hat{\sigma}_{b_1}}. \tag{12.11}$$

Or, as illustrated in Section 8.17, one may construct a confidence interval for β_1 and see whether the hypothesized value of β_1 is contained within the interval. A $(1 - \alpha)$ confidence interval for β_1 with $n - 2$ df is:

$$b_1 - t\hat{\sigma}_{b_1} \leq \beta_1 \leq b_1 + t\hat{\sigma}_{b_1} \tag{12.12}$$

EXAMPLE. Test H_0: $\beta_1 = 0$ for the data of Table 12.2.
From previous computations we know that

$$b_1 = 3.1 \quad \text{and} \quad \hat{\sigma}_{b_1} = \sqrt{0.0633} = 0.252.$$

Then,

$$t = \frac{b_1}{\hat{\sigma}_{b_1}} = \frac{3.1}{0.252} = 12.32.$$

Using a significance level of 0.05, the critical value of t with 3 df is $t = 3.182$, and H_0 is rejected. Similarly, the 95 percent confidence interval is:

$$b_1 - t\hat{\sigma}_{b_1} \leq \beta_1 \leq b_1 + t\hat{\sigma}_{b_1}$$
$$3.1 - 3.182(0.252) \leq \beta_1 \leq 3.1 + 3.182(0.252)$$
$$2.30 \leq \beta_1 \leq 3.90,$$

and since zero is not contained in the interval, the hypothesis $\beta_1 = 0$ would be rejected.

Estimation of $E(Y)$ Having established that Y is linearly related to X (by the rejection of H_0: $\beta_1 = 0$), the analyst will usually wish to exploit the relationship when estimating the dependent variable. We shall consider the estimation of two quantities: (1) the mean of the Y population, $E(Y_h)$, for some given value of X, X_h; and (2) the value of the dependent variable for a new observation, $Y_{h(new)}$, for some given value of X, X_h. An example of the first would be to estimate the mean exam score for the population of students whose QPA $= 2.0$. By our model,

$$E(Y_h) = \beta_0 + \beta_1 X_h;$$

hence, if β_0 and β_1 were known, $E(Y_h)$ could be estimated *without error*. Since only estimators, b_0 and b_1, are known, $E(Y_h)$ must be estimated. The point estimator of $E(Y_h)$, $\hat{Y}_h$, is given by:

$$\hat{Y}_h = b_0 + b_1 X_h.$$

No problems so far. The problem arises when interval estimates of $E(Y_h)$ are desired. Note the $\hat{Y}_h$ is a linear combination of two random variables, b_0 and b_1; hence $\hat{Y}_h$ is also a random variable. It can be shown that

$$\hat{Y}_h \sim N[E(Y_h), \sigma_{Y_h}]$$

where

$$\sigma_{\hat{Y}_h}^2 = \sigma_\varepsilon^2 \left[\frac{1}{n} + \frac{(X_h - \bar{X})^2}{\sum X^2 - n\bar{X}^2} \right].$$

As usual, the variance must be estimated:

$$\hat{\sigma}_{\hat{Y}_h}^2 = \hat{\sigma}_\varepsilon^2 \left[\frac{1}{n} + \frac{(X_h - \bar{X})^2}{\sum X^2 - n\bar{X}^2} \right]. \tag{12.13}$$

Our estimators are unbiased for $E(\hat{Y}_h) = E(Y_h)$ and $E(\hat{\sigma}_{\hat{Y}_h}^2) = \hat{\sigma}_{\hat{Y}_h}^2$. Although Equation (12.13) shall not be derived, an examination of its components reveals nothing surprising. The magnitude of $\hat{\sigma}_{\hat{Y}_h}^2$, reflected in the width of the confidence interval, is:

(1) Directly related to $\hat{\sigma}_\varepsilon^2$; as the variance of the error term (i.e., the effect of excluded variables) increases, the less precise will be estimates of $E(Y_h)$;

(2) Inversely related to sample size, n; and

(3) Directly related to $(X_h - \bar{X})^2$. In other words, our most precise estimates are made in the "center" of the data, if $X_h = \bar{X}$ this term is zero; but as X_h moves away from $\bar{X}$ the estimates become less precise.

A confidence interval for $E(Y_h)$ would be constructed using:

$$\hat{Y}_h - t\hat{\sigma}_{\hat{Y}_h} \le E(Y_h) \le \hat{Y}_h + t\hat{\sigma}_{\hat{Y}_h}. \tag{12.14}$$

EXAMPLE. Using the data of Table 12.2, construct a 95 percent confidence interval for $E(Y)$ when

a. $X = 4$

b. $X = 2$.

a. Using previous computations, a point estimate of $E(Y_4)$ is given by

$$\hat{Y}_4 = 2.8 + 3.1(4)$$
$$= 2.8 + 12.4 = 15.2,$$

and the variance of $\hat{Y}_4$ is:

$$\hat{\sigma}^2_{\hat{Y}_4} = \hat{\sigma}^2_\varepsilon \left[\frac{1}{n} + \frac{(X_h - \bar{X})^2}{\sum X^2 - n\bar{X}^2} \right]$$

$$= 0.633 \left[\frac{1}{5} + \frac{(4 - 2)^2}{10} \right]$$

$$= 0.633 \left[\frac{1}{5} + \frac{4}{10} \right]$$

$$= 0.633(0.6) = 0.38,$$

and

$$\hat{\sigma}_{\hat{Y}_4} = \sqrt{0.38} = 0.62.$$

Substituting in Equation (12.14) yields

$$15.2 - (3.182)(0.62) \le E(Y_4) \le 15.2 + (3.182)(0.62)$$
$$13.2 \le E(Y_4) \le 17.2.$$

b. Similarly,

$$\hat{Y}_2 = 2.8 + 3.1(2) = 9.0;$$
$$\hat{\sigma}^2_{\hat{Y}_2} = 0.633 \left[\frac{1}{5} + \frac{(2 - 2)^2}{10} \right]$$
$$= 0.633(0.2) = 0.127$$
$$\hat{\sigma}_{\hat{Y}_2} = \sqrt{0.127} = 0.36.$$

Then,

$$9.0 - 3.182(0.36) \le E(Y_2) \le 9.0 + 3.182(0.36)$$
$$7.9 \le E(Y_2) \le 10.1.$$

Note that the width of the confidence interval in (b) is more narrow than that of (a); this is because in (b) $X_h = \bar{X}$ while in (a) the point of interest moved away from the center of the data.

Estimation of a
New Observation

The second quantity we wish to estimate is $Y_{h(new)}$. For example, if a student with a QPA of 1.5 missed the exam in Quantitative II and is taking the exam late, what estimate does one make of the score for this individual? Note, we are *not* estimating the mean score of the population with a QPA of 1.5, rather the score of an individual observation *within* that population. The model

$$Y_h = \beta_0 + \beta_1 X_h + \varepsilon_h$$

clearly shows that now, even if β_0 and β_1 were known, Y_h could not be predicted with certainty because of the existence of ε_h. Hence, an estimate of $Y_{h(new)}$

involves not only the uncertainty associated with locating the mean of the population to which it belongs (i.e., not knowing β_0 and β_1), but an additional source of variation within the population. Thus,

$$\sigma^2_{\hat{Y}_{h(new)}} = [\text{Variation associated with estimating } E(Y_h)] +$$
$$[\text{Variation within the population whose mean is } E(Y_h)]$$

$$= \sigma^2_{\hat{Y}_h} + \sigma^2_{\varepsilon}$$

$$= \sigma^2_{\varepsilon}\left[\frac{1}{n} + \frac{(X_h - \bar{X})^2}{\sum X^2 - n\bar{X}^2}\right] + \sigma^2_{\varepsilon}$$

$$= \sigma^2_{\varepsilon}\left[1 + \frac{1}{n} + \frac{(X_h - \bar{X})^2}{\sum X^2 - n\bar{X}^2}\right],$$

and its estimator is given by

$$\hat{\sigma}^2_{\hat{Y}_{h(new)}} = \hat{\sigma}^2_{\varepsilon}\left[1 + \frac{1}{n} + \frac{(X_h - \bar{X})^2}{\sum X^2 - n\bar{X}^2}\right]. \tag{12.15}$$

Whether estimating $Y_{h(new)}$ or $E(Y_h)$, the point estimator is the same: $b_0 + b_1 X_h$; only the confidence interval needs to be modified to reflect the differing variances of the estimators:

$$\hat{Y}_h - t\hat{\sigma}_{\hat{Y}_{h(new)}} \leq Y_{h(new)} \leq \hat{Y}_h + t\hat{\sigma}_{\hat{Y}_{h(new)}}. \tag{12.16}$$

EXAMPLE. Construct a 95 percent confidence interval for $Y_{h(new)}$ given $X = 4$.

The point estimate is:

$$\hat{Y}_4 = 2.8 + 3.1(4) = 15.2. \quad \text{Central tendency}$$

The estimated variance of $Y_{4(new)}$, using Equation (12.15), is

$$\hat{\sigma}^2_{\hat{Y}_{4(new)}} = 0.633\left[1 + \frac{1}{5} + \frac{(4-2)^2}{10}\right]$$

$$= 0.633[1.6] = 1.01,$$

and

$$\hat{\sigma}_{\hat{Y}_{4(new)}} = \sqrt{1.01} = 1.00. \quad \text{Dispersion}$$

The new confidence interval is then given by Equation (12.16):

$$15.2 - 3.182(1.00) \leq Y_{4(new)} \leq 15.2 + 3.182(1.00)$$
$$12.0 \leq Y_{4(new)} \leq 18.4.$$

Note that the width of the 95 percent confidence interval for $Y_{4(new)}$ is 6.4, while for $E(Y_4)$ the width was 4.0. This is as expected for predictions for $Y_{h(new)}$ involve an additional element of uncertainty, the variation within the distribution of Y values.

12.7 OTHER TOPICS

The computational burden of regression analysis is not readily apparent when dealing with a small number of observations whose values are "nice" numbers such as 1, 4, 15, etc. (i.e., as in our continuing example of this chapter).

However, when n increases and/or the values are numbers such as 2,147 or 56.32, the calculations of $\sum X^2, \sum XY, \hat{\sigma}_\varepsilon^2$, etc;, become quite tedious. Computers and "canned" programs remove the computational burden, enabling the analyst to spend relatively more time in gathering data, checking model assumptions, and interpreting the results of the analysis. Figure 12.5 is an example of computer output using a regression package with time-sharing (teletype) system.

FIGURE 12.5: SAMPLE
COMPUTER OUTPUT
USING DATA OF
TABLE 12.2

NOW, READ EACH ROW IN A FREE FORMAT *
```
0   3
1   5
2  10
3  12
4  15
```

DO YOU WISH TO PRINT THE DATA JUST READ IN *
NO

DO YOU WISH TO CHANGE SOME VALUES
NO

SPECIFY THE DEPENDENT VARIABLE
2

SPECIFY THE INDEPENDENT VARIABLE
1

INTERCEPT . . . .   2.80000 b_0
REGRESSION COEFFICIENT . . . .   3.10000 b_1

STD. ERROR OF REG. COEF. . . . . $\hat{\sigma}_{b_1}$ 0.252
COMPUTED T-VALUE . . . .   12.318 $b_1/\hat{\sigma}_{b_1}$

CORRELATION COEFFICIENT . . . . . . r 0.990
STANDARD ERROR OF ESTIMATE . . .   0.796 $\hat{\sigma}_\varepsilon$

ANALYSIS OF VARIANCE FOR THE REGRESSION

0 SOURCE OF VARIATION	D.F.	SUM OF SQ.	MEAN SQ.	F VALUE
ATTRIBUTABLE TO REGRESSION	1	96.100	96.100	151.739
DEVIATION FROM REGRESSION	3	1.900	0.633	
TOTAL	4	98.000		

The major effort required of the user is the "reading" (typing) in of the data. The values that were computed earlier are annotated.

The "analysis of variance for the regression" is of no concern to those whose study of regression analysis will not go beyond the simple linear case. In such an instance, it is merely an alternate (but equivalent) procedure for testing $H_0: \beta_1 = 0$. However, the ANOVA procedure is basic to multiple regression and will be discussed in the following chapter.

Correlation

The correlation coefficient is a standard part of the output of most computer packages. We shall see that the correlation coefficient is used to describe the *degree of relationship* between variables.

Recall that $SSE = \sum(Y_i - \hat{Y}_i)^2$ measures the variability of the observed values about the regression equation. Similarly, $\sum(Y_i - \bar{Y})^2$ measures variability about the sample mean $\bar{Y}$; in regression analysis this quantity is termed the total sum of squares:

$$SST = \sum(Y_i - \bar{Y})^2. \tag{12.17}$$

SST can be zero only if all Y_i are the same value; SSE can be zero only if all Y_i fall perfectly on the regression line, $b_0 + b_1 X_i$. It can be shown that:

$$SST \geq SSE.$$

The equality holds only if the fitted regression line has $b_1 = 0$; in which case $b_0 = \bar{Y}$ and $\hat{Y}_i = b_0 = \bar{Y}$. Therefore we see that $0 \leq SSE \leq SST$. If SSE is zero, we have a perfect relationship between X and Y. If $SSE = SST$, we have *no* relationship since $b_1 = 0$ for such an occurrence. Hence, the use of SSE and SST is suggested as a measure of the degree of relationship. The numerator of Equation (12.18) defines a quantity termed the regression the *coefficient of determination:*

$$r^2 = \frac{SST - SSE}{SST}, \tag{12.18}$$

and $r^2 = 0$ indicates no relationship, $r^2 = 1$ indicates a perfect relationship, and values in between reflect varying degrees of relationship. Figure 12.6 illustrates cases where $r^2 = 1$ and $r^2 = 0$. On the left, all points fall on the line and $SSE = 0$. On the right, $\hat{Y} = b_0 + b_1 X = b_0 + 0X = b_0$ (or $\bar{Y}$) and $SSE = SST$. The numerator of Equation (12.18) defines a quantity termed the regression sum of squares:

$$SSR = SST - SSE. \tag{12.19}$$

As such, SSR shows the reduction in variability of the Y_i when measured about the regression equation ($\hat{Y}_i$) instead of their mean ($\bar{Y}$). It follows from our knowledge of SSE that $0 \leq SSR \leq SST$. Formula (12.18) can be rewritten as

$$r^2 = \frac{SSR}{SST}. \tag{12.20}$$

The coefficient of determination can be interpreted as the proportion of variation in $Y(SST)$ that is "explained" by $X(SSR)$. Then, for example, if $r^2 = 1$, all the variation in Y is "explained" by X.

FIGURE 12.6: SCATTER DIAGRAM ILLUSTRATING DATA WITH $r^2 = 1$ AND $r^2 = 0$

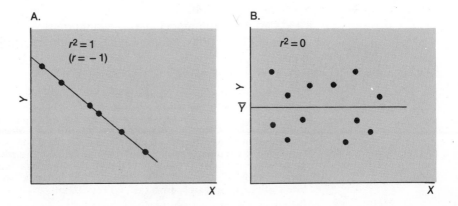

The *coefficient of correlation*, r, is merely the square root of the coefficient of determination, and it is given the sign of b_1. Hence, $-1 \le r \le +1$, and one hears such terms as "negatively correlated;" i.e., $r < 0$. Negatively correlated simply means *increasing* values of X are associated with *decreasing* values of Y. The coefficient of correlation can be directly computed:

$$r = \frac{\sum(X_i - \bar{X})(Y_i - \bar{Y})}{\sqrt{\sum(X_i - \bar{X})^2 \sum(Y_i - \bar{Y})^2}}$$

$$= \frac{\sum X_i Y_i - n\bar{X}\bar{Y}}{\sqrt{[\sum X_i^2 - n\bar{X}^2][\sum Y_i^2 - n\bar{Y}^2]}}. \tag{12.21}$$

The proper sign for r, i.e., the sign of b_1, will naturally result when using Equation (12.21).

EXAMPLE. Compute the coefficient of correlation for the data of Table 12.2.

Using Formula (12.21),

$$r = \frac{\sum XY - n\bar{X}\bar{Y}}{\sqrt{(\sum X^2 - n\bar{X}^2)(\sum Y^2 - n\bar{Y}^2)}}$$

$$= \frac{121 - (5)(2)(9)}{\sqrt{[30 - (5)(2)^2][503 - (5)(9)^2]}}$$

$$= \frac{31}{\sqrt{(10)(98)}}$$

$$= \frac{31}{\sqrt{980}} = \frac{31}{31.30} = 0.99.$$

Thus, X and Y have a high, positive correlation. (Note the result above appears on the sample computer output of Figure 12.5.) Since $r^2 = (0.99)^2 = 0.98$, we can also say that 98 percent of the variation in Y is "explained" by X.[4]

**12.8
PRECAUTIONARY
NOTES**

The proper utilization of regression analysis entails far more than a set of numbers mechanically obtained from the computer. The trained analyst is always aware of the model *and* its assumptions, and is fully cognizant of the technique's limitations. Before leaving this subject, cautions for the novice are in order.

---

[4] The results of this chapter are applicable whether the X_i are assumed to be fixed constants or assumed to be the observed values of a random variable. Another model, the correlation model, assumes both X and Y are random variables *and* specifies a joint distribution (the bivariate normal) for the two variables. However, to quote Miller and Wichern: "Simple linear regression analysis then yields all the information of a correlation analysis and in addition (a) indicates how the variables X and Y are related and (b) is applicable regardless of whether X is taken to be a fixed (controllable) variable or a random variable. That is, correlation may be viewed as a subset of regression analysis." (Robert B. Miller, and Dean W. Wichern, *Intermediate Business Statistics* [New York: Holt, Rinehart and Winston, 1977], p. 221).

1. Although one of the major goals of regression analysis is prediction, the regression equation should not be used for estimation of values outside the range of those in the sample observations; such a practice is termed *extrapolation*. Consider, for example, the scatter diagram of height (Y) versus age (X) for the nineteen students of our Quantitative II class (Figure 12.7).

FIGURE 12.7: SCATTER DIAGRAM OF HEIGHT VERSUS AGE FOR 19 STUDENTS

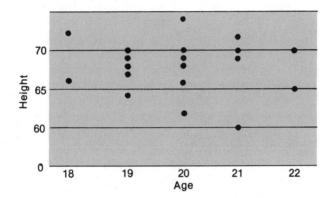

We note no discernible pattern in the observations, and if the hypothesis $\beta_1 = 0$ is tested, the test result is to accept H_0 ($t = -0.346$). In words we conclude that no relationship exists between X and Y, and for predictive purposes the model $Y_i = \beta_0 + \varepsilon_i$ is as useful as a model containing the variable age. So far so good. But the fearless prognosticator is going to be embarrassed when predictions are made for ten-year-olds. Is this a typical example of the dangers of extrapolation? No. The real-world examples are far less obvious.

2. The rejection of H_0: $\beta_1 = 0$ suggests that we accept the alternate hypothesis that X and Y are linearly related. However, this should not be construed as to imply a *cause-and-effect* relationship. It could be that X "causes" Y, or vice versa. Or it could be that both are being caused by some other factor. For example, height and weight of elementary-school children are related but which, if either, is the causal variable? Or are both being caused by age?

3. One should not "blindly" accept the summary statistics of a regression analysis. An especially enlightening example by Anscombe[5] yields the following results:

Number of observations . . . . $n = 11$
Mean of the dependent variable. . . . . $\bar{Y} = 7.5$
Mean of the independent variable. . . . . $\bar{X} = 9.0$
Regression equation. . . . . $Y = 3 + 0.5x$

---

[5] F. J. Anscombe, "Graphs in Statistical Analysis," *The American Statistician*, vol. 27, no. 1 (February 1973), pp. 17–21. The data summary and Figure 12.8 are reproduced with permission.

Estimated standard deviation of b_1. . . . . $\hat{\sigma}_{b_1} = 0.118$
t test for H_0: $\beta_1 = 0$. . . . . $t = 4.24$
Error sum of squares . . . . $SSE = 13.75$
Total sum of squares . . . . $SST = 41.25$
Coefficient of determination . . . . $r^2 = 0.667$

One mentally visualizes the data associated with the above in something similar to Figure 12.8(A), and, in fact, the data of that scatter diagram do yield the above. However, the data associated with panels B, C, and D of Figure 12.8 also yield identical results. In B, it appears that the linear model is inappropriate, and a curvilinear relationship is suggested. In panel C, if one observation were deleted, the remaining points would all fall on a line and $SSE = 0$ and $r^2 = 1$. The "eccentric" observation in panel C is termed an *outlier*, which is defined as an observation whose residual is much greater than for other observations. When an outlier is observed, one should check for measurement or recording error. In panel D, if one observation were deleted, b_1 would be zero, and $SSE = SST$ and $r^2 = 0$; the slope of the regression equation is entirely determined by one observation.

FIGURE 12.8: SCATTER
DIAGRAMS OF VARIOUS
DATA SETS YIELDING
IDENTICAL REGRESSION
OUTPUT

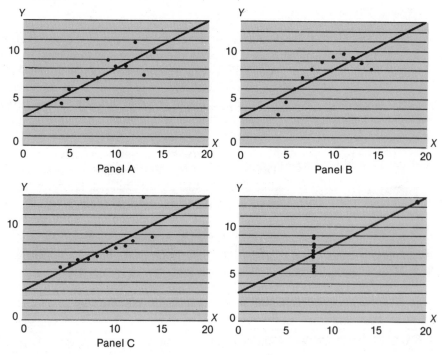

4. Finally one must be careful not to equate "statistical" significance with "practical" significance. If the aim of a regression study is to establish precise estimates (confidence intervals) for Y through use of X, the applicability of the study can be evaluated only through constructing such estimates and

comparing their precision to the user's requirements. Rejection of $H_0 : \beta_1 = 0$, i.e., establishing that a statistically significant relationship exists, does not mean that precise prediction (practical significance) will follow. However, in exploratory research, the goal of the study may be to determine what factors are associated with the dependent variable (e.g., smoking and lung cancer). In this instance, rejection of $H_0 : \beta_1 = 0$ is of practical significance for it aids in directing the analyst's future research efforts.

Other pitfalls encountered in regression analysis include non-independence of error terms (especially when dealing with time-series data), a nonconstant variance of the Y populations for different values of X (the technical term here is heteroskedasticity), and the non-normality of the Y populations. Examples and remedial measures are found in the references.

12.9 TERMS AND FORMULAS

Coefficient of correlation (r)
Coefficient of determination (r^2)
Deterministic relationship
Error mean square (MSE)
Error sum of squares (SSE)
Estimated error variance ($\hat{\sigma}_\varepsilon^2$)
Estimated value of the dependent variable ($\hat{Y}_i$)
Extrapolation
Least squares estimators (b_0, b_1)
Mean of a Y population ($E(Y_h)$)
Observed value of the dependent variable (Y_i)
Outlier
Regression equation ($\hat{Y}_i = b_0 + b_1 X_1$)
Residual ($e_i = Y_i - \hat{Y}_i$)
Scatter diagram
Simple linear regression model ($Y_i = \beta_0 + \beta_1 X_i + \varepsilon_i$)
Statistical relationship
Standard error of estimate ($\hat{\sigma}_\varepsilon$)
Total sum of squares (SST)
Value of Y for a new observation ($Y_{h(\text{new})}$)

Regression equation
$$\hat{Y}_i = b_0 + b_1 X_i$$

Least squares estimators
$$b_1 = \frac{\sum XY - n\bar{X}\bar{Y}}{\sum X^2 - n\bar{X}^2}$$

$$b_0 = \frac{\sum Y - b_1 \sum X}{n}$$

Error sum of squares
$$SSE = \sum (Y_i - \hat{Y}_i)^2$$

Error mean square

$$MSE = \hat{\sigma}_\varepsilon^2$$

$$= \frac{SSE}{n-2}$$

$$= \frac{\sum Y^2 - b_0 \sum Y - b_1 \sum XY}{n-2}$$

Standard error of estimate

$$\hat{\sigma}_\varepsilon = \sqrt{MSE}$$

Estimated variance of b_1

$$\hat{\sigma}_{b_1}^2 = \frac{\hat{\sigma}_\varepsilon^2}{\sum X^2 - n\bar{X}^2}$$

Test statistic for $H_o: \beta_1 = 0$

$$t = \frac{b_1}{\hat{\sigma}_{b_1}}$$

Estimated variance of $\hat{Y}_h$

$$\hat{\sigma}_{\hat{Y}_h}^2 = \hat{\sigma}_\varepsilon^2 \left[\frac{1}{n} + \frac{(X_h - \bar{X})^2}{\sum X^2 - n\bar{X}^2} \right]$$

Estimated variance of $\hat{Y}_{h(new)}$

$$\hat{\sigma}_{\hat{Y}_{h(new)}}^2 = \hat{\sigma}_\varepsilon^2 \left[1 + \frac{1}{n} + \frac{(X_h - \bar{X})^2}{\sum X^2 - n\bar{X}^2} \right]$$

Total sum of squares

$$SST = \sum (Y_i - \bar{Y})^2$$

Regression sum of squares

$$SSR = SST - SSE$$

Coefficient of determination

$$r^2 = \frac{SSR}{SST}$$

Coefficient of correlation

$$r = \frac{\sum XY - n\bar{X}\bar{Y}}{\sqrt{(\sum X^2 - n\bar{X}^2)(\sum Y^2 - n\bar{Y}^2)}}$$

QUESTIONS

A.   True or False

F   1.   Unless $Y_i = \hat{Y}_i$ for all $i = 1, 2, \ldots, n$, the regression model is invalid.

T   2.   The standard error of estimate can never be negative.

F   3.   The regression coefficient, b_1, can never be negative.

T   4.   The regression model assumes that the probability distributions of Y given various values of X have identical standard deviations.

F   5.   Under the assumptions of our model, b_1 is a t-distributed random variable.

F   6.   Given some X_h, a 90% confidence interval for $Y_{h(new)}$ is more narrow than a 90% confidence interval for $E(Y_h)$.

T   7.   Given some level of confidence, the most narrow confidence interval for $E(Y_h)$ will occur when $X_h = \bar{X}$.

T   8.   For our model, acceptance of $H_0: \beta_1 = 0$ implies no linear relationship between X and Y.

F   9.   Sales and advertising expenditures are an example of a deterministic relationship.

F   10.   One would expect IQ and QPA to be negatively correlated.

B. Fill in

1. The expected value of b_1 is ___b_1___.

2. The t distribution used to construct confidence intervals for $E(Y_h)$ has ___$n-2$___ degrees of freedom.

3. A measure of the variation of the observed values (Y_i) about the regression equation would be ___$\hat{\sigma}_\varepsilon^2$ ($or \hat{\sigma}_\varepsilon^2$ or SSE)___.

4. A pioneer in regression analysis was ___Galton___.

5. The regression model assumes that ε_i is a random variable with a ___normal dist.___ distribution.

6. A synonym for the estimated error variance ($\hat{\sigma}_\varepsilon^2$) is ___error mean square MSE___.

7. In order to test hypotheses about β_1, one would use the ___t___ distribution.

8. The maximum value of SSE is ___SST___.

9. Given ten observations (X_i, Y_i) but for which all X_i are the same, SST equals ___SSE___.

10. The expected value of ε_i is ___0___.

EXERCISES

1. Given the simple linear regression model

$$Y_i = \beta_o + \beta_1 X_i + \varepsilon_i,$$

define realistic variables X and Y for which you would expect:
a. β_1 to be positive.
b. β_1 to be negative.
c. β_1 to be zero.

2. Given the following data:

X	Y
1	12
2	11
4	9
6	8
7	5

b) $\hat{y} = 13.16 - 1.04x$

c) .64

a. Plot the scatter diagram.
b. Determine the regression equation and plot.
c. Compute the estimated error variance.

3. Refer to Problem 2.
a. Test $H_0: \beta_1 = 0$ at the 0.05 level.
b. Construct a 95 percent confidence interval for β_1.

4. Refer to Problem 2.
a. Given $X = 4$, construct a 90 percent confidence interval for $E(Y_h)$.
b. Given $X = 2$, construct a 90 percent confidence interval for $Y_{h(new)}$.

$8.16 \le E(Y_4) \le 9.8$

5. Compute the coefficient of determination for the data of Problem 2.

$8.89 \le y_2 \le 13.27$

6. A regression equation for a set of data has a y-axis intercept of 10 and passes through the point (20, 90).

$\hat{y} = 10 + 4x$

    a. Determine the regression equation.

$e_j = 85 - 82 = 3$

    b. Given the observation ($X_j = 18$, $Y_j = 85$), determine the value of e_j (i.e., $Y_j - \hat{Y}_j$).

7. The credit union at United University has a large number of members who make deposits via payroll deduction. A sample of 50 such members yields the following data:

$$Y = \text{amount of payroll deduction (hundreds of dollars)}$$
$$X = \text{annual salary (hundreds of dollars)}$$

and preliminary calculations show:

$$\hat{Y}_i = -2 + 0.05 X_i$$
$$SSE = 6$$
$$SST = 20$$
$$X \text{ ranges from 70 to 140}$$
$$\sum X^2 - n\bar{X}^2 = 4800$$
$$\bar{X} = 100$$

    a. Compute the estimated error variance.

    b. Test $H_0: \beta_1 = 0$ at the 0.01 level.

.837 8. Compute the coefficient of correlation for Problem 7.

9. Refer to Problem 7.

    a. For payroll deduction members whose salary is \$12,000 (i.e., $X_h = 120$), construct a 95 percent confidence interval for the mean deduction of this population.

    b. Given that John Dran is on payroll deduction and his salary is \$10,000, develop a 90 percent confidence interval for his annual deduction.

    c. Compute a point estimate for $E(Y_h)$ when $X_h = 30$; if an unrealistic number results, what could be the problem?

10. TV Shack, a national firm with hundreds of retail stores, recently investigated the relationship between expenditures for supplies (Y) and sales (X) for 25 stores in 1976. X and Y are both in thousands of dollars. Previous calculations show:

$$\hat{Y} = 2 + 0.015X$$
$$\hat{\sigma}_\varepsilon = 0.5$$
$$\bar{X} = 300$$
$$\sum X^2 - n\bar{X}^2 = 10,000$$

$t = \dfrac{b_1}{\hat{\sigma}_{b_1}} = 3.00$ Reject

    a. Do the data give evidence that there is a linear relationship between X and Y?

    b. The corporate staff had earlier predicted that for a sales increase of \$100, an \$0.80 increase in supply expenditures should be expected, do the data support this? (i.e., test $H_0: \beta_1 = 0.008$).

$t = 1.4$

11. See Problem 10. A store with 1976 sales of $300,000 had supplies expense of $9,000, should the store manager be commended, reprimanded, or ignored in this regard?

12. Professor Gordon is saddled with two problems: large class sizes and low student ratings. The former takes time away from his research while the latter has him in trouble with the dean. Professor Gordon believes that his low ratings are due to his large classes (e.g., he can't learn the students' names, there is less discussion, etc.). He samples ten classes in the School of Business in order to test his hypothesis.

X Class Size	Y Class Rating of Instructor
10	4.2
20	4.3
20	3.1
20	3.6
30	4.0
30	4.1
30	3.2
40	4.0
40	3.0
40	3.1
50	3.0

a. Plot the scatter diagram.
b. Determine the regression equation.
c. Test $H_0: \beta_1 = 0$ at the 0.10 level.

13. The personnel manager of the Normal Company has gathered data for 1976 concerning age (X) and days of sick-leave (Y) for a random sample of 14 salaried employees.

X	Y
22	7
25	6
25	9
30	2
32	4
35	2
35	1
35	0
40	3
45	2
46	1
50	4
55	15
60	6

a. Plot the data.
b. Determine and interpret the regression equation.
c. Test for a linear relationship between X and Y at the 0.10 level.

14. Gilvan Inc. has compiled data on its ten sales territories for the past year. The data are X (population in millions) and Y (sales in millions of dollars), and an observation represents a territory. Preliminary calculations show:

$$n = 10 \qquad \sum XY = 459$$
$$\overline{X}55\sum X = 55 \qquad \sum X^2 = 385$$
$$6.8 \; \sum Y = 68 \qquad \sum Y^2 = 560$$

1.135+1.03X *a.* Determine and interpret the regression equation.
Reject 8.44 *b.* Test $H_0: \beta_1 = 0$ at the 0.05 level.
.897 *c.* Compute r^2.

15. The Internal Revenue Service of Tralfamadore is seeking to establish standards to aid its auditors in selecting returns to undergo a complete audit. A sample of audited returns for 1976 is given below:

X Income ($ thousands)	Y Interest Expense ($ hundreds)
4.1	2.0
5.0	4.0
7.0	5.0
10.2	6.0
11.0	9.0
11.0	8.0
15.0	10.0
17.5	14.0
20.0	16.0
30.0	29.0

a. Plot the data.
b. Determine and interpret the regression equation.
c. Test for the significance of a linear relationship between income and interest expense.

16. The Tralfamadorian I.R.S. has established a regression equation for X (income in $ thousands) and Y (deduction claimed for gifts and charities in $ hundreds):

$$\hat{Y} = 0.02 + 0.42X$$

a) $t = 2.80$
Reject

b)
$4.062 \le E(Y_{10}) \le 4.378$

c) $3.287 \le E(Y_{12}) \le 6.833$

and

$$\hat{\sigma}_\varepsilon = 0.60$$
$$n = 100$$
$$\overline{X} = 10$$
$$\sum X^2 - n\overline{X}^2 = 16.$$

a. Test $H_0: \beta_1 = 0$ at the 0.01 level.
b. Determine the 99 percent confidence interval for the mean (gifts and charities) deduction of the population whose income is $10,000.
c. Chester Tight had an income of $12,000 and claimed a deduction of $3,000 for gifts and charities; would you recommend an audit of his return?

17. The data of Table 12.1 yield the following:

$$n = 19 \qquad \sum XY = 2{,}901.7$$
$$\sum X = 35.5 \qquad \sum Y = 1{,}527$$
$$\sum X^2 = 70.83 \qquad \sum Y^2 = 124{,}561$$

    a. Determine the regression equation.
    b. Compute the estimated error variance.
    c. Compute the estimated variance of b_1.
    d. Test the hypothesis of no linear relationship between QPA and Exam Score.
    e. Compute a point estimate of the score of an individual with a QPA = 2.0.

18. Recall that $\hat{\sigma}_x^2 = \sum (X_i - \bar{X})^2/(n-1) = (\sum X^2 - n\bar{X}^2)/(n-1)$. Use this and the formulas for b_1 and r given in this chapter to show that:

$$b_1 = \frac{\hat{\sigma}_y}{\hat{\sigma}_x} \cdot r.$$

REFERENCES

Anscombe, F. J. "Graphs in Statistical Analysis." *The American Statistician*, vol. 27, no. 1 (February 1973), pp. 17–21.

Barrett, James P. "The Coefficient of Determination—Some Limitations," *The American Statistician*, vol. 28, no. 1 (February 1974), pp. 19–20.

Draper, N. R. and Smith, H. *Applied Regression Analysis*, New York: John Wiley & Sons, Inc., 1966.

Neter, John, and Wasserman, William *Applied Linear Statistical Models*. Homewood, Ill.: Richard D. Irwin, Inc., 1974.

Stanley, Julian C., and Glass, Gene V. "An Algebraic Proof that the Sum of the Squared Errors in Estimating Y from X via b_1 and b_0 is Minimal," *The American Statistician*, vol. 23, no. 1 (February 1969), pp. 25–26.

13

Extensions of Regression Analysis

**Optional material.

13 EXTENSIONS OF REGRESSION ANALYSIS

Inquire into the length and shortness of men's
lives, according to the times, countries, climates,
and places in which they were born and lived . . .
And according to their parentage and family . . .
their food, diet, manner of living, exercise, and
the like, with regard to the air in which they
live and dwell.

Francis Bacon
History of Life and Death (1623)

13.1
INTRODUCTION

The previous chapter dealt with simple linear regression, the adjective simple denoting that a single independent variable was considered in the analysis. In many instances such a model will prove adequate for a given problem and provide the user with estimates of sufficient precision. In other instances frequently encountered in business and economics, a more sophisticated model must be employed to meet the requirements of precision or to answer more subtle questions concerning the relationships among variables or both.

The construction of more sophisticated models often involves the inclusion of more than one independent variable. For example, as Bacon notes above the

> ## FRANCIS BACON (1561—1626)
>
> Bacon was born in York House, a mansion on the river Thames in London. His father served as Lord Keeper of the Seal to Queen Elizabeth, one of the highest positions in the realm. Ambitious, talented, and determined to follow in the footsteps of his father, Bacon was frustrated for years until his star began to rise under King James I with an appointment as Solicitor General in 1607. He was eventually named Lord Chancellor (1618), but was impeached by Parliment in 1621. (Bacon: "He that never climbed never fell.") Bacon is remembered here as a philosopher of science. His *Novum Organum* was a summons to the scientific method emphasizing experimentation, observation, and induction. This was a bold call in an age when learning largely consisted of rehashing old books and philosophies. Bacon, although making no great scientific discoveries, fostered a spirit of inquiry and examination, helped to free men from antiquity and superstition, and encouraged knowledge aimed toward the betterment of mankind. Bacon's death is representative of his philosophy: he died of pneumonia contracted while studying the preservation of meat by freezing (in snow), and on his deathbed he wrote to a friend, "As for the experiment itself, it succeeded excellently well." Modern science, both physical and social, remains indebted to the chief spokesman for an age of reason.

"length and shortness of men's lives" would likely be statistically related to a number of variables such as climate, diet, exercise, etc. A model of length of life that considered only one of these variables would possess rather poor predictive capabilities (even assuming $H_0: \beta_1 = 0$ would be rejected). Why? Because the error term, representing the combined effect of all other variables, would be quite large.

In a business example, one would consider unit sales of a product to be statistically related to its price. True, but would not sales also be related to advertising expenditures, number of competitors, economic conditions, etc? In this case, the simple linear model would be less precise than a more elaborate model containing additional independent variables. By explicit consideration of several relevant variables, the analyst can better explain and predict values of the dependent variable. This extension has analogies in our study of ANOVA in Chapter 11. There the movement from the completely randomized design to the randomized block design was an effort to isolate a second factor (the block effect) and thus decrease the magnitude of the error term.

The major concern of this chapter is the development of multiple linear regression analysis. It is called multiple because more than one independent variable will be included in the model; linear because the model will be of the form: $Y_i = \beta_0 + \beta_1 X_{i1} + \beta_2 X_{i2} + \cdots + \varepsilon_i$. Other refinements will be added as we move through the chapter. Before turning to these subjects, however, simple linear regression must be reexamined in conjunction with the F test, a concept essential to further study of regression analysis.

13.2 THE F TEST
IN REGRESSION

Analysis of variance and the F distribution were studied in Chapter 11, and Chapter 12 introduced the various sums of squares utilized in simple linear regression. To recapitulate, the total sum of squares,

$$SST = \sum (Y_i - \bar{Y})^2, \tag{13.1}$$

measures the variability of the dependent variable about its mean. The greater SST, the greater is the variability among the Y observations. The error sum of squares,

$$SSE = \sum (Y_i - \hat{Y}_i)^2, \tag{13.2}$$

measures the variation of the Y observations about the fitted regression line. The smaller SSE, the closer the observed values lie to the regression line; if SSE equals zero, the observed values all lie directly on the line.

Now since the regression line is obtained using the least squares procedure (i.e., b_0 and b_1 are chosen so as to minimize SSE), SST represents an upper bound for SSE. Why? Because the line $Y_i = \bar{Y}$, in which case $b_0 = \bar{Y}$ and $b_1 = 0$, will yield an SSE equal to SST (see Figure 12.6), and if *any* other line yields a lesser SSE, the least squares procedure will find it. Hence,

$$SST \geq SSE.$$

The difference, if any, between SST and SSE is termed the regression sum of squares, *SSR*. Analogous to the partitioning of the sum of squares in Chapter 11, we have in regression

$$SST = SSE + SSR \qquad (13.3)$$

The interpretation of *SSR* is just the *opposite* of *SSE*. What would $SSR = SST$ indicate?

If computations were done by hand, the sums of squares would be obtained using

$$SST = \sum Y^2 - n\bar{Y}^2; \qquad (13.4)$$

$$SSE = \sum Y^2 - b_0 \sum Y - b_1 \sum XY; \qquad (13.5)$$

$$SSR = SST - SSE. \qquad (13.6)$$

As developed in Chapter 11, any sum of squares has associated with it a number of degrees of freedom. *SST* has $n - 1$ degrees of freedom, one degree of freedom being lost because of a constraint on the deviations associated with the mean $\bar{Y}$. *SSE* has $n - 2$ degrees of freedom, two being lost due to constraints associated with b_0 and b_1. The degrees of freedom for *SSR* can be obtained by the difference in degrees of freedom for *SST* and *SSE*:

$$(n - 1) - (n - 2) = 1.$$

Degrees of freedom, like sums of squares, are additive. Further, a sum of squares divided by its degrees of freedom yields a mean square; the mean square regression (*MSR*) and mean square error (*MSE*) then being:

$$MSR = SSR/1; \qquad (13.7)$$

$$MSE = SSE/(n - 2). \qquad (13.8)$$

In the previous chapter, we were given that $E(MSE) = \sigma_\varepsilon^2$. The expected value of *MSR* can be shown to be

$$E(MSR) = \sigma_\varepsilon^2 + \beta_1^2 \sum (X_i - \bar{X})^2.$$

Now if $\beta_1 = 0$, then *MSE* and *MSR* are both estimators of the error variance σ_ε^2 (they are also *independent* estimators), and the ratio *MSR/MSE* would follow an F distribution with $v_1 = 1$ and $v_2 = n - 2$. Hence, the F distribution may be used to test $H_0: \beta_1 = 0$. If H_0 is true, the computed value of *MSR/MSE* is an observed value of an F distributed random variable. If H_0 is false, then the second term of Equation (13.9) is some *positive* value and the expected value of *MSR* is greater than σ_ε^2. Therefore *large* values of *MSR/MSE* indicate rejection of the null hypothesis, i.e., the critical region is entirely in the right tail. The

TABLE 13.1: ANOVA TABLE FOR SIMPLE LINEAR REGRESSION

Source of Variation	SS	v	MS	F
Regression	SSR	1	$MSR = SSR/1$	MSR/MSE
Error	SSE	$n - 2$	$MSE = SSE/(n - 2)$	
Total	SST	$n - 1$		

computational results of the F test are usually displayed in an ANOVA table of the form of Table 13.1.

EXAMPLE. The data and requisite computations from Section 12.4 are reproduced below. Use the F distribution to test $H_0: \beta_1 = 0$ at the 0.05 level of significance.

TABLE 13.2: DATA AND
PRELIMINARY
CALCULATIONS FOR
EXAMPLE PROBLEM

(1) X	(2) Y	(3) XY	(4) X^2	(5) Y^2
0	3	0	0	9
1	5	5	1	25
2	10	20	4	100
3	12	36	9	144
4	15	60	16	225
$\sum X = 10$	$\sum Y = 45$	$\sum XY = 121$	$\sum X^2 = 30$	$\sum Y^2 = 503$

Then, using Equation (13.4)

$$SST = \sum Y^2 - n\bar{Y}^2$$
$$= 503 - 5(9)^2 = 503 - 405 = 98.0.$$

SSE was computed (as was MSE) in Section 12.5:

$$SSE = 1.9,$$

and then,

$$SSR = 98.0 - 1.9 = 96.1.$$

The ANOVA table for regression is:

Source of Variation	SS	v	MS	F
Regression	96.1	1	96.1	$\dfrac{96.1}{0.633} = 151.74$
Error	1.9	3	0.633	
Total	98.0	4		

The critical value of $F_{1, 3, 0.05} = 10.1$; $H_0: \beta_1 = 0$ is rejected; and we conclude $\beta_1 \neq 0$. A decision rule for the p-value reaches the same conclusion, since $P(F \geq 151.74 | H_0 \text{ true}) < 0.01$ and $\alpha = 0.05$.

The student may now reconcile the above with the ANOVA table of Figure 12.5, the sample computer output. In Chapter 12, we stated that the ANOVA approach was an alternate but equivalent procedure to using the t distribution for testing $H_0: \beta_1 = 0$, and the above illustrates that both arrive at the same decision: rejection of H_0. The connection between the two tests is quite specific: the computed F value (151.74) is equal to the *square* of the computed t value (12.32 as obtained in Section 12.6). It can also be shown that the *square* of a t

distributed random variable with v_K degrees of freedom is F distributed with $v_1 = 1$ and $v_2 = v_K$. Hence, at any given level of significance (α), the F test of $H_0: \beta_1 = 0$ is algebraically equivalent to the t test; remember though that *all* the critical region is in the right tail for F while t is a two-tailed test.

Why then do we need the t test? The F test? In simple linear regression the t test is quite sufficient for testing $H_0: \beta_1 = 0$. In fact, the t test is more versatile since one-tailed hypotheses (e.g., $H_0: \beta_1 \geq 0$) or hypotheses of the form $H_0: \beta_1 = 5$ can be tested by means of t but not F. The F test does have certain unique applications in simple linear regression, but they are not introduced in this text. For our purposes, the F test is introduced because of its usefulness in testing certain hypotheses in multiple regression (where there will be no alternative t test).

EXAMPLE. Given a data set of 22 observations with the following computations:

$$\sum X^2 - n\bar{X}^2 = 50; \qquad b_1 = 2;$$
$$SST = 1,200; \qquad SSE = 1,000.$$

Use both the F and t distributions to test $H_0: \beta_1 = 0$ at the 0.01 level of significance.

The ANOVA table is given as:

Source of Variation	SS	v	MS	F
Regression	200	1	200	4.00
Error	1,000	20	50	
Total	1,200	21		

SSR was obtained as the difference of SST and SSE, and since $n = 22$, the degrees of freedom are all easily obtained. The critical $F_{1, 20, 0.01} = 8.10$, and $H_0: \beta_1 = 0$ is accepted.

The estimated variance of b_1 is, using Equation (12.8),

$$\hat{\sigma}_{b_1}^2 = \frac{\hat{\sigma}_\varepsilon^2}{\sum X^2 - n\bar{X}^2}$$

$$= \frac{MSE}{\sum X^2 - n\bar{X}^2} = \frac{50}{50} = 1.0,$$

and hence $\hat{\sigma}_{b_1} = \sqrt{\hat{\sigma}_{b_1}^2} = 1.0$. Then t is computed as

$$t = \frac{b_1 - 0}{\hat{\sigma}_{b_1}} = \frac{2.0}{1.0} = 2.0.$$

The critical value of t for a *two-tailed* test at $\alpha = 0.01$ and $v = 20$ is $t = 2.845$. As with the F test, $H_0: \beta_1 = 0$ is accepted. (Note again that the computed F is the square of the computed t value, and that this relationship also holds for their respective critical values.)

13.3 MULTIPLE REGRESSION

The basic principles of simple linear regression may now be extended to a regression model with several independent variables. The method of least squares, t and F tests, etc., all have counterparts in multiple regression, and they will be introduced as we proceed. First, the model.

The general form of the multiple linear regression model is

$$Y_i = \beta_0 + \beta_1 X_{i1} + \beta_2 X_{i2} + \cdots + \beta_K X_{iK} + \varepsilon_i, \qquad (13.9)$$

where

Y_i = Value of the dependent variable for the ith observation;
X_{i1} = Value of the first independent variable for the ith observation;
$\vdots$
X_{iK} = Value of the Kth independent variable for the ith observation;
$\beta_0, \beta_1, \ldots, \beta_K$ = Parameters;
ε_i = Value of the error term for the ith observation.

Conceptually, the above is a straightforward extension of the simple linear regression model, only more independent variables are included. The assumptions of the model, testing, and estimation techniques similarly will not require a new conceptual orientation. For $K = 2$, the estimated regression equation is a plane "fitted" to the observed data to minimize the error sum of squares, $\sum(Y_i - \hat{Y}_i)^2$ (see Figure 13.1). However, the computations involved in multiple regression are very laborious. And when there are more than two independent variables, an analysis without the aid of a computer is impractical. Accordingly

FIGURE 13.1: GRAPH OF THE REGRESSION EQUATION (A PLANE) FOR TWO INDEPENDENT VARIABLES

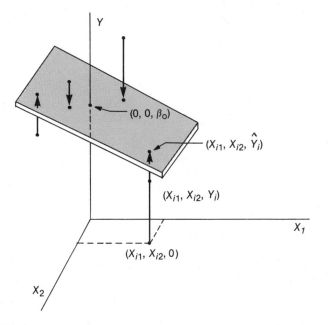

we shall initially restrict ourselves to the two-independent variable case. This enables us to illustrate all the requisites of multiple regression without becoming mired in computational details. (An analogy: If you were given 3,000 five-digit numbers and told the sum of these numbers, you clearly understand what has been done; given adequate time you could reproduce the result. In this chapter we want the student to understand what calculations are necessary and why, but to minutely reproduce the calculations for a six-variable model would be extravagant.)

For an example of a two-variable model, consider Stereo Ltd., a retailer of stereo components with numerous stores throughout the Midwest. Stereo Ltd. desires to develop a regression model for predicting store sales, which they plan to use in locating new stores. The corporate sales staff believes that store sales can be predicted on the basis of (1) number of competitors within a five-mile radius of store location, a trading zone, and (2) population within this zone. The model is then

$$Y_i = \beta_0 + \beta_1 X_{i1} + \beta_2 X_{i2} + \varepsilon_i,$$

where

Y = Store sales ($ ten thousands);
X_1 = Number of competitors in trading zone;
X_2 = Population in trading zone (thousands).

Values for the two independent variables can be ascertained through market research and data published by the Bureau of the Census. Annual sales by store is, of course, available in corporate headquarters for existing stores. In order to estimate the model, a random sample of existing stores (say $n = 10$) will be selected, and the 1979 data obtained for each observation. Then, using the least-squares method a regression equation can be computed of the form

$$\hat{Y}_i = b_0 + b_1 X_{i1} + b_2 X_{i2}.$$

As in simple linear regression, $\hat{Y}_i$ is a point estimate of the *mean* sales of stores that are characterized by both (1) X_{i1} competitors in the trading zone and (2) a population of X_{i2}. The quantities b_0, b_1, and b_2 are the least-squares estimators of β_0, β_1, and β_2; i.e., they are chosen in such a way as to minimize the sum of the squared deviations:

$$\sum(Y_i - \hat{Y}_i)^2 = \sum[\hat{Y}_i - (b_0 + b_1 X_{i1} + b_2 X_{i2})]^2 = \sum e_i^2.$$

The parameters β_1 and β_2 are usually termed regression coefficients (some texts prefer *partial* regression coefficients), and *the b_1 and b_2 are the estimated regression coefficients.* The regression coefficients are often explained as the change in the dependent variable associated with a unit change in the respective independent variable while all other independent variables are held constant. In the Stereo Ltd. model, then, β_2 could be interpreted as the change in expected store sales given a unit (in this instance 1,000 people since the data are in thousands) change in X_2, assuming X_1 is held constant. While such an interpretation is useful initially, some modification will later be required.

The assumptions of the model are similar to those for simple linear regression. They are: (1) the error term, ε_i, is a normally distributed random variable with a mean of zero and variance σ_ε^2 for all $i = 1, 2, \ldots, n$, and (2) the error term for some observation, say ε_i, is independent of the error term for any other observation, say ε_j. These assumptions imply that for any combination of X's there exists a normal distribution of Y values, each such distribution having the same variance, and with mean given by

$$E(Y) = \beta_0 + \beta_1 X_1 + \beta_2 X_2.$$

Note that these assumptions are not necessary to "fit" the regression equation (i.e., to calculate the b_i), but they are required for the estimation and inferential procedures to be introduced.

In simple linear regression, the least squares procedure of minimizing $\sum(Y_i - \hat{Y}_i)^2$ involved simultaneously solving the two normal equations for b_0 and b_1 (see Section 12.4). With two independent variables, the values of b_0, b_1, and b_2 are obtained by simultaneous solution of three normal equations:

$$\begin{aligned}
b_0 n &+ b_1 \sum X_1 &+ b_2 \sum X_2 &= \sum Y; \\
b_0 \sum X_1 &+ b_1 \sum X_1^2 &+ b_2 \sum X_1 X_2 &= \sum X_1 Y; \\
b_0 \sum X_2 &+ b_1 \sum X_1 X_2 &+ b_2 \sum X_2^2 &= \sum X_2 Y.
\end{aligned}$$

If one *must* obtain a solution by hand (and thus forego the speed and accuracy of the computer), one faces a formidable task. However, the solution may be simplified if the various summations required (e.g., $\sum X_{i1}$) are obtained after expressing each value as a deviation from its mean (e.g., $X_{i1} - \bar{X}_1$). Then, instead of dealing with rather cumbersome notation such as $\sum(X_{i1} - \bar{X}_1)(X_{i2} - \bar{X}_2)$, define:

$$\begin{aligned}
SS_Y &= \sum(Y - \bar{Y})^2 = \sum Y^2 - n\bar{Y}^2; \\
SS_1 &= \sum(X_1 - \bar{X}_1)^2 = \sum X_1^2 - n\bar{X}_1^2; \\
SS_2 &= \sum(X_2 - \bar{X}_2)^2 = \sum X_2^2 - n\bar{X}_2^2; \\
SP_{Y1} &= \sum(Y - \bar{Y})(X_1 - \bar{X}_1) = \sum X_1 Y - n\bar{X}_1\bar{Y}; \\
SP_{Y2} &= \sum(Y - \bar{Y})(X_2 - \bar{X}_2) = \sum X_2 Y - n\bar{X}_2\bar{Y}; \\
SP_{12} &= \sum(X_1 - \bar{X}_1)(X_2 - \bar{X}_2) = \sum X_1 X_2 - n\bar{X}_1\bar{X}_2.
\end{aligned} \qquad (13.10)$$

Then, b_0, b_1, and b_2 are obtained by substituting in the following:

$$b_1 = \frac{SP_{Y1}(SS_2) - SP_{Y2}(SP_{12})}{SS_1(SS_2) - (SP_{12})^2};$$

$$b_2 = \frac{SP_{Y2}(SS_1) - SP_{Y1}(SP_{12})}{SS_1(SS_2) - (SP_{12})^2}; \qquad (13.11)$$

$$b_0 = \bar{Y} - b_1\bar{X}_1 - b_2\bar{X}_2.$$

We reiterate that if the number of observations is large or the number of independent variables is greater than two, solutions without computer assistance become quite impractical.

EXAMPLE. The staff of Stereo Ltd. has taken a random sample of ten company stores and obtained the data given in Table 13.3 with preliminary computations.

TABLE 13.3: DATA FOR STEREO LTD. EXAMPLE

Store	Sales Y	Competitors X_1	Population X_2
1	6	4	34
2	15	3	92
3	12	2	75
4	9	3	36
5	17	1	78
6	5	5	8
7	11	4	23
8	16	2	69
9	9	3	10
10	10	3	25
Sum	110	30	450
Mean	11	3	45

Next, the various sums of squares and sums of cross-products are computed as shown in Table 13.4.

TABLE 13.4: CALCULATIONS FOR STEREO LTD. DATA

Y^2	X_1^2	X_2^2	X_1Y	X_2Y	X_1X_2
36	16	1,156	24	204	136
225	9	8,464	45	1,380	276
144	4	5,625	24	900	150
81	9	1,296	27	324	108
289	1	6,084	17	1,326	78
25	25	64	25	40	40
121	16	529	44	253	92
256	4	4,761	32	1,104	138
91	9	100	27	90	30
100	9	625	30	250	75
1,358	102	28.704	295	5,871	1,123

Then, the "corrected" sums of squares and cross-products are calculated. For example,

$$SS_Y = \sum(Y - \bar{Y})^2 = \sum Y^2 - n\bar{Y}^2$$
$$= 1,358 - (10)(11)^2$$
$$= 148.$$
$$SP_{Y1} = \sum(Y - \bar{Y})(X_1 - \bar{X}_1) = \sum X_1 Y - n\bar{X}_1\bar{Y}$$
$$= 295 - (10)(3)(11)$$
$$= -35.$$

Similarly,

$$SS_1 = 12;$$
$$SS_2 = 8,454;$$
$$SP_{Y2} = 921;$$
$$SP_{12} = -227.$$

Finally, the regression coefficients are calculated by substituting in Formulas 13.11. Hence,

$$b_1 = \frac{SP_{Y1}(SS_2) - SP_{Y2}(SP_{12})}{SS_1(SS_2) - (SP_{12})^2}$$

$$= \frac{(-35)(8454) - (921)(-227)}{(12)(8454) - (-227)^2}$$

$$= -1.739,$$

and

$$b_2 = 0.062.$$

Then,

$$b_0 = \bar{Y} - b_1\bar{X}_1 - b_2\bar{X}_2$$
$$= 11 - (-1.739)(3) - (0.062)(45)$$
$$= 13.417.$$

The multiple regression equation for the Stereo Ltd. data is then

$$\hat{Y} = 13.417 - 1.739X_1 + 0.062X_2.$$

The above indicates that when the number of competitors in the trading zone (X_1) increases by one, a *decrease* in sales of 1.739 (or $17,390) is expected, assuming other variables are unchanged. Similarly, as the population in the trading zone (X_2) increases by one (1,000 persons), Stereo Ltd. expects an *increase* in store sales of 0.062 (or $620), assuming other variables are unchanged. Nevertheless, the mere "fitting" of the regression equation, however arduous the task, is but one stage of the analysis. Remember: the least squares method may be applied to *any* set of data. For example, the Stereo Ltd. sales could have been regressed against X_1 (miles to nearest airport) and X_2 (inches of rainfall in 1979), and a regression equation thus obtained. But obviously such a model would possess little merit. How does the analyst evaluate a model? By statistical inference and observation of the model's predictive capabilities. We now turn to these matters.

13.4 INFERENCE IN MULTIPLE REGRESSION

The preceding section dealt with using the method of least squares to determine estimates of the model parameters β_0, β_1, and β_2. Various tests concerning the model and its parameters have been developed, but the assumptions underlying the model are prerequisite to their validity. We will present two of the tests most frequently employed in regression analysis: (1) a test of the hypothesis that *all* $\beta_i = 0$ for $i = 1, \ldots, K$ and (2) a test of the hypothesis that a particular $\beta_i = 0$.

Overall Test

The initial test usually performed in a regression analysis is a test for the significance of the relationship between the dependent variable and the entire set of K independent variables, i.e., an *overall* test. The null hypothesis is of general form

$$H_0: \beta_1 = \beta_2 = \cdots = \beta_K = 0.$$

The alternate hypothesis is

$$H_1: \text{Not all } \beta_i = 0 \ (i = 1, 2, \ldots, K).$$

If the analyst fails to reject the null hypothesis, this implies that a useful linear model based upon these K independent variables (or any subset thereof) does not exist. The analyst would then discard the model and, perhaps, attempt to find other independent variables that are linearly related to Y or investigate the possibility of nonlinear relationships. The test of this hypothesis employs the F distribution and is a modification of the F test introduced in the beginning of this chapter.

In Section 13.2, the F test in simple linear regression required computation of the various sums of squares, the mean squares, and finally the ratio MSR/MSE. The procedure in multiple regression requires the same quantities; first we must find SSR and SSE.

$$
\begin{aligned}
SST &= \sum (Y_i - \bar{Y})^2 \\
&= \sum Y^2 - n(\bar{Y})^2; \\
SSE &= \sum (Y_i - \hat{Y}_i)^2; \\
SSR &= SST - SSE.
\end{aligned}
\tag{13.12}
$$

The above appear identical to those previously given; the difference, of course, is that now the $\hat{Y}_i = b_0 + b_1 X_{i1} + \cdots + b_K X_{iK}$, i.e., the predicted values of Y, are obtained from the *multiple* regression equation. The above sums of squares can also be obtained using the regression coefficients and the "corrected" sums of squares and cross-products. The following present less of a computational burden:

$$
\begin{aligned}
SST &= SS_Y; \\
SSR &= b_1(SP_{Y1}) + b_2(SP_{Y2}); \\
SSE &= SST - SSR.
\end{aligned}
\tag{13.13}
$$

The degrees of freedom associated with SSE and SSR are modified to reflect additional constraints associated with estimating the b_i. As before SST has $n - 1$ degrees of freedom, but now SSE has df $= n - K - 1$. The degrees of freedom for $SSR = (n - 1) - (n - K - 1) = K$. The mean squares are given by

$$
\begin{aligned}
MSR &= SSR/K, \\
MSE &= SSE/(n - K - 1),
\end{aligned}
\tag{13.14}
$$

and the test statistic is

$$F = MSR/MSE. \qquad (13.15)$$

If H_0 is true, MSR/MSE is an observed value of an F distributed random variable with $v_1 = K$ and $v_2 = n - K - 1$. If H_0 is false, we expect MSR to be *greater* than MSE; hence, as in simple linear regression, the critical region is entirely in the right tail. Hence, the decision rule indicates that H_0 is rejected if the computed F is greater than the critical F or, equivalently, if $P(F \geq$ computed value $|H_0$ true$) < \alpha$; otherwise accept H_0. An ANOVA table in the form of Table 13.5 is used to display the computational results.

TABLE 13.5: ANOVA
TABLE FOR MULTIPLE
REGRESSION

Source of Variation	SS	v	MS	F
Regression	SSR	K	$MSR = SSR/K$	MSR/MSE
Error	SSE	$n - K - 1$	$MSE - SSE/(n - K - 1)$	
Total	SST	$n - 1$		

EXAMPLE. Using the Stereo Ltd. data, test the hypothesis that $\beta_1 = \beta_2 = 0$.

We first require the sums of squares; for illustration purposes we shall employ Formula 13.12; SST is easily obtained from our previous calculations as $SS_Y = 148$. To obtain SSE, the values of the Y_1 must be calculated by using the values of X_{i1} and X_{i2} in the regression equation:

$$\hat{Y}_i = b_0 + b_1 X_{i1} + b_2 X_{i2}$$
$$= 13.417 - 1.739 X_{i1} + 0.062 X_{i2}.$$

Since for the first observation, $X_1 = 4$ and $X_2 = 34$,

$$\hat{Y}_1 = 13.417 - 1.739(4) + 0.062(34)$$
$$= 8.576.$$

Similarly,

$$\hat{Y}_2 = 13.417 - 1.739(3) + 0.062(92)$$
$$= 13.909$$

The remaining computations are summarized in Table 13.6.

TABLE 13.6:
COMPUTATIONS FOR
ERROR SUM OF
SQUARES

i	Y_i	$\hat{Y}_i$	$Y_i - \hat{Y}_i$	$(Y_i - \hat{Y}_i)^2$
1	6	8.576	−2.576	6.636
2	15	13.904	1.096	1.201
3	12	14.589	−2.589	6.703
4	9	10.432	−1.432	2.051
5	17	16.514	0.486	0.236
6	5	5.218	−0.218	0.048
7	11	7.887	3.113	9.691
8	16	14.217	1.783	3.179
9	9	8.820	0.180	0.032
10	10	9.750	0.250	0.062
			$SSE =$	29.839

Finally,

$$SSR = SST - SSE$$
$$= 148 - 29.839$$
$$= 118.161.[1]$$

The degrees of freedom for SSR and SSE are $K = 2$ and $n - K - 1 = 7$ respectively. Computing the mean squares:

$$MSR = \frac{SSR}{K} = \frac{118.161}{2} = 59.080$$

$$MSE = \frac{SSE}{N - K - 1} = \frac{29.839}{7} = 4.263.$$

The computations are presented in Table 13.7.

TABLE 13.7: ANOVA
TABLE FOR STEREO LTD.

Source of Variation	SS	v	MS	F
Regression	118.161	2	59.080	13.860
Error	29.839	7	4.263	
Total	148	9		

At the 0.05 level of significance, the critical value of F with $v_1 = 2$ and $v_2 = 7$ is 4.74. Since the computed value of F is 13.860, we reject $H_0: \beta_1 = \beta_2 = 0$. The overall F test has thus provided encouragement to the analyst that the model under consideration may prove useful, and one would now proceed to further evaluation.

Test on a Specific Regression Coefficient

If the analyst rejects the null hypothesis that all β_i $(i = 1, \ldots, K)$ equal zero, then he concludes that one or more of the population regression coefficients is different from zero. However, the overall test does not indicate which *individual* coefficients are significant. Naturally, one desires to test the individual coefficients in order to delete nonsignificant variables and thus simplify the model. Such a test can employ either the t or F distributions with equivalent results. We shall computationally illustrate only the F test; the formulae involving the t test are quite complex unless matrix algebra is used. We shall term this test the *marginal F* test; some authors refer to the *incremental* or *partial F* test. Note that the marginal F test would not ordinarily be performed if the null hypothesis of the overall test were accepted.

---

[1] As an exercise, let's use Formula 13.13 to compute SSR:

$$SSR = b_1(SP_{Y1}) + b_2(SP_{Y2})$$
$$= -1.739(-35) + 0.062(921)$$
$$= 60.865 + 57.102$$
$$= 117.967.$$

The slight difference in the two results for SSR is due to rounding error and no matter for concern.

To test the significance of an individual coefficient, the marginal F test considers two models: (1) a model containing *all* the variables, and (2) a model with all variables *except* the one to be tested for significance. The gist of the test is to determine the contribution made (as measured by a reduction in the error sum of squares) by adding a particular variable to a model that originally did not contain such a term.

Consider the Stereo Ltd. example. There a regression equation of the form

$$\hat{Y}_i = b_0 + b_1 X_{i1} + b_2 X_{i2}$$

was computed. Subsequently the various sums of squares, MSE, etc. were computed and $H_0\colon \beta_1 = \beta_2 = 0$ was rejected. Now we ask: Is the above regression equation significantly better in predicting Y, store sales, than one in which X_1, number of competitors, is the only independent variable? To answer this question we must first examine a model containing only X_1, and then compare it to the model containing both X_1 and X_2. Now adding a variable to a model cannot cause SSE to increase, at worst the error sum of squares would be unchanged.[2] The reduction in SSE attributable to X_2 is the difference between SSE for a model with X_1 alone and SSE for a model with both X_1 and X_2; symbolically,

$$SSR(X_2 | X_1) = SSE(X_1) - SSE(X_1, X_2). \tag{13.16}$$

To test $H_0\colon \beta_2 = 0$ versus $H_1\colon \beta_2 \neq 0$, given X_1 is already in the model, one would compute the test statistic

$$F = \frac{SSE(X_1) - SSE(X_1, X_2)}{SSE(X_1, X_2)/(n-3)}, \tag{13.17}$$

where $(n-3)$ are the degrees of freedom associated with SSE for a model containing both X_1 and X_2, and the numerator has df $= 1$. If the null hypothesis is true, the test statistic above would be an observed value of a F distributed random variable with $v_1 = 1$ and $v_2 = n - 3$. The rejection region is again in the right tail; small values of the test statistic indicating that little reduction in SSE has occurred.

EXAMPLE. The two-variable model of sales for the Stereo Ltd. stores has been obtained. Given a single variable model with X_1 (number of competitors), does the addition of X_2 (population) to the model prove significant? I.e., test $H_0\colon \beta_2 = 0$.

To test the above hypothesis using the F distribution, we require the error sum of squares for models with (1) X_1 alone and (2) both X_1 and X_2. The latter we obtain from our previous computations (Table 11.7) as

---

[2] Although SSE cannot increase as variables are added to a model, this does not hold for MSE. From Equation (13.14), $MSE = SSE/(n - K - 1)$; note that as K (the number of independent variables) increases, the denominator (degrees of freedom) decreases. Hence, if the reduction in SSE is trivial as variables are added, MSE can increase. Since MSE is the estimated error variance, we should like to minimize its value. The point here is that we should not haphazardly add variables to a model under the pretense that if they don't help, at least no harm is done.

$$SSE(X_1, X_2) = 29.839.$$

To obtain $SSE(X_1)$, we would develop a regression model with X_1 alone. Assuming you don't need the practice, use the computer printout of Figure 13.2, where we observe:

$$SSE(X_1) = 45.917.$$

Using Equation (13.17) with $n = 10$, the test statistic is

$$F = \frac{SSE(X_1) - SSE(X_1, X_2)}{SSE(X_1, X_2)/(10 - 3)}$$

$$= \frac{45.917 - 29.839}{(29.839)/(7)}$$

$$= \frac{16.078}{4.263} = 3.772.$$

If $H_0: \beta_2 = 0$ were true, the above would represent an observed value of an F distributed random variable with $v_1 = 1$ and $v_2 = 7$. At the 0.10 level of significance, the critical value is $F = 3.59$. Thus the null hypothesis is rejected and the addition of X_2 provides a significant reduction in SSE beyond that of a model with X_1 alone.

FIGURE 13.2: OUTPUT FOR REGRESSION ANALYSIS OF SALES (Y) AND NUMBER OF COMPETITORS (X_1)

INTERCEPT	19.74998
REGRESSION COEFFICIENT	−2.91667
STD. ERROR OF REG. COEF.	0.692
COMPUTED T-VALUE	−4.217
CORRELATION COEFFICIENT	−0.831
STANDARD ERROR OF ESTIMATE . . .	2.396

ANALYSIS OF VARIANCE FOR THE REGRESSION

0 SOURCE OF VARIATION	D.F.	SUM OF SQ.	MEAN SQ.	F VALUE
ATTRIBUTABLE TO REGRESSION	1	102.083	102.083	17.786
DEVIATION FROM REGRESSION	8	45.917	5.740	
TOTAL	9	148.000		

Figure 13.3 is a reproduction of the printout of the SPSS package that employs the F distribution to test $H_0: \beta_i = 0$. The computed F value for X_2 (population, denoted in the lower left of the table by POP) is 3.785; the difference between this and our previous result is due to rounding error. Note that we may also test the significance of the adding X_1 (Competitors, denoted in the table by COMPET) to a model containing only X_2; in this case the F value is 4.196.

To test the significance of any particular variable X_i for any $i = 1, \ldots, K$ (although we state the hypothesis in term of its regression coefficient β_i) in a regression model of K independent variables, compute

$$\frac{SSE(\text{All } K \text{ variables except } X_i) - SSE(\text{All } K \text{ variables})}{SSE(\text{All } K \text{ variables})/(n - K - 1)}. \tag{13.18}$$

FIGURE 13.3: MULTIPLE REGRESSION OUTPUT USING THE SPSS PACKAGE

MULTIPLE REGRESSION RUN USING CARD INPUT AND RAW DATA   07/18/77   PAGE 4
FILE STEREO   (CREATION DATE = 07/18/77)   LTD, RANDOM SAMPLE OF STORES
*********************************** M U L T I P L E   R E G R E S S I O N ******************   VARIABLE LIST 1
                                                                                               REGRESSION LIST 1
DEPENDENT VARIABLE.. SALES
VARIABLE(S) ENTERED ON STEP NUMBER 1..   COMPET

MULTIPLE R	0.83051	ANALYSIS OF VARIANCE	DF	SUM OF SQUARES	MEAN SQUARE	F
R SQUARE	0.68975	REGRESSION	1.	102.08333	102.08333	17.78584
ADJUSTED R SQUARE	0.65097	RESIDUAL	8.	45.91667	5.73958	
STANDARD ERROR	2.39574					

--------------- VARIABLES IN THE EQUATION --------------- ------- VARIABLES NOT IN THE EQUATION -------

VARIANCE	B	BETA	STD ERROR B	F	VARIABLE	BETA IN	PARTIAL	TOLERANCE	F
COMPET	−2.916667	−0.83051	0.69159	17.786	POP	0.47041	0.59242	0.49206	3.785
(CONSTANT)	19.75000								

****

VARIABLE(S) ENTERED ON STEP NUMBER 2..   POP

MULTIPLE R	0.89367	ANALYSIS OF VARIANCE	DF	SUM OF SQUARES	MEAN SQUARE	F
R SQUARE	0.79864	REGRESSION	2.	118.19852	59.09926	13.88169
ADJUSTED R SQUARE	0.74111	RESIDUAL	7.	29.80148	4.25735	
STANDARD ERROR	2.06334					

--------------- VARIABLES IN THE EQUATION --- ------------ ------- VARIABLES NOT IN THE EQUATION -------

VARIABLE	B	BETA	STD ERROR B	F	VARIABLE	BETA IN	PARTIAL	TOLERANCE	F
COMPET	−1.739278	−0.49525	0.84912	4.196					
POP	0.6224083D-01	0.47041	0.03199	3.785					
(CONSTANT)	13.41700								

MAXIMUM STEP REACHED $\hat{Y} = 13.1417 - 1.739(x) + .6224(x_2)$

If $H_0: \beta_i = 0$ is true, the above will follow the F distribution with $v_1 = 1$ and $v_2 = n - K - 1$. Again the critical region is in the right tail. The student should be careful to understand that the test of $H_0: \beta_i = 0$ is a marginal test; it measures the value of X_i in reducing the error sum of squares beyond that of a model containing the other $K - 1$ independent variables. The test result of $H_0: \beta_i = 0$ will likely vary depending upon the nature and number of independent variables in the model.

Consider the Stereo Ltd. example with $Y = $ sales and independent variables

$X_1 = $ number of competitors in trading zone;

$X_2 = $ population in trading zone;

$X_3 = $ number of registered vehicles in trading zone.

In a two variable model with X_1 and X_2, X_2 was significant in reducing SSE beyond that of X_1 alone. Now consider the three-variable model with X_1, X_2, and X_3 and test $H_0: \beta_2 = 0$. I.e., Would the addition of X_2 to a model already containing X_1 and X_3 significantly reduce the error sum of squares and hence improve our predictions concerning Y? Answer: Likely not, because X_2 and X_3 are both attempting to measure the same thing: potential store patronage. And once X_2 or X_3 is included in the model, the other would be superfluous. Note further that we would expect both $H_0: \beta_2 = 0$ and $H_0: \beta_3 = 0$ to be accepted in a model of X_1, X_2, and X_3, but we recognize that only X_2 or X_3 should be deleted. The moral of this is that the F test for individual coefficients should be used to delete only one variable at a time; do not summarily remove all variables whose test statistic is not significant. To summarize this succinctly, we quote

Cramer ("Significance Tests and Tests of Models in Multiple Regression," *The American Statistician*, vol. 26, no. 4 [October 1972], p. 26):

> To say that a regression coefficient . . . is significant, one must specify explicitly or implicitly a model except when the variables are uncorrelated; only in the case of uncorrelated variables is the significance of a coefficient independent of what other variables are included in the model.

We might add that the X_i are rarely uncorrelated in business and economic applications of regression analysis.

As mentioned earlier, the marginal test of $H_0: \beta_i = 0$ can be performed using the t distribution as well as the F test just presented. The two tests yield equivalent results, as did their counterparts in simple linear regression. Some computer packages will include both tests, while others will include only one test. Those including both do so primarily for the convenience of the user. The *interpretation* of the t test is quite straight forward; one merely compares the test statistic

$$t = \frac{b_i}{\hat{\sigma}_{b_i}} \tag{13.19}$$

with a critical value from a t distribution with $(n - K - 1)$ degrees of freedom. The *computation* of the test statistic is not so straight forward. The estimated standard error, $\hat{\sigma}_{b_i}$, has a very complex formula (unless expressed in matrix algebra), and will not be presented. Computer routines employing the t test will ordinarily include the estimated standard error as well as the test statistic for all variables in the model. Figure 13.4 is an example of the output of a regression program using the t test. In Figure 13.4, variables 2 and 3 are X_1 and X_2 respectively. Note that the estimated standard errors, $\hat{\sigma}_{b_i}$, are reported and that the computed t values are the square roots of the corresponding F values in Figure 13.3. The student must remember that this t test is equivalent to the marginal F test and, as such, tests the significance of a variable *in a particular model*. The test result may change when different models are entertained.

FIGURE 13.4: MULTIPLE REGRESSION OUTPUT USING t TO TEST H_0: $\beta_i = 0$

VARIABLE	REG. COEF.	STD. ERROR COEF.	COMPUTED T	BETA COEF.
2	−1.73928	0.84912	−2.04834	−0.49525
3	0.06224	0.03199	1.94557	0.47041

INTERCEPT	13.41698		
MULTIPLE CORRELATION	0.89367	(ADJUSTED R =	0.87947)
STD. ERROR OF ESTIMATE	2.06334	(ADJUSTED SE =	2.18850)

ANALYSIS OF VARIANCE FOR THE REGRESSION

0 SOURCE OF VARIATION	D.F.	SUM OF SQ.	MEAN SQ.	F VALUE
ATTRIBUTABLE TO REGRESSION	2	118.199	59.099	13.882
DEVIATION FROM REGRESSION	7	29.801	4.257	
TOTAL	9	148.000		

There are other inferential procedures in multiple regression that the interested student may wish to pursue in a more advanced text. Included in these procedures would be a test of, for example, $H_0: \beta_2 = \beta_4 = 0$ in a five-variable model. This test uses the F distribution and enables the analyst to test the significance of a subset of variables from the set comprising the model. Finally, as in simple linear regression, confidence intervals for means or new observations, or both, may be computed for given values of the independent variables in the model. Computations necessary for such confidence intervals are complex and would ordinarily not be attempted by hand. Many computer packages include an option to compute such estimates should they be desired by the analyst.

**13.5 QUALITATIVE VARIABLES

The regression models studied thus far have considered only *quantitative* variables. In this and the previous chapter such variables included sales, population, number of competitors, income, QPA, age, class size, etc.; these variables possess values determined through some measuring or counting process. However, many variables of interest to the analyst are *qualitative* (or categorical) in nature; these variables assume states through some classification process. Examples of such variables are sex, race, marital status, region (Northeast, Midwest, etc.), profession (lawyer, physician, etc.), and class instructor (Professors Duncan, Dow, etc.). The analyst may wish to consider qualitative variables as their inclusion in the model could prove significant in reducing the error sum of squares and providing more precise estimates. Using qualitative variables in regression requires the construction of *dummy* variables (sometimes termed indicator variables). The definition and use of dummy variables is now our concern.

As an initial example consider a marketing application. Citizens' Bank has constructed a simple linear regression model with Y (monthly charges on the bank's Credi-Card) and X_1 (customer's monthly income). The observations obtained represent a random sample from its population of 4,000 cardholders and their June transactions. A staff member has suggested that sex (male, female) of cardholder is an important determinant of card usage and should be included in the model. The creation of the necessary dummy variable may be accomplished in a variety of ways, but the easiest method is to use a coding system of 0's and 1's. Then, we could define X_2 (sex) as having the value one if cardholder is male and zero if not male, i.e., female. The coding process for six cardholders is illustrated below.

Observation	Sex	X_2
1	Male	1
2	Male	1
3	Female	0
4	Female	0
5	Male	1
6	Female	0

Given the numerical values for X_2 and the corresponding data for Y and X_1, the calculation of the regression equation and the inferential procedures would be accomplished as set forth in the previous sections—i.e., the inclusion of dummy variables in a regression model does not require any changes in the computational routines. (Nevertheless, we shall later present a numerical illustration.) The Credi-Card model would then have X_1 (income) and X_2 (sex) as independent variables and be of the form

$$\hat{Y} = b_0 + b_1 X_1 + b_2 X_2.$$

We note that b_2, like any regression coefficient, may be positive or negative. How does one interpret b_2? First note that for females $X_2 = 0$, and

$$\hat{Y} = b_0 + b_1 X_1 + b_2 \quad (0)$$
$$= b_0 + b_1 X_1 \quad \text{(females)}.$$

For males, $X_2 = 1$ and

$$\hat{Y} = b_0 + b_1 X_1 + b_2 \quad (1)$$
$$= (b_0 + b_2) + b_1 X_1 \quad \text{(males)}.$$

Using the dummy variables has thus given us *two* regression lines (see Figure 13.5). The lines are parallel (i.e., have the same slope b_1), but different intercepts. Hence, b_2 is interpreted as the effect of belonging to the male category versus the female category. Note also that this effect is constant over the entire range of incomes.

FIGURE 13.5:
REGRESSION ANALYSIS
WITH A QUANTITATIVE
AND A QUALITATIVE
VARIABLE

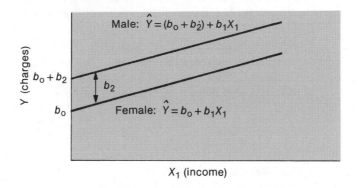

How would a qualitative variable with more than two categories be included in a regression model? First, a general rule: *a qualitative variable with q categories requires the creation of q − 1 dummy variables*. The variable SEX with two categories thus required only one dummy variable. Suppose that in a study concerning professional schools we desired to include the qualitative variable major (law, engineering, or business). To include this variable in a regression model requires the creation of $q - 1 = 3 - 1 = 2$ dummy variables.[3] The numerical values for both dummy variables will consist of 0's and 1's. Define:

$$X_1 = 1 \text{ if business major}$$
$$\quad 0 \text{ if not business major;}$$

$$X_2 = 1 \text{ if law major}$$
$$\quad 0 \text{ if not law major.}$$

_____

[3] Note that including qualitative variables with numerous categories will require many dummy variables. For example, suppose we desired to regress faculty salary (Y) against age, school, and rank. If there were nine schools in the university (education, business, humanities, etc.) and five ranks (lecturer, instructor, assistant professor, etc.), the *two* qualitative variables would require $9 - 1 = 8$ plus $5 - 1 = 4$ or *twelve* dummy variables.

The coding process for five students would be:

Observation	Major	X_1	X_2
1	Business	1	0
2	Engineering	0	0
3	Law	0	1
4	Business	1	0
5	Engineering	0	0

Suppose the model was Y (starting salary) with X_1, X_2, and X_3 (QPA) as independent variables. The regression equation would be of the form

$$\hat{Y} = b_0 + b_1 X_1 + b_2 X_2 + b_3 X_3.$$

Analogous with previous results, the above yields three regression lines:

$$\hat{Y} = b_0 + b_3 X_3 \quad \text{(engineering major)};$$
$$\hat{Y} = (b_0 + b_1) + b_3 X_3 \quad \text{(business major)};$$
$$\hat{Y} = (b_0 + b_2) + b_3 X_3 \quad \text{(law major)}.$$

Thus, b_1 measures the effect of business major versus engineering major, the latter being coded with all zeroes (i.e., $X_1 = X_2 = 0$ for engineers). Similarly, b_2 measures the effect of law major versus engineering major. Note that the effect of major on starting salary is constant over all QPAs in this model.

EXAMPLE. A staff member of Stereo Ltd. believes that store sales are significantly different (he thinks higher) when a campus is located within the store's trading zone. College students, in his opinion, represent a very lucrative market for stereo equipment. Accordingly for the ten sample stores previously studied, he finds that stores 2, 3, 5, 7, 8 and 10 have a campus within their trading zones. He then developes the dummy variables X_3 (CAMPUS) as follows:

$$X_3 = 1 \text{ if campus within trading zone}$$
$$0 \text{ if not.}$$

Then, values for X_3 are:

Observation	Campus	X_3
1	No	0
2	Yes	1
3	Yes	1
4	No	0
5	Yes	1
6	No	0
7	Yes	1
8	Yes	1
9	No	0
10	Yes	1

Then, using the above and data from Table 13.3, a regression model for sales (Y), with X_1 and X_3 as independent variables, can be obtained. The results are given in the printout of Figure 13.6.

The regression equation is

$$\hat{Y} = 14.636 - 1.970X_1 + 3.788X_3.$$

Using the (marginal) t test for $H_0: \beta_3 = 0$, the computed t value is 2.730, and the hypothesis is rejected at the 0.05 level of significance. The student might note that, as measured by SSE, X_1 and X_3 yield a better model for sales than X_1 and X_2 (compare Figures 13.4 and 13.6). Of course, the analyst would probably try a model with all three independent variables included, but we leave that as an exercise for the student.

FIGURE 13.6:
REGRESSION RESULTS
WITH Y (SALES), X_1
(COMPETITORS), AND X_3
(CAMPUS)

VARIABLE	REG. COEF.	STD. ERROR COEF.	COMPUTED T	BETA COEF.
1	−1.96970	0.62061	−3.17383	−0.56087
3	3.78788	1.38772	2.72958	0.48236

INTERCEPT	14.63636	
MULTIPLE CORRELATION	0.92180	(ADJUSTED R = 0.91155)
STD. ERROR OF ESTIMATE	1.78255	(ADJUSTED SE = 1.89068)

ANALYSIS OF VARIANCE FOR THE REGRESSION

0 SQUARE OF VARIATION	D.F.	SUM OF SQ.	MEAN SQ.	F VALUE
ATTRIBUTABLE TO REGRESSION	2	125.758	62.879	19.789
DEVIATION FROM REGRESSION	7	22.242	3.177	
TOTAL	9	148.000		

This review of dummy variables has served only to introduce the student to their utility in regression analysis. Three other uses of dummy variables deserve comment. (1) In our models with dummy variables, the effect of some category (e.g., law major) was constant across the entire range of the quantitative variable (e.g., QPA). If law students' salaries increase more rapidly with QPA than engineering students, then the effect is not constant but varies depending upon QPA, and the regression equations should appear as in Figure 13.7. This case is beyond our goals for this text but can be resolved using dummy variables. (2) Dummy variables are also used to test for the equality of two regression

FIGURE 13.7:
REGRESSION
EQUATIONS WHEN THE
EFFECT OF MAJOR IS
NOT CONSTANT OVER
QPA

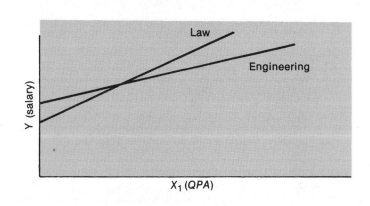

equations, say to compare an Ohio bank's study of credit-card usage with that of a California bank. We assume that the variables employed in each study are the same (e.g., income, age, and sex). (3) Finally, dummy variables are used in developing piece-wise linear regression equations. For example in a study of variable costs (Y) versus lot size (X), the slope (i.e., marginal cost) may be distinctly different after X increases beyond some point because of the firm's ability to then buy raw materials in carload lots at lower prices. For these and other applications, consult the book by Neter and Wasserman or that by Kerlinger and Pedhazur as cited in this chapter's references.

13.6 THE COEFFICIENT OF MULTIPLE DETERMINATION

Section 12.7 defined the coefficient of determination, r^2, for simple linear regression. In multiple regression, a similar statistic exists: the *coefficient of multiple determination, R^2*.

$$R^2 = \frac{SSR}{SST}$$

$$= 1 - \frac{SSE}{SST}. \tag{13.20}$$

R^2 shows the proportion of the variability of the dependent variable Y that is "explained" by the multiple regression equation. As before, this coefficient will assume a value between zero and one. When $R^2 = 1$, the observed values Y_i and predicted values $\hat{Y}_i$ are equal for $i = 1, 2, \ldots, n$.

The *coefficient of multiple correlation* is defined as the positive square root of R^2:

$$R = \sqrt{R^2}. \tag{13.21}$$

In Chapter 12, r was positive or negative depending upon the sign of b_1. In multiple regression, R is defined as a positive number since the various regression coefficients may assume different signs, e.g., b_1 could be positive while b_2 is negative. Both R and R^2 frequently appear on the output of computer regression programs.

As variables are added to a model, there is usually a decrease in SSE (it will never increase) with a corresponding increase in R^2; this occurs regardless of the relevance of the additional variables to the study. In fact, if $n \leq (K + 1)$, a perfect fit will be obtained no matter how ludicrous the hypothesized relationship. For example, in a model of QPA(Y), height (X_1) and weight (X_2), $R^2 = 1$ if only $K + 1 = 3$ observations are used. Accordingly, one may encounter the *adjusted* coefficient of multiple determination (or correlation). This statistic adjusts R^2 *downward* when the number of observations (n) does not greatly exceed the number of independent variables (K).

$$R_a^2 = 1 - \frac{(n - 1)}{(n - K - 1)} \frac{SSE}{SST} \tag{13.22}$$

When n is much larger than K, R^2 and R_a^2 will be approximately equal.

13.7 OTHER TOPICS

Chapter 12 concluded with a number of warnings for the novice in simple linear regression. The reader was cautioned about the dangers of extrapolation, blind acceptance of computer printouts, statistical versus practical significance, and the use of regression to infer a cause-and-effect relationship. These caveats are equally applicable in multiple regression, and Section 12.8 deserves a thorough review.

Several additional problems are also encountered in multiple regression applications. For example, in Chapter 12, we observed the utility of a scatter diagram in verifying the validity of the linear model and its assumptions. But when there are several independent variables, a plot of the observed data is no longer possible. However, other graphic analyses are possible. In particular, a plot of the residuals ($e_i = Y_i - \hat{Y}_i$) versus the predicted values ($\hat{Y}_i$) is often used to detect abnormalities in the model. The reader should consult Anscombe, Draper and Smith, or Neter and Wasserman for a complete discussion on the examination of residuals.

Multicollinearity is another problem encountered in multiple regression. Multicollinearity exists when the independent variables are themselves related, a very common occurrence in business and economic applications. If multicollinearity is present, the regression coefficients will shift in magnitude (occasionally even change signs depending on the variables included in the model For instance, we have constructed several different models using the Stereo Ltd. example, all of which contained X_1 (competitors). Table 13.8 shows that the regression coefficient for X_1 has varied from -1.470 to -2.917 depending on the model. This serves to warn us against interpreting a regression coefficient too literally. Yes, given a unit change in X_1 and holding other independent variables in the model constant, b_1 does show the effect on Y. Practically, variables related to X_1 do not remain constant while X_1 changes. Extreme multicollinearity may also cause significant rounding errors in computer routines. One solution to the problem sometimes suggested is to drop all but one of the interrelated variables. In summary, multicollinearity will cause the regression coefficients to be unstable (1) from sample to sample for the same model or (2) from model to model for the same data (as in Table 13.8). Nevertheless, this condition generally does not affect the prediction aspects of regression analysis.

TABLE 13.8: VALUES OF b_1 FOR VARIOUS MODELS OF Y (SALES) WITH X_1 (COMPETITORS), X_2 (POPULATION), AND X_3 (CAMPUS)

Variables Included	b_1	Source
X_1	-2.917	Figure 13.2
X_1, X_2	-1.739	Figure 13.3
X_1, X_3	-1.970	Figure 13.6
X_1, X_2, X_3	-1.470	(Not illustrated)

Finally, another topic worthy of further study is that of selecting the "best" regression equation. In applied work, the analyst may have 10, 20, even 30 or more independent variables as "candidates" for inclusion in the regression model. How does one screen the variables in an efficient manner? Fortunately, several computer routines are available which are invaluable in larger-scale problems. (1) *Backward elimination* begins with a model containing all the independent variables. Then the marginal F (or t) test is used to delete the variable with the *smallest F* value *if* this value is *below* the predetermined critical value of F. When a variable is dropped, a *new* regression equation is found using the remaining variables, and the marginal F test is used again. The process continues until the variables left in the model are all significant. (2) The *forward selection* procedure adds variables to the model. The first variable to enter is the one with the *largest F* value if this value is *above* the predetermined critical value of F. Then, *given that this variable is in the model*, another variable is entered using the marginal F test as an entry criterion. The process continues until no excluded variable's entry significantly reduces *SSE*. (3) The *stepwise* procedure is probably the most popular of selection routines. It combines the forward selection procedure with the F test on variables previously included in the model. For example, with the addition of a sixth independent variable, the variable that entered on the second "step" may no longer be significant. (Remember that the significance of a variable is dependent on what other variables are included in the model.) If a variable entered on earlier step later becomes insignificant, the stepwise procedure deletes such a variable. The process terminates when no excluded variable has a marginal F above the critical F value and no included variable has a marginal F below the critical F value.

Backward elimination, forward selection, and stepwise are well-known selection procedures, and still other procedures are also in existence. Some computer packages for statistical analysis (e.g., SAS) allow the user to choose the selection procedure to be employed. Sophisticated programs and reams of printout are no substitute for familiarity with the phenomenon under study and an understanding of the principles of regression analysis. We should note that the various procedures do not necessarily lead to the same model, and personal judgment remains an important part of model construction. Given a choice, the authors recommend the stepwise procedure.

13.8 TERMS AND FORMULAS

Multiple regression

Estimated regression coefficients $(b_1, b_2, \ldots)$

Overall F test

Marginal (or partial) F test

Dummy (or indicator) variables

Multicollinearity

Backward elimination

Forward selection

Stepwise procedure

Multiple regression equation for K independent variables

$$\hat{Y}_i = b_0 + b_1 X_{i1} + b_2 X_{i2} + \cdots + b_K X_{iK}$$

Total sum of squares

$$SST = \sum (Y_i - \bar{Y})^2$$

Error sum of squares

$$SSE = \sum(Y_i - \hat{Y}_i)^2$$

Regression sum of squares

$$SSR = SST - SSE$$

Mean squares, degrees of freedom, and test statistic for simple linear regression

See Table 13.1

Test statistic for overall F test (H_0: $\beta_1 = \beta_2 = \cdots = \beta_K = 0$)

$$F = \frac{MSR}{MSE}$$

$$= \frac{SSR/K}{SSE/(n - K - 1)}$$

Test statistic for marginal F test (H_0: $\beta_i = 0$)

$$F = \frac{\left(\begin{array}{c} SSE(\text{All } K \text{ variables except } X_i) \\ - SSE(\text{All } K \text{ variables}) \end{array} \right)}{SSE(\text{all } K \text{ variables})/(n - K - 1)}$$

Coefficient of multiple determination

$$R^2 = \frac{SSR}{SST}$$

$$= 1 - \frac{SSE}{SST}$$

Coefficient of multiple correlation

$$R = \sqrt{R^2}$$

Adjusted coefficient of multiple determination

$$R_a^2 = 1 - \frac{(n - 1)SSE}{(n - K - 1)SST}$$

QUESTIONS

A. True or False

1. The minimum value of SSE is zero.
2. Tests in regression analysis using the F distribution have a one-tailed rejection region.
3. The overall test of H_0: $\beta_1 = \beta_2 = \cdots = \beta_K$ employs the t-distribution.
4. SSE cannot increase as variables are added to a regression model.
5. The result of the marginal test H_0: $\beta_2 = 0$ may vary depending upon what variables are included in the model.
6. In a regression model with seven independent variables, the marginal test accepted H_0 for four variables; these four should be dropped from the model.
7. In the marginal test, the t or F distributions may be used with equivalent results.
8. Zodiac sign is a qualitative variable.
9. Age is a qualitative variable.

~~F~~ 10. The regression coefficient for a dummy variable must be positive.

~~F~~ 11. $R_a^2 \geq R^2$.

B. Fill in

1. In the marginal test of $H_0: \beta_3 = 0$, the computed $t = 4.0$; the F value for this test would be $(4)^2 = 16$.

2. The null hypothesis of the overall F test is of the form: $H_0\ \beta_1 = \beta_2 = \beta_K = 0$

3. _Dummy variable_ are required when the analyst desires to include qualitative data.

4. Given seven observations and six independent variables, the coefficient of multiple determination will equal _1.00_.

5. Name two variable-selection procedures: _Stepwise, Forward_
   _____.

6. _Multicolinear_ describes a situation where the independent variables are themselves related.

7. If $Y_i = \hat{Y}_i$ for $i = 1, 2, \ldots, n$, then SSR equals _SST_.

8. The _Marginal_ test is used to determine significance of a variable in a specific model.

9. If make of car (Ford, Chrysler, AMC, or GM) was to be considered, _Three_ dummy variables would be required.

10. If $K = 10$, $n = 20$, and $R^2 = 0.90$, $R_a^2 = $ _.789_.

EXERCISES

1. In the following situations, specify two or more quantitative (independent) variables that you would expect to be useful in a multiple regression model for:
   a. A marketing director for a national diaper service company predicting store revenue per year.
   b. An oil company predicting annual volume of gas sold at a service station.
   c. A university board of trustees predicting faculty salaries.

2. In the above problem, specify a qualitative variable that you would expect to be useful in the models.

3. For Problem 12 in the previous chapter, test $H_0: \beta_1 = 0$ using the F test and the 0.10 level of significance.

4. Given a simple linear regression model with 23 observations, $SST = 1,736$, and $SSR = 972$, test $H_0: \beta_1 = 0$ at the 0.05 level.

*$F = 4.32$
Reject*

5. Given a simple linear regression model with $SSE = 406$, $SST = 3,790$, and $n = 30$,
   a. Determine SSR.
   b. Determine MSR and MSE.
   c. Test $H_0: \beta_1 = 0$ at the 0.05 level.

6. The following results were calculated from a data set of five observations using the simple linear regression model:

$$b_0 = 20.4 \qquad \sum X^2 - n\bar{X}^2 = 50.0$$
$$SSR = 1,352.0$$
$$b_1 = 5.2 \qquad SST = 1,480.0$$

a. Determine SSE.
b. Compute the estimated error variance.
c. Use the t distribution to test $H_0: \beta_1 = 0$ at the 0.05 level.
d. Use the F distribution to test $H_0: \beta_1 = 0$ at the 0.05 level.
e. Square the test statistic in part c and compare it to the test statistic in part d.

7. The following data result from a linear regression model with two independent variables: $n = 30$; $SST = 2,980$; $SSE = 344$.
a. Determine SSR, MSR, and MSE.
b. Test $H_0: \beta_1 = \beta_2 = 0$ at the 0.05 level.
c. If one rejects the hypothesis of part b, would the *marginal* tests of $H_0: \beta_1 = 0$ and $H_0: \beta_2 = 0$ both necessarily indicate rejection?

8. Suppose a distributor of hospital supplies does business in three states: New Jersey, New York, and Pennsylvania.
a. How many dummy variables would be necessary to include the qualitative variable "state" in a regression analysis?
b. Define the necessary dummy variables and indicate their values for an observation from New York.

9. The chancellor's office of a state university system took a random sample of graduate students from two of the system's campuses. A regression model was developed of the form

$$\hat{Y} = 1.12 + 0.53X_1 - 0.18X_2,$$

where

Y = graduate grade point average;
X_1 = undergraduate grade point average;
X_2 = 0 if Playton campus, 1 if Tech campus.

a. What is the point estimate of a graduate student at the Tech campus whose undergraduate grade point average was 2.12?
b. Sketch the regression equation(s).

10. A large real estate fim is using regression analysis to study the selling price of homes. Their model will include such variables as square footage, age, size of lot, etc. If the firm is concerned with homes in four large subdivisions (Forest Park, Tanglewood, Coventry, and Vesthaven), how many dummy variables are necessary to include "subdivision" in their model? Specify the values of the dummy variables for a house in Coventry.

11. In Problem 12 of the previous chapter, Professor Gordon performed a regression analysis of class rating (Y) and class size. Suppose that he

teaches at an urban university with night classes, and five of his sample classes met at night. This new information as well as the previous data is given below.

Class Size	Night Class	(Y) Class Rating of Instructor
10	No	4.2
20	No	4.3
20	Yes	3.1
20	Yes	3.6
30	No	4.0
30	No	4.1
30	Yes	3.2
40	No	4.0
40	Yes	3.0
40	No	3.1
50	Yes	3.0

a. Construct a dummy variable (X_2) to reflect whether the class meets day or night.
b. Obtain the regression equation with both class size (X_1) and night (X_2) included as independent variables.
c. Test $H_0: \beta_1 = \beta_2 = 0$ using the overall (F) test at the 0.05 level.
d. Perform the marginal test of $H_0: \beta_2 = 0$ at the 0.05 level.
e. With the regression equation of part b, predict the class rating of an instructor with 28 students in a night class.

12. The Tralfamadorian I.R.S. (Problem 15, Chapter 12) was seeking methods to help its auditors select tax returns to audit. One of their employees suggested the number of dependents claimed on the return be included as an independent variable. The data are given below:

X_1 Income ($ thousands)	X_2 Dependents	Y Interest Expense ($ hundreds)
4.1	2	2.0
5.0	5	4.0
7.0	1	5.0
10.2	4	6.0
11.0	2	9.0
11.0	2	8.0
15.0	3	10.0
17.5	1	14.0
20.0	4	16.0
30.0	3	29.0

a. Determine the values of b_0, b_1 and b_2 for the regression equation $\hat{Y} = b_0 + b_1 X_1 + b_2 X_2$.
b. Determine the values of SST, SSR, and SSE.
c. Compute the coefficient of multiple determination, R^2.
d. Test $H_0: \beta_1 = \beta_2 = 0$ using the overall F test.

    e. In a model containing only X_1, $SSE = 16.90$. Use this information to do a marginal test on X_2, given X_1 is in the model.

13. In Problem 7 of the previous chapter, the credit union at United University had used simple linear regression with fifty observations where:

Y = amount of payroll deduction ($ hundreds);
X = annual salary ($ hundreds);
$SSE = 6.0$;
$SST = 20.0$

The credit union manager thought a model that included age of the employee would be an improvement. He determined the age of the 50 employees in his sample, and a multiple regression analysis with both X_1 (salary) and X_2 (age) had an error sum of squares of 2.2.

    a. Determine $SSE(X_1)$ and $SSE(X_1, X_2)$.
    b. Determine $SSR(X_2 | X_1)$.
    c. Determine $SST(X_1)$ and $SST(X_1, X_2)$.
    d. Test $H_0: \beta_1 = \beta_2 = 0$ at the 0.05 level.
    e. Test $H_0: \beta_2 = 0$ when X_1 is included in the model (the marginal test).
    f. Determine the coefficient of multiple determination, R^2.

14. For report purposes, regression equations are sometimes written in the following form:

$$\hat{Y} = 17.20 + 9.61X_1 + 35.73X_2 - 3.22X_3 + 14.01X_4.$$
$$(6.93) \qquad (5.27) \qquad (0.79) \qquad (12.95)$$

where the numbers in parentheses are the estimated standard errors of the regression coefficients. For example, $\hat{\sigma}_{b_1} = 5.27$. Assume $n = 20$.

    a. Use the above to do the marginal tests of $H_0: \beta_i = 0$ at the 0.05 level for $i = 1, 2, 3, 4$.
    b. Based on the above, if a single independent variable could be dropped from the model, which one, if any, would you drop?
    c. Could you now recommend a "final" model containing two independent variables?
    d. For the model above, $R^2 = 0.78$; compute the adjusted coefficient of determination.
    e. Predict Y when $X_1 = 2.10$, $X_2 = 4.12$, $X_3 = 2.79$ and $X_4 = 0.78$.

15. A financial analyst for an investments firm was trying to identify undervalued stocks in the electric utilities group. He selected a random sample of 20 such stocks listed on the New York Stock Exchange and determined values for:

$\hat{Y}$ = Price/earnings ratio;
X_1 = Yield (%);
X_2 = Five-year earnings per share growth rate (%);
X_3 = Current ratio;
X_4 = Percent nuclear generation.

He summarized his results with the following equation (estimated standard errors in parentheses):

$$\hat{Y} = 6 + 0.134X_1 + 0.056X_2 + 0.191X_3 - 0.045X_4$$
$$\phantom{\hat{Y} = 6 + }(0.036)\quad (0.011)\quad (0.157)\quad (0.012)$$
$$(R^2 = 0.84;\ SST = 245.3).$$

a. Test $H_0: \beta_1 = \beta_2 = \beta_3 = \beta_4 = 0$ at the 0.05 level.
b. Perform the marginal test of $H_0: \beta_i = 0$ for each independent variable at the 0.05 level.
c. Compute the adjusted coefficient of determination.
d. United Edison's common stock has an 8 percent yield and a growth rate of 10 percent. The company's current ratio is 2.2 and 40 percent of its power is from nuclear generation. What is the point estimate of the stock's price/earnings ratio?

16. The personnel director of a large service company developed a multiple regression equation using a random sample of 30 white-collar workers where

Y = Score on a job satisfaction test;
X_1 = Age (years);
X_2 = Annual salary ($ thousands);
X_3 = Sex (0 if male, 1 if female).

Using all three independent variables, $SSE = 2,310$ and $SST = 5,832$.
a. Determine SSR.
b. Determine the adjusted coefficient of multiple determination, R_a^2.
c. Test $H_0: \beta_1 = \beta_2 = \beta_3 = 0$ at the 0.05 level.
d. Suppose the director had a regression model using only X_1 (age) and X_2 (salary) and $SSE(X_1, X_2) = 2,405$. Is X_3 (sex) significant in a model containing X_1 and X_2? (Hint: Use Formula 13.18.)

REFERENCES

Anscombe, F. J. "Graphs in Statistical Analysis." *The American Statistician*, vol. 27, no. 1 (February 1973), pp. 17–21.

"Beyond the Prima Facie Case in Employment Discrimination Law: Statistical Proof and Rebuttal." *Harvard Law Review*, vol. 89 (1975), pp. 387–422.

Cramer, Elliot M. "Significance Tests and Test of Models in Multiple Regression." *The American Statistician*, vol. 26, no. 4 (October 1972), pp. 26–30.

Crocker, Douglas C. "Some Interpretations of the Multiple Correlation Coefficient." *The American Statistician*, vol. 26, no. 2 (April 1972), pp. 31–33.

Draper, N. R., and Smith, H. *Applied Regression Analysis*. New York: John Wiley & Sons, Inc., 1966.

Finkelstein, Michael O. "Regression Models in Administrative Proceedings." *Harvard Law Review*, vol. 86 (1973), pp. 1442–75.

Geary, R. C., and Lesser, C. E. V. "Significance Tests in Multiple Regression." *The American Statistician*, vol. 22, no. 1 (February 1968), pp. 20–21.

Harris, Richard J. *A Primer of Multivariate Statistics*. New York: Academic Press, 1975.

Kerlinger, Fred N., and Pedhazur, Elazar J. *Multiple Regression in Behavioral Research*. New York: Holt, Rinehart and Winston, Inc., 1973.

Neter, John, and Wasserman, William *Applied Linear Statistical Models*. Homewood, Ill.: Richard D. Irwin, Inc., 1974.

Nie, Norman., et al. *SPSS: Statistical Package for the Social Sciences*. New York: McGraw-Hill, 1975.

14

Time Series Analysis

14 TIME SERIES ANALYSIS

I know of no way of judging the
the future but by the past.

Patrick Henry (1775)

**14.1
INTRODUCTION**

Production, sales, and profit are some of the goals of private business. In the pursuit of these goals, management must plan for the future by making forecasts or predictions. Valid forecasts can be made only after management has a thorough understanding of how various conditions affect their business operations. Much of this understanding may be gained by observing the past. Careful examination of past experience reveals much about what must be considered in making forecasts and planning future operations. A retail business may examine the pattern of sales over several years in order to plan orders for the coming holiday season. Management may also consider how various economic factors such as inflation, unemployment, interest rates, etc., affect retail sales. This may require a study of the past, comparison with the present, and prediction for the future.

Analyzing information for previous time periods is the subject of *time series analysis.*

> A *time series* is the measurement of a variable taken over time.

One time series, the U.S. National Debt for 1951–79, is portrayed in Figure 14.1. Measures of a variable taken over time can be studied in order to describe the movement of the variable over time and to assist in forecasting future values of the variable. This chapter covers the major concepts in time series analysis—the description of various movements in a time series and the process of predicting future observations. Patrick Henry spoke of the process of "judging the future . . . by the past" as he addressed the Second Virginia Convention on March 23, 1795. A few minutes later he spoke the immortal words ". . . give me liberty or give me death."

**14.2
COMPONENTS OF
TIME SERIES**

Four major components, or movements, of time series may be identified.

> The components of time series are:
>
> Trend (T)
> Cycle (C)
> Seasonal (S)
> Irregular (I).

FIGURE 14.1: THE U.S.
NATIONAL DEBT FOR
FISCAL YEARS 1951–1979
(1978 AND 1979 ARE
ESTIMATES)

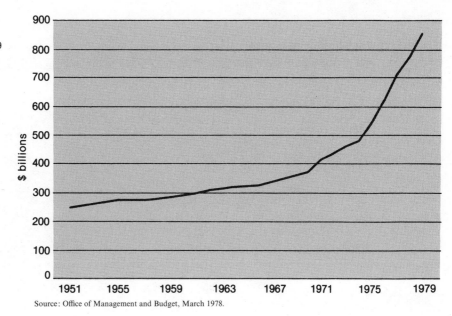

Source: Office of Management and Budget, March 1978.

The most widely used model of a time series, the *multiplicative model*, describes the values, *Y*, of a series as the product of these four components. Thus

$$Y = T \cdot C \cdot S \cdot I.$$

The *classical approach* to time series analysis involves the decomposition of a series into these four components. This subject is taken up later in the chapter. First, these four components are discussed.

The *trend* component of a time series is the overall average movement over a long period of time. The trend of the U.S. National Debt over the years 1951–79 is described by the smooth curve shown as a dashed line in Figure 14.2. Another time series, the purchasing power of the U.S. dollar from 1950 through 1977, and its trend (the dashed line) is given in Figure 14.3. The trend of a time series generally either increases or decreases over time. This may be described mathematically by a *trend equation* expressing the movement of the variable *Y* over time (*X*), i.e., *Y* = *f*(*X*). The form of the equation may be linear or nonlinear and may indicate either a direct or inverse relationship.

The *cycle* component of a time series is a wavelike, repetitive movement fluctuating about the trend of the series. In Figure 14.4 the quarterly profits (after taxes) per dollar of sales for all U.S. manufacturing corporations fluctuate above and below the curve representing the trend. The length of time before the cyclical movement begins to repeat itself, the *period* of the cycle, often varies throughout the time series. One period of 5 years may be followed by a 10-year period, a 20-year period, etc. The cycle component refers to repetitive movements with periods greater than one year. The *amplitude*, or height of the fluctuations above and below the trend, may also vary among periods.

FIGURE 14.2: THE U.S.
NATIONAL DEBT FOR
FISCAL YEARS 1951–1979
(1978 AND 1979 ARE
ESTIMATES)

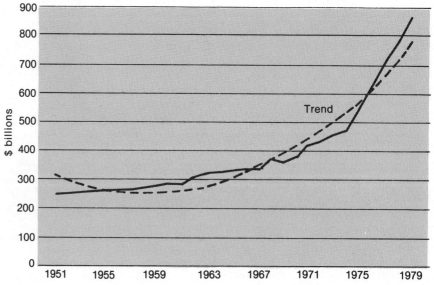

Source: Office of Management and Budget, March 1978.

FIGURE 14.3:
PURCHASING POWER OF
THE U.S. DOLLAR,
1950–1977 (MEASURED
BY THE CONSUMER
PRICE INDEX, 1967 = 100)

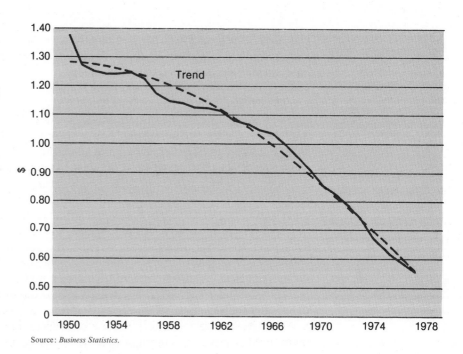

Source: *Business Statistics.*

FIGURE 14.4: QUARTERLY PROFITS (AFTER TAXES) PER DOLLAR OF SALES FOR ALL
U.S. MANUFACTURING CORPORATIONS, 1947–1976

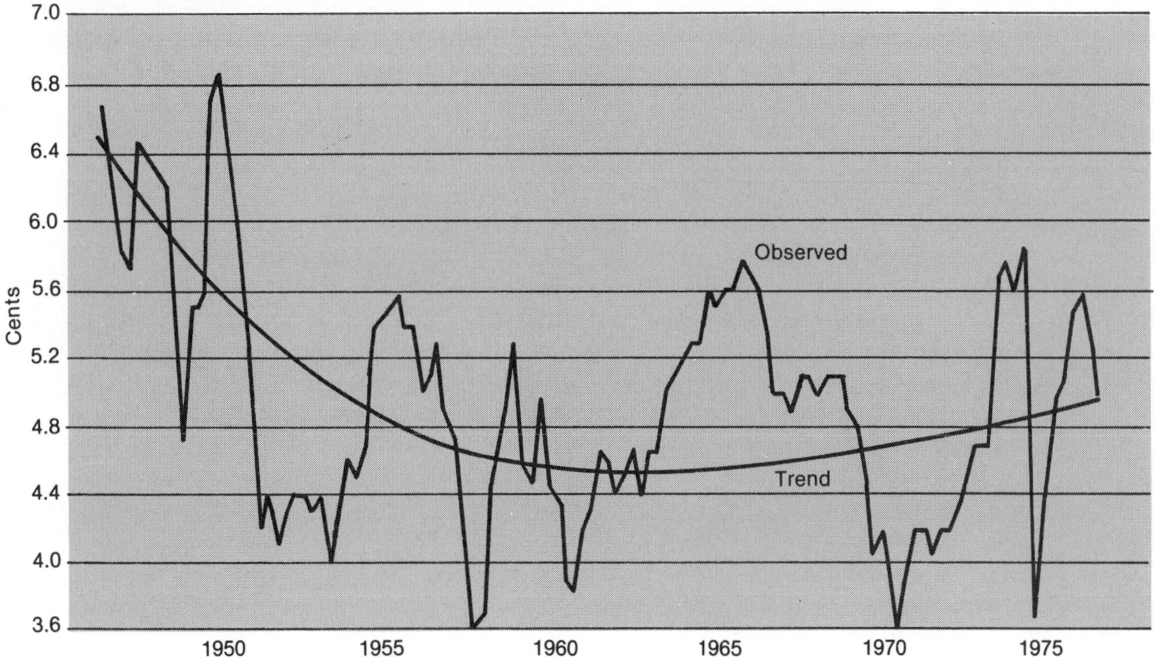

Source: *Business Statistics.*

Cycles may be attributed to internal factors of an organization such as inventory policies and to external factors such as weather, economic conditions, government policy, etc. Identification of cycles is important in detecting turning points in a series and in making intermediate-term forecasts.

The *seasonal* component of a time series is a repetitive, fluctuating movement occurring within a time period of one year or less. Monthly values of magazine advertising are shown in Figure 14.5. Notice that peak values of advertising generally occur in the months of April, May, October, and November, and the lowest levels of advertising occur in July, August, and January. There is a fluctuating movement between high and low values twice each year. Further, the pattern is repeated year after year at about the same points in time. Seasonal movements tend to be more regular and predictable than cycles.

Seasonal movements, also called *seasonal variation*, are measured by indexes. These may describe changes on a quarterly, monthly, weekly, or any other basis within one year. Knowledge of the seasonal pattern enables managers to plan for short-term fluctuations. Seasonal indexes are used to adjust data for seasonal variation so that measures of change in a series are not influenced by seasonal patterns. Finally, a seasonal indexes are useful in short-term forecasting.

The *irregular* component of a time series refers to the movements in the series that cannot be classified as either trend, cycle, or seasonal. In this sense, irregular

FIGURE 14.5: U.S. MAGAZINE ADVERTISING COST BY MONTHS, JANUARY 1973–JULY 1978

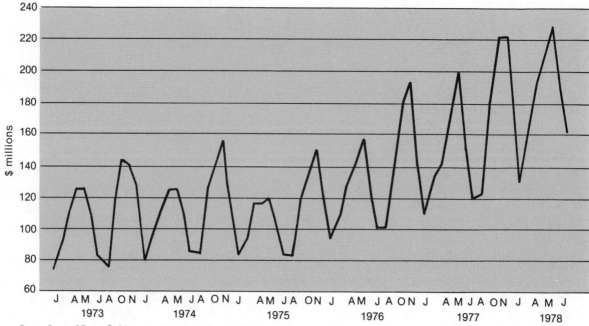

Source: *Survey of Current Business.*

variation is a residual measure, what is left over after isolating the other components. Events such as strikes, wars, fires, storms, etc., may affect a time series, but not in a predictable fashion. Hence, a strike at National Motors may produce a sharp drop in sales because of the work stoppage. Sales of a weapons manufacturer may rise rapidly for a short time during a conflict between nations.

Because of its random nature, irregular variation does not lend itself to measurement and prediction.

Smoothing is a fundamental approach to measuring time series. Smoothing methods are introduced in the next section and are followed by other procedures for measuring and using the trend, cycle, and seasonal components.

14.3 SMOOTHING CONCEPTS

Smoothing attempts to remove the effect of irregular or random variation on the values of time series. Smoothing out the irregular variation gives a clearer indication of the underlying movement in a series. Smoothing techniques may be applied to series measured over any time period—annual, quarterly, monthly, weekly, etc.

Monthly sales of Rapid Computing's System 2000 are shown in Figure 14.6. One smoothing technique, *moving averages*, is used to obtain the dashed and dotted lines. These two lines show smoothed values of monthly sales. The effects of irregular movements are minimized to eliminate erratic fluctuations and produce a smoother movement of the series. The two major *smoothing techniques* are *moving averages* and *exponential smoothing*.

FIGURE 14.6: MONTHLY SALES (y_t) OF RAPID COMPUTING'S SYSTEM 2000 AND MOVING AVERAGES ($\bar{y}_t$)

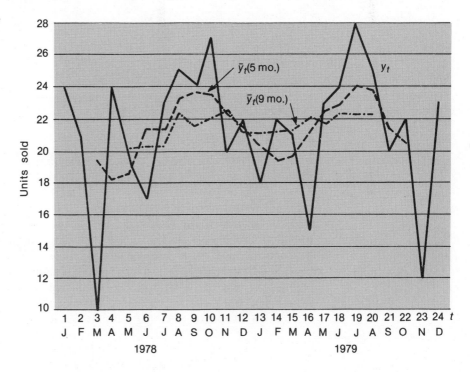

14.4 MOVING AVERAGES

Moving averages are smoothed values of time series that result from averaging the values of the series over successive time intervals. The dashed line in Figure of 14.6 is a five-month moving average of the sales data. Average values of sales for successive five-month intervals are computed in Table 14.1, Column (4). Letting y_t represent the observed value of the series at time t, average (arithmetic mean) sales for the first five months are

TABLE 14.1: MOVING AVERAGES OF MONTHLY SALES FOR RAPID COMPUTING'S SYSTEM 2000

		(1) t	*(2)* Units Sold y_t	*(3)* 5-Month Moving Total	*(4)* 5-Month Moving Average	*(5)* 9-Month Moving Total	*(6)* 9-Month Moving Average
1978	J	1	24				
	F	2	21				
	M	3	10	98	19.6		
	A	4	24	91	18.2		
	M	5	19	93	18.6	187	20.8
	J	6	17	108	21.6	190	21.1
	J	7	23	108	21.6	189	21.0
	A	8	25	116	23.2	201	22.3
	S	9	24	119	23.8	195	21.7
	O	10	27	118	23.6	198	22.0
	N	11	20	111	22.2	202	22.4
	D	12	22	109	21.8	194	21.6
1979	J	13	18	103	20.6	192	21.3
	F	14	22	98	19.6	192	21.3
	M	15	21	99	19.8	193	21.4
	A	16	15	105	21.0	198	22.0
	M	17	23	111	22.2	196	21.8
	J	18	24	115	23.0	200	22.2
	J	19	28	120	24.0	190	21.1
	A	20	25	119	23.8	192	21.3
	S	21	20	107	21.4		
	O	22	22	102	20.4		
	N	23	12				
	D	24	23				

$$\tfrac{1}{5}(y_1 + y_2 + y_3 + y_4 + y_5) = \tfrac{1}{5}(24 + 21 + 10 + 24 + 19)$$
$$= \tfrac{1}{5}(98) = 19.6.$$

Since the average is used to represent the interval of five months, the average value is centered at the middle of the interval (the moving total and moving average are listed opposite the third month ($t = 3$) in Table 14.1 and the average is plotted on the third month in Figure 14.6). If $\bar{y}_t$ is the average for time t, the third month average is $\bar{y}_3 = 19.6$. Successive intervals of time for computing moving averages are established by shifting the interval forward by one time period. This amounts to simply incrementing all values of t by one. Thus the average for the second interval of five months is

$$\bar{y}_4 = \tfrac{1}{5}(y_2 + y_3 + y_4 + y_5 + y_6)$$
$$= \tfrac{1}{5}(21 + 10 + 24 + 19 + 17)$$
$$= \tfrac{1}{5}(91) = 18.2.$$

Continuing, the third interval of five months is centered on time period $t = 5$ and has an average

$$\bar{y}_5 = \tfrac{1}{5}(y_3 + y_4 + y_5 + y_6 + y_7).$$

The computation of a moving average may be generalized as follows.

A *moving average* for an interval of n time periods centered on time t is

$$\bar{y}_t = \frac{1}{n}(y_{t-(n-1)/2} + \cdots + y_{t-1} + y_t + y_{t+1} + \cdots + y_{t+(n-1)/2})$$

where y_t is the value of the series at time t and n is an odd number.

An odd number of time periods is usually used to determine a moving average so that the computed average is centered on a point in time for which there is an observed value y_t. If the number of time periods is even, the computed average is centered between two observed values of y_t. Odd numbers enable comparisons to be made of observed and smoothed values, while even numbers do not.

A large number of time periods is more effective in smoothing than a small number. The nine-month moving average computed in Table 14.1, Column (6), and shown in Figure 14.6 as a dotted line, is more effective in smoothing out erratic fluctuations in sales than the five-month moving average. An examination of the five- and nine-month moving averages points out a disadvantage of the moving average techniques for measuring a time series. There is not a computed average for all time periods for which observed values y_t exist. A value of $\bar{y}_1$, $\bar{y}_2$, $\bar{y}_{23}$, and $\bar{y}_{24}$ cannot be obtained for the five-month moving average and the nine-month average does not provide $\bar{y}_t$ for the first and last four time periods.

Moving averages are more useful for studying the historical movement of time series than for forecasting. The method presented in the next section is a more appropriate forecasting tool.

14.5 EXPONENTIAL SMOOTHING

A second smoothing technique, exponential smoothing, is a very popular approach to short-term forecasting. *Exponential smoothing* determines a smoothed value of a time series by computing an exponentially weighted average of the current and all previous values of the series. A weighting system is used to assign the desired degree of importance to the current value and the previous values. The current value is usually assigned a larger weight than previous values. To describe the process, let y_t be the observed value of the series at time t, w is the weight assigned to the current value where $0 < w < 1$, and S_t is the smoothed value at time t. Exponential smoothing begins by setting the smoothed value at the first time period equal to the current value. Thus for $t = 1$,

$$S_1 = y_1.$$

The smoothed value for the second period is the weighted average of the current value y_2 and the previous value y_1,

$$S_2 = wy_2 + (1 - w)y_1.$$

The third period smoothed value weights the current value y_3 by w, the second period value by $w(1 - w)$ and the first period by $(1 - w)^2$.

$$S_3 = wy_3 + w(1 - w)y_2 + (1 - w)^2 y_1.$$

For the fourth and fifth periods, the smoothed values are

$$S_4 = wy_4 + w(1 - w)y_3 + w(1 - w)^2 y_2 + (1 - w)^3 y_1$$

and

$$S_5 = wy_5 + w(1 - w)y_4 + w(1 - w)^2 y_3 + w(1 - w)^3 y_2 + (1 - w)^4 y_1.$$

This shows that the smoothed value for the current period is an exponentially weighted average of the observed value for the current period and all previous periods. The weighting system changes exponentially as the smoothing process moves forward in time.

While the above explains the basis of exponential smoothing, a more efficient process for obtaining values of S_t is actually used. It can be shown algebraically that, for all time periods beyond the first, the

$$\begin{pmatrix} \text{Current} \\ \text{smoothed} \\ \text{value} \end{pmatrix} = w \begin{pmatrix} \text{Current} \\ \text{observed} \\ \text{value} \end{pmatrix} + (1 - w) \begin{pmatrix} \text{Previous} \\ \text{smoothed} \\ \text{value} \end{pmatrix}. \qquad (14.1)$$

Symbolically, this may be written as

$$S_t = wy_t + (1 - w)S_{t-1}. \qquad (14.2)$$

TABLE 14.2: EXPONENTIALLY SMOOTHED VALUES OF RAPID COMPUTING'S SYSTEM 2000 SALES			*(1)* t	*(2)* y_t	*(3)* S_t $(w = 0.1)$	*(4)* S_t $(w = 0.4)$
1978	J		1	24	24.00	24.00
	F		2	21	23.70	22.80
	M		3	10	22.33	17.68
	A		4	24	22.50	20.21
	M		5	19	22.15	19.73
	J		6	17	21.64	18.64
	J		7	23	21.78	20.38
	A		8	25	22.10	22.23
	S		9	24	22.29	22.94
	O		10	27	22.76	24.56
	N		11	20	22.48	22.74
	D		12	22	22.43	22.44
1979	J		13	18	21.99	20.66
	F		14	22	21.99	21.20
	M		15	21	21.89	21.12
	A		16	15	21.20	18.67
	M		17	23	21.38	20.40
	J		18	24	21.64	21.84
	J		19	28	22.28	24.30
	A		20	25	22.55	24.58
	S		21	20	22.30	22.75
	O		22	22	22.27	22.45
	N		23	12	21.24	18.27
	D		24	23	21.42	20.16

To demonstrate that Equation 14.2 is equivalent to the previous procedure, determine S_3 from Equation 14.2:

$$S_3 = wy_3 + (1 - w)S_2.$$

Now apply Equation 14.2 to determine S_2 in the above.

$$S_3 = wy_3 + (1 - w)[wy_2 + (1 - w)S_1].$$

Since $S_1 = y_1$, the above is

$$S_3 = wy_3 + w(1 - w)y_2 + (1 - w)^2 y_1,$$

the same as the earlier result.

Equation 14.2 is the standard mechanism for determining exponentially smoothed values of a time series. It is efficient because only the previous smoothed value must be retained in computing a smoothed value for the next time period. This reduces data storage requirements tremendously.

Monthly sales for Rapid Computing's System 2000 are given in Column (2) of Table 14.2. Smoothed values computed by Equation 14.2 are given in Column (3) using a weight of $w = 0.1$ for the current observed value. Setting $S_1 = y_1$ for the first time period, the smoothed values are:

$$S_1 = 24.00$$
$$S_2 = 0.1(21) + 0.9(24.00) = 23.70$$
$$S_3 = 0.1(10) + 0.9(23.70) = 22.33$$
$$\vdots$$
$$S_{24} = 0.1(23) + 0.9(21.24) = 21.42.$$

The smoothed values are shown in Figure 14.7 as a dotted line. The exponential smoothing technique has reduced the erratic fluctuations in the series.

An important consideration in exponential smoothing is the choice of the weight assigned to the current observed value. This weight w is called the *smoothing constant*. The exponential smoothing procedure reacts quickly to changes in the current and recent observed values if w has a high value. The danger in using high values of w is that the smoothing procedure reacts too quickly to erratic fluctuations in the series. Smoothed values of the sales series using a smoothing constant of $w = 0.4$ are given in Column (4) of Table 14.2. The values plotted as a dashed line in Figure 14.7 show a larger reaction to fluctuations in the series than with $w = 0.1$. Extremely small values of w may cause the smoothing procedure to react too slowly to changes in the series. Since the purpose of smoothing is to cancel out the effects of extreme fluctuations, relatively small values of w are generally used. Values of w between 0.1 and 0.3 are common. In practice, the analyst generally experiments with different values of w to select that most appropriate for a given application.

The basic approach to forecasting with exponential smoothing suggests that the smoothed value S_t for the most recent time period be used as the forecast for any future time period. For example, if Rapid Computing prefers a smoothing constant $w = 0.1$, the sales forecast for January 1980 is the smoothed value for

FIGURE 14.7: MONTHLY
SALES (y_t) OF RAPID
COMPUTING'S SYSTEM
2000 AND SMOOTHED
VALUES (S_t)

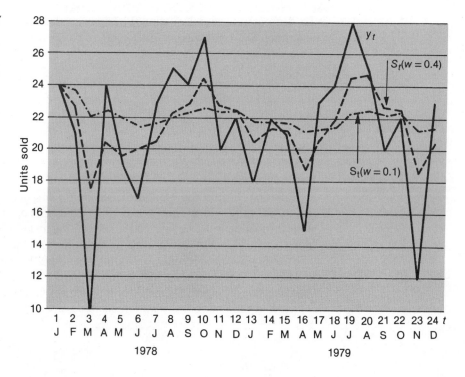

December 1979, 21.42. The basic approach of using the most recent smoothed value ignores any trend and/or seasonal effects that may be present. Advanced methods of exponential smoothing adjust for trend and seasonal effects but require that additional data (more than the most recent smoothed value) be carried forward in the analysis.

We now turn to the classical methods of decomposing a time series into the four basic components of trend, cycle, seasonal, and irregular.

14.6 TREND ANALYSIS

The *trend* of a time series is the overall average movement over a long period of time. Trend, in this sense, is a part of our everyday language. We speak of the trend in new car prices, the trend in unemployment, clothing trends, etc. By measuring trend, we provide a statement of the long-term direction of movement in some variable. The purposes of measuring trend include the following:

1. *To describe historical movements.* Descriptions of trend provide useful comparisons. A company can use the trend of sales over the past two decades to see how the company has performed relative to its industry. An individual might examine how changes in family income compare with changes in the cost of living. Government may compare the pace of tax revenues with public expenditures.

2. *For forecasting and prediction.* Government may find that some cuts in public expenditures become necessary when forecasted tax revenues do not

provide the required funds. Planning for the future is a key to successful business operations. Corporate planning departments and other organizational units must make and use forecasts of key operating characteristics such as sales, profits, raw material availability and cost, etc.

3. *To eliminate the effect of trend in a time series.* The basic method for isolating the cycle component is to remove the trend factor from the original data. Thus trend measurement is useful in studying other components of time series.

Trend is measured by a mathematical equation describing the long-term movement of a variable Y over time (X), i.e., $Y = f(X)$. The *trend equation* is established from historical values of the series measured over a long period of time. At least 15 years of historical data are preferred. With shorter time periods, the resulting trend may be unduly affected by short-term fluctuations and random occurrences. Annual data are generally used for trend measurement. Data measured over intervals of time shorter than a year are not pertinent to measuring the average movement over a long period of time. Annual data also eliminate any seasonal influence; the series is then composed of only three factors, $Y = T \cdot C \cdot I$.

Occasionally it is appropriate to edit the values of time series prior to measuring the trend or any other component. Adjusting the data for price changes is perhaps the most common editing procedure. Dollar sales measured over time may be influenced by changes in prices. If we want to describe the trend of dollar sales separately from changes in price, the values of the original series, measured in current dollars, may be deflated by an appropriate price index as described in Chapter 3. A trend equation established from the series of real dollar sales provides a truer picture of the growth or decline in sales. Adjustments for price changes are almost always appropriate in measuring trends in series such as income and wages.

Data are also edited for population changes. As an example, consider the concern in recent years over the trend in cigarette smoking. The question of interest is to determine whether people are smoking more than before, or less. Data are generally available on total usage, such as the number of cartons or packages sold. But if population is increasing, total usage could be increasing when the average person smokes the same as, or even less than, before. Adjusting the data for population provides a measure of usage in per capita terms, which can then be analyzed to detect trends in smoking.

14.7
LEAST SQUARES
TREND EQUATIONS

In describing the trend of a time series by an equation, we seek the best description possible of the average movement over a long period of time. The method of least squares was introduced in Chapter 12 as a procedure for determining a line of best fit for a regression equation. A simple regression equation describes the average relationship between two variables, Y and X. A trend equation may be viewed as a special case of a regression equation where the independent variable measures time. Using values of variable Y observed over time, variable X, a trend equation may be established. The equation provides values of trend, $\hat{Y}$ (read "Y-hat"), over time. The mathematical properties of a least-squares trend equation are:

1. $\sum(Y - \hat{Y}) = 0$;
2. $\sum(Y - \hat{Y})^2 = $ a minimum.

This indicates that the deviations of the observed values of Y about the trend values $\hat{Y}$ sum to zero. The sum of the squared deviations is a minimum for a given form of trend equation. For example, if a linear trend equation is determined by the method of least squares, then no other linear equation will have a smaller sum of squared deviations of the observed values about the trend values. Since a line of best fit is defined as the equation that has the smallest value for $\sum(Y - \hat{Y})^2$, the same property is also useful for comparing different forms of equations.

Even though a trend equation may be viewed as a special case of a regression equation, the inferences about $Y|X$ for regression equations are not valid for trend equations. In regression analysis, the conditional distribution of Y for a given value of X is assumed normal. The observed Y for a given X is a random observation from this conditional distribution. Successive values of $Y|X$ are also independent. In time series, the observed value in a given year is not a random observation from a conditional distribution. Also, a time series does not meet the independence assumption. This simply means that the confidence statements made about a regression equation are not valid for trend equations. But the least squares method for fitting a trend equation to a time series is entirely appropriate.

Once any preliminary editing of data is complete, the analyst proceeds to fit an equation to the data.

The steps in fitting a trend equation are:

1. Plot the data.
2. Select the best model (form of equation).
3. Use the least squares method to fit the desired model to the data.

A plot of the series over time provides a picture of the general movement of the series. The plot must be examined in order to decide on the model, or form of equation, that best represents the average movement over time. First, determine whether the series is increasing or decreasing over time. Then decide if the trend appears to be linear or nonlinear. It may be helpful to sketch a freehand line as a subjective approximation of trend. If possible, decide on the model that appears best to describe the trend, and fit this to the data.

In some cases, the best model cannot be chosen solely on the basis of looking at the plotted data. Then further analysis is performed to select the best fitting model. This is discussed following the presentation of three widely used models, the linear, quadratic, and exponential.

14.8 LINEAR
TREND EQUATION

EXAMPLE. Reliable Tire Company produces a radial automobile tire called "The Gripper." Annual sales of this tire for 1965–79 are shown in Table 14.3, Column (2). The time series is plotted in Figure 14.8. The average movement of the series over the 15-year period appears to be best described by a straight line. Using the method of least squares, fit a linear trend to the data.

TABLE 14.3: ANNUAL
SALES OF "THE
GRIPPER" TIRE

(1)	(2) Unit Sales (000) Y	(3) X	(4) XY	(5) X^2	(6) Trend Values $\hat{Y}$	(7) $Y - \hat{Y}$	(8) $(Y - \hat{Y})^2$
1965	23	−7	−161	49	23.6	−0.6	0.36
1966	28	−6	−168	36	27.6	0.4	0.16
1967	34	−5	−170	25	31.6	2.4	5.76
1968	35	−4	−140	16	35.6	−0.6	0.36
1969	40	−3	−120	9	39.6	0.4	0.16
1970	41	−2	−82	4	43.6	−2.6	6.76
1971	48	−1	−48	1	47.6	0.4	0.16
1972	50	0	0	0	51.6	−1.6	2.56
1973	58	+1	58	1	55.6	2.4	5.76
1974	60	+2	120	4	59.6	0.4	0.16
1975	61	+3	183	9	63.6	−2.6	6.76
1976	68	+4	272	16	67.6	0.4	0.16
1977	72	+5	360	25	71.6	0.4	0.16
1978	76	+6	456	36	75.6	0.4	0.16
1979	80	+7	560	49	79.6	0.4	0.16
	774	0	1,120	280		0	29.60

The least squares method for fitting the two-variable linear equation

$$\hat{Y} = b_0 + b_1 X$$

is introduced in Section 12.4. The least squares estimators of b_0 and b_1 are

$$b_1 = \frac{\sum XY - n\bar{X}\bar{Y}}{\sum X^2 - n\bar{X}^2};$$ (14.3)

$$b_0 = \frac{\sum Y - b_1 \sum X}{n}.$$ (14.4)

The equations for these estimators may be simplified for trend equations by coding the independent variable in a time series. The successive values of time in Column (1) differ by increments of one. Time increments are regular and predictable, whereas the independent variable in regression analysis is not. This enables the variable time to be coded so that Equations (14.3) and (14.4) may be simplified. The coded values of the variable time are labeled X in Column (3) of Table 14.3. Note that the column total is zero. This coding simplifies Equations (14.3) and (14.4).

FIGURE 14.8: ANNUAL
SALES OF "THE
GRIPPER" TIRE

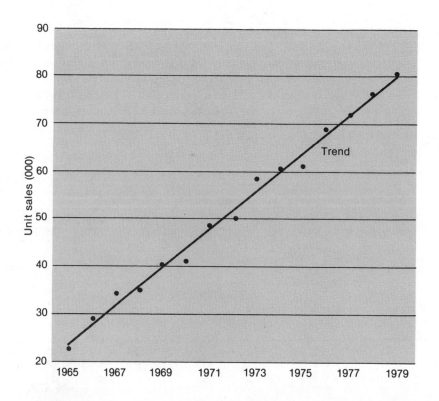

The least squares estimators b_0 and b_1 in a linear trend equation

$$\hat{Y} = b_0 + b_1 X$$

are

$$b_0 = \frac{\sum Y}{n} \quad \text{and} \quad b_1 = \frac{\sum XY}{\sum X^2} \tag{14.5}$$

when the variable time, X, is coded so that $\sum X = 0$, therefore $\bar{X} = 0$.

    The values of the variable time are coded as follows. If the number of time periods is odd (as in Table 14.3), assign the middle period a value of zero. As you move away the middle period, assign values of $-1, -2, -3, \ldots$ to earlier time periods and assign values of $+1, +2, +3, \ldots$ to later time periods. If the time series contains an even number of time periods, first identify the two middle time periods. Assign the earlier a value of -1; assign the later a value of $+1$. Moving backward in time away from the -1 value, assign values of -3, $-5, -7$, etc. Moving forward in time away from the $+1$ value, assign $+3, +5$, $+7$, etc. These coding techniques reflect the original time measures since the change in the value of X is constant over time. For an odd number of time periods, a change of one X-unit represents one time period. For an even number of time periods, one X-unit represents half the original time period.

Solution. The necessary computations for using equation (14.5) to find b_0 and b_1 are provided in Columns (4) and (5) of Table 14.3 Substituting,

$$b_0 = \frac{\sum Y}{n} = \frac{774}{15} = 51.6;$$

$$b_1 = \frac{\sum XY}{\sum X^2} = \frac{1120}{280} = 4.0.$$

The trend equation is $\hat{Y} = 51.6 + 4.0X$.

A trend equation describes the average movement over time. This statement of trend is interpreted in terms of b_0 and b_1, the parameters of a linear equation. If $X = 0$, then the value of the equation $\hat{Y} = 51.6$. For this example, $X = 0$ in 1972. The trend is described as follows. "Average annual sales, centered in 1972, are 51,600 and annual sales increase by an average of 4,000 units per year." Values of trend for any time period are found by substituting the coded value X. Using the subscript i to represent time for (Y_i, X_i), the trend value in 1969 is

$$\hat{Y}_{69} = 51.6 + 4.0X_{69}$$
$$= 51.6 + 4.0(-3) = 39.6.$$

Trend values for 1965–79 are shown in Column (6) of Table 14.3. These trend values provide the straight line in Figure 14.8. The last two columns of the table illustrate the least-squares properties. Column (7) verifies that $\sum(Y - \hat{Y}) = 0$. Column (8) provides $\sum(Y - \hat{Y})^2 = 29.6$. This quantity varies for different data and different forms of equations. For this problem, no other linear equation provides a smaller $\sum(Y - \hat{Y})^2$ than 29.6. In determining a trend equation, it is not necessary to compute these quantities. The least squares properties are met for a linear equation whenever Equation (14.5) is used. The computations are provided here only for illustrative purposes.

The form of equation selected for describing the trend of a time series reflects the way the values of the series (Y) change over time. If, on the average, Y changes by a constant amount for each unit change in X, then a linear equation best represents the trend. If the average change in Y is not constant, then some form of nonlinear equation provides the best description of trend. Two basic forms of nonlinear equations, the quadratic and exponential, are presented next.

14.9 QUADRATIC TREND EQUATION

A quadratic trend equation has the form

$$\hat{Y} = b_0 + b_1X + b_2X^2.$$

The graph of a quadratic equation is called a *parabola*. The quadratic equation can be used to describe four different types of nonlinear trends. As shown in Figure 14.9, the value of $\hat{Y}$ in a quadratic equation may, for each unit change in X, be increasing by either an increasing or decreasing amount (panels A and B); or decreasing by either an increasing or decreasing amount (panels C

FIGURE 14.9: FOUR
TYPES OF MOVEMENT
FOR QUADRATIC
EQUATIONS
$\hat{Y} = b_0 + b_1 X + b_2 X^2$

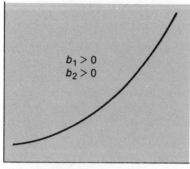

$b_1 > 0$
$b_2 > 0$

A. Y increasing by an increasing
amount

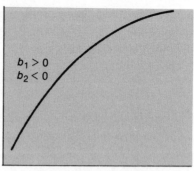

$b_1 > 0$
$b_2 < 0$

B. Y increasing by an decreasing
amount

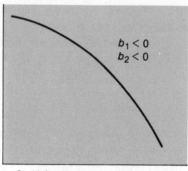

$b_1 < 0$
$b_2 < 0$

C. Y decreasing by an increasing
amount

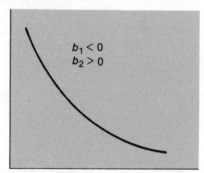

$b_1 < 0$
$b_2 > 0$

D. Y decreasing by a decreasing
amount

and D). When the variable time is coded so that $\sum X = 0$, the least squares estimators[1] b_0 and b_2 are found by simultaneous solution of the two equations

$$\sum Y = nb_0 + b_2 \sum X^2; \tag{14.6}$$
$$\sum X^2 Y = b_0 \sum X^2 + b_2 \sum X^4. \tag{14.7}$$

The estimator b_1 is

$$b_1 = \frac{\sum XY}{\sum X^2}. \tag{14.8}$$

---

[1] The method described in Section 12.4 is used to find the following least squares normal equations for fitting the quadratic equation.

$$\sum Y = nb_0 + b_1 \sum X + b_2 \sum X^2$$
$$\sum XY = b_0 \sum X + b_1 \sum X^2 + b_2 \sum X^3$$
$$\sum X^2 Y = b_0 \sum X^2 + b_1 \sum X^3 + b_2 \sum X^4$$

When $\sum X = 0$, all summations involving odd powers of X are zero, so that these reduce to Equations 14.6–14.8.

The purchasing power of the U.S. dollar for 1950–77 was portrayed earlier in Figure 14.3. A glance at Figure 14.3 indicates that trend of the series is non-linear and has the general shape described in panel C, Figure 14.9. Table 14.4 provides data for this series along with the computations required for fitting a quadratic equation. Note the different coding of X for an even number of time periods. Substituting into Equations (14.6) and (14.7),

$$28.525 = 28b_0 + 7308b_2;$$
$$7144.661 = 7308b_0 + 3,427,452b_2.$$

Solving simultaneously,

$$b_0 = 1.07032;$$
$$b_2 = -0.00020.$$

Substitution into Equation (14.8) provides

$$b_1 = -0.01365.$$

TABLE 14.4: PURCHASING POWER OF THE DOLLAR, 1950–1977 MEASURED BY THE CONSUMER PRICE INDEX (1967 = 100)		(1) Purchasing Power of the Dollar* Y	(2) X	(3) XY	(4) X^2	(5) X^2Y	(6) X^4
1950		$1.387	−27	−37.449	729	1011.123	531,441
1951		1.285	−25	−32.125	625	803.125	390,625
1952		1.258	−23	−28.934	529	665.482	279,841
1953		1.248	−21	−26.208	441	550.368	194,481
1954		1.242	−19	−23.598	361	448.362	130,321
1955		1.247	−17	−21.199	289	360.383	83,521
1956		1.229	−15	−18.435	225	276.525	50,625
1957		1.186	−13	−15.418	169	200.434	28,561
1958		1.155	−11	−12.705	121	139.755	14,641
1959		1.145	−9	−10.305	81	92.745	6,561
1960		1.127	−7	−7.889	49	55.223	2,401
1961		1.116	−5	−5.580	25	27.900	625
1962		1.104	−3	−3.312	9	9.936	81
1963		1.091	−1	−1.091	1	1.091	1
1964		1.076	1	1.076	1	1.076	1
1965		1.058	3	3.174	9	9.522	81
1966		1.029	5	5.145	25	25.725	625
1967		1.000	7	7.000	49	49.000	2,401
1968		0.960	9	8.640	81	77.760	6,561
1969		0.911	11	10.021	121	110.231	14,641
1970		0.860	13	11.180	169	145.340	28,561
1971		0.824	15	12.360	225	185.400	50,625
1972		0.798	17	13.566	289	230.622	83,521
1973		0.752	19	14.288	361	271.472	130,321
1974		0.678	21	14.238	441	298.998	194,481
1975		0.621	23	14.283	529	328.509	279,841
1976		0.587	25	14.675	625	366.875	390,625
1977		0.551	27	14.877	729	401.679	531,441
		28.525	0	−99.725	7308	7144.661	3,427,452

* Source: *Business Statistics.*

The quadratic equation describing the trend in purchasing power is

$$\hat{Y} = 1.07032 - 0.01365X - 0.00020X^2.$$

As before, values of trend are found by substituting the appropriate X-value into the equation. To illustrate, the trend in 1975 is

$$\hat{Y}_{75} = 1.07032 - 0.01365X_{75} - 0.00020X_{75}^2$$
$$= 1.07032 - 0.01365(23) - 0.00020(23)^2$$
$$= 0.651.$$

Both the trend and the original series are shown in Figure 14.10.

FIGURE 14.10:
PURCHASING POWER OF
THE DOLLAR, 1950–1977,
MEASURED BY THE
CONSUMER PRICE INDEX
(1967 = 100)

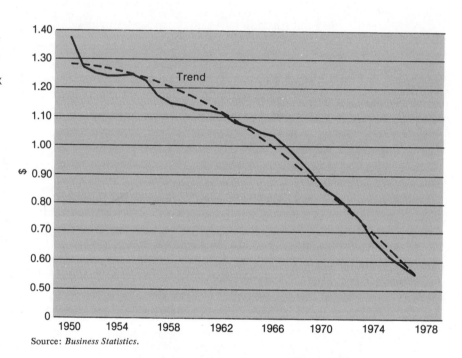

Source: *Business Statistics*.

14.10
EXPONENTIAL
TREND EQUATION

The exponential trend equation

$$\hat{Y} = b_0 b_1^X \tag{14.9}$$

is used to describe trend when the changes in the Y-values approximate a constant percentage over time. Consider the changes in Y over time t below. From Column (2), Y is clearly increasing over time. Column (3) indicates that Y is increasing by an increasing amount. A quadratic equation was used to describe this type of trend in Figure 14.9, panel A. But further examination in Column (4) shows that Y changes by a constant percentage.

(1)	(2)	(3) Absolute Change from Previous Value	(4) Percentage Change from Previous Value
t	Y		
1	100	—	—
2	140	40	40
3	196	56	40
4	274.4	78.4	40
5	384.16	109.76	40
6	537.824	153.664	40
7	752.9536	215.1296	40

In this sense, the exponential trend may be viewed as a special case of the quadratic trends described in Figure 14.9, panels A and D. Panels A and D are illustrative of trends that may or may not reflect constant percentage changes in Y. If the average change in Y approximates a constant percentage, then the exponential equation is used; if not, the quadratic equation is used.

To determine whether the quadratic or exponential trend is best, it may be helpful to plot the data on a semilogarithmic chart. This chart has a logarithmic scale on the vertical axis and an arithmetic scale on the horizontal axis. If successive values of Y change by a constant percentage, the series forms a straight line on a semilogarithmic chart. The values of Y from above are plotted on an arithmetic chart (panel A) and on a semilogarithmic chart (panel B) in Figure 14.11.

Computations involving the exponential equation are expressed in logarithms. Taking the logarithm of both sides of the equation

$$\hat{Y} = b_0 b_1^X$$

yields

$$\log \hat{Y} = \log b_0 + \log b_1 \cdot X. \tag{14.10}$$

FIGURE 14.11: ARITHMETIC AND SEMILOGARITHMIC CHARTS

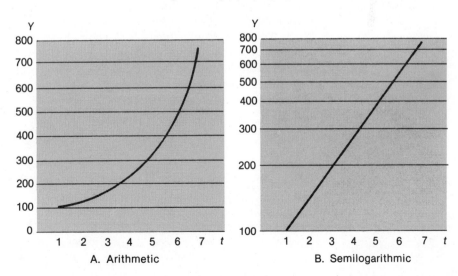

A. Arithmetic   B. Semilogarithmic

TABLE 14.5:
PRODUCTION OF STANIK
SHAVERS, 1966–1980

	(1) Production (thousand units) Y	(2) X	(3) log Y	(4) X log Y	(5) X²
1966	320	−7	2.5051	−17.5357	49
1967	339	−6	2.5302	−15.1812	36
1968	362	−5	2.5587	−12.7935	25
1969	381	−4	2.5809	−10.3236	16
1970	405	−3	2.6075	−7.8225	9
1971	430	−2	2.6336	−5.2672	4
1972	451	−1	2.6542	−2.6542	1
1973	475	0	2.6767	0	0
1974	510	1	2.7076	2.7076	1
1975	542	2	2.7340	5.4680	4
1976	573	3	2.7582	8.2746	9
1977	610	4	2.7853	11.1412	16
1978	644	5	2.8089	14.0445	25
1979	680	6	2.8325	16.9950	36
1980	725	7	2.8603	20.0221	49
		0	40.2337	7.0751	280

This is a linear equation in terms of $\log \hat{Y}$ and the original values of X. The least squares estimators $\log b_0$ and $\log b_1$ are[2]

$$\log b_0 = \frac{\sum \log Y}{n}; \tag{14.11}$$

$$\log b_1 = \frac{\sum X \log Y}{\sum X^2}. \tag{14.12}$$

Annual production of safety razors for 1966–80 by an American firm, Stanik, is listed in Column (1) of Table 14.5. The approximate trend on an arithmetic chart, shown in Figure 14.12, is increasing by an increasing amount. When plotted on a semilogarithmic chart, Figure 14.13, the data fall almost perfectly in a straight line, indicating that the exponential trend model is appropriate. The required computations are given in Table 14.5. A table for logarithms is provided in Appendix L. Substituting into Equations (14.11) and (14.12),

$$\log b_0 = \frac{40.2337}{15} = 2.6822;$$

$$\log b_1 = \frac{7.0751}{280} = 0.0253.$$

---

[2] The least squares normal equations are

$$\sum \log Y = n \log b_0 + \log b_1 \cdot \sum X;$$
$$\sum X \log Y = \log b_0 \cdot \sum X + \log b_1 \cdot \sum X^2.$$

If $\sum X = 0$, these reduce to Equations (14.11) and (14.12).

FIGURE 14.12:
PRODUCTION OF STANIK
SHAVERS, 1966–1980

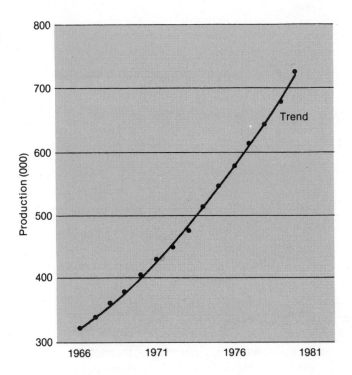

FIGURE 14.13:
SEMILOGARITHMIC
CHART OF STANIK
SHAVER PRODUCTION,
1966–1980

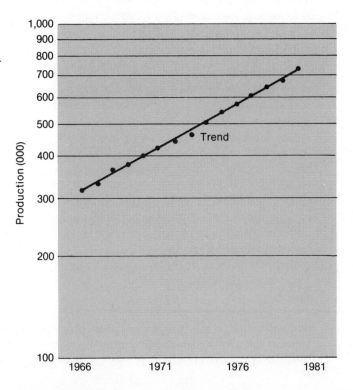

The exponential equation, expressed in terms of log Y, is

$$\log \hat{Y} = 2.6822 + 0.0253X. \tag{14.13}$$

Trend values are found in this form of equation by substituting the appropriate X-value and finding the antilog of log $\hat{Y}$. Finding the value of trend in 1970,

$$\log \hat{Y}_{70} = 2.6822 + 0.0253X_{70}$$
$$= 2.6822 + 0.0253(-3)$$
$$= 2.6063.$$

The trend value is

$$\hat{Y}_{70} = \text{antilog } 2.6063$$
$$= 404 \text{ thousand units.}$$

The straight line shown in Figure 14.13 and the curve in Figure 14.12 represent the computed trend values.

The constant percentage change measured by an exponential trend is found by expressing the equation in original units of $\hat{Y}$ and X, rather than in logarithms of $\hat{Y}$. Finding antilogs in Equation (14.13), the exponential trend equation, in the form $\hat{Y} = b_0 b_1^X$, is

$$\hat{Y} = (481)(1.06)^X.$$

The constant percentage change is found by substituting the value of b_1 in the expression $100(b_1 - 1)$. For the above equation, this is $100(1.06 - 1) = 6$. This indicates that the trend values for production of Stanik Shavers increase by 6 percent per year. If there is a constant percentage decrease over time, the expression $100(b_1 - 1)$ produces a negative value. For example, the trend equation

$$\hat{Y} = (200)(0.93)^X$$

provides a 7 percent decrease in Y each year, since $100(0.93 - 1) = -7$.

The exponential trend is the best form of trend equation if, on the average, changes in Y approximate a constant percentage. When an exponential trend is suspected, plotting the data on a semilogarithmic chart provides a preliminary indication of the adequacy of the exponential model. An exponential trend should be used if the approximate trend on a semilogarithmic chart is linear.

14.11 SELECTING THE BEST MODEL

By plotting the data, the analyst can usually determine which model, or form of equation, best describes the long-term movement over time. If the approximate trend is linear on an arithmetic chart, the linear model is best. If the approximate trend appears as a parabola on an arithmetic chart, the quadratic model *may* be best. If the same data approximate a straight line on a semilogarithmic chart, the exponential model is best; otherwise the quadratic is preferable.

In some cases the analyst may be undecided, after examining the plotted data, as to which model is best. These cases require further analysis in order to select the best model. The following procedure may be used. Fit an equation to the data for each model that may be appropriate. Compute $\sum(Y - \hat{Y})^2$ for each model. The best model for describing trend has the smallest $\sum(Y - \hat{Y})^2$. This comparison is based on the least squares property that defines a line of best fit, i.e., $\sum(Y - \hat{Y})^2$ is a minimum. Fitting a given form of equation by the least squares method guarantees that the best fitting equation *of that form* will result. But if we are uncertain about which form of equation is best, we can compare the best fitting linear equation with the best fitting quadratic equation, etc. This requires the computation of more than one form of equation for a given time series and determining $\sum(Y - \hat{Y})^2$ for each form of equation.

A computer routine for fitting equations by the least squares method is the most efficient way to perform the analysis suggested here. To illustrate this comparison, a computer routine was used to verify which model is best for the illustrations in Sections 14.9 and 14.10. The results, given in Table 14.6, indicate that the quadratic equation provides a better fit than the exponential equation for the data on the purchasing power of the dollar. For production of Stanik Shavers, the exponential is better.

TABLE 14.6: COMPUTED VALUES OF $\sum(Y - \hat{Y})^2$ FOR TWO TREND MODELS

	Model	
Illustration (reference)	*Quadratic*	*Exponential*
Purchasing power of the dollar	0.02	0.17
(Section 14.9)		
Production of Stanik Shavers	85.3	74.4
(Section 14.10)		

14.12 USE OF THE COMPUTER

The major computer packages for performing statistical computations contain procedures for trend analysis. The use of the computer in fitting a trend equation is illustrated in this section. The data on annual sales of "The Gripper" tire are used here so that the computer results may be compared with the manual computations in Section 14.8. The SAS computer package is used here; other packages provide similar output. The data on annual sales, along with the coded values of variable X—Columns (2) and (3) of Table 14.3—constitute the input for the computer analysis.

A plot of a time series is examined first in order to decide on the form of equation to fit to the data. The computer-generated plot is given in Figure 14.14. Since the approximate trend appears linear, a linear equation is fitted by the method of least squares. The SAS general linear models procedure provides the results shown in Figure 14.15. Note that the equation, $\hat{Y} = 51.6 + 4.0X$, is the same as determined in Section 14.8. For each observed value X, the trend value Y, labeled the "predicted value", is computed. The "residual" is the difference "observed value" minus "predicted value", or $(Y - \hat{Y})$. The least squares property $\sum(Y - \hat{Y}) = 0$, is labeled the "sum of residuals". The "sum of squared residuals" is the $\sum(Y - \hat{Y})^2 = 29.6$, which was determined in Section 14.8.

Since the analyst must select the form of equation to fit to the data, the plotted data must be carefully examined. If the analyst is uncertain about the form of equation, more than one model could be used; the sum of squared residuals can be compared to select the best-fitting equation.

FIGURE 14.14:
COMPUTER PLOT OF
TIME SERIES DATA

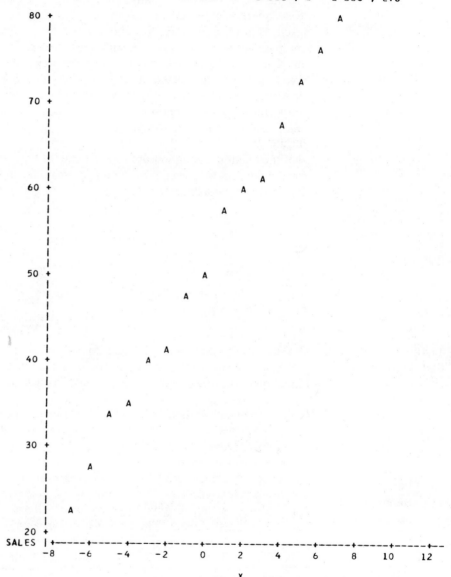

S T A T I S T I C A L   A N A L Y S I S   S Y S T E M

PLOT OF X*SALES   LEGEND: A = 1 OBS , B = 2 OBS , ETC

FIGURE 14.15:
COMPUTER-GENERATED
TREND ANALYSIS

S T A T I S T I C A L   A N A L Y S I S   S Y S T E M

GENERAL LINEAR MODELS PROCEDURE

DEPENDENT VARIABLE: SALES

PARAMETER	ESTIMATE
INTERCEPT	51.60000000
X	4.00000000

OBSERVATION	OBSERVED VALUE	PREDICTED VALUE	RESIDUAL
1	23.00000000	23.60000000	-0.60000000
2	28.00000000	27.60000000	0.40000000
3	34.00000000	31.60000000	2.40000000
4	35.00000000	35.60000000	-0.60000000
5	40.00000000	39.60000000	0.40000000
6	41.00000000	43.60000000	-2.60000000
7	48.00000000	47.60000000	0.40000000
8	50.00000000	51.60000000	-1.60000000
9	58.00000000	55.60000000	2.40000000
10	60.00000000	59.60000000	0.40000000
11	61.00000000	63.60000000	-2.60000000
12	68.00000000	67.60000000	0.40000000
13	72.00000000	71.60000000	0.40000000
14	76.00000000	75.60000000	0.40000000
15	80.00000000	79.60000000	0.40000000

SUM OF RESIDUALS	0.00000000
SUM OF SQUARED RESIDUALS	29.60000000

**14.13
FORECASTING**

Forecasting is predicting future observations of time series. Traditional methods of forecasting begin with trend analysis. We have seen in the previous sections how to describe the trend of a series by an equation. Once this equation has been established, values of trend for future time periods are obtained by simply extending the present trend forward in time. Consider the trend equation for annual sales of "The Gripper" tire, established in Section 14.8:

$$\hat{Y} = 51.6 + 4.0X$$

where

$X = 0$ at 1972,
X-unit is one year,
Y is unit sales (000).

The information stated below the equation is necessary for using the equation. First, the "origin" of the equation is stated. The *origin* is the point in time where the coded value of X equals zero. This is stated here as the year 1972. In time series analysis, data are usually centered on the middle of the time period under consideration. Thus the exact origin of this trend equation is the middle of 1972, or July 1, 1972. Second, the statement of the X-unit of the equation indicates that a one-unit change in variable X measures one year in time. This can be seen from the values of X in Column (3) of Table 14.3. Third, the unit of measurement for the values of Y is stated.

To project the present trend forward to a future date, substitute into the equation the number of X-units from the origin to the future date. The trend value for annual sales of "The Gripper" tire in 1985 is determined as follows.

1. Find the number of X-units from the origin to the future date.

$$
\begin{array}{lr}
\text{Future date} & 1985 \\
\text{Origin} & \underline{1972} \\
& 13 \text{ years}
\end{array}
$$

$$
\begin{array}{lr}
\text{Multiply by the number} & \\
\quad \text{of } X\text{-units per year} & \underline{\times\ 1} \\
& 13\ X\text{-units}
\end{array}
$$

2. Substitute the value of X into the equation.

$$\hat{Y}_{85} = 51.6 + 4.0(13)$$
$$= 103.6.$$

Step 1 indicates that X would have a value of 13 if the coding for time in Column (3), Table 14.3, is extended to 1985. Step 2 provides a trend value of 103,600 units for annual sales in 1985. The same procedure is used for other dates and other series. The extension of the trend to 1985 is illustrated graphically by line segment (a) in Figure 14.16.

FIGURE 14.16: ANNUAL SALES OF "THE GRIPPER" TIRE WITH FORECASTS FOR 1985

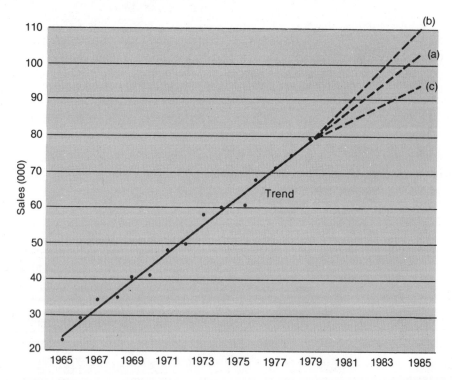

EXAMPLE. The trend equation for the purchasing power of the dollar, established in Section 14.9, is

$$\hat{Y} = 1.07032 - 0.01365X - 0.00020X^2$$

where

$X = 0$ at 1963–64,
X-unit is $\frac{1}{2}$ year,
Y is purchasing power ($).

Determine the trend value for purchasing power in 1982.

Solution. The statement of origin indicates that $X = 0$ at a point in time halfway between 1963 and 1964 (see Column (2) of Table 14.4). The point in time half-way between July 1, 1963, and July 1, 1964, is January 1, 1964. Thus the origin could be stated as $X = 0$ at January 1, 1964. To determine trend in 1982, follow the two steps suggested above.

1. Find the number of X-units from the origin to the future date.

Future date	July 1, 1982
Origin	January 1, 1964
	$18\frac{1}{2}$ years
Multiply by the number of X-units per year	$\times\ 2$
	37 X-units

2. Substitute the value of X into the equation.

$$\hat{Y}_{82} = 1.07032 - 0.01365(37) - 0.00020(37)^2$$
$$0.29147.$$

The trend value indicates that the purchasing power of a dollar in 1982 is $0.29, relative to 1967.

Extension of the present trend into the future is called *extrapolation*. This may or may not provide a valid forecast. Forecasting must incorporate the uncertainties of the future and an assessment of how possible future occurrences might affect a time series. Consider the trend in annual sales of "The Gripper" tire. If a major recession occurs between 1979 and 1985, the present trend in tire sales, as well as other series, may not continue. Estimates of future sales might be revised downward from the extrapolated value. A different economic picture, on the other hand, may suggest revising estimates upward. In practice, a forecast may incorporate trend descriptions, expert opinion, and managerial judgement. This subjective process may produce a range of values, rather than a single value, as a forecast. Line segments (b) and (c) of Figure 14.16 are illustrative of what may be called "optimistic" and "pessimistic" forecasts for 1985. Forecasts are sometimes generated for a variety of economic conditions. Because of the uncertainty of the future, forecasting based on extrapolation is limited to intermediate terms of no more than three to five years for most time series.

Determination of the best-fitting equation does not guarantee that trend extrapolation will provide a valid forecast. The equation is fitted to historical data. Because conditions may change, a different equation might be more appropriate for forecasting than for describing the historical data.

While traditional forecasting methods center on trend analysis, other prediction procedures also are employed. Regression equations, presented in Chapters 12 and 13, are often useful. To predict future values of a variable through regression analysis, a regression equation relating the time series, the dependent variable, to one or more independent variables is established. As an example, a regression equation relating automobile tire sales, over time, to a variable such as personal income might be used to predict sales of "The Gripper" tire. An estimate of personal income for 1985 could be substituted into the regression equation to produce a predicted value of tire sales in 1982. Multiple regression allows the time series to be related to more than one independent variable.

Other methods for forecasting can be found in the references at the end of the chapter.

14.14 CYCLICAL VARIATION

The cycle component of a time series is a wavelike repetitive movement fluctuating about the trend of the series. Examination of cycles is important for detecting turning points in a series and in making short-term forecasts. A measure of the cyclical movement can be obtained from the observed values and trend values of a series. If annual data are used to establish a trend equation, the observed values (Y) of the series are products of the trend (T), cycle (C), and irregular (I) components. If the observed values are expressed as a *percentage of the trend* values,

$$\frac{Y}{\hat{Y}} \cdot 100 = \frac{T \cdot C \cdot I}{T} \cdot 100 = C \cdot I \cdot 100, \qquad (14.14)$$

FIGURE 14.17:
PERCENTAGES OF
TREND FOR THE
PURCHASING POWER OF
THE DOLLAR, 1950–1977

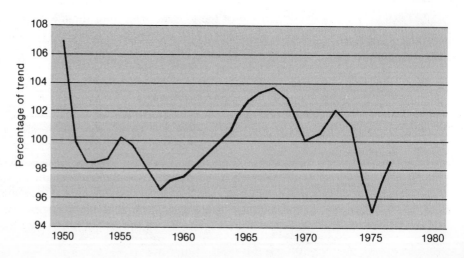

a measure of the cyclical component, with some irregular influence, is obtained. The percentage-of-trend values, $(Y/\hat{Y}) \cdot 100$, are referred to as *cyclical relatives*.

Percentage-of-trend values for the purchasing power of the dollar (originally given in Table 14.4) are determined in Table 14.7. The trend values are determined from the trend equation established in Section 14.9. The percentages of trend are plotted in Figure 14.17. This shows that the series is generally below the trend (100) for 1951–62 and above the trend for 1963–73. The period 1951–73 represents a complete cycle. Even though the period and amplitude of cycles are not usually constant, knowledge of the cycle is helpful in predicting turning points above and below the trend of a series. This knowledge could assist in adjusting projected trend values for forecasting purposes. The average length of the cycle could be determined by using more annual data.

TABLE 14.7:
DETERMINING
PERCENTAGES OF
TREND FOR THE
PURCHASING POWER OF
THE DOLLAR

	(1) Observed Value* Y	(2) Trend Value $\hat{Y}$	(3) Percentage of Trend $(Y/\hat{Y}) \cdot 100$
1950	1.387	1.295	107.1
1951	1.285	1.288	99.8
1952	1.258	1.280	98.3
1953	1.248	1.270	98.3
1954	1.242	1.258	98.7
1955	1.247	1.245	100.2
1956	1.229	1.231	99.8
1957	1.186	1.214	97.7
1958	1.155	1.197	96.5
1959	1.145	1.177	97.3
1960	1.127	1.156	97.5
1961	1.116	1.134	98.4
1962	1.104	1.109	99.5
1963	1.091	1.084	100.6
1964	1.076	1.056	101.9
1965	1.058	1.028	102.9
1966	1.029	0.997	103.2
1967	1.000	0.965	103.6
1968	0.960	0.931	103.1
1969	0.911	0.896	101.7
1970	0.860	0.860	100.0
1971	0.824	0.821	100.4
1972	0.798	0.781	102.2
1973	0.752	0.740	101.6
1974	0.678	0.697	97.3
1975	0.621	0.652	95.2
1976	0.587	0.606	96.9
1977	0.551	0.558	98.7

* Source: *Business Statistics.*

14.15 SEASONAL VARIATION

The seasonal component of a time series is a repetitive, fluctuating movement occurring within a time period of one year or less. Seasonal variation is produced by a variety of conditions, but they are often classified as either (1) natural or (2) man-made phenomena. Weather is the most obvious natural phenomenon that produces seasonal fluctuations. Retail sales of various clothing vary according to the weather associated with the seasons of the year. Sales of snowski equipment and automobile snow tires reach a peak during winter months. Man-made conditions that produce seasonal fluctuations include customs and traditions such as those in the holiday season surrounding Christmas and New Years. Retail stores generally have peak sales around this time. New car sales are generally highest in the fall, when new models are introduced. Tax accounting firms and tax preparation services are busiest during March and April because of the April 15 deadline for filing federal income tax returns.

There are three important reasons for studying seasonal variation. First, identification of seasonal patterns provides an understanding and explanation of short-term fluctuations. Second, knowledge of seasonal movements is useful in short-term forecasting, e.g., predicting product sales by month over the next year or two. Third, seasonal variation may be measured for the purpose of eliminating it from the series. The presence of seasonal patterns indicates what changes are expected in a series. In order to determine whether actual changes are more or less than expected, data must be adjusted for seasonal variation. Government publications such as *Business Statistics*, *Business Conditions Digest*, and *Survey of Current Business* contain measures of many economic indicators that are adjusted for seasonal variation, e.g., retail sales, unemployment, orders for new plant and equipment, etc.

Seasonal variation is measured by an index. The basic data for determining seasonal indexes are usually monthly or quarterly measurements. At least six years of data are preferable. The major method for computing seasonal indexes is the *ratio-to-moving average*. The method begins by computing centered 12-month (or 4-quarter) moving averages for the data available. The moving averages represent the trend-cycle components. Since the original data contain all four time series components, ratios of the original data (Y) to the moving averages produce measures of the seasonal-irregular components. Symbolically, the ratio-to-moving average is

$$\frac{Y}{T \cdot C} = \frac{T \cdot C \cdot S \cdot I}{T \cdot C} = S \cdot I. \tag{14.15}$$

The $S \cdot I$ ratios, multiplied by 100, are called *seasonal relatives*. These seasonal relatives are then averaged to minimize the irregular influence, which results in the seasonal indexes.

EXAMPLE. A major electric power company wishes to study the seasonal pattern of electricity usage in all-electric homes. Using a large number of these homes as a data base, a time series of the average kilowatt hours (KWH) used per month for 1974–79 was established. This is illustrated in Figure 14.18; the numerical data are provided in Columns (1) and (2) of Table 14.8.

Solution. Computations for determining seasonal indexes by the ratio-to-moving average method are given in Tables 14.8 and 14.9. Centered 12-month moving averages and seasonal relatives are computed in Table 14.8. The seasonal relatives are averaged to produce the seasonal indexes in Table 14.9. Details of the procedure and discussion of the results follow.

FIGURE 14.18: AVERAGE KILOWATT HOURS USED PER MONTH IN ALL-ELECTRIC
SINGLE-FAMILY DWELLINGS, 1974–1979

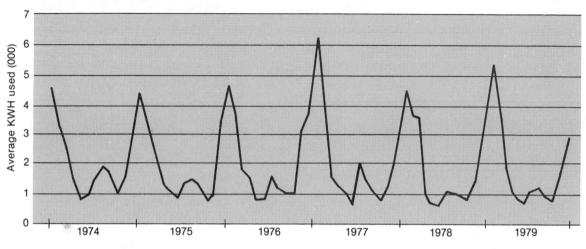

TABLE 14.8: AVERAGE KWH USED AND RATIOS-TO-MOVING AVERAGES, 1974–1979

(1) Date	(2) KWH Used	(3) 12-Month Moving Total	(4) Centered 24-Month Moving Total	(5) Centered Moving Average*	(6) Seasonal Relatives†	(7) Deseasonalized KWH Used‡
1974 J	4520					1831
F	3370					1751
M	2560					2257
A	1510					2217
M	910					1932
J	1020					2481
		24,840				
J	1530		49,580	2065.83	74.06	2045
		24,740				
A	1850		49,600	2066.67	89.52	2617
		24,860				
S	1710		49,680	2070.00	82.61	2855
		24,820				
O	1120		49,600	2066.67	54.19	2440
		24,780				
N	1670		49,860	2077.50	80.39	2183
		25,080				
D	3070		50,060	2085.83	147.18	1881
		24,980				
1975 J	4420		49,810	2077.50	212.76	1791
		24,830				
F	3490		49,320	2055.00	169.83	1813
		24,490				
M	2520		48,670	2027.92	124.27	2222
		24,180				
A	1470		48,050	2002.08	73.42	2159
		23,870				
M	1210		47,270	1969.58	61.43	2569
		23,400				

TABLE 14.8
(*continued*)

(1)		*(2)*	*(3)*	*(4)*	*(5)*	*(6)*	*(7)*
				Centered			
			12-Month	*24-Month*	*Centered*		
		KWH	*Moving*	*Moving*	*Moving*	*Seasonal*	*Deseasonalized*
Date		*Used*	*Total*	*Total*	*Average**	*Relatives†*	*KWH Used‡*
	J	920		47,090	1962.08	46.89	2238
			23,690				
	J	1380		47,690	1987.08	69.45	1845
			24,000				
	A	1510		48,310	2012.92	75.02	2136
			24,310				
	S	1400		48,060	2002.50	69.91	2337
			23,750				
	O	810		47,610	1983.75	40.83	1765
			23,860				
	N	1200		47,360	1973.33	60.81	1569
			23,500				
	D	3360		47,020	1959.17	171.50	2059
			23,520				
1976	J	4730		47,400	1975.00	239.49	1917
			23,880				
	F	3800		47,520	1980.00	191.92	1974
			23,640				
	M	1960		46,990	1957.92	100.11	1728
			23,350				
	A	1580		47,030	1959.58	80.63	2320
			23,680				
	M	850		49,420	2059.17	41.28	1805
			25,740				
	J	940		52,010	2167.08	43.38	2287
			26,270				
	J	1740		54,070	2252.92	77.23	2326
			27,800				
	A	1270		55,980	2332.50	54.45	1796
			28,180				
	S	1110		56,210	2342.08	47.39	1853
			28,030				
	O	1140		55,910	2329.58	48.94	2484
			27,880				
	N	3260		56,070	2336.25	139.54	4261
			28,190				
	D	3890		56,220	2342.50	166.06	2384
			28,030				
1977	J	6260		56,520	2355.00	265.82	2536
			28,490				
	F	4180		57,270	2386.25	175.17	2171
			28,780				
	M	1810		57,580	2399.17	75.44	1596
			28,800				
	A	1430		57,340	2389.17	59.85	2100
			28,540				
	M	1160		55,240	2301.67	50.40	2463
			26,700				
	J	780		52,090	2170.42	35.94	1898
			25,390				
	J	2200		49,150	2047.92	107.43	2941
			23,760				
	A	1560		47,620	1984.17	78.62	2206
			23,860				

TABLE 14.8
(continued)

(1) Date	(2) KWH Used	(3) 12-Month Moving Total	(4) Centered 24-Month Moving Total	(5) Centered Moving Average*	(6) Seasonal Relatives†	(7) Deseasonalized KWH Used‡
S	1130		49,610	2067.08	54.67	1886
		25,750				
O	880		51,200	2133.33	41.25	1917
		25,450				
N	1420		50,560	2106.67	67.40	1856
		25,110				
D	2580		50,200	2091.67	123.35	1581
		25,090				
1978 J	4630		49,230	2051.25	225.72	1876
		24,140				
F	4280		47,880	1995.00	214.54	2223
		23,740				
M	3700		47,420	1975.83	187.26	3263
		23,680				
A	1130		47,380	1974.17	57.24	1659
		23,700				
M	820		47,540	1980.83	41.40	1741
		23,840				
J	760		48,490	2020.42	37.62	1849
		24,650				
J	1250		49,980	2082.50	60.02	1671
		25,330				
A	1160		50,460	2102.50	55.17	1641
		25,130				
S	1070		49,020	2042.50	52.39	1786
		23,690				
O	900		47,640	1985.00	45.34	1961
		23,950				
N	1560		48,050	2002.08	77.92	2039
		24,100				
D	3390		48,270	2011.25	168.55	2077
		24,170				
1979 J	5310		48,350	2014.58	263.58	2152
		24,180				
F	4080		48,660	2027.50	201.23	2119
		24,480				
M	2260		49,060	2044.16	110.56	1993
		24,580				
A	1390		49,150	2047.91	67.87	2041
		24,570				
M	970		49,390	2057.91	47.14	2059
		24,820				
J	830		49,350	2056.25	40.36	2019
		24,530				
J	1260					1684
A	1460					2065
S	1170					1953
O	890					1939
N	1810					2366
D	3100					1900

* Column (4) divided by 24.
† Column (2) divided by Column (5), times 100.
‡ Column (2) divided by the seasonal index, S (see Table 14.9).

The centered 12-month moving averages of average KWH used are computed in Columns (3)–(5) of Table 14.8. The basic computation of a moving average is described in Section 14.4. Moving averages are computed for the purpose of determining seasonal relatives by expressing the original data as a ratio to the moving averages (Equation 14.15). Twelve-month moving averages provide trend-cycle measures because seasonal fluctuations averaged over a year tend to cancel one another out. The moving averages must be centered on a point in time corresponding to the original data (months) so that the ratio $Y/(T \cdot C)$ produces $S \cdot I$. Moving averages computed for an even number of time periods are not centered on the original values. Thus the 12-month moving averages must be adjusted to accomplish this. Totals for successive 12-month intervals are determined in Column (3). These totals are centered between the original values. In Column (4), successive pairs of 12-month moving totals are added. This centers the 24-month moving totals on the original monthly values. Dividing by 24 produces, in (Column (5), centered moving averages. Ratios-to-moving averages are determined by dividing the values in Column (2) by the values in Column (5). These ratios, when multiplied by 100, are the seasonal relatives provided in Column (6).

The seasonal relatives from Column (6), Table 14.8, are classified in Table 14.9 by month and year. The six seasonal relatives for any one month differ slightly because of the influence of irregular variation. The irregular component causes some values of $S \cdot I$ to be too large and others too small. A modified mean of the $S \cdot I$ for a given month is computed to minimize the irregular influence. Modified means are computed by eliminating extremely large and small values and then computing an arithmetic mean of the remaining values. The modified means for each month in Table 14.9 are computed by eliminating the largest and smallest value (marked through) and computing an arithmetic mean of the four remaining values. The modified means must usually be adjusted to force the total of the 12 monthly values to 1200. This is accomplished in Table 14.9 by multiplying each modified mean by 1200/1181.02, the ratio of the desired total to the existing total. The results are the 12 monthly indexes for average KWH used.

Seasonal indexes are index numbers exactly as described in Chapter 3. Their base value is 100. If seasonal variation is measured by months, a seasonal index of 100 reflects a typical or average monthly value of the series. For months where the series has larger than average values, the index is larger than 100. Smaller-than-average values are represented by indexes below 100. Since the average value of the index is 100, the sum of twelve monthly indexes is 1200. This is the reason for the adjustment in the previous paragraph. If seasonal variation is measured by quarterly indexes, the sum of the four indexes must be 400.

Let us see how these seasonal indexes could be used to accomplish the three purposes for measuring seasonal variation. First, seasonal indexes are useful for understanding and explaining short-term fluctuations. The seasonal indexes in Table 14.9 show that, for all-electric residential homes, more electricity is used in the cold winter months of December, January, and February than at

TABLE 14.9: COMPUTATION OF SEASONAL INDEXES FOR AVERAGE KWH USED, 1974–1979

	Jan.	Feb.	Mar.	Apr.	May	Jun.	Jul.	Aug.	Sep.	Oct.	Nov.	Dec.	Total
1974	—	—	—	—	—	—	74.06	~~89.52~~	~~82.61~~	~~54.19~~	80.39	147.18	
1975	~~212.76~~	~~169.85~~	124.27	73.42	~~61.43~~	~~46.89~~	69.45	75.02	69.91	~~40.83~~	~~60.81~~	~~171.50~~	
1976	239.49	191.92	100.11	~~80.63~~	~~41.28~~	43.38	77.23	~~54.45~~	~~47.39~~	48.94	~~139.54~~	166.06	
1977	~~265.82~~	175.17	~~75.44~~	59.85	50.40	~~35.94~~	~~107.43~~	78.62	54.67	41.25	67.40	~~123.35~~	
1978	225.72	~~214.54~~	~~187.26~~	~~57.24~~	41.40	37.62	~~60.02~~	55.17	52.39	45.34	77.92	168.55	
1979	263.58	201.23	110.56	67.87	47.14	40.36							
Modified mean .	242.93	189.44	111.65	67.05	46.31	40.45	73.58	69.60	58.99	45.18	75.24	160.60	1181.02
Seasonal index, S	246.8	192.5	113.4	68.1	47.1	41.1	74.8	70.7	59.9	45.9	76.5	163.2	1200.0

other times in the year. Electricity usage is lowest in the months of May and June. This could be because as spring ends, little or no heat is required, and the weather is not yet hot enough to require much air conditioning. But usage rises in July and August (air conditioning for high temperatures) before declining in September and October (months in between heavy air conditioning and heating needs). This pattern of usage applies only to this class of customer (all-electric single family homes) in this geographical area. Peak demand at electric utilities throughout most of the U.S. generally occurs in summer months rather than winter. This is because sources of heating (gas, oil, etc.) other than electricity exist, yet there are few alternatives to electricity for air conditioning.

A second purpose for seasonal indexes is for short-term forecasting. When annual forecasts are made, quarterly or monthly forecasts can be obtained by using the seasonal indexes. Suppose the electric power company referred to in this section has estimated total usage by all classes of customers for the next year. Assume that next year's annual usage for customers in all-electric single family dwellings is predicted to be 30,000 KWH per home. Then monthly usage per home is expected to average 2,500 KWH (30,000 ÷ 12). The forecast of usage in a given month is the product of the seasonal index (expressed as a proportion) for that month and average monthly usage. The average KWH forecast for January of next year is (2.468)(2,500) = 6,170. The January index indicates that January usage is 246.8 percent of average monthly usage, or 6,170 KWH.

The third purpose for measuring seasonal variation is to eliminate its influence on the data. The impact of seasonality in a time series is eliminated by dividing the original values of the series (Y) by the seasonal indexes (S) and multiplying by 100. This is

$$\frac{Y}{S} \cdot 100 = \frac{T \cdot C \cdot S \cdot I}{S} \cdot 100 = T \cdot C \cdot I. \qquad (14.16)$$

The resulting values are basically trend-cycle measures. Data adjusted for seasonal variation are often called *deseasonalized data*. Deseasonalized average KWH used is listed in Column (7) of Table 14.8. These values are obtained by dividing the values in column (2) by the appropriate seasonal indexes (Table 14.9) and multiplying by 100. The January 1974 value of 1,831, or (4520 ÷ 246.8) · 100, represents average KWH used if the seasonal factor were not present. Both the original and deseasonalized series are shown in Figure 14.19. The original series, which contains seasonality, fluctuates more than the deseasonalized series, which does not contain seasonality. Many of the major time series published by U.S. government agencies are provided in both unadjusted and deseasonalized form.

Elimination of seasonal influence is also important in determining whether changes in the values of a series are important. Suppose a retailer sells 4,320 units of an item in the third quarter and 6,700 in the fourth quarter. Have sales increased more than expected from the third to the fourth quarter if the seasonal index is 96 in the third quarter and 140 in the fourth? Sales adjusted for seasonal variation in the third quarter are 4,500, or (4,320 ÷ 96) · 100. Deseasonalized sales for the fourth quarter are also 4,500, or (6,700 ÷ 140) · 100. Assuming no

FIGURE 14.19: ORIGINAL AND DESEASONALIZED AVERAGE KWH USED, 1974–1979

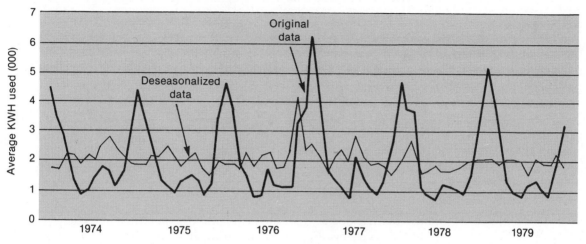

14.16 SUMMARY

trend-cycle influence, sales are exactly what should be expected. In practice, the measurement of seasonal variation and adjustment for seasonality is not expected to produce results this precise. But the results do provide valid indicators of the effect of seasonality.

A time series is the measurement of a variable taken over time. Time series are studied in order to understand and explain movements over time, to make comparisons over time, and to make predictions or forecasts. The four components of time series are trend (T), cycle (C), seasonal (S), and irregular (I). The multiplicative model for time series describes the values of the series (Y) as a product of these components, $Y = T \cdot C \cdot S \cdot I$.

Trend is the overall average movement over a long period of time. Trend is used to describe historical movements and for making forecasts and predictions. Trend is also used to determine the cycle component. Cycle is a wavelike repetitive movement fluctuating about the trend of the series. Identification of cycles is important in detecting turning points in a series and in making intermediate-term forecasts. Any fluctuating repetitive movement with a period greater than one year is considered cycle. Seasonal refers to repetitive, fluctuating movements with periods of one year or less. Knowledge of seasonal patterns is useful for short-term forecasting and for adjusting data for seasonal variation. Identifying seasonal patterns helps to understand the fluctuations in a series. Irregular is the term applied to any movements in a series that cannot be identified as trend, cycle, or seasonal. Strikes, fires, floods, and wars are some examples of events occurring irregularly that affect time series.

Time series analysis seeks to describe and use these four components. Classical methods to describe or measure these components are based on decomposing the series into their individual components. Some components are isolated by removing the influence of other components. Smoothing techniques seek to remove irregular variation from the series. Moving averages provide smoothed values by averaging the values of the series over successive time intervals. Exponential smoothing, a second smoothing technique, determines smoothed values by computing an exponentially weighted average of the current and all previous values of a series.

The trend component of a time series is described by an equation. The least squares method is used to fit an equation of a given mathematical form to the series. The form of equation is chosen by examining the plotted data. A linear equation, $\hat{Y} = b_0 + b_1 X$, is best if the approximate trend appears to form a straight line on an arithmetic chart. If the approximate trend appears as a parabola on an arithmetic chart, the quadratic equation, $\hat{Y} = b_0 + b_1 X + b_2 X^2$, may be best. For parabolas where the approximate trend is increasing by an increasing amount over time, or decreasing by a decreasing amount, plot the data on a semilogarithmic chart. If the plotted data approximates a straight line on a semilogarithmic chart, the exponential equation $\hat{Y} = b_0 b_1^X$ is best; otherwise the quadratic is preferable. A linear trend indicates that the trend values are changing by a constant amount. An exponential trend indicates that trend values change by a constant percentage. For the quadratic trend, the trend values either increase or decrease by either an increasing or decreasing amount. Many computer routines are available to assist in fitting trends to time series.

Extrapolation is the process of extending the present trend into the future. The extrapolated trend value may or may not be a valid forecast. Forecasting must consider factors that may affect a series in the future and whether the present trend will continue. Forecasting should include managerial judgment, economic indicators, etc.

Trend may be used to measure the cyclical component of a time series. When annual data consisting of $T \cdot C \cdot I$ are expressed as a percentage of trend (T), the result is a measure of cyclical variation $(C \cdot I)$. After multiplying by 100, the resulting values are called cyclical relatives.

Seasonal variation is measured by an index. Using monthly (or quarterly) data, the ratio-to-moving average method begins with the computation of a centered 12-month (or 4-quarter) moving average. This represents the trend-cycle $(T \cdot C)$ components. Expressing the original data $(T \cdot C \cdot S \cdot I)$ as a ratio to the moving average, and multiplying by 100 provides seasonal relatives $(S \cdot I)$. A modified mean of the seasonal relatives is used to determine the seasonal indexes.

An understanding of the movements in a time series enables managers to more effectively plan future operations. Forecasting is perhaps the most frequent objective of studying a time series. Understanding past fluctuations and editing data are other reasons for examining time series.

14.17 TERMS AND FORMULAS

Amplitude	Forecast
Arithmetic chart	Irregular component
Cycle component	Least squares method
Cyclical relatives	Moving average
Exponential smoothing	Origin
Extrapolation	Percentage of trend

Period

Ratio-to-moving average

Residuals

Seasonal component

Seasonal relatives

Semilogarithmic chart

Smoothing

Smoothing constant

Time series

Trend component

Trend equation

Multiplicative model
$$Y = T \cdot C \cdot S \cdot I$$

Exponentially smoothed value
$$S_t = wy_t + (1 - w)S_{t-1}$$

Moving average

$$\bar{y}_t = \frac{1}{n}(y_{t-(n-1)/2} + \cdots + y_{t-1} + y_t + y_{t+1} + \cdots + y_{t+(n-1)/2})$$

Formulas to determine trend equations when $\sum X = 0$.

1. Linear: $\hat{Y} = b_0 + b_1 X$

$$b_0 = \frac{\sum Y}{n}$$

$$b_1 = \frac{\sum XY}{\sum X^2}$$

2. Quadratic: $\hat{Y} = b_0 + b_1 X + b_2 X^2$
   for b_0 and b_2:

$$\sum Y = nb_0 + b_2 \sum X^2$$
$$\sum X^2 Y = b_0 \sum X^2 + b^2 \sum X^4$$

$$b_1 = \frac{\sum XY}{\sum X^2}$$

3. Exponential: $\hat{Y} = b_0 b_1^X$
   Convert to log $\hat{Y} = \log b_0 + (\log b_1) \cdot X$

$$\log b_0 = \frac{\sum \log Y}{n}$$

$$\log b_1 = \frac{\sum X \log Y}{\sum X^2}$$

QUESTIONS

True or False

1. A quadratic trend equation exhibits a constant percentage change in Y for each unit change in X.

2. Deseasonalized values of a time series result when the original values are divided by the seasonal index and multiplied by 100.

3. The seasonal component of a time series is not present when the series is measured on an annual basis.

4. The irregular component of a time series is measured by an index.

5. Cyclical fluctuations are more regular and predictable than seasonal movements.

6. If two trend models are compared, the best-fitting has the smallest $\sum(Y - \hat{Y})^2$.

7. Moving averages are more useful for forecasting than for studying the historical movements in a time series.

8. One advantage of exponential smoothing is that only the current smoothed value must be retained for computing next period's smoothed value, thereby reducing data storage requirements.

9. Small values of the smoothing constant cause the smoothing procedure to react slowly to changes in a series.

10. An exponential equation plots as a straight line on a semilogarithmic chart.

Fill in

1. Retail inventories of domestic new passenger cars at the end of December, 1976 were, according to the *Survey of Current Business*, 1,465 thousand units. The seasonally adjusted figure for that date was 1,554 thousand units. The December seasonal index is _____ .

2. An exponential trend equation, expressed in logarithms of $\hat{Y}$, is log $\hat{Y} = 2.8754 + 0.0218X$. This equation exhibits a _____ percent change in $\hat{Y}$ for a unit change in X.

3. The trend equation for unit production at a factory is $\hat{Y} = 36,000 + 200X$, where the origin is 1968 and the X-unit is one year. The trend in unit production will reach 40,000 in the year _____ .

4. The trend equation for sales, in thousand dollars, for a travel firm sponsoring Carribean cruises, is log $\hat{Y} = 2.3871 + 0.0691X$, where the origin is 1970 and the X-unit is one year. The predicted value of sales in 1980 is _____ thousand dollars.

5. The forecast of 1982 retail sales for Gilvan, Ltd., is $96,000. The seasonal index for second quarter sales is 125. The second quarter retail sales forecast is _____ .

6. Which of the four types of movements in a time series is illustrated by each of the following?
   a. A recession.
   b. An increase in beer sales in summer months.
   c. A decrease in the number of families living on U.S. farms.
   d. The declining enrollment in secondary schools.
   e. A strike by steelworkers.

EXERCISES

1. Quarterly profits (after taxes) per dollar of sales for all U.S. manufacturing corporations over 1947–76 are given in Figure 14.4. The data, measured in cents, are provided below for the ten years 1967–76. Determine five-quarter moving averages of the data for 1967–76. Compare the moving averages with the original data by plotting both on the same chart.

		Quarter		
Year	I	II	III	IV
1967	5.0	5.0	4.9	5.1
1968	5.1	5.0	5.1	5.1
1969	5.1	4.9	4.8	4.5
1970	4.1	4.2	4.0	3.6
1971	4.0	4.2	4.2	4.1
1972	4.2	4.2	4.3	4.5
1973	4.7	4.7	4.7	5.7
1974	5.8	5.6	5.9	4.9
1975	3.8	4.4	5.0	5.1
1976	5.5	5.6	5.3	5.0

Source: *Survey of Current Business.*

2. Refer to the data in Exercise 1, quarterly profits (after taxes) per dollar of sales for all U.S. manufacturing corporations, 1967–76.
   a. Use exponential smoothing to determine smoothed values of quarterly profits. Set the first smoothed value equal to the observed value (first quarter of 1967). Use a smoothing constant $w = 0.1$ in Equation (14.2).
   b. Compare the smoothed and original values by plotting both on the same chart.
   c. What is your estimate of profits per dollar of sales for the first quarter, 1977?

3. The production of primary aluminum in the U.S. for 1968–77 is given below.
   a. Fit a linear trend equation by the method of least squares.
   b. Determine the predicted value of trend for 1982 and 1985.
   c. Determine the cyclical relatives for 1968–77 as percentage-of-trend values; prepare a chart of the cyclical relatives.

Year	Production (thousands of short tons)
1968	3255
1969	3793
1970	3976
1971	3925
1972	4122
1973	4529
1974	4903
1975	3879
1976	4251
1977	4539

Source: *Survey of Current Business.*

4. A retailer's unit sales of an increasingly popular item are given below for 1964–79.
   a. Find the linear trend equation.
   b. Interpret the equation.
   c. Find the value of trend in 1982.

Year	Sales (hundreds)
1964	30
1965	32
1966	35
1967	36
1968	38
1969	38
1970	37
1971	39
1972	40
1973	45
1974	46
1975	46
1976	47
1977	49
1978	49
1979	52

5. The U.S. national debt for 1951–79 is shown in Figure 14.1 at the beginning of the chapter. Data for this series are given below.

   a. Fit a quadratic trend equation to the data.

   b. When will the U.S. national debt reach one trillion dollars, if the present trend continues?

Year	National Debt ($ billions)
1951	254
1952	258
1953	265
1954	271
1955	274
1956	273
1957	272
1958	280
1959	288
1960	291
1961	293
1962	303
1963	311
1964	317
1965	323
1966	329
1967	341
1968	370
1969	367
1970	383
1971	410
1972	437
1973	468
1974	486
1975	544
1976	632
1977	709
1978	778*
1979	866*

* 1978 and 1979 are estimates.

Source: Office of Management and Budget, March 1978.

6. The *Y*-values shown below are used in Section 14.10 as an illustration of a constant 40 percent increase in *Y* over time. Determine the exponential trend equation for these data and verify that it reflects this constant percentage change.

Time t	Y
1	100
2	140
3	196
4	274.4
5	384.16
6	537.824
7	752.9536

7. The *Survey of Current Business* provides the average retail price of automobile gasoline (regular grade, excluding taxes) based on selected service stations in 50–55 cities. The data are given below for 1960–77.
   a. Plot the data and fit a quadratic trend equation.
   b. Find the trend value in 1981.
   c. What factors must be considered in forecasting gasoline price?

Year	Price (dollars per gallon)
1960	0.210
1961	0.205
1962	0.204
1963	0.201
1964	0.200
1965	0.208
1966	0.216
1967	0.226
1968	0.230
1969	0.239
1970	0.246
1971	0.252
1972	0.244
1973	0.269
1974	0.404
1975	0.455
1976	0.474
1977	0.507

8. U.S. per capita personal Income and the Consumer Price Index for 1962–77 are given below.
   a. Use the CPI to adjust per capita personal income for price level changes and fit the appropriate trend equation to the result, real per capita personal income.
   b. Predict per capita personal income for 1980 by
      (1) using the equation to determine a trend value and
      (2) adjusting the trend value by the preliminary CPI (obtain from library) for 1980.

Year	Per Capita Personal Income ($)	Consumer Price Index (1967 = 100)
1962	2,370	90.6
1963	2,458	91.7
1964	2,590	92.9
1965	2,770	94.5
1966	2,986	97.2
1967	3,170	100.0
1968	3,436	104.2
1969	3,708	109.8
1970	3,943	116.3
1971	4,164	121.3
1972	4,493	125.3
1973	4,980	133.1
1974	5,428	147.7
1975	5,861	161.2
1976	6,403	170.5
1977	7,019	181.5

Source: *Survey of Current Business.*

9. The Securities and Exchange Commission prohibited fixed minimum commissions on stock transactions on May 1, 1975. The New York Stock Exchange opposed this move because, in part, they feared that the decline in the number of brokerage firms holding NYSE membership would accelerate. The number of NYSE member firms for 1968–77 is given below

   a. Determine the least-squares equation that best describes trend in NYSE membership.

   b. Do the data indicate that membership declined more rapidly in 1975–77 than before? (For further information, see the source of the data: *The Wall Street Journal*, November 24, 1978, editorial page.)

   c. Predict NYSE membership in 1980.

	NYSE Member Firms
1968	646
1969	622
1970	572
1971	577
1972	558
1973	523
1974	508
1975	494
1976	481
1977	473

10. U.S. production of gasoline for 1962–76 is given below.

   a. Use the method of least squares to fit the best form of trend equation to these data.

b. Find the value of this trend in 1985.

c. When will the trend value reach 3,000 million barrels?

Year	Production (millions of barrels)
1962	1,583
1963	1,625
1964	1,661
1965	1,704
1966	1,793
1967	1,846
1968	1,940
1969	2,028
1970	2,105
1971	2,203
1972	2,320
1973	2,402
1974	2,338
1975	2,394
1976	2,517

Source: *Survey of Current Business.*

11. Determine the appropriate trend equation for the shipments of office equipment (S.I.C. code 357: office, computing, and accounting machines) for 1958–78). Predict trend in 1981.

Year	Shipments ($ millions)
1958	2,079
1959	2,406
1960	2,783
1961	2,971
1962	3,229
1963	3,503
1964	3,859
1965	4,265
1966	5,936
1967	5,732
1968	6,215
1969	7,420
1970	7,693
1971	6,909
1972	8,605
1973	10,054
1974	12,179
1975	11,528
1976	13,724
1977	15,791
1978	18,277*

* Estimated.

Source: *U.S. Department of Commerce.*

12. Refer to Exercise 11, shipments of office equipment for 1958–78. Estimate trend by determining five-year moving averages of the annual data. What are the disadvantages of using the method of moving averages to estimate trend?

13. The U.S. population in the "Under 5" age group is given below for the years 1961–74.
    a. Plot the data.
    b. Determine the best fitting trend equation; plot on the chart prepared in (a).
    c. Interpret the trend equation.
    d. Find the trend value in 1980. Use a library source to compare the actual value with this trend value.

Year	U.S. Population under 5 (thousands)
1961	20,522
1962	20,469
1963	20,342
1964	20,165
1965	19,824
1966	19,208
1967	18,563
1968	17,913
1969	17,376
1970	17,156
1971	17,174
1972	17,006
1973	16,714
1974	16,304

Source: *Economic Report of the President*, 1975.

14. Refer to the data in Exercise 9, the number of New York Stock Exchange member firms for 1968–77. Use exponential smoothing to obtain estimates of trend, the smoothed values of the series. Set the smoothed value for 1968 equal to the observed value and use a smoothing constant $w = 0.2$. What is your predicted value for 1978?

15. Contracts and orders for plant and equipment over 1960–76 are given below.
    a. Use a computer routine to fit quadratic and exponential equations to the data.
    b. Determine which equation provides the best fit by comparing the $\sum(Y - \hat{Y})^2$ for each.
    c. Find the 1979 trend value for each equation.

Year	Contracts and Orders for Plant and Equipment ($ billions)
1960	$ 40.62
1961	41.28
1962	45.11
1963	49.49
1964	56.74
1965	63.19
1966	75.32
1967	71.70
1968	98.62
1969	111.21
1970	98.90
1971	107.98
1972	124.38
1973	156.85
1974	176.85
1975	154.79
1976	182.87

Source: *Survey of Current Business.*

16. Refer to the data in Exercise 1, quarterly profits (after taxes) per dollar of sales for all U.S. manufacturing corporations, 1967–76. Determine the four quarterly indexes of seasonal variation by the moving average method. Use centered four-quarter moving averages to estimate the trend-cycle component.

17. Monthly magazine advertising cost for January 1973 through July 1978 is illustrated in Figure 14.5. The actual data, rounded to the nearest million dollars, are given below.

    a. Determine the monthly seasonal indexes of advertising cost by the ratio-to-moving average method; use centered 12-month moving averages to estimate the trend-cycle component.

	U.S. Magazine Advertising Cost ($ millions)					
	1973	1974	1975	1976	1977	1978
January.	73	81	82	93	112	130
February	90	98	95	109	136	160
March	110	112	118	130	154	194
April	126	126	119	145	177	213
May.	126	127	120	159	200	231
June.	110	111	105	123	151	188
July	81	86	83	101	119	163
August	77	85	82	101	122	—
September	119	126	118	143	173	—
October	144	143	135	182	221	—
November	140	158	152	194	222	—
December	118	119	120	142	178	—

Source: *Survey of Current Business.*

b. Find the value of advertising cost, adjusted for seasonal variation, in July 1978.

18. Monthly sales of Vangil Enterprises for 1974–79 are given below. Determine the monthly seasonal indexes by the ratio-to-moving average method (use centered 12-month moving averages to estimate the trend-cycle component).

		Sales ($000)			*Sales ($000)*
1974	J.	119	1977	J.	175
	F	86		F	137
	M.	59		M.	74
	A	35		A	61
	M.	23		M.	50
	J.	32		J.	38
	J.	49		J.	106
	A	63		A	74
	S	58		S	55
	O.	36		O.	39
	N.	54		N.	57
	D.	80		D.	86
1975	J.	115	1978	J.	143
	F	106		F	140
	M.	77		M.	128
	A	51		A	49
	M.	42		M.	37
	J.	37		J.	38
	J.	55		J.	64
	A	64		A	59
	S	59		S	56
	O.	30		O.	41
	N.	44		N.	62
	D.	89		D.	110
1976	J.	115	1979	J.	145
	F	103		F	121
	M.	61		M.	65
	A	52		A	50
	M.	31		M.	43
	J.	34		J.	36
	J.	74		J.	60
	A	53		A	65
	S	46		S	52
	O.	47		O.	37
	N.	84		N.	55
	D.	101		D.	90

19. A department store forecasts 1981 annual sales of $500,000. Seasonal indexes of sales for each quarter of the year are: I, 81; II, 103; III, 122; IV, 94. Determine the sales forecast for each quarter of 1981.

20. In the second quarter of 1979, the director of sales of the Commonwealth Company argued with the vice president of finance over the significance of changes in sales during the first quarter. Sales rose from $750,000 in January to $900,000 in February and to $1,200,000 in March. The director

of sales claimed that the increases resulted from a new sales program. The vice president of finance said the increases were attributable to seasonal variation. The seasonal index for January is 95, February 115, and March 110. Who is right? Why?

21. Frabastat production totaled 90,000 units in 1979. If the April seasonal index is 130, estimate Frabastat production for the month of April.

22. Seasonal indexes of sales for C. J. Spenny, Inc., are 110 in September and 98 in October. Describe the percentage change in sales from September to October.

23. A skateboard manufacturer reports August sales of 15,000 units in 1980 and an August seasonal index of 112. Estimate annual sales for 1980.

24. Use sources in your library (e.g., *Business Statistics, Survey of Current Business*) to collect annual data on a time series assigned by your instructor. Fit the appropriate trend equation and forecast the value of the series for next year.

REFERENCES

Mendenhall, William, and Reinmuth, James E. *Statistics for Management and Economics*, 3d ed. North Scituate, Mass.: Duxbury Press, 1978.

Miller, Robert B., and Wichern, Dean W. *Intermediate Business Statistics*. New York: Holt, Rinehart and Winston, Inc., 1977.

Nelson, Charles R. *Applied Time Series Analysis for Managerial Forecasting*. San Francisco: Holden-Day, Inc., 1973.

Neter, John; Wasserman, William; and Whitmore, G. A. *Applied Statistics*. Boston: Allyn and Bacon, Inc., 1978.

Wheelwright, Steven C., and Madridakis, Spyros *Forecasting Methods for Management*, 2d ed. New York: John Wiley & Sons, Inc., 1977.

Wonnacott, Thomas H., and Wonnacott, Ronald J. *Introductory Statistics for Business and Economics*, 2d ed. New York: John Wiley & Sons, Inc., 1977.

15

Nonparametric Methods

**Optional material.

15 NONPARAMETRIC METHODS

The dolphin's strength deserts
him on dry land.

                    Ion

Many of the procedures for testing hypotheses and estimating parameters previously discussed are termed parametric methods. Parametric methods can be loosely defined as those requiring stringent assumptions about the distribution of the population. For example, in using the t distribution to test $H_0: \mu = \mu_0$, one assumes that the population is normally distributed. And, in using the F distribution and ANOVA to test for the equality of means, the populations are assumed normal with equal variances. In both instances above, the validity of the test is dependent upon the appropriateness of the assumptions. When the assumptions underlying a statistical test are violated, the test (like the dolphin on dry land) may lose much of its strength.

Nonparametric (or distribution-free) statistics provide techniques useful to the analyst who is unwilling or unable to make the assumptions required of parametric methods. Several nonparametric tests, although not identified as such at the time, have been already introduced. Most notably, procedures resting on the central limit theorem and the goodness-of-fit test and test for independence using the chi-square distribution are nonparametric. Note, however, that testing $H_0: \sigma^2 = \sigma_0^2$ with the chi-square distribution assumes a normally distributed population and, hence, is a parametric method. For most of the parametric statistical tests discussed earlier in the text, a nonparametric counterpart has been developed. In this chapter, we shall review four tests using nonparametric methods. The utility of these and other nonparametric tests is best appreciated, however, when one is familiar with the concepts of measurement.

15.2 LEVELS OF MEASUREMENT

The selection of a statistical technique is influenced by the level of measurement possessed by the data to be analyzed. For example, if the data are the *weights* of five individuals, then computation of the mean weight, $\bar{X}$, of the group yields a meaningful result. However, if the data are *sex* for five individuals (say three males and two females), the mean sex is undefined. Suppose that instead of classifying sex by the labels (names) "male" or "female," one classifies sex by the labels "zero" or "one." The distinction is maintained with either pair of labels, and "a rose by any name . . . ." In the latter case, one might be tempted to compute a mean since the data would then consist of three zeroes and two ones; however, the resultant $\bar{X}$ is without meaning. The procedures that are valid on the variables weight and sex are different because their levels of measurement are different.

Levels of measurement are usually defined by the four classifications identified by S. S. Stevens: nominal, ordinal, interval, and ratio. The *nominal scale*, "weakest" of the four, is simply classifying each element by a label (name and/or number). Examples of the nominal scale would include sex (male, female) or major (accounting, economics, marketing, etc.). If the labels are numbers, it must be remembered that they are arbitrarily chosen and do not possess the properties of "real" numbers, for example, their sum has no meaning.

The *ordinal scale* is the next level of measurement. Here, the elements can be arranged in a sensible order, e.g., best to worst, smallest to largest, darkest to brightest. Data measured on an ordinal scale are usually given in ranks. For example, the finishers of a marathon could be ranked in order of finish 1, 2, 3, . . . The *order* of such numbers is meaningful, i.e., 2 took longer to finish than 1, 7 took longer than 4. The *difference* in ordinal numbers is *not* meaningful; the marathon winner could have beaten 2 by ten seconds, while 2 beat 3 by eight minutes. Similarly, university faculty can be ranked as instructors (1), assistant professors (2), associate professors (3), and full professors (4). Such numbers do not imply that two instructors equal one assistant professor.

The data are measured on an *interval scale* when the elements can be distinguished, ordered, *and* possess a meaningful difference. Unlike the nominal and ordinal scales, data at this level can be viewed as truly numerical and subject to the various arithmetic operations. The interval scale, however, does *not* have a true zero, i.e., complete absence of the characteristic being measured. The classic example is the measurement of temperature. The Fahrenheit and Celsius scales are equally valid measurements of temperature (heat). However, the freezing point of water is defined as 0° C but 32° F; thus, on either scale, the zero is arbitrary and does not mean the total absence of heat. The interval scale does mean that the difference (interval) in heat between 40° and 50° F is the same as between 80° and 90° F. We reiterate this is not a property of the lower scales of measurement.

The *ratio scale* has the properties of the interval scale *plus* a fixed, inherently defined zero. Examples would include measurements of area or volume. The term "ratio" is employed because ratio comparisons are meaningful. For example, a container of ten cubic feet has twice the volume of one with five cubic feet. Note this kind of comparison is not appropriate for interval data, e.g., 100° F is not "twice as hot" as 50° F.

The four levels of measurement have been presented in order of increasing precision, and any level can be transformed to that of a *lower* level, e.g., ratio to ordinal. The concept of measurement is important in statistics because lower levels of measurement restrict one's freedom in analysis of the data. For example, $\bar{X}$ and S^2 have no meaning when computed from data at the nominal or ordinal levels. Hence the parametric procedures concerned with an hypothesis test or estimation of μ are not applicable. One of the advantages of nonparametric methods is their applicability to data measured at *less* than the interval scale; in particular, the four methods presented in this chapter can all be used with ordinal (or higher scaled) data.

As mentioned in the introductory section, nonparametric methods also require "milder" assumptions than those associated with parametric statistics. In summary then, one may select a nonparametric test because the population assumptions are thought untenable or the level of measurement is inadequate for parametric methods.

When the analyst has data for which the parametric assumptions and required level of measurement are met, what loss is entailed by choosing the nonparametric counterpart? Such a question is usually answered by studying the large-sample efficiency of the competing tests. Large-sample efficiency, also known as asymptotic relative efficiency (ARE), is the ratio of the sample sizes required by each of the two tests to yield the same level of power.

Table 15.1 shows the large-sample efficiency for the tests of this chapter relative to their parametric counterparts. These ratios were computed by assuming that the populations being sampled were normally distributed. The efficiency of test A to test B is given by n_b/n_a. If the quotient is less than one, test B is more efficient; it requires a smaller sample to yield the same power. Note that although the nonparametric tests in Table 15.1 are all less efficient that their parametric counterparts, the ratios are fairly close to one. For small samples or when the parametric assumptions are not met, a nonparametric test could be more efficient than its parametric competitor.

TABLE 15.1: EFFICIENCY OF SELECTED NONPARA-METRIC TESTS RELATIVE TO THEIR PARAMETRIC COUNTERPARTS WHEN PARAMETRIC ASSUMPTIONS ARE MET

Nonparametric Test	Counterpart Parametric Test	Large-Sample Efficiency
Wilcoxon signed rank	Student's t	0.955
Mann-Whitney-Wilcoxon	Student's t	0.955
Kruskal-Wallis	F-test (one-way ANOVA)	0.955
Spearman rank correlation	Pearson's r or t-test of b	0.912

15.3 RANKED DATA

The nonparametric methods of this chapter are all based on ranks. Hence, the *required* level of measurement is ordinal or better; if better (interval or ratio), the data are converted to ranks. The rank is assigned to interval or ratio data by arranging the n sample observations in *ascending* order; the smallest number is given rank 1, the next smallest rank 2, . . . , and the largest rank n.

A mathematical theorem of some use in later calculations says the sum of integers $1, 2, \ldots, n$ is given by

$$1 + 2 + 3 + \cdots + n = \frac{n(n + 1)}{2}. \tag{15.1}$$

For example, given the rank of four observations, the rank sum is $1 + 2 + 3 + 4 = (4)(5)/2 = 10$.

When dealing with ranked data, a problem with ties occasionally arises. For example, how shall the observations 2, 5, 7, 7, and 10 be ranked? The procedure usually recommended is the *midrank method*. This method assigns the same rank to each observation in a tied set; this rank is simply the average of the ranks being occupied by the ties. Hence, 2, 5, 7, 7, and 10 would be ranked 1, 2, 3.5, 3.5, and 5 (3.5 is the average of ranks 3 and 4). Using the midrank procedure does *not* change the rank sum from the $n(n + 1)/2$ value.

FRANK WILCOXON (1892–1965)

Frank Wilcoxon was born in Ireland to American parents. He was reared and educated in the Northeast, receiving his Ph.D. in physical chemistry from Cornell in 1924. His contributions as a chemist were many, especially in the area of insecticides. Here, of course, he is recognized for his work in statistics. Two tests presented in this chapter were developed by Wilcoxon and bear his name. Both were published in 1945 and fostered an increased interest in nonparametric methods. The major part of his career was with the American Cyanamid Company, but in 1960 Wilcoxon joined Florida State University as a Distinguished Lecturer. There he was a popular instructor despite giving no partial credit for incorrect answers. He was an outdoorsman, accomplished guitarist, and avid bicyclist. Wilcoxon, in his sixties, was also one of the first American owners of a Honda motorcycle.

EXAMPLE. Given the following hourly wages, convert the observations to ranked data and compute the rank sum.

Wage: $2.90, 3.10, 3.10, 4.00, 4.20, 4.50, 4.50, 4.50, 5.12.

Note that we have a tie in ranks 2 and 3 and also in ranks 6, 7, and 8. Using the midrank method rank $(2 + 3)/2 = 2.5$ will be assigned to each of the $3.10 values, and rank $(6 + 7 + 8)/3 = 7$ will be assigned to each of the $4.50 values. The data above would then correspond to the following ranks:

Rank: 1, 2.5, 2.5, 4, 5, 7, 7, 7, 9.

The rank sum can be obtained by summing the ranks above, but it is easier to use our new theorem:

$$\text{Rank sum} = \frac{n(n + 1)}{2}$$

$$= \frac{9(10)}{2} = 45.$$

15.4 WILCOXON SIGNED RANK TEST

The Wilcoxon signed rank test is the nonparametric counterpart of Student's t test of $H_0: \mu = \mu_0$. The hypothesis for the Wilcoxon test, however, is usually stated in terms of the population median, M; this convention shall be followed here. (In Chapter 2, Md was used for the median; since the mean and mode are not referred to in this chapter, we adopt the more convenient M for median.) The Wilcoxon test assumes (1) the data represent a random sample drawn from a continuous, symmetric population and (2) measurement on an ordinal or higher scale. Note that the assumptions of Student's t are those of Wilcoxon, *plus* a *particular* continuous, symmetric distribution (the normal) is assumed.

The computation of the test statistic for the Wilcoxon signed rank test is as follows:

> 1. Find the difference between each observed value, X_i, and the hypothesized median, M_0.
> 2. If any $X_i - M_0 = 0$, omit this observation and reduce the sample size accordingly.
> 3. Rank the differences in order of *absolute* magnitude; the smallest absolute difference will be given rank one.
> 4. If there are ties in the differences, assign ranks using the midrank method.
> 5. Give each rank a positive or negative sign according to the sign of its $X_i - M_0$; e.g., if $X_i - M_0 < 0$, then its signed rank is negative.
> 6. Compute the sum of the positive ranks, symbolized by T_+, *or* the sum of the negative ranks, T_- (given one, the other can be found by using $T_+ + T_- = n(n + 1)/2$).

The test statistic will be either T_+ or T_-; Table 15.2 gives the appropriate one for the various null hypotheses. The null hypothesis will be *rejected* if the test statistic is equal to or less than the critical value given in Appendix J. Before explaining the rationale underlying the test, an example would be helpful.

TABLE 15.2: APPROPRIATE TEST STATISTIC FOR VARIOUS NULL HYPOTHESES WITH THE WILCOXON SIGNED RANK TEST

Null Hypothesis	Test Statistic
$H_0: M \geq M_0$	T_+
$H_0: M \leq M_0$	T_-
$H_0: M = M_0$	Smaller of T_+ and T_-

EXAMPLE. A hospital insurance underwriter is investigating the length of stay for diabetes patients at Mercy Hospital. From national data it is known that the median length of stay for diabetes patients is 6.0 days. A random sample of eleven patients' records shows the following length of stays.

Length (days): 3.0, 5.0, 6.0, 6.5, 7.5, 8.0, 8.0, 9.5, 10.0, 11.5, 12.5.

At the 0.025 level of significance, test the hypothesis that Mercy Hospital's median length of stay for diabetes patients is equal to or less than 6.0 days; i.e.,

$$H_0: \quad M \leq 6.0$$
$$H_1: \quad M > 6.0$$
$$\alpha = 0.025.$$

The computations required to obtain the test statistic are shown in Table 15.3. Note the third observation has a zero difference and will be ignored, reducing n from eleven to ten; also note the midrank method was applied to the tie in ranks four and five. From Table 15.2, the test statistic will be T_- and equals:

$$T_- = 6 + 2 = 8.$$

TABLE 15.3 COMPUTATIONS REQUIRED FOR EXAMPLE

X_i	$X_i - 6.0$	$|X_i - 6.0|$	Rank	Sign
3.0	-3.0	3.0	6	$-$
5.0	-1.0	1.0	2	$-$
6.0	0	0	—	
6.5	0.5	0.5	1	$+$
7.5	1.5	1.5	3	$+$
8.0	2.0	2.0	4.5	$+$
8.0	2.0	2.0	4.5	$+$
9.5	3.5	3.5	7	$+$
10.0	4.0	4.0	8	$+$
11.5	5.5	5.5	9	$+$
12.5	6.5	6.5	10	$+$

As a check, we can compute

$$T_+ = 1 + 3 + 4.5 + 4.5 + 7 + 8 + 9 + 10 = 47,$$

and verify using Fomula 15.1:

$$T_- + T_+ = 8 + 47$$
$$= \frac{n(n+1)}{2} = \frac{10(11)}{2} = 55.$$

The critical value is given in Appendix J as $T = 8$ (the 0.025 level and $n = 10$). The decision rule is reject H_0 if the test statistic is equal to or less than the critical value; therefore $H_0: M \leq 6.0$ is rejected.

The rationale of the Wilcoxon signed rank test is rather easily explained. Recall, by definition of the median, an observation is equally likely to be above or below the median value. And, if the distribution is symmetrical, any difference $(X_i - M)$ of a given magnitude, and therefore any rank, is equally likely to be positive as negative. For example, if $n = 6$ and $H_0: M = M_0$ is true, the sign attached to each rank $1, 2, \ldots, 6$ is equally likely to be positive as negative. One possible sample outcome is:

Rank:	1	2	3	4	5	6
Sign:	$+$	$+$	$+$	$+$	$+$	$+$

In this case, $T_- = 0$ and $T_+ = 21$; the probability of this result is $(\frac{1}{2})^6 = 0.0156$. Similarly, it can be shown $P(T_+ = 0) = 0.0156$. The probability of the smaller of T_+ and T_- equaling zero is then $2(0.0156) = 0.0312$. Hence, if testing at the 0.05 level and a T of zero is observed with $n = 6$, $H_0: M = M_0$ is rejected (or the p-value of 0.0312 may be cited). Note, of the $(2)^6 = 64$ possible ranking outcomes, only one yields a T_- of zero, that shown above. However, $T_- = 3$ could arise because rank 3 is negative, or because ranks 1 and 2 are negative. For samples not tabulated in Appendix J ($n > 25$), the distribution of both T_+ and T_- can be considered normal with

$$\mu_T = \frac{n(n+1)}{4}, \tag{15.2}$$

$$\sigma_T = \sqrt{\frac{n(n+1)(2n+1)}{24}}, \tag{15.3}$$

and

$$Z = \frac{T - \mu_T}{\sigma_T} \tag{15.4}$$

is the test statistic and is used as in previous chapters. As before, if $X_i - M_0 = 0$, it is dropped; and the midrank method is used for tied ranks.[1]

EXAMPLE. Continuing the previous illustration, a sample of *thirty* diabetes patients is obtained to test $H_0: M \leq 6.0$ versus $H_1: M > 6.0$. Computations similar to that of Table 15.3 yield no differences of zero and $T_- = 100$. Test H_0 at the 0.05 level.

Given $n = 30$, we compute μ_T and σ_T using Formulas 15.2 and 15.3.

$$\mu_T = \frac{n(n + 1)}{4} = \frac{30(31)}{4} = 232.5;$$

$$\sigma_T = \sqrt{\frac{n(n + 1)(2n + 1)}{24}}$$

$$= \sqrt{\frac{30(31)(61)}{24}} = 48.6.$$

Computing Z,

$$Z = \frac{T - \mu_T}{\sigma_T}$$

$$= \frac{100 - 232.5}{48.6} = -2.73.$$

For a one-tailed test at 0.05 level, the critical values of $Z = -1.65$. Hence $H_0: M \leq 6.0$ is rejected.

**Wilcoxon Test for Paired Samples

The analyst may, by design or accident, encounter data that *appear* to be the "ordinary" two-sample case as illustrated in Sections 9.5 and 9.6. This case assumes that the observations are independent both within and between the two samples. The data would typically be of the form:

Sample 1	Sample 2
$260(X_{11})$	$230(X_{12})$
$180(X_{21})$	$169(X_{22})$
$\vdots$	$\vdots$
$150(X_{n_1 1})$	$138(X_{n_2 2})$

Suppose, however, that the data represent a random sample of patients that were given a particular diet plan and X_{11} = weight of patient one before diet

---

[1] Two "refinements" are possible that somewhat improve the test when using the normal approximation: (1) T, in the absence of ties, can assume only integer values, and a correction for continuity is sometimes used, and (2) when there are many ties, an adjustment in the computation of the standard deviation will improve the approximation. The interested reader should consult Gibbons as cited in this chapter's references.

and X_{12} = weight of patient one after one month on the diet plan. Now a person's weight after the diet is *not* independent of his weight before dieting. For example, X_{11} = 260 pounds might be for a forty-year-old male and X_{21} = 180 could be the weight for a thirty-year-old female. When the observations are not independent between samples, the methods of Section 9.5 and 9.6 (or the following Section 15.5) are no longer appropriate.

The example above illustrates what is termed *paired or matched samples.* Such data arise in comparative experiments when the measurements are the "before and after" on the same experimental unit. Another example would be the job satisfaction scores of the same group of managers before and after a corporate reorganization. Similarly, the sales of a product in a certain group of stores could be analyzed before and after a special promotional campaign. Paired samples can also occur when the two measurements are made on different experimental units, but where the units are still matched in some way. For example, a study could involve identical twins with a measurement being obtained on each.

Given paired observations, one procedure for testing $H_0: \mu_1 - \mu_2 = 0$ or $H_0: M_1 - M_2 = 0$ is simply a variation on the Wilcoxon one-sample test of $H_0: M = M_0$. This procedure reduces the two related samples to a single sample by an analysis of the *differences*

$$D_i = X_{i1} - X_{i2}.$$

Then, we test $H_0: M_1 - M_2 = 0$ by testing $H_0: M_D = 0$, i.e., that the median of the population of differences is zero. The test assumes that the D_i are a random sample from a population of differences that is continuous and symmetric. After obtaining the D_i one proceeds exactly as in the previous one sample case. (Any D_i equal to zero is dropped, and nonzero ties are resolved with the midrank method.)

EXAMPLE. A marketing analyst at the Second National Bank is investigating male/female differences in check-writing activity for *joint accounts* at the bank. His random sample of eight joint accounts yields the following.

Account	Male-Signed Checks	Female-Signed Checks
1	5	14
2	12	8
3	8	10
4	15	10
5	17	6
6	4	14
7	6	3
8	3	19

(Note that because of the pairing by account, the observations are not independent between samples.) At the 0.05 level of significance test $H_0: M_1 - M_2 = 0$ versus $H_1: M_1 - M_2 \neq 0$.

A test of H_0 above is equivalent to testing $H_0: M_D = 0$. The computations required to obtain the test statistic are shown in Table 15.4. There is no $D_i - M_0$ column as in Table 15.3 since M_0 (the hypothesized median) equals 0.

Account	Male-Signed Checks	Female-Signed Checks	D_i	$\lvert D_i \rvert$	Rank	Sign
1	5	14	-9	9	5	$-$
2	12	8	4	4	3	$+$
3	8	10	-2	2	1	$-$
4	15	10	5	5	4	$+$
5	17	6	11	11	7	$+$
6	4	14	-10	10	6	$-$
7	6	3	3	3	2	$+$
8	3	19	-16	16	8	$-$

From Table 15.2, the test statistic is the smaller of T_+ and T_-.

$$T_+ = 3 + 4 + 7 + 2 = 16;$$
$$T_- = 5 + 1 + 6 + 8 = 20.$$

Note that $T_+ + T_- = n(n + 1)/2 = 36$, as required. The critical value is given in Appendix J as $T = 4$. The decision rule is to reject H_0 if the test statistic (in this case T_+) is equal to or less than the critical value. Therefore, H_0 is not rejected, and we conclude that there is no difference in male/female check-writing activity for joint accounts.

We finally note the rationale, large-sample approximation (Formulas 15.2–15.4), and the procedure for one-tailed tests for paired samples is the same as for the one-sample case. For example, if we desired to test $H_0: M_1 - M_2 \geq 0$, H_0 would be rejected if T_+ were too small (see Table 15.2).

15.5 MANN-WHITNEY-WILCOXON TEST

The previous section analyzed data where the measurements occur in pairs. Our attention is now turned to the case of two mutually independent samples, where each observation is independent of all other observations (both within and between samples). In Section 9.6, the t test of $H_0: \mu_1 - \mu_2 = 0$ was introduced. The assumptions of this test are that two mutually independent, random samples are drawn from normally distributed populations having equal variances. The nonparametric counterpart of the t test is the Mann-Whitney-Wilcoxon (MWW) test.[2] Although we shall state the hypotheses in terms of the medians, MWW is equally valid when the test is on means μ_1 and μ_2. While the MWW test assumes the populations have identical "shapes" and hence equal variances, it does *not* assume they are normally distributed; indeed, the test does not even assume symmetrical populations. Because of these less stringent assumptions and because the data need be only at the ordinal level,

---

[2] The MWW test may appear in various forms and under different names (e.g., the Mann-Whitney U test or the Wilcoxon rank sum test). These tests yield equivalent results when applied to the same data, but the student should not utilize another table of critical values without verifying that the test statistic is that defined here.

the MWW test may be employed when application of the t test would be inappropriate.

The test statistic, T_1, for MWW is quite easily obtained; the sample data consists of n_1 observations from the first population and n_2 observations from the second population.

1. The $n_1 + n_2$ observations are pooled and arrayed in *ascending* order.
2. Ranks are assigned to all observations in the array; the smallest value is given rank one and the largest is rank $n_1 + n_2$.
3. The midrank method is used in case of ties.
4. The test statistic, T_1, is simply the sum of the ranks assigned to the n_1 observations from population one.

An intuitive explanation of the test can be given without great difficulty. If $H_0: M_1 - M_2 = 0$ were true, the smallest observation (that assigned rank 1) is equally likely to have come from population one as from population two. Similarly, *any* given rank is equally likely from either population. Consider the following:

Observation's Population	Observation's Rank
1	1
1	2
1	3
1	4
2	5
2	6
2	7
2	8

Here, $T_1 = 1 + 2 + 3 + 4 = 10$, the *smallest possible* value of T_1 when $n_1 = 4$. Such an ordering would certainly not support the null hypothesis. Note the *largest* possible value of $T_1 = 5 + 6 + 7 + 8 = 26$. Intuitively we would expect T_1 to be "around" 18 (the average of 10 and 26). In fact, when H_0 is true, the expected value of T_1 is given by

$$E(T_1) = \frac{n_1(n_1 + n_2 + 1)}{2}. \tag{15.5}$$

What is the probability of obtaining $T_1 = 10$ when $n_1 = n_2 = 4$? There are $_8C_4 = 8!/4!4! = 70$ possible rankings that could be assigned to the eight observations, and if H_0 is true all 70 are equally likely. Only one, that shown above, would yield a $T_1 = 10$ (and none can yield a smaller value), hence $P(T_1 = 10) = P(T_1 \leq 10) = \frac{1}{70} = 0.014$. We later see in Appendix K that a value of $T_1 = 10$ would be significant.

Acceptance or rejection of the null hypothesis is ascertained by using Appendix K to find the appropriate critical value as shown in Table 15.5. The null hypothesis is rejected as also indicated in Table 15.5.

TABLE 15.5:
APPROPRIATE CRITICAL
VALUES AND DECISION
RULES FOR NULL
HYPOTHESES WITH
THE MANN-WHITNEY-
WILCOXON TEST; $E(T_1) =$
$n_1(n_1 + n_2 + 1)/2$

Null Hypothesis	Critical Values	Reject H_0 When
$H_0: M_1 - M_2 \geq 0$	T_L	$T_1 \leq T_L$
$H_0: M_1 - M_2 \leq 0$	T_U	$T_1 \geq T_U$
$H_0: M_1 - M_2 = 0$	T_L and T_U	$T_1 \leq T_L$ or $T_1 \geq T_U$

EXAMPLE. Given $n_1 = n_2 = 6$ and $T_1 = 26$, determine the critical value and accept or reject $H_0: M_1 - M_2 \geq 0$ at $\alpha = 0.05$.

The null hypothesis is in doubt when the ranks assigned to observations from population one tend to be small. Using Table 15.5 and Appendix K, the critical value is $T_L = 28$. I.e., the probability of obtaining a value of T_1 less than or equal to that observed is less than 0.05, and H_0 is rejected.

EXAMPLE. Given $n_1 = 4, n_2 = 9$ and $T_1 = 39$, determine the critical value and test $H_0: M_1 - M_2 = 0$ at the 0.05 level. The critical values from Table 15.5 and Appendix K are

$$T_L = 15 \quad \text{and} \quad T_U = 41,$$

and H_0 would *not* be rejected.

Appendix K is valid only for $n_1 \leq n_2$, i.e., T_1 must be computed for the *smaller* sample. If $n_1 > n_2$, could we still use Appendix K to obtain the critical values? Yes, since we can simply designate the smaller sample as having come from population one. A detailed example is now in order.

EXAMPLE. A marketing analyst at Amoko Oil is investigating gasoline buying patterns. In particular, she is interested in the quantity of gas bought when the customer elects "full service" versus "self-service." In the latter, the gas is pumped by the customer rather than by the service station attendant. Letting the full-service group be population one, test $H_0: M_1 - M_2 \geq 0$ versus $H_1: M_1 - M_2 < 0$ with the following data at the 0.05 level.

$ Purchase (full service):   1.00, 1.00, 3.00, 4.20, 6.75

$ Purchase (self-service):   2.00, 5.00, 7.00, 7.00, 9.30

The sample data are first pooled and then arrayed. Next, ranks are assigned, and the midrank method is used for ties. This procedure is shown in Table 15.6;

TABLE 15.6: REQUIRED
COMPUTATIONS FOR
EXAMPLE

Value	Rank	Population 1
1.00	1.5	x
1.00	1.5	x
2.00	3	
3.00	4	x
4.20	5	x
5.00	6	
6.75	7	x
7.00	8.5	
7.00	8.5	
9.30	10	

the last column is for convenience in identifying the ranks for computation of T_1. Summing to obtain T_1,

$$T_1 = 1.5 + 1.5 + 4 + 5 + 7 = 19.$$

From Table 15.5, the appropriate critical value is;

$$T_L = 19,$$

and H_0 is rejected. The sample data support the *alternate* hypothesis and indicate that the median purchase is *lower* for full service customers than for self-service customers.

For larger samples not tabulated in Appendix K, the sampling distribution of T_1 may be approximated by the normal distribution with parameters μ_{T_1} (which is the $E(T_1)$ given by Formula 15.5) and

$$\sigma_{T_1} = \sqrt{\frac{n_1 n_2 (n_1 + n_2 + 1)}{12}}. \tag{15.6}$$

As in the previous case, T_1 is computed for the *smaller* sample. The test statistic,

$$Z = \frac{T_1 - \mu_{T_1}}{\sigma_{T_1}}, \tag{15.7}$$

is employed in the "usual" manner.[3]

EXAMPLE. Continuing the Amoko Oil illustration, suppose a sample of $n_1 = 12$ and $n_2 = 15$ yields a $T_1 = 128$. Test $H_0: M_1 - M_2 \geq 0$ versus H_1: $M_1 - M_2 < 0$ at the 0.05 level.

Using Formulas (15.4)–(15.6), we compute

$$\mu_{T_1} = \frac{n_1 (n_1 + n_2 + 1)}{2}$$

$$= \frac{12(12 + 15 + 1)}{2} = 168;$$

$$\sigma_{T_1} = \sqrt{\frac{n_1 n_2 (n_1 + n_2 + 1)}{12}}$$

$$= \sqrt{\frac{(12)(15)(12 + 15 + 1)}{12}} = \sqrt{420} = 20.5,$$

and

$$Z = \frac{T_1 - \mu_{T_1}}{\sigma_{T_1}} = \frac{128 - 168}{20.5} = -1.95.$$

_____

[3] As noted in the large-sample approximation for the Wilcoxon signed rank test, a correction exists for (1) continuity and (2) numerous tied ranks. Similar corrections somewhat improve the normal approximation for the MWW test but shall not be presented here. Gibbons is again the recommended reference.

The critical value is -1.64, so the null hypothesis is rejected. (The p-value = 0.0256.) See Figure 15.1.

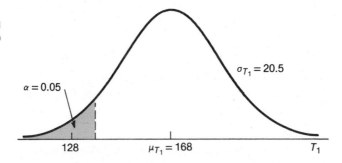

$\sigma_{T_1} = 20.5$

$\alpha = 0.05$

128     $\mu_{T_1} = 168$ T_1

15.6 KRUSKAL-WALLIS TEST

The Kruskal-Wallis test is the nonparametric counterpart of the F test (one-way ANOVA) of Chapter 11. As such, it is an extension of the Mann-Whitney-Wilcoxon test to more than two populations. The F test requires that the K populations each be normally distributed with equal variances. The Kruskal-Wallis test requires only that the K populations be continuous and have the same shape; the test remains completely valid for skewed population distributions. Further, Kruskal-Wallis can be used with ordinal data. Both tests, of course, require mutually independent random samples.

The hypotheses for the Kruskal-Wallis test are:

$$H_0: \quad M_1 = M_2 = \cdots = M_k;$$
$$H_1: \quad \text{not all } M_j \text{ are equal.}$$

The computation of the test statistic, H, follows a procedure very similar to in the Mann-Whitney-Wilcoxon test.

1. The $n_1 + n_2 + \cdots + n_k = n$ observations are pooled and arrayed in ascending order.

2. Ranks are assigned to all observations (smallest is rank 1); in case of ties, the midrank method is used.

3. The sum of the ranks assigned to each group, R_j, is computed; e.g., $R_1 =$ sum of the ranks assigned to the sample observations from population one.

4. The test statistic, H, is computed using

$$H = \frac{12}{n(n+1)} \left[\sum^k \frac{R_j^2}{n_j} \right] - 3(n+1). \tag{15.8}$$

For samples of even moderate size, the distribution of H is approximately chi-square with $(k-1)$ degrees of freedom. The approximation is satisfactory when each $n_j > 5$. For very small samples, some $n_j \le 5$, the exact distribution of H has been tabulated; e.g., see Siegel as cited in the references. We shall, however, confine our discussion to samples where the approximation holds. The decision rule will be: *Reject* the null hypothesis if the observed value of the test statistic *exceeds* the (right-tail) critical value of the chi-square distribution with $(k - 1)$ degrees of freedom.[4]

EXAMPLE. In another aspect of her study on gasoline-buying patterns, the marketing analyst at Amoko Oil is investigating differences by methods of payment. In this case, she wishes to determine whether there are differences in the median purchase (in dollars) when the method of payment is (1) Amoko credit card, (2) other credit cards, and (3) cash. Test:

$$H_0: \quad M_1 = M_2 = M_3$$
$$H_1: \quad \text{not all } M_j \text{ are equal}$$

at the 0.10 level of significance using the following data.

$ Purchase (Amoko Credit Card):   3.00, 11.50, 5.00, 7.00, 10.00, 9.40

$ Purchase (other credit cards):   4.00, 7.50, 9.50, 5.00, 8.20, 10.25

$ Purchase (cash):   5.00, 1.00, 2.00, 6.50, 4.50, 1.00

First, we arrange the observations in order of magnitude within each sample; this is not necessary but simplifies the assignment of ranks. Next, we assign ranks to all $n = 18$ observations considering them to be a single sample, and using the midrank method for ties. Then, we obtain the sum of the ranks for each sample (see Table 15.7).

TABLE 15.7:
COMPUTATION OF
THE R_j FOR THE
KRUSKAL-WALLIS TEST.

Amoko Credit Card		Other Credit Cards		Cash	
Observation	*Rank*	*Observation*	*Rank*	*Observation*	*Rank*
3.00	4	4.00	5	1.00	1.5
5.00	8	5.00	8	1.00	1.5
7.00	11	7.50	12	2.00	3
9.40	14	8.20	13	4.50	6
10.00	16	9.50	15	5.00	8
11.50	18	10.25	17	6.50	10
	$R_1 = 71$		$R_2 = 70$		$R_3 = 30$

Then, using Equation (15.8) to compute H:

$$H = \frac{12}{n(n + 1)} \left[\sum \frac{R_j^2}{n_j} \right] - 3(n + 1)$$

$$= \frac{12}{18(19)} \left[\frac{71^2}{6} + \frac{70^2}{6} + \frac{30^2}{6} \right] - 3(19)$$

$$= 63.40 - 57 = 6.40.$$

---

[4] When numerous ties are observed in the sample data, a correction exists to adjust the computed value of the test statistic, H. The correction usually has little effect and will not be presented here.

The critical value of the statistic is that of chi-square with $k - 1 = 3 - 1 = 2$ degrees of freedom; at the 0.10 level from Appendix H,

$$\chi^2_{2,\,0.10} = 4.61$$

The observed value of 6.40 exceeds the critical value, and H_0 is rejected. The data indicate significant differences in the median purchase of gasoline among the three methods of payments.

The rationale of the Kruskal-Wallis test is best explained by examining an alternate form of Equation (15.8):

$$H = \frac{12}{n(n + 1)} \sum \frac{[R_j - n_j(n + 1)/2]^2}{n_j}. \tag{15.9}$$

Under H_0, it can be shown that $E(R_j) = n_j(n + 1)/2$, the second term in brackets of Equation (15.9). When the ranks assigned to a sample from one of the populations tend to be all high or all low (as in the cash group, population three, of the example), the difference between the observed and expected R_j will be large. Since the differences in observed R_j and expected R_j are *squared*, large values of H indicate large discrepancies between the R_j's and those expected under H_0. Thus, *large* values of H (the one-tailed rejection region) suggest rejection of the null hypothesis.

15.7 SPEARMAN'S RANK CORRELATION COEFFICIENT

Questions concerning independence, or the existence of a relationship between variables, have been addressed in various sections of this text. For example, in Chapter 10 we studied the use of the chi-square distribution to test if two variables of classification were independent. In Chapter 12, we used regression analysis to test for a linear relationship between X and Y. This section introduces a method of measuring and testing the degree of association between the two variables measured at the ordinal level: the Spearman rank correlation coefficient. As before, the assumptions associated with application of the rank correlation coefficient are less stringent than those of its parametric counterpart. Since many variables of interest in business applications are not measured above the ordinal level (e.g., equality of employment references, individuals' overall health, prestige of law firms, etc.), rank correlation is a very useful technique.

In computing the Spearman rank correlation coefficient, r_s, the data consist of n observations with a pair of measurements (X_i, Y_i) for each observation.

1. If the measurements for a variable are at the interval or ratio scale, the data are converted to ranks. The midrank method is used for ties.

2. Let U_i denote the rank of X_i and V_i denote the rank of Y_i for the ith observation.

3. Denote the difference between ranks for the ith observation as

$$D_i = U_i - V_i. \tag{15.10}$$

4. Compute the rank correlation coefficient:

$$r_s = 1 - \frac{6 \sum D_i^2}{n(n^2 - 1)} \tag{15.11}$$

When the data appear as:

$$\text{Observation:} \quad 1 \quad 2 \quad 3 \quad 4$$
$$\text{X rank:} \qquad 1 \quad 2 \quad 3 \quad 4$$
$$\text{Y rank:} \qquad 1 \quad 2 \quad 3 \quad 4$$

there is complete agreement in the rankings. In the above case $D_i = 0$ for all i, and, in Formula (15.11) with $\sum D_i^2 = 0$, we see $r_s = 1$. If the Y rankings for the above had been:

$$Y \text{rank:} \quad 4 \quad 3 \quad 2 \quad 1,$$

there is complete disagreement. In this case, $\sum D_i^2$ would assume its *maximum* value of $n(n^2 - 1)/3$ and $r_s = -1$. (The student should verify this result for the data above.) Thus, when $|r_s|$ is near one, an association between X and Y is suggested. When $|r_s|$ is near zero, independence or no association between X and Y is suggested.

EXAMPLE. A sales manager for a life insurance company has ranked ten of the company's sales representatives according to their physical attractiveness; he considered neatness, attire, "good looks," etc. in developing his rankings. Rank 1 was assigned to the individual considered least attractive. He then obtained the sales volume (Y) achieved by each in 1979. Compute the rank correlation coefficient for the data given below.

TABLE 15.8: DATA AND REQUIRED COMPUTATIONS FOR EXAMPLE

Sales Representative	Attractiveness $X_i = U_i$	Sales ($ thousands), Y_i	Sales Rank V_i	$D_i = U_i - V_i$	D_i^2
1	7	175	2	5	25
2	2	190	3	−1	1
3	4	250	6	−2	4
4	9	150	1	8	64
5	3	300	8	−5	25
6	10	270	7	3	9
7	6	210	4	2	4
8	1	310	9	−8	64
9	5	400	10	−5	25
10	8	240	5	3	9
					$\sum D_i^2 = \overline{230}$

The sample data are given in the first three columns of Table 15.8. Since Y was not given in ranks, the fourth column is required. The last two columns illustrate the computation of $\sum D_i^2$. Then substituting in Equation (15.11):

$$r_s = 1 - \frac{6 \sum D_i^2}{n(n^2 - 1)}$$

$$= 1 - \frac{6(230)}{10(10^2 - 1)}$$

$$= 1 - \frac{1380}{990} = 1 - 1.39 = -0.39$$

Note that if one of the rankings were reversed (say rank 1 were most attractive), the *sign* but not the absolute value of r_s would be changed.

While the value of the rank correlation coefficient has utility as a descriptive measure of association between the two sets of ranks in the sample, the analyst may wish to test a hypothesis about the association of X and Y in the population. We shall illustrate a test of:

$$H_0: \quad X \text{ and } Y \text{ are not associated;}$$
$$H_1: \quad X \text{ and } Y \text{ are associated.}$$

For $n \geq 10$, the test statistic is:

$$t = \frac{r_s}{\sqrt{(1 - r_s^2)/(n - 2)}}, \tag{15.12}$$

which, under H_0, is t distributed with $(n - 2)$ degrees of freedom. For small n, the above is invalid, and tables of the sampling distribution of r_s must be consulted.[5]

EXAMPLE. Using the previous example's results,

$$n = 10, \qquad r_s = -0.39,$$

test H_0: X and Y are not associated at the 0.10 level of significance.

Substituting in Equation (15.11),

$$t = \frac{r_s}{\sqrt{(1 - r_s^2)/(n - 2)}}$$

$$= \frac{-0.39}{\sqrt{(1 - (-.39)^2)/8}}$$

$$= \frac{-0.39}{\sqrt{0.106}} = \frac{-0.39}{0.326} = -1.20$$

With $(n - 2) = 8$ degrees of freedom, the critical values of $t = \pm 1.86$, and H_0 is not rejected. The observed value of r_s is *not* large enough to indicate an association between attractiveness and sales performance.

15.8 SUMMARY

This chapter introduced a number of nonparametric methods with useful applications to problems in business. Nonparametric methods are also referred to as distribution-free methods or "weak assumption" statistics. As we have seen, the latter is perhaps the most descriptive name. For every technique in

---

[5] When $n < 10$, the reader could consult Gibbons for the appropriate tables. Gibbons also reviews (1) the one-tailed test concerning the rank correlation coefficient and (2) a correction for ties; we shall not pursue these matters here.

this chapter, the assumptions for valid application were weaker than those required for the parametric counterpart discussed in earlier chapters. Despite the milder assumptions, we saw in Table 15.1 that the nonparametric techniques were all quite efficient when compared to their parametric counterparts. As a final bonus, one may. observe that the computations associated with non-parametric methods are simple and easily done. (For example, compare the computations of Kruskal-Wallis with those of one-way ANOVA.)

After a review of the concepts of measurement, rank data were discussed, and the midrank method for handling ties was given. Recall that all of the procedure use rank data for the computation of the test statistic. The Wilcoxon signed-rank test was used to test hypotheses about a population median (or mean) and for tests with paired data. The Mann-Whitney-Wilcoxon test (MWW) was used in the two-independent-samples case to compare medians. The Kruskal-Wallis test was an extension of MWW to the k-sample case. Finally, Spearman's rank correlation coefficient, r_s, was introduced.

This chapter serves as an introduction to nonparametrics; many other useful procedures have been developed. For example, the Friedman test is a non-parametric alternative to the two-way ANOVA (randomized block design) of Chapter 11. Other useful procedures, such as the runs test, have no parametric equivalent. For a further discussion, consult one of the texts devoted wholly to nonparametrics; Gibbons is especially recommended.

15.9 TERMS AND FORMULAS

Levels of measurement (nominal, ordinal, interval, and ratio)

Asymptotic relative efficiency (ARE)

Midrank method

Wilcoxon signed rank test (test statistic: T_+ or T_-)

Mann-Whitney-Wilcoxon test (test statistic: T_1)

Kruskal-Wallis test (test statistic: H)

Spearman rank correlation coefficient (r_s)

Sum of integers $1, 2, \ldots, n$ (rank sum)	$\dfrac{n(n + 1)}{2}$
Expected value of T	$\mu_T = \dfrac{n(n + 1)}{4}$
Standard deviation of T	$\sigma_T = \sqrt{\dfrac{n(n + 1)(2n + 1)}{24}}$
Expected value of T_1	$E(T_1) = \dfrac{n_1(n_1 + n_2 + 1)}{2}$
Standard deviation of T_1	$\sigma_{T_1} = \sqrt{\dfrac{n_1 n_2(n_1 + n_2 + 1)}{12}}$

Kruskal-Wallis test statistic

$$H = \frac{12}{n(n+1)} \left[\sum \frac{R_j^2}{n_j} \right] - 3(n+1)$$

Spearman rank correlation coefficient

$$r_s = 1 - \frac{6\sum D_i^2}{n(n^2-1)}$$

Rank correlation test statistic

$$t = \frac{r_s}{\sqrt{(1-r_s^2)/(n-2)}}$$

QUESTIONS

A. True or False

1. Measurements on an interval scale may be transformed into measurements on the ordinal level.

2. Measurements on a nominal scale may be transformed into measurements on the ratio scale.

3. Only large values of H in the Kruskal-Wallis test discredit the null hypothesis.

4. A fixed, inherently defined zero is a property of the interval scale.

5. The parametric counterpart to the Mann-Whitney-Wilcoxon test is simple linear regression.

6. A nonparametric test is always less efficient than its parametric counterpart.

7. If the midrank method is used for ties, the sum of the ranks is no longer equal to $n(n+1)/2$.

8. The Wilcoxon signed rank test assumes the population is symmetrically distributed.

9. The Kruskall-Wallis test assumes the k populations have the same shape.

10. The test statistic for MWW, T_1, is computed for the smaller sample.

B. Fill in

1. If a tie is observed for ranks 10, 11, and 12, the midrank method would assign a rank of _____ to each of the three observations.

2. Temperature is an example of measurement at the _____ level.

3. The parametric counterpart to the Kruskal-Wallis test is _____ _____.

4. With large n_j, the distribution of H (the Kruskall-Wallis test statistic) is approximately _____.

5. The sum of integers 1, 2, 3, ..., 39, and 40 is _____. Hint: Use formula (15.1).

6. In the MWW test with $n_1 = 5$ and $n_2 = 6$, the smallest possible value of $T_1 = $ _____.

7. For large samples, the sampling distribution of T_1 may be approximated by the _____ distribution.

8. If the X and Y ranks are in complete disagreement, $r_s =$ _____.

9. Married, never married, or divorced is an example of a measurement using the _____ scale.

10. In the Wilcoxon signed rank test, any observation with $X_i - M_0 = 0$ is _____.

EXERCISES

1. Use the midrank method to assign ranks when there are ties for:
   a. Ranks 8 and 9.
   b. Ranks 14, 15, and 16.
   c. Ranks 6, 7, 8, and 9.

2. Determine the critical value(s) of the test statistic for the Wilcoxon signed rank test where
   a. $H_0: M = M_0$, $n = 20$, and $\alpha = 0.05$.
   b. $H_0: M \geq M_0$, $n = 17$, and $\alpha = 0.01$.
   c. $H_0: M \leq M_0$, $n = 13$, and $\alpha = 0.025$.

3. Determine the critical value(s) of the test statistic for the Mann-Whitney-Wilcoxon test where
   a. $H_0: M_1 - M_2 \geq 0$, $n_1 = 9$, $n_2 = 10$, and $\alpha = 0.025$.
   b. $H_0: M_1 - M_2 = 0$, $n_1 = 4$, $n_2 = 6$, and $\alpha = 0.10$.
   c. $H_0: M_1 - M_2 \leq 0$, $n_1 = 5$, $n_2 = 7$, and $\alpha = 0.05$.

4. Several complaints have been directed to a hospital administrator concerning the operation of the emergency room. The major complaint concerns the waiting time before seeing a physician. The administrator pulls a sample of records and observes the following lengths of time (in hours) from registration to initiation of care by a physician: 1.1, 1.7, 2.4, 0.2, 3.0, 0.9, 1.5, 1.6, 2.4, and 0.1. Test $H_0: M \leq 1.0$ with the Wilcoxon signed rank test at the 0.025 level.

5. A random sample of eight students in a large freshman chemistry class had the following grades on the midterm exam: 91, 73, 62, 88, 81, 64, 72, 79. Test the hypothesis that the median score for the class is 75. Use the Wilcoxon signed rank test at the 0.05 level of significance.

6. A campus recruiter, extolling the virtues of his company, says that a new graduate could expect to double his salary in five years. A sample of graduates hired five years ago showed their salaries had multiplied by the following amounts: 1.8, 2.9, 1.4, 1.9, 2.3, 1.4, 1.7, 2.5, and 2.0. Does this evidence support the recruiter's claim? Use the Wilcoxon signed rank test with $\alpha = 0.025$.

7. Klean, a laundry detergent, recently had a change in packaging and product formula. These changes were accompanied by a special promotional campaign. The data below represent sales in ten stores before and after the changes.

Store	Sales before Changes	Sales after Changes
1	10	14
2	17	25
3	26	22
4	29	25
5	13	20
6	19	29
7	22	22
8	31	34
9	16	17
10	20	29

Use the Wilcoxon test for paired sample data to determine if the changes led to increased sales. Let $\alpha = 0.025$.

8. A company psychologist gave a test to measure anxiety to the employee and spouse of families about to be transferred to a new location. The scores for eight families are given below (higher scores indicate greater anxiety):

Family	Employee	Spouse
1	41	50
2	30	25
3	54	48
4	49	55
5	36	25
6	20	22
7	30	35
8	40	37

Do the data indicate any difference in the level of anxiety between employee and spouse prior to the relocation? Use the Wilcoxon test for paired samples at the 0.05 level.

9. A large operator of a taxi service has begun replacing the original equipment tires (bias-belted) with radial tires. Radials, among other claimed benefits, are supposed to increase gas mileage. The following data has been collected for nine taxis that were changed to radials:

Taxi	Miles per Gallon with Bias Tires	Miles per Gallon with Radial Tires
1	8.2	9.1
2	12.0	12.4
3	13.0	12.9
4	10.1	10.4
5	9.6	9.8
6	13.9	14.5
7	11.2	11.2
8	9.4	10.3
9	12.3	12.7

Does it appear that gas mileage is increased with radial tires? Use the Wilcoxon test for paired samples at the 0.025 level of significance.

10. A toy manufacturer has developed two sets of assembly instructions for its new wagon. (There are alternative satisfactory ways to put the wagon together.) In order to determine which set is most easily followed, the manufacturer gives the wagon to 20 second-grade children, with the parents understanding that assembly time must be recorded. Half the wagons have Set A instructions, the other half Set B. The following assembly times (in minutes) are reported.

$T_1 = 72$
$T_o = 138$

Reject

Set A	Set B
15	34
32	48
27	22
50	28
22	36
31	39
28	53
18	51
30	46
25	70

Do the data indicate a significant difference in assembly times? Use Mann-Whitney-Wilcoxon and the 0.05 level of significance.

11. In Problem 10, suppose in another experiment with a first-grade class of 30 children, 14 received wagons with Set A instructions and 16 with Set B. Given that $T_1 = 152$, test again the hypothesis of Problem 10 at the 0.05 level.

12. Initial starting salaries for a random sample of recent graduates in accounting and engineering were obtained, then pooled and ranked as below (rank one is highest salary):

$T_1 = 53$
$H_o = $ accept

Accounting	Engineering
2	1
5	3
6	4
8	7
9	11
10	12
13	14

Is there a significant difference in accounting and engineering starting salaries? Use MWW and the 0.05 level of significance.

13. Professor Jones, a jogging enthusiast, believes that the sport is helpful in schoolwork by promoting both physical and mental health. In particular, he thinks joggers have a higher QPA than nonjoggers. A random sample

of students in the Business School were asked if they jogged regularly; their responses and their QPA from school files are tabulated as:

Nonjoggers' QPA	Joggers' QPA
1.60	2.02
2.00	1.43
1.12	1.56
1.25	1.83
2.40	2.45
1.36	
1.42	

Use the Mann-Whitney-Wilcoxon test to determine if joggers' QPA is significantly greater than nonjoggers'. Test at the 0.05 level.

14. The vice president of human resources at Omni, Inc., administered a job-satisfaction test to random sample of middle managers in the Sales, Manufacturing, and Research and Development divisions of the company. He obtained the following scores (higher scores indicate greater job satisfaction):

Sales	Manufacturing	R&D
60	51	65
70	47	83
55	63	77
43	74	78
74	56	69
82	71	80
	62	

Do the data indicate any difference in job satisfaction among the three divisions? Use the Kruskal-Wallis test at the 0.05 level of significance.

15. A major motel chain annually buys hundreds of color televisions. Last year the firm bought several sets from three major manufacturers and had detailed records kept on their repair costs over the year for a sample of each brand. The following data were obtained.

Repair Costs in Dollars		
Brand A	Brand B	Brand C
$8	$10	$14
25	17	28
15	43	10
7	29	10
22	37	34
30	42	25
	35	

Use the Kruskal-Wallis test to determine whether there is a significant difference in repair costs among the three brands. Let $\alpha = 0.05$.

16. A large steel mill operates three shifts of workers; the plant manager believes there is a significant difference in output among the shifts. A sample of production reports for each shift was obtained, and the following data are from the pooled rankings (rank one assigned to highest production).

	Shift 1	Shift 2	Shift 3
	$n_1 = 10$	$n_2 = 10$	$n_3 = 10$
	$R_1 = 150$	$R_2 = 110$	$R_3 = 205$

$H = 5.86$
Reject

Test the hypothesis of no difference in production levels among the three shifts: $H_0: M_1 = M_2 = M_3$. Let $\alpha = 0.10$.

17. Given a rank correlation coefficient of $r_s = -0.19$ and $n = 14$, test the hypothesis of no association between the two sets of ranks at the 0.01 level.

18. Julius Feast, a well-known wine connoiseur, recently published his rankings of ten California red wines in the "moderate" price range. The following gives his rankings and the recent price per bottle.

$T_c = 1.75$
accept

Wine	Feast's Rank	Price
A	4	$5.25
B	7	6.00
C	10	7.00
D	1	8.50
E	2	8.00
F	5	6.25
G	9	5.10
H	8	5.40
I	3	6.75
J	6	7.20

Compute the rank correlation coefficient for the data above. Test the hypothesis of no association at the 0.05 level.

19. A number of accounting majors were recently interviewed by one of the "big eight" firms. The following represents the personnel director's ranking of the candidates and his assessment of the quality of the candidate's college (rank one is best):

Candidate	Interview Rank	Rank of College
1	10	6
2	3	7.5
3	2	3
4	9	9
5	5	1
6	6	7.5
7	1	2
8	4	4.5
9	8	10
10	7	4.5

Compute the rank correlation coefficient and test the hypothesis of no association at the 0.05 level.

REFERENCES

Bradley, James "Nonparametric Statistics." In Kirk, Roger E., ed., *Statistical Issues: A Reader for the Behavioral Sciences.* Belmont, Cal.: Brooks/Cole Publishing Company, 1972, pp. 329–38.

Bradley, Ralph A. and Hollander, Myles "Frank Wilcoxon." In Kruskal, William H., and Judith M. Tanur, eds., *International Encyclopedia of Statistics.* New York: The Free Press, 1978, pp. 1245–50.

Daniel, Wayne W. *Applied Nonparametric Statistics.* Boston: Houghton Mifflin Company, 1978.

Gibbons, Jean Dickinson *Nonparametric Methods for Quantitative Analysis.* New York: Holt, Rinehart & Winston, 1976.

Noether, Gottfried E. *Introduction to Statistics: A Nonparametric Approach.* Boston: Houghton Mifflin Company, 1976.

Siegel, Sidney *Nonparametric Statistics for the Behavioral Sciences.* New York: McGraw-Hill Book Company, 1956.

Statistics in Business

"The time has come," the Walrus said,
"To talk of many things:
Of shoes—and ships—and sealing wax—
Of cabbages—and kings—
And why the sea is boiling hot—
And whether pigs have wings."

Lewis Carroll

16.1 CONCEPTS REVISITED

The basic concepts of statistics are presented in the preceding chapters. A brief overview of these concepts is given in this section of the current chapter. Then a look at the practice of statistics in today's business world is provided. These selected applications of statistics give further insight into the ways in which statistical information and procedures are currently used to assist in business decision making.

Statistical methods are used for description and inference. Characteristics of data are described by computed measures such as the arithmetic mean, median, and standard deviation. Various types of tables and charts are used to summarize and present data. For situations characterized by uncertain events, probability may be used to describe the degree of uncertainty. In addition, probability provides the foundation for statistical inference. The two branches of statistical inference are estimation and testing. Estimation procedures provide ways for predicting the value of some population parameter, such as the mean, based on sample information. Testing provides information for decisions by comparing sample occurrences with hypotheses. The p-value of a hypothesis test provides the probability of certain sample results based on the assumption that the null hypothesis is true. When sample results are surprising, we must question the validity of the null hypothesis. Concepts such as regression analysis are extensions of the basic descriptive and inference principles. A regression equation is descriptive when used to "describe" the relationship between variables. If the regression equation is used to obtain predicted values of the dependent variable, then the regression analysis is inferential. Both point and interval estimates of the dependent variable are possible; testing is used to determine the significance of the relationship.

As in any field, an introduction provides a look only at the surface. The surface of statistics is provided throughout Chapters 1–15. The statistician must have a much deeper knowledge of these concepts, as well as those more advanced. Managers must be able to understand the basic concepts in order to communicate with quantitative specialists, such as statisticians, and to effectively use the kind of statistical information these specialists provide. With this brief review, we now turn to selected applications of these concepts. In the words of Lewis Carroll, "The time has come ... to talk of many things ...." Specifically, we wish to provide a flavor for the application of statistics in business today.

16.2 PERSONNEL MANAGEMENT

Personnel directors and managers have witnessed a tremendous change in their duties and responsibilities during the last decade. In earlier years, personnel decisions such as employee selection, performance appraisal, and salary and promotion recommendations were made on the basis of the internal policies of a company or government office.

These practices were considered, for the most part, the private affairs of the organizations. But the Civil Rights Act of 1964 and its revision in 1972 changed all that. Now federal agencies such as the Department of Labor and the Equal Employment Opportunity Commission (EEOC) issue guidelines that must be adhered to for many personnel decisions. Compliance with the various regulations has increased the volume of data that must be recorded and analyzed. Personnel testing programs are closely scrutinized for fairness and validity. Even though statistical analysis has been used for test validation for some time, compliance regulations have markedly increased the need for this analysis. Two illustrations follow that demonstrate the importance of statistical concepts in the personnel management function.

Personnel Information Systems

Shortly after computers became standard office equipment, many organizations found that an endless stream of reports were being generated for which few, if any, individuals had any use. Personnel managers could obtain foot-high stacks of printouts listing all employees by salary, job classification, etc. Such reports are helpful, but they are far from adequate for reporting the type of information often required by EEOC and other guidelines. Cross-tabulation reports of employees by sex, race, age, and salary are typical requirements. Rather than reacting to demands by compliance agencies, the effective personnel administrator spends time designing reports that not only meet federal guidelines but alert management to patterns, trends, and exceptions on employee data. These not only alert management to compliance difficulties prior to investigation, but allow management to insure that internal policies and objectives are being met.

Some examples of frequency distributions, histograms, and scatter diagrams that provide personnel information are given in Figures 16.1–16.5.[1] Figure 16.1 provides frequency distributions, actual and relative, of a company's workforce cross-classified by race and EEO job category. A glance at this shows, for example, that 13.33 percent of the 45 managers are Hispanic Americans and also that 16.89 percent of the entire workforce of 444 is of this race. Thus, the

---

[1] From Carlton W. Dukes, "The EEO Report: Make Your Computer Work for You," *Management World*, vol. 6, no. 2 (December 1977), pp. 13–16. Figures 16.1 through 16.5 are reproduced with permission.

FIGURE 16.1

EEO Status
Percent of Workforce by Job Level

EEOCODE			RACE			
	WHITE NON-HISPANIC	BLACK NON-HISPANIC	ASIAN OR PACIFIC ISLANDER	HISPANIC AMERICAN	NATIVE AMERICAN (ALASKAN)	TOTALS
MANAGERS	19 42.22%	10 22.22%	8 17.78%	6 13.33%	2 4.44%	45 100.00%
PROFESSIONAL	32 46.38%	15 21.74%	11 15.94%	8 11.59%	3 4.35%	69 100.00%
TECHNICAL	28 36.36%	19 24.68%	9 11.69%	14 18.18%	7 9.09%	77 100.00%
SALESWORK	41 43.16%	25 26.32%	12 12.63%	15 15.79%	2 2.11%	95 100.00%
CLERICAL	44 37.29%	30 25.42%	14 11.86%	22 18.64%	8 6.78%	118 100.00%
SKILLCRAFT	1 8.33%	5 41.67%	2 16.67%	4 33.33%	0 0.00%	12 100.00%
OPERATORS	10 52.63%	5 26.32%	1 5.26%	3 15.79%	0 0.00%	19 100.00%
UNSKILLED	0 0.00%	2 40.00%	1 20.00%	2 40.00%	0 0.00%	5 100.00%
SERVICEWORK	0 0.00%	1 25.00%	1 25.00%	1 25.00%	1 25.00%	4 100.00%
TOTALS	175 39.41%	112 25.23%	59 13.29%	75 16.89%	23 5.18%	444 100.00%

proportion of managers who are Hispanic Americans is not substantially different from the proportion of the workforce made up of this race. Figure 16.2 is the frequency distribution of employees by race from the last row of Figure 16.1. Asterisks are used to make a histogram of the frequency distribution, which allows a visual comparison of employees by race. In Figure 16.3, the payrate totals for each race are shown. Looking at Figures 16.2 and 16.3 together shows, for example, that Asian or Pacific Islanders make up 13.29 percent of the workforce, and their payrate is 14.84 percent of the total.

FIGURE 16.2

Distribution of Workforce

RACE	FREQUENCY		* = 5
WHITE NON-HISPANIC	175 39.41%	.***********************************	
BLACK NON-HISPANIC	112 25.23%	.**********************	
ASIAN OR PACIFIC ISLANDER	59 13.29%	.************	
HISPANIC AMERICAN	75 16.89%	.***************	
NATIVE AMERICAN (ALASKAN)	23 5.18%	.*****	
TOTAL	444		

FIGURE 16.3

Distribution of Salaries

```
        RACE      PAYRATE                                              *= 5000

        WHITE
        NON-      149071.0000   .**************************-*
        HISPANIC       39.70%

        BLACK
        NON-       88326.0000   .******************
        HISPANIC       23.52%

        ASIAN OR
        PACIFIC    55712.0000   .***********
        ISLANDER       14.84%

        HISPANIC   60529.0000   .************
        AMERICAN       16.12%

        NATIVE
        AMERICAN   21824.0000   .****
        (ALASKAN)       5.81%

        TOTAL     375462.0000
```

Monthly salaries are plotted against EEO job categories in Figure 16.4. The job categories 1–9 are specified, in order, in Figure 16.1. This shows, for example, that eight clerks (Category 5) make as much as four of the managers (Category 1). Even one unskilled worker (Category 8) makes as much as these four managers. This type of chart can be used to make sure that salaries are in line with job classifications and to identify exceptions. Exceptions can be investigated to determine whether there are valid reasons for them. Monthly salaries are plotted against age in Figure 16.5. The mean age and monthly salary are plotted as horizontal and vertical lines. Using the concept of a normal distribution, rectangles represent the number of standard deviations away from the mean. This can be used to identify exceptional wages based on age; for example, those more than one standard deviation away from the mean. Other types of tables and charts could be used for statistical description of employee characteristics. The efficient personnel manager will develop the kind of procedures that best serves the organization's needs.

FIGURE 16.4

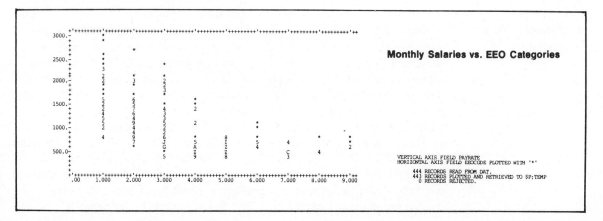

Monthly Salaries vs. EEO Categories

FIGURE 16.5

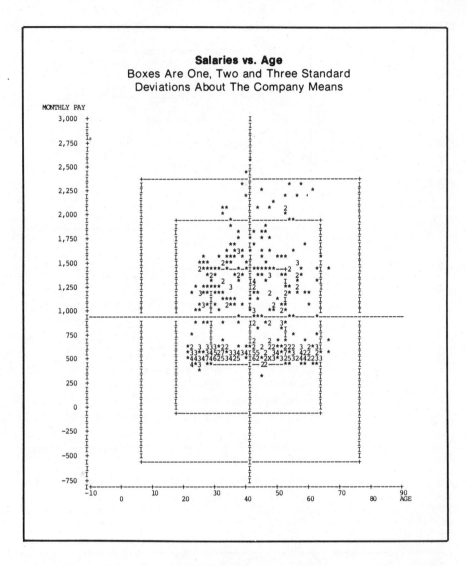

Employment Test
Validation

Employment tests for certain jobs are standard procedure in many organizations. Aptitude tests for various skills are most common but some organizations use tests as a part of their employee selection procedure for professional occupations. And of course, colleges and universities require tests as part of their entrance requirements. The object of such tests is to predict future performance as an employee or student. Aptitude test scores are used to predict productivity; SAT scores are used to predict grade point averages. Predictions are made on the basis of regression equations. First, tests are constructed to measure special skills required on the job. Before actual use for employee selection, the tests must be validated. Current employees can be used as a data base for performance and test scores. A regression equation is constructed from these data and verified for statistical significance. The test must also be validated for fairness to different groups of applicants, for example, different race classifications such as minority and nonminority. The test may be used for employee selection once it has been demonstrated as racially unbiased and a valid predictor of performance.

As an illustration, Feild et al. examined some employment tests for potential use in selecting production workers engaged in the construction of boxboard containers.[2] The purpose of the study was to select from several available tests those that were racially unbiased and valid predictors of good employees for identical production activities in boxboard container construction. Two geographically distant manufacturing plants of a large, national paper company were chosen for the study. Job analyses for the activities performed by the workers were combined with interviews of supervisors to determine the essential job skills required. The essential skills were: (a) taking measurements accurate to 16ths of an inch; (b) comprehending specifications for boxboard construction from computer printouts; (c) using basic shop arithmetic; and (d) accepting and following oral instructions. Four tests were examined for measuring the skills: (1) a specially developed Rule Measurements Test, which required the use of a ruler for taking various measurements of a three-dimensional figure; (2) a specially developed Specifications Sheet Test to measure reading and comprehension of computer printouts of box specifications; (3) a commercially available Numerical Test of basic arithmetic calculations and reading of charts and diagrams; and (4) a commercially available Oral Directions Test of ability to comprehend spoken instructions.

The company wished to know which, if any, of these tests could be used as predictors of two criteria, performance and productivity, for both minority and nonminority employees. Performance was measured by a subjective rating assigned by the supervisor. An eight-point, verbally defined rating scale was used to rate each employee on six characteristics: (a) speed of work; (b) quality of work; (c) care in handling company property; (d) ability to handle different jobs; (e) ability to work without supervision; and (f) attitude toward supervisor. Employee productivity was measured by the number of boxboard containers constructed in conformity with minimum quality standards. Data on performance, productivity, and results of the four tests were collected for 100 employees. The minority members of the group consisted of 42 blacks and 10 Mexican-Americans. The rest of the group consisted of 48 white employees.

Tests for differences in mean age and education of the two groups revealed no statistically significant differences. Differences in job tenure were significant. There were no significant differences between the mean scores of the two groups for performance rating and productivity. For the four predictor tests, no significant differences between mean scores for the two groups were found on the Rule Measurements and Specifications Sheet tests. Mean scores were significantly different for the two commercially prepared tests. Individual tests

[2] Hubert S. Feild, Gerald A. Bayley, and Susan M. Bayley, "Employment Test Validation for Minority and Nonminority Production Workers," *Personnel Psychology*, vol. 30, no. 1 (Spring 1977), pp. 37–46.

of significance of the relationship of each predictor (test) to each criterion (performance and productivity) were performed. All predictors except the Oral Directions test were significantly related to each criterion.

Since mean scores between minorities and nonminorities on the two criteria and the four predictors were not significantly different, the criteria and tests are not considered racially biased. Thus, the tests can be used without regard to race. With this in mind, the two groups were combined and stepwise multiple regression was used to select the best predictors of the two criteria. The Rule Measurements and Specifications tests were significant predictors for both criteria; the other tests were not. To simplify the use of the tests in employment offices, the scores of the two tests were summed and regressed with each criteria. This composite predictor proved a significant predictor (p-value < 0.001) of both performance and productivity.

The statistical investigation revealed no racial bias in the tests and established a composite score, the sum of two test scores, as a valid predictor of performance and productivity. This validates the company's use of the tests for selection of future employees. The statistical investigations in this study are typical of those performed in employee selection tests to determine whether they are racially unbiased and valid predictors of successful employees. The major statistical procedures used are hypothesis tests for differences in means and regression analysis.

16.3 MARKETING

Extensive statistical analysis is frequently used today to provide information for marketing decisions. This includes advertising and promotional strategies, sales forecasting, new product decisions, pricing and product mix decisions, and consumer behavior studies. Some examples follow.

Marketing Strategy

Successful marketing strategy depends on an understanding of the major factors contributing to sales. Funk et al., investigated the development of effective marketing programs for beef in retail food chains.[3] A retail demand function for beef was established from data collected in Toronto food chains each week from January 1974 through May 1975. The demand function established through multiple regression analysis was designed to predict weekly beef sales for each chain. The independent variables considered in the analysis were the chain's beef prices, competitors' beef prices, the chain's prices of substitute items (pork, chicken, veal, and lamb), competitors' prices of substitute items, the chain's local newspaper advertising for all meats, the competitors' local newspaper advertising for all meats, and some seasonal factors. The significant predictors of a chain's weekly beef sales were the chain's beef prices, prices of substitute items, and advertising. Advertising and pricing of competitors were not significant.

---

[3] T. F. Funk, Karl D. Meilke, and H. B. Huff, "Effects of Retail Pricing and Advertising on Fresh Beef Sales," *American Journal of Agricultural Economics*, vol. 59, no. 3 (August 1977), pp. 533–37.

The study indicates that aggregate beef sales for a chain are price elastic, implying that a chain may increase total revenue by lowering prices, provided other factors remain unchanged. Advertising elasticity was positive but less elastic than price. Thus, some trade-offs between price and advertising may be possible, depending on the costs of each. Pork advertising had a negative effect on beef sales, while other meat advertising had a positive effect.

This study illustrates how store managers can conduct demand studies to assist in the selection of basic marketing strategy. The results are useful in determining the effect of price and advertising changes on sales. An understanding of this relationship can assist in increasing sales and profits.

Consumer Behavior

Marketing researchers spend considerable effort in trying to understand how consumers react to various kinds of advertising. Some advertisements rely on humor, some on a "scientific" image, while others emphasize either a masculine, macho image or a feminine image.

Some promotional campaigns feature attractive female models to call attention to the product. A study reported in 1977 investigated the male reaction to female models in advertisements.[4] Two hypotheses were investigated. First, the researchers believed that an ad containing a decorative female model would be recognized better by males than an ad without a model. Second, they believed that the female model would not facilitate recognition of brand names in the ads. The study focused only on print advertising. Two hundred widely varying ads were selected from current popular magazines, some with decorative models and some without. Male undergraduates at Purdue University participated in the study during the Spring semester, 1976. Four groups of students viewed 50 different magazine ads, projected on a large screen, for 15 seconds each. Two groups viewed ads with decorative models and two groups viewed ads without models.

Two different recognition tests were given after the initial viewing. A recognition test for the entire ad consisted of exposing subjects to 50 ads. For each group, 25 of the "old" ads were mixed with 25 "new" ads. Subjects were scored on their ability to identify each ad as old or new. The brand name recognition test consisted of 25 "old" brand names mixed with 25 "new" brand names, all without photographs. A similar scoring system was used for this test. An analysis of variance procedure was used to analyze the data. The results verified the hypotheses. Recall of ads was significantly better for ads with decorative models, but the presence of decorative models did not significantly influence identification of the brand of product involved.

A similar type of study was conducted to determine how product name influences consumer perception of a product.[5] The study was based on the

[4] Robert W. Chestnut, with Charles C. LaChance and Amy Lubitz, "The 'Decorative' Female Model: Sexual Stimuli and The Recognition of Advertisements," *Journal of Advertising*, vol. 6, no. 4 (Fall 1977), pp. 11–14.

[5] Hershey H. Friedman and William S. Dipple, Jr., "The Effect of Masculine and Feminine Brand Names on the Perceived Taste of a Cigarette," *Decision Sciences*, vol. 9, no. 3 (July 1978), pp. 467–71.

society stereotype that men are strong and women are weak. Since cigarette ads often exhibit strength or weakness, this product was selected for the study. The major hypothesis was that consumers would allow the usual male/female stereotype to influence their ratings of a cigarette, given either a masculine or a feminine name. A cigarette not yet commercially available was used in the study. No brand name or recognizable markings were on the cigarettes or packages. A pretest was conducted to select masculine and feminine names. On a 1-to-10 feminine-masculine scale, the name "April" had a mean score of 1.46, while the name "Frontiersman" had a mean score of 9.42.

A sample of 200 smokers, 100 male and 100 female, was used in the study. The subjects had essentially similar demographic characteristics. After trying the cigarette, the subjects were told that a company was interested in marketing the cigarette under the name "April" or "Frontiersman" (50 males and 50 females were told one name and the others were given the second name). The subjects then rated the cigarette on seven characteristics:

1. Bland flavor/Rich flavor.
2. Hot tasting/Cool tasting.
3. Weak taste/Strong taste.
4. Harsh/Mild.
5. Unenjoyable/Enjoyable.
6. Masculine/Feminine.
7. Definitely would not purchase/Definitely intend to purchase.

Each characteristic was assigned a number from 1 to 8, going from the left option to the right option.

Analysis of variance and other statistical test procedures were used to analyze the results. Many of the results were predictable. Men preferred cigarettes named "Frontiersman," while women preferred the identical cigarette when named "April." The women rated "Frontiersman" as having blander flavor, hotter and stronger taste, harsher, less enjoyable, and more masculine than "April," even though both groups smoked identical cigarettes. Differences between mean scores for the two brands given by women were significant (p-value < 0.05) for six of the characteristics. Only two characteristics showed significant differences in mean scores given by men. Thus, it appeared that women may be more prone to stereotyping than men.

The study indicates that sex stereotypes of our society influence consumer perception of products. This strengthens the importance of pretesting brand names of new products. Careful attention should be paid to the sex of the target market and the perceived gender of the brand name.

16.4   FINANCE

The modern financial manager relies heavily on statistical analysis for assistance in financial planning and decision making. Financial forecasts, for example, include predictions of sales and inventories and projections of both the need for funds and the cost of funds. Regression analysis, exponential smoothing, and decision theory analysis are some of the statistical procedures used in financial forecasting. Some of the basic measures from descriptive statistics are used to describe various financial characteristics. Decision theory analysis is widely used for capital budgeting and investment decisions. The risk associated with a possible investment is often measured by either the standard deviation or coefficient of variation of the distribution of possible returns. The systematic risk of a given stock, as represented by its beta coefficient, is determined from a linear regression equation relating the stock's rate of return to the market rate of return. The following discussion provides additional insight into the role of statistics in some financial decisions.

Capital Investment Decisions

The AIL Division of Cutler-Hammer, Incorporated, used decision theory analysis to assist in considering an opportunity for capital investment in a new product.[6] The AIL Division of Cutler-Hammer, an electrical corporation with annual sales of about $400 million, operates primarily in the defense market and has annual sales of about $100 million. In 1974, an independent inventor offered AIL the opportunity to invest in a new, patented flight-safety system. The product was still in the development stage, but it was sufficiently developed to allow evaluation of probable costs. The immediate decision for AIL was whether to purchase a six-month option on the patent rights at a cost of $50,000. Securing the option would allow AIL to further investigate costs and the probable markets for the product. In addition, AIL would have the opportunity to enter into a license agreement with the inventor at the end of this period. Since the inventor wished to offer the patent rights to other companies if AIL was not interested, a decision was required in two weeks. A task group was assigned to investigate the opportunity and to recommend a plan of action.

FIGURE 16.6: AIL'S DECISION PROBLEM

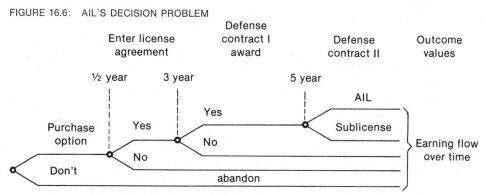

Source: Reproduced with permission from Jacob Ulvila, Rex V. Brown, and Karle S. Packard, "A Case in On-Line Decision Analysis for Product Planning," *Decision Sciences*, vol. 8, no. 3 (July 1977), p. 602.

The task group, together with outside consultants, used a decision theory analysis framework for their investigation. The major considerations are outlined in the tree diagram shown in Figure 16.6. If AIL purchased the option, they

---

[6] Jacob Ulvila, Rex V. Brown, and Karle S. Packard, "A Case in On-Line Decision Analysis for Product Planning," *Decision Sciences*, vol. 8, no. 3 (July 1977), pp. 598–615.

would spend about $150,000 investigating the patent's validity and researching the market. At the end of six months, they would have to decide whether to enter into a license agreement with the inventor. This would require a $300,000 front-end royalty payment to the inventor. AIL would then spend about $500,000 over $2\frac{1}{2}$ years developing and marketing the product for a particular defense application. Thus it would be three years before they would know whether an initial two-year defense contract could be obtained for the product. If a contract was awarded, a $250,000 manufacturing facility would be required. First-year production should produce net earnings of about $2.5 million. At the end of the initial two-year contract, a second two-year contract could be awarded either to AIL (net earnings $5 million) or to another contractor who would become a sub-licensee of AIL (net earnings $3 million). Net earnings beyond that time were estimated at either $1 million or at $425,000, depending on whether AIL or a sub-licensee was the primary contractor.

The task force then assessed the probabilities of the possible events and the present value of the flows of earnings, discounted at an annual pre-tax rate of 20 percent. The expected present value of purchasing the six-month option was about $85,000. The task force and consultants then used this information and model for further analysis. This amounted to "fine-tuning" of the model; its basic structure remained unchanged. AIL finally decided to reject the option and to adopt a wait-and-see attitude as to how the product would develop. They could seek a sub-license at a later time in the same manner as other contractors. Management of AIL was very pleased at the type of information the decision theory analysis afforded during the two weeks' time available for the decision.

Decision theory analysis is used to support investment decisions in many large corporations today. The use of decision theory analysis by E.I. duPont de Nemours and Company, The Pillsbury Company, General Electric Company, Ford Motor Company, and Inmont Corporation has been reported by Rex V. Brown, a consultant specializing in decision theory analysis.[7] In 1966, General Electric required that investment proposals for more than $500,000 be accompanied by a probabilistic assessment of the rate of return and other related data. More than 500 studies using decision theory analysis were conducted over the next four years. The company maintains a large library of computer programs for performing the analysis.

Financial Forecasting

Financial forecasts are based on predictions of sales. Forecasts of cash flow, inventory levels, profit, etc., can be determined by applying standard ratios to a sales forecast. Since the sales forecast is the base for financial forecasts, extreme care should be used in its preparation. Methods for forecasting sales vary widely. Some sales forecasts are made simply by a "hunch" or "guess" involving a single person or pooling the opinions of a group of people. While such subjective

---

[7] See Rex V. Brown, "Do Managers Find Decision Theory Useful?," *Harvard Business Review* (May–June 1970), pp. 78–89; and Rex V. Brown, Andrew S. Kahr, and Cameron Peterson, *Decision Analysis for the Manager* (New York: Holt, Rinehart and Winston, 1974), chap. 7.

approaches are satisfactory in some situations, well-defined statistical procedures are generally preferred. William M. Sharp, a budgetary and forecasting analyst at the Dictaphone Corporation, suggests that no matter what forecasting method is used, it is important to know the average forecast error.[8] The accuracy of a forecast can be anticipated by knowing the average forecast error. Typical measures of forecast error include the average absolute difference in the observed and predicted values, the average percentage difference, or simply the standard error of estimate. Sharp reports that Dictaphone uses subjective methods, traditional trend analysis, and, more recently, exponential smoothing in forecasting.

Three different departments at Southern New England Telephone Company prepare sales forecasts.[9] The Comptroller Department forecasts total company revenue from toll messages and from business and residence main stations, based on current customers, rates, and services. Operations Plans, an engineering group, projects short- and long-term growth for residential business customers for use in construction planning. The Marketing Department makes forecasts in its research on new products and services and proposed rate changes. The Marketing Department regularly uses five forecasting techniques for various purposes. (1) The Delphi Method is usually used to forecast sales of a new product to large commercial users. The method involves using sales consultants as a panel of experts to provide feedback about their customers' acceptance of new equipment. (2) User Surveys are designed to predict the effect of rate cases and to get feedback on new products and services that are not technically complex. (3) Historical Analogy involves comparing the introduction of a new product or service with the results of a similar product or service introduced at an earlier time. The forecast for the new equipment is based on what was observed in the past for similar equipment. (4) Multiple Regression Analysis relates sales of one type of service or equipment to other variables. To illustrate, the company determined that sales of their newer Totalphone equipment was related to sales of Touchtone phones, Princess phones, and auxiliary services. Past changes in these independent variables assisted in predicting sales of Totalphones. (5) The fifth method, New Product Trials, is the least used. In addition to being expensive, trial use of new products and services is subject to rigid utility regulations.

Financial analysis, including forecasting, lies at the heart of corporate planning models. A survey of corporations thought to be either using, developing, or considering some form of corporate planning model was conducted.[10] Of 346 respondents, 73 percent were either using or developing a model, 15 percent were planning to develop a model, and 12 percent had no plans for a model. Of those currently using a corporate planning model, 80

[8] William M. Sharp, "Financial Forecasting: Trend and Probability Controls," *Management World*, vol. 4, no. 5 (May 1975), pp. 12–16.

[9] Kevin McCrohan, "Forecasting Business Needs in the Telephone Market," *Industrial Marketing Management*, vol. 7, no. 2 (1978), pp. 109–13.

[10] Thomas H. Naylor and Horst Schauland, "A Survey of Users of Corporate Planning Models," *Management Science*, vol. 22, no. 9 (May 1976), pp. 927–37.

percent have modeled the financial structure of their business. Further, financial analysis dominates these firms' applications of corporate planning models. This includes cash flow analysis, budgeting, investment, and various types of forecasts. The relative frequency of specific applications of corporate planning models, for those firms with models, is summarized in Table 16.1. Many of these applications involve some type of financial forecasting, particularly for sales and revenue. Simpler forecasting techniques are used more frequently in these corporate models than the more complex techniques. Table 16.2 provides the percentage of use of various techniques for those firms having corporate models.

TABLE 16.1:
APPLICATIONS OF
CORPORATE MODELS

Applications	Percentage
Cash flow analysis	65
Financial forecasting	65
Balance sheet projections	64
Financial analysis	60
Pro forma financial reports	55
Profit planning	53
Long-term forecasts	50
Budgeting	47
Sales forecasts	41
Investment analysis	35
Marketing planning	33
Short-term forecasts	33

Source: Reproduced with permission from Thomas H. Naylor and Horst Schuland, "A Survey of Users of Corporate Planning Models," *Management Science*, vol. 22, no. 9 (May 1976), p. 930.

TABLE 16.2:
FORECASTING
TECHNIQUES USED IN
CORPORATE MODELS

Forecasting Technique	Percentage
Growth rate	50
Linear time trend	40
Moving average	22
Exponential smoothing	20
Nonlinear time trends	15
Adaptive forecasting	9
Box-Jenkins	4

Source: Reproduced with permission from Thomas H. Naylor and Horst Schuland, "A Survey of Users of Corporate Planning Models," *Management Science*, vol. 22, no. 9 (May 1976), p. 933.

The growth in usage of statistical and other formal analysis for financial planning will continue as more financial managers become familiar and comfortable with these procedures. The increased availability and reduced cost of computing equipment also promotes the efficient development and use of financial models.

16.5 REAL ESTATE   The real estate industry has, in recent years, been a frequent user of multiple regression analysis for property appraisal. Multiple regression analysis is used to predict property value as a function of several characteristics of the property. Many private apprisal firms and governmental assessment offices now routinely use multiple regression analysis in the appraisal function. Generally, a regression model is constructed, tested, and verified for each different type of property. Thus, one model might be used for suburban single-family dwellings, another for apartment buildings, and other models for office buildings, shopping centers, etc. Model building begins by collecting data on the dependent variable and all independent variables. As an example, a model to predict the selling price of a single-family dwelling might include such independent variables as size of house (square feet); number of bedrooms; number of bathrooms; whether the house has a garage, fireplace, swimming pool; the age of the house; and the location (measured by some desirability rating). Using the appropriate regression model, the most important predictors of selling price are determined by the statistical significance of each independent variable.

The development of regression models by the San Mateo County, California Assessor's Office for predicting the selling price of apartment buildings is described by George Gipe, a property analyst for the assessor's office.[11] Stepwise regression analysis was used to construct a multiple linear regression equation to predict the selling price of 134 different apartment properties. Results for one of the regression models developed is provided in Table 16.3.

TABLE 16.3:
REGRESSION EQUATION
FOR SELLING PRICE
OF APARTMENT
PROPERTIES

Variable	Coding (units)	Coefficient	Marginal F Test
Total land area	Square feet	− 1.862	25.1
Desirability rating	1 thru 5	6503.949	16.6
Neighborhood 3	0 or 1	−13315.789	4.2
Effective year built	Year − 1900	480.747	10.5
Below average construction quality	0 or 1	−10705.801	8.5
Twenty-five or more units	0 or 1	27394.254	7.7
Total living area	Square feet	9.600	30.1
Total number of bedrooms	Actual number	− 2000.603	13.2
Swimming pool	0 or 1	−25946.391	14.4
Total monthly income	Dollars	76.048	93.0

Number of sales 134
Regression constant . . . . −36759.895
Adjusted R squared 0.983
Sample mean of $Y(\bar{Y})$ 124535.0
Sample standard deviation of Y 96509.4
Standard error of the estimate 12007.6

Source: Adapted, with permission, from Table 1, p. 29 of George W. Gipe, "Developing a Multiple Regression Model for Multi-Family Residential Properties," *The Real Estate Appraiser*, vol. 42, no. 3 (May–June 1976 issue), published by Society of Real Estate Appraisers, Chicago, Illinois, pp. 28–33. No further reproduction is authorized.

---

[11] Reprinted, with permission, from the May–June, 1976 issue of *The Real Estate Appraiser*, published by Society of Real Estate Appraisers, Chicago, Ill. No further reproduction is authorized.

Some of the ten independent variables are quantitative, while others are qualitative and must be coded numerically for use in the regression equation. The mean selling price of the 134 properties is $124,535; the standard deviation is $96,509.40. The coefficient of determination for the multiple regression equation is 0.983. Thus the equation explains 98.3 percent of the variance in selling price.

Gipe cautions that the assumptions of multiple regression analysis must be throughly investigated for each application. When the assumptions are met and a valid model is established, the result is most valuable. The initial cost of developing a model may be relatively high, but operating costs are quite reasonable. Computing costs for a typical appraisal have been estimated at between $3 and $10. Today's low cost of minicomputers and widely available time-sharing arrangements bring the development of valuation models within the reach of almost all organizations. Once an equation is established, predictions may be made with only a small hand-held calculator.

16.6 ACCOUNTING

The accounting profession, much like that of financial analysts, has adopted many statistical procedures in recent years. In fact, accounts and financial analysts share many responsibilities in the areas of financial planning, budgeting and forecasting. Two applications of statistical methods to accounting are discussed in this section.

Cost Estimation and Control

There are many uses for regression analysis in cost accounting. For example, predictions of the annual maintenance and repair costs of similar machines might be obtained from a simple regression equation relating costs to another variable such as hours of usage, sales volume, or age of the machine. If it is important to consider all three variables, then a multiple regression equation could be used. Simple linear regression analysis is particularly useful for describing the fixed and variable portion of some expense item. In the equation $\hat{Y} = b_0 + b_1 X$, where, for example, Y measures expenses and X measures sales volume ($\$$), b_0 is the estimated fixed expense and b_1 is the variable expense per dollar of sales.

A method for using multiple regression analysis to estimate unit costs associated with various products was described in *The Accounting Review*.[12] The article describes a repair shop where adding machines, typewriters, and calculators were overhauled. An estimate of the direct cost per unit of each machine was desired. Data were available on the number of adding machines, typewriters, and calculators repaired each week, as well as the total number of hours allocated to machine overhaul each week. In week one, for example, five adding machines, seven typewriters, and eight calculators were overhauled. A total of 64 hours were spent in machine overhaul that week. If direct labor-hours per unit could be allocated to each machine, a labor cost per hour could

---

[12] Paul R. McClenon, "Cost Finding through Multiple Correlation Analysis," *The Accounting Review*, vol. 38, no. 3 (1963), pp. 540–47.

be used to convert this figure to direct labor costs per unit for the overhaul of each type machine. The following approach was taken.

Construct a regression equation $\hat{Y} = b_0 + b_1X_1 + b_2X_2 + b_3X_3$, where total hours in overhaul per week (Y) is a function of the number of adding machines overhauled that week (X_1), the number of typewriters overhauled that week (X_2), and the number of calculators overhauled that week (X_3). The objective is to establish the estimates b_0, b_1, b_2, and b_3. The estimate b_1, for example, is the average direct labor-hours for overhauling an adding machine. If $b_1 = 1.5$ and six adding machines are repaired in a week, then the estimated direct labor-hours involved in adding machine repair that week is $(1.5)(6) = 9$. The value b_0 is an estimate of indirect time or "overhead" hours per week. Thus, the equation for overhaul labor hours per week is

$$\text{Total hours} = \text{Indirect hours} + \quad \text{Direct hours,} \quad \text{or}$$
$$\hat{Y} = \quad b_0 \quad + b_1X_1 + b_2X_2 + b_3X_3.$$

The solution given for an illustrative example is $\hat{Y} = 7.88 + 1.46X_1 + 2.41X_2 + 4.02X_3$. The estimated direct labor hours are 1.46 per adding machine, 2.41 per typewriter, and 4.02 per calculator. An average of 7.88 indirect labor-hours or "overhead" is allocated to this activity.

Before using multiple regression analysis in situations such as that described here, caution must be exercised to investigate the necessary assumptions and to validate the model. Statistical validity can be assured through tests of significance. The standard error of estimate and the multiple coefficient of determination provide additional indicators of the reliability of the results.

Auditing

The auditor's primary function is to investigate financial records, procedures, and statements and to give an opinion on the appropriateness of the procedures used and the fairness of the financial statements as they portray the financial condition of an organization. This attestation provides credibility to the financial statements, which are used by creditors, shareholders, and other investors who provide capital to a corporation. Accounts receivable, accounts payable, inventories, etc., are some of the records that must be examined. The purpose of auditing these records is to determine whether any errors exist in the statements of these accounts. In the past, auditors relied on an examination of the entire list of accounts in order to determine whether errors exist and, if so, the number and amount of the errors. Auditors have, in recent years, relied more on statistical sampling for these audits. Sampling is less expensive and less time consuming since fewer accounts must be examined in detail. Often the sample provides more accuracy than a complete examination. Many people and a large amount of time are involved in auditing all accounts. Thus the probability of human error increases because of the lack of training for some employees and the boredom that comes with repetitious work. Since a sample involves fewer items, there is less chance for boredom and, since fewer people are required, properly trained personnel may be used. An application of sampling in auditing follows.

The Chesapeake and Ohio Railroad Company (C & O) conducts periodic audits of waybills indicating the amount of revenue due on interline, less-than-carload, freight shipments.[13] When shipments travel over several railroads, the total freight revenue is divided among the railroads. The amount due each railroad is computed from a waybill that gives details about the goods shipped, route, charges, etc. C & O wished to determine whether an accurate division of total revenue could be made on the basis of a sample at a substantial savings in cost. They selected all waybills over the Pere Marquette district of the C & O and another railroad, during a six-month period, for the study. Approximately 23,000 waybills were involved, with charges varying from $2.00 to $200.00. Because the variability in waybill charges was large, a stratified sampling procedure was selected. The population of waybills was divided into more homogeneous strata based on the dollar amount, as shown in Table 16.4. Then the proportion of each stratum that should be sampled was decided on the basis of the variability within strata. A larger proportion of the stratum is selected as the variability of revenue due C & O increases. Random sampling is used to select waybills within each stratum.

TABLE 16.4:
CHESAPEAKE AND
OHIO STRATIFIED
SAMPLING PLAN

Stratum	Waybills with Charges	Proportion to Be Sampled
1	$0 and 5.00	1%
2	5.01 and 10.00	10
3	10.01 and 20.00	20
4	20.01 and 40.00	50
5	40.01 and over	100

Source: Reproduced with permission from p. 206 of John Neter, "How Accountants Save Money by Sampling," in Judith M. Tanur et al., eds., *Statistics: A Guide to the Unknown* (San Francisco: Holden-Day, Inc., 1972), pp. 203–11.

A total of 2,072 of the 22,984 waybills in the population (9 percent) was selected for the sample. For each, the amount of freight due C & O was calculated. Then an estimate of the amount due for each stratum in the population was obtained, and these values were added to estimate the amount due C & O for the entire population of waybills. In order to assess the accuracy of the sampling procedure, revenue due C & O was also calculated from the entire population of waybills. The results were:

Amount due C & O based on census: $64,651

Amount due C & O based on sample: 64,568

Difference: $ 83

This shows that a sample of 9 percent of the waybills provided an estimate of the revenue due within $83 of that obtained from a complete examination of

---

[13] John Neter, "How Accountants Save Money by Sampling," in Judith M. Tanur et al. eds., *Statistics: A Guide to the Unknown* (San Francisco: Holden-Day, Inc., 1972), pp. 203–11.

the population. Since the complete count costs $5,000 and the sample costs only $1,000, a tremendous cost savings is possible through sampling. Assuming the complete count is accurate, it is not worth $4,000 to uncover an error of $83. And, in repeated samples, the errors against and in favor of C & O will probably balance one another.

Because of the advantages of statistical sampling for performing audits, accountants have developed many specialized sampling procedures for different purposes. These procedures are all based on the sampling concepts presented Chapter 6. Sampling plans used in auditing include variables sampling, attribute sampling, acceptance sampling, difference sampling, ratio sampling, and dollar-unit sampling.

16.7 BANKING

In the previous sections, some of the ways in which statistical analysis is used in various functional areas of a business organization have been presented. These functions exist in the firms of all major industries, including the banking industry. For example, the auditing departments of many banks use sampling procedures for verification of account balances, for checks on internal control, for transactions testing, etc. Loan departments may use regression analysis to develop a profile of the worthy loan applicant. The investment department may use a model based on decision theory analysis as an aid in choosing among alternative investments. Trend analysis, exponential smoothing, and regression models are useful in forecasting the volume of demand deposits, etc. And the personnel department may use significance testing to validate the fairness and predictive ability of instruments used in recruiting, hiring, and salary and promotion decisions. Two specific applications of statistical concepts in banking are presented next. Since banks share many of the characteristics of other businesses, similar applications of these concepts are also found in other organizations.

Evaluating the Quality of Service

Since a bank is a service organization, its success is partially dependent on the quality of service provided. One measure of a bank's service is the efficiency with which customers are served as they arrive at a teller station, either inside the bank or at a drive-in window. If too few teller stations are open during peak periods, long lines of customers may form. This increases the time between the arrival of a customer and the departure after business has been transacted. The bank faces a potential loss of customers if long waiting lines and lengthy waiting times occur frequently. Opening additional teller stations may reduce customer waiting time, but at an increased cost of providing additional service. Thus the bank must make a trade-off between short waiting times and increased costs. Probability theory may assist in analyzing this situation.

Probability is the basis of a mathematical model called *queuing theory*. Any situation where individual units may wait in line for some processing or service can be formulated as a queuing model. The components of a queuing model are shown in Figure 16.7. A queuing system has two major components—the waiting line, or queue, and the service facility. For a bank, customers arrive to conduct business, e.g., cash checks and make deposits, at a teller station, which is the service facility. If the teller(s) is busy, the customer must wait in line, the queue. Eventually the customer gets to the head of the line and receives the desired service at the teller station. Queuing analysis is used to study the system

FIGURE 16.7:
COMPONENTS OF A
QUEUING SYSTEM

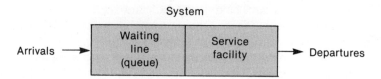

so that an appropriate level of service may be provided at a reasonable cost. The queuing study begins by determining the probability distribution for the arrival and service components of the queuing system. The Poisson probability distribution (Chapter 5) is often used to model the number of customers arriving during specified time intervals. When probability distributions are identified for the arrival and service processes, and cost measures are applied, trade-offs in reduced waiting time and increased cost can be examined. The bank may decide on a level of service that specifies that 90 percent of the customers will not have to wait in line more than two minutes.

This approach was taken in a study to evaluate the expansion of drive-in window facilities at American Exchange Bank and Trust Company in Norman, Oklahoma.[14] The bank was concerned about loss of customers due to poor service. Due to space limitations at the bank, two options for expansion were available: an automated "robo" system or an increase in the number of teller windows. The existing drive-in facility contained three teller stations. Enough space existed to add two teller stations or four "robo" stations. The "robo" system consisted of a speaker and pneumatic tube arrangement that allowed remote processing of banking transactions, while requiring only about half the space of a teller window. The objective stated by bank officials was to expand these facilities to assure that customer waiting time would be limited to four minutes. Cost considerations were given secondary importance by the bank officials. The bank operated by opening an additional drive-in window when four or more cars were waiting.

Data were collected in order to determine the appropriate distributions of time between arrivals and service. Once a model was established, various combinations of teller stations and "robo" stations were explored. A recommendation was made that the existing drive-in window system be expanded to five lanes by adding two "robo" stations to the existing three teller windows. According to the results of the waiting line model, there would be only a 5 percent

[14] B. L. Foote, "A Queuing Case Study of Drive-In Banking," *Interfaces*, vol. 6, no. 4 (August 1976), pp. 31–37.

chance that customers would have to wait more than four minutes. The bank constructed the system as suggested, and later reported that the system operated as indicated. The system constructed is shown in Figure 16.8.

FIGURE 16.8: DRIVE-IN
SYSTEM ADOPTED

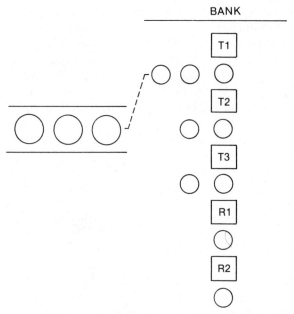

Source: Reproduced with permission from p. 35 of B. L. Foote, "A Queuing Case Study of Drive-In Banking," *Interfaces*, vol. 6, no. 4 (August 1976), Fig. III on p. 35.

Because service organizations are sensitive to customer feelings, there is a tendency to staff customer-service areas for peak traffic periods. This practice may be expensive and inefficient if the personnel must be retained through slack periods. In branch banking, queuing analysis has been used to determine optimal combinations of full-time and part-time employees needed to meet desired service standards.[15] This involves applying cost measures to various staffing combinations to determine the least-cost arrangement that meets the desired service criteria. Central Federal Savings of San Diego has been applying queuing theory for this type of analysis since 1975. The annual staff payroll and fringe benefit savings and cost avoidance have amounted to more than $270,000. First Federal Savings of Phoenix, Arizona, achieved annual savings of about $50,000 since beginning to use this queuing approach to branch staffing in 1976. Other banks reporting similar savings as a result of the same approach include Women's Savings Bank of Cleveland, Ohio; Central Federal Savings of Nassau County, Long Beach, N.Y.; and Lincoln Savings Bank, Brooklyn, N.Y.

---

[15] Peter J. Oeth, "Cut Cost, Not Service," *Savings and Loan News*, vol. 99, no. 5 (May 1978), pp. 58–62.

Estimating Demand
Deposit Potential

Banking has become increasingly competitive in recent years. Liberalized branch banking laws, bank mergers and acquisitions, and the establishment of new banking institutions are some of the factors influencing the increased competition. Correspondingly, the risk associated with establishing new branch offices has increased. In order to assess the possible benefit of establishing a new branch office location, an estimate of the share of deposit potential is required. This estimate is usually viewed as a percentage of the total demand deposit potential for the locality. Traditional methods of estimating the total demand deposit potential for a locality involve use of the Federal Deposit Insurance Corporation's (FDIC) county estimates of deposits per capita. These estimates are broadly classified and therefore may be inappropriate for many branch location possibilities. Further, there is no measure of the accuracy of the estimates; much of the method is subjective, even though logical and intuitively correct.

Regression analysis has proven a desirable method for estimating deposit potential.[16] This approach was taken in a study for a large commercial bank in Philadelphia that wished to expand into new localities. The study developed a regression model to predict, for a locality, demand deposit account balance (Y) as a function of per capita income (X). The rationale for this is as follows. Since demand deposits are non-interest-bearing assets, the rational individual (or family) attempts to maintain the smallest balance necessary to meet short-term expenditures. And because short-term expenditures are largely determined by personal income, demand deposit account balances should be significantly influenced by personal income. To establish the regression model, data were collected on the average personal demand deposit account balances by postal ZIP codes for current accounts at the bank. Data were also available on per capita income by Minor Civil Divisions (MCD). Since the MCD boundaries matched ZIP code boundaries, the data could be paired. Thirty-seven sample observations resulted, with per capita income ranging from $4668 to $10,655 and average personal demand deposit account balances ranging from $650 to $1993.

The best-fitting regression model was the exponential equation, $\hat{Y} = b_0 b_1^X$. The following results were obtained:

$$b_0 = 308.895;$$
$$b_1 = 1.0001447;$$
$$\text{Standard error of } b_1 = 0.1228;$$
$$\text{p-value for } b_1 = 0.001;$$
$$r^2 = 0.66.$$

Thus the regression equation appears quite useful. The output of the model, $\hat{Y}$, is an average demand deposit account balance for an individual in a given locality. The bank wished to estimate the total demand deposit potential for the entire population in a given locality. By determining the number of house-

---

[16] William D. Miller and Kenneth F. Carlin, "Estimating Demand Deposit Potential," *The Bankers Magazine*, vol. 161, no. 6 (November–December 1978), pp. 76–81.

holds (NH) in the locality and an estimate of the average number of demand deposit accounts per household (APH), the bank could estimate total demand deposit potential (TDDP) by the following relation:

$$TDDP = (NH)(APH)(\hat{Y}).$$

16.8 SURVEYING PUBLIC OPINION

Samples are taken for the purpose of making inferences about the population. Many samples are taken for the purpose of measuring public opinion. Manufacturers conduct surveys and hire market research organizations to conduct surveys of consumer preference. Politicians and government agencies use polls and surveys to judge public opinion. Television networks rely on ratings to determine fees for advertising and to decide whether a show should stay on the air. The results of these various surveys and polls are used to make decisions that, in turn, influence our lives daily. The following discussion introduces some of the issues in public opinion polling.

Television Ratings[17]

The television ratings are perhaps the most familiar, and probably the most controversial, polls of public opinion. The "Nielsen" ratings are used by television networks to justify charging advertisers up to $150,000 for a one-minute commercial on a top-rated, prime-time program. The ratings, which are widely publicized in newspapers and magazines, purport to tell us which shows we like most. And then we find that our favorite show has been cancelled—because of low ratings! These are only a few of the frustrations expressed by network executives, producers, advertisers, advertising agencies, and the public. The frustration probably arises more from a lack of understanding of the ratings than from serious questions about the methods involved.

The A. C. Nielsen Company is probably the largest market research organization. It operates in 22 countries and earns about a third of a billion dollars a year. Its major business is market surveys at the retail level, which are designed to determine which products, brands, packages and sizes are being purchased. Manufacturers and advertising agencies pay Nielsen and other market research organizations for this information. The ratings are a small part of the company's business, accounting for only 11 percent of earnings. Yet the ratings have made the company famous. Other companies, such as Arbitron, provide information about television viewing, but Nielsen is the leader.

Television ratings are done by the Media Research Division of the company. Two types of information about television viewing are collected, each in different ways. "Ratings" are determined from audimeters attached to televisions in a sample of households. A random sample of 1,170 homes with TVs, geographically spread throughout the United States, are chosen from a Census Bureau master list. Field representatives interview the families selected and try to persuade them to join the ratings panel. For those who do, the company pays

---

[17] Based on the following sources: David Chagall, "Can You Believe the Ratings?", *TV Guide*, vol. 26, no. 25 (June 25, 1978), pp. 2–4 ff.; David Chagall, "The Viewers Who Go Uncounted," *TV Guide*, vol. 26, no. 26 (July 1, 1978), pp. 20–22 ff.; Michael Wheeler, "The Nielsen 1200," Chapter 10 of *Lies, Damn Lies, and Statistics* (New York: Liveright, 1976); *What the Ratings Really Mean* (Chicago: A. Nielsen Company, 1964).

$25 for allowing the audimeter to be hooked up and another $25 for each of the five years of the contract. The company also pays half of all TV-repair costs during that period. The device compiles records of all TV watched on a 24-hour basis; the information is received by the company via special telephone lines. Because of malfunctions, TVs out for repair, family vacations, etc., information is usually received from about 1,000 of the audimeters at any one time.

The "ratings" number simply reflects the percentage of sample households that had a show tuned in for at least six minutes. The top-ranked show in the Fall of 1977, " Laverne and Shirley," had a rating of 31.6. This means that 31.6 percent of the approximately 1,000 households watched this program. The remainder were either watching other programs or not watching TV. It is the use of the result of the sample of 1,170 households to judge the viewing habits of 73 million that raises questions from some people. Nielsen reports that the sampling error for a show ranked 20, "Welcome Back, Kotter" in the fall of 1977, is 1.3. Thus, with 68 percent confidence, "Kotter" ranked between 18.7 and 21.3. Even though Nielsen reports his results with 68 percent confidence, sampling theory indicates that with 95 percent confidence, "Kotter" ranks between 17.4 and 22.6, $20 \pm 2(1.3)$. In the fall of 1977, 32 of the 40 top-ranked shows were within these limits. While the standard error of a rating may be small, the difference in ratings by successive ranks is also small.

The second measure of television viewing is the "share." Rating is based on all households having audimeters, while share is based only on those households actually watching TV. The audience share is simply the proportion of households watching a given program out of all households in a sample who are watching TV. Shares are determined from logbooks, or diaries, kept by a family for one week. A sample of 2,200 families is selected nationally for this purpose. The family records in the diary when they watched TV and what programs were watched. When the diaries are returned, Nielsen tabulates the audience share of a show and its competitors. If, for example, the Super Bowl has an 85 percent audience share, this means that 85 percent of all families watching television during that time on Sunday afternoon were tuned to the game. Fifteen percent were watching other programs.

Critics of the Nielsen ratings argue that the ratings do not accurately reflect all segments of the population. Most agree that the ratings reflect what is referred to as "Middle America." High-income and low-income groups are under-represented in proportion to their numbers in the United States. The ratings panel (audimeter families) has a lower proportion of blacks and Spanish-speaking people than in the population as a whole. This is not to say that the samples are purposely biased. Many families refuse to participate in the survey, for one reason or another. But the fact remains that the ratings on shows that appeal to certain groups may be affected. Many places where TV is watched are not possible sample sites. These include hotels, hospitals, prisons, military bases, and colleges and universities. Nielsen admits that people living alone are under-represented. Almost half are over 65; in 1977, they accounted for 23 percent of all households but only 17 percent of the ratings panel. Their preferences in viewing habits may differ from other families.

Despite the criticisms and readily admitted problems, the ratings do serve as general indicators of viewing preference, and that is the purpose for which they were designed. They must be interpreted as sample measures with the standard statistical properties. Television networks, the major ratings users, have been approached with alternatives to the Nielsen ratings many times. The fact that these suggestions have been turned down is an indication of the networks' satisfaction with the present system.

Pollsters, the Vox Populi[18]

Whether or not pollsters are the "voice of the people," their influence is far-reaching. Newspapers, magazines, radio, and television issue frequent reports on recent polls conducted by the well-known organizations of Gallup, Harris, and Roper. They report on such issues as the changes in the public's support of the president, attitudes on current wars (Vietnam earlier), the pardoning of Richard Nixon and Patty Hearst, the 55 mph speed limit, the cost of living, etc. The publicity accompanying these surveys have made pollsters famous. Yet most, like the A. C. Nielsen Company, have commercial market research as their major business activity. Political polling brings the publicity that attracts corporate clients. And since the methods used in political and commercial polls are generally the same, it is easy to conduct both. Many pollsters also incorporate questions for their corporate clients into their political polls.

Commercial market research is primarily concerned with reporting past sales of products and predictions of customer acceptance for products under consideration. Commercial polling is extremely big business today. Gallup charges about $300 for each question presented to one segment of a sample. If 15,000 people were each asked 25 questions, the cost could easily reach $20,000. The best-known pollsters have regular report services they sell to clients. The Roper Organization grosses over $150,000 annually from its *Roper Reports*, published ten times a year. About 45 organizations pay $3,600 annually for the basic service. Forty-five firms pay $25,000 annually for the *Harris Perspective*, which grosses a million dollars a year for the company. The *Public Opinion Index*, published by Opinion Research Corporation of Princeton, New Jersey, provides 24 issues annually to 50 subscribers paying $6,000 yearly. The firm of Daniel Yankelovich publishes two reports. Eighty organizations each pay $11,500 to receive the once-a-year *Yankelovich Monitor*, a report on changing values and life styles in the U.S. *Corporate Priorities*, a study of public attitude toward big business, quality and safety of products, social responsibility of business, etc., goes for $23,000. In addition to regular reports, pollsters also conduct special surveys at the request of clients. Several years ago Alan R. Nelson Research of New York did an exhaustive study of the value of various athletes' endorsements of products. This was requested by 20 national companies and advertising agencies, who each paid $6,000.

---

[18] Based primarily on Michael Wheeler, *Lies, Damn Lies, and Statistics* (New York: Liveright, 1976).

Not all market surveys lead to satisfied clients and happy endings. Albert Sindlinger, a specialist in automotive studies, conducted a survey early in 1957 for Ford Motor Company that predicted 200,000 units of a proposed new car would be purchased in its first year of introduction. Ford planned production accordingly but by September, when the car was introduced, the economy had turned down. Sindlinger revised his estimate of first year sales to 54,000. Even though this proved correct, Ford was extremely unhappy. The car, named the Edsel, was eventually scrapped.

Many of the major pollsters have strong interests in political polling, for reasons beyond the acknowledgment that the fame of such polls brings corporate business. Politicians are also strongly attracted to polling. By the seventies, every major political candidate had a pollster. The pollster collected data on the public acceptance of the candidate and reported the candidate's changing fortune in the public's eye. Polls also influence public opinion of candidates by the attention the polls receive. This also assists with campaign contributions, endorsements and media coverage. Louis Harris conducted private political polls in 240 different campaigns between 1956 and 1963. He was John Kennedy's pollster for the 1960 Presidential race. Pat Caddell served as George McGovern's pollster in the 1972 presidential race and as Edward Kennedy's pollster in his 1976 senatorial race.

Political polling is based on random sampling. Sample areas throughout the United States are chosen so that a community's chance of being selected is proportional to its share of total population. Several stages may be involved in the selection process. States may be selected first, counties second, etc. Interviewers then question a selected number of households within each sample area. A typical Harris national political poll is based on about 100 sample areas and 1,200 to 1,500 interviews. Gallup uses 150 to 360 sample areas in collecting about the same number of interviews.

The sampling error in the Gallup and Harris national polls is usually 3 or 4 percent. If, for example, a candidate is reported to have 48 percent of the vote at a given time, his actual percentage may range between 42 and 54 percent, with 95 percent confidence (using a sampling error of 3 percent). Most of the major polls rarely publicize the sampling error, but pollsters complain that reporters fail to understand the statistical implications of the percentages. Perhaps pollsters should make a better effort to educate their clients and the public on the statistical interpretations of their findings.

Most of the major firms continually conduct research to improve polling methods. The way in which questions are phrased by interviewers and identification of the voting population are two areas that plague pollsters. Of those eligible to vote in the United States, only about two-thirds are registered. Fewer than that actually vote. A major study conducted in 104 cities during 1960 showed that the percentage of eligible voters who are registered to vote ranged from 32 to 96 percent.[19] The percentage of eligible voters who actually voted

---

[19] Edward Tufte, "Registration and Voting," in Judith M. Tanur et al., eds., *Statistics: A Guide to the Unknown* (San Francisco: Holden-Day, Inc., 1972), pp. 153–61.

ranged from 24 to 85 percent. The correlation between voting turnout and registration was 0.88. Multiple regression was used to estimate the importance of several factors on the registration rate. Convenience of registration, longer registration office hours, and availability of more offices were directly related to registration rates as was the closeness of recent elections. Socioeconomic variables such as age, race, and education also influenced registration rates. These findings played a role in the passage of the Voting Rights Act of 1965, which was designed to simplify registration procedures and eliminate certain obstacles to registration.

16.9 SUMMARY

The purpose of this book is to introduce you to the world of statistics. Basic procedures in statistical inference and description have been presented. Having equipped you with a knowledge of these concepts, we have sought in this final chapter to give you a sampling of their use in a variety of situations. The applications presented here are not intended to be exhaustive. No doubt you can think of others. Our intent is to give you insight into *why* there is a need for statistics in today's business world.

Obviously there are many business situations, much more complex than those presented here, where statistical methods are used. Often, methods beyond the scope of this text are required. But advanced methods are based on the same principles covered in this text. Familiarity with these basic concepts equips business managers with the knowledge to recognize situations where statistical analysis is helpful and to incorporate the results of statistical studies into managerial decisions. This familiarity includes the ability to effectively communicate your special needs to statisticians and other quantitative specialists and to be on guard for inappropriate uses of statistical methods.

Finally, let us remind you that there is a difference between *probability* and *outcome*. Even the best managerial judgment combined with sound statistical analysis occasionally produces a less than optimal result. Our best decision may have a high probability of producing success, but an outcome with a small probability of occurrence may result. This is the essence of decision making under conditions of uncertainty. An understanding of the basic concepts helps to avoid the conclusion reached in the accompanying cartoon.

© 1974 United Feature Syndicate, Inc.

REFERENCES

Chagall, David "Can You Believe the Ratings?" *TV Guide*, vol. 26, no. 25 (June 25, 1978), pp. 2–4 ff.

_____ "The Viewers Who Go Uncounted." *TV Guide*, vol. 26, no. 26 (July 1, 1978), pp. 20–22 ff.

Kirk, Roger E., ed. *Statistical Issues: A Reader for the Behavioral Sciences.* Belmont, Cal.: Wadsworth Publishing Company, 1972.

Mansfield, Edwin, ed. *Elementary Statistics for Economics and Business: Selected Readings.* New York: W. W. Norton and Company, Inc., 1970.

Peters, William S., ed. *Readings in Applied Statistics.* Englewood Cliffs, N.J.: Prentice-Hall, Inc., 1969.

Reichard, Robert S. *The Figure Finaglers.* New York: McGraw-Hill Book Company, 1974.

Tanur, Judith M., et al., eds. *Statistics: A Guide to the Unknown.* San Francisco: Holden-Day, Inc., 1972.

Wheeler, Michael *Lies, Damn Lies, and Statistics.* New York: Liveright, 1976.

APPENDIXES

AND

ANSWERS TO ODD-NUMBERED QUESTIONS AND EXERCISES

APPENDIX A: NOTATION, SYMBOLS, AND SUMMATION OPERATIONS

Notation and symbols are used in mathematics as shorthand expressions for mathematical concepts and statements. In this section the notation and symbols that are used extensively in statistics are reviewed.

Symbols such as X and Y are used as *variable names* to indicate different groups of data. For example, the symbol X could be the variable name for the number of cars sold per day at an automobile dealership; the symbol Y might be a variable name for the closing price each day of a certain stock. *Subscripts* for variables are used to identify individual observations of a variable. Suppose the automobile dealership sold three cars on one day, seven on the second day, and four on the third day. This can be symbolically represented with the subscripted variable X_i where the subscript i identifies a particular day. Thus X_1 represents the number of cars sold on the first day, X_2 the number sold on the second day, and X_3 the number sold on the third day.

The symbol $\sum$, capital Greek letter sigma, represents summation in mathematics. The expression

$$\sum_{i=1}^{n} X_i$$

represents the summation of the observations of variable X for the subscripts from 1 to n. The expression may be expanded for the individual values of i so that

$$\sum_{i=1}^{n} X_i = X_1 + X_2 + \cdots + X_n.$$

The total number of cars sold in three days at the automobile dealership described above may be expressed as

$$\sum_{i=1}^{3} X_i = X_1 + X_2 + X_3.$$

Substituting the values of X_i,

$$\sum_{i=1}^{3} X_i = 3 + 7 + 4 = 14.$$

Many statistical expressions involve summation operations. The following rules are useful in performing these operations.

Rule 1. If the values of a variable X_i are each multiplied by a constant c, then

$$\sum_{i=1}^{n} cX_i = c \sum_{i=1}^{n} X_i.$$

The sum of a constant times a variable equals the constant times the sum of the variable. If $c = 3$, $X_1 = 2$, $X_2 = 5$, and $X_3 = 6$, then

$$\sum_{i=1}^{3} cX_i = 3(2) + 3(5) + 3(6)$$

$$= c \sum_{i=1}^{3} X_i = 3(2 + 5 + 6) = 3(13) = 39.$$

Rule 2. If a constant c is added to itself n times, then

$$\sum_{i=1}^{n} c = nc.$$

The sum of a constant is n times the constant. If the constant 4 is added 3 times, then

$$\sum_{i=1}^{3} 4 = 4 + 4 + 4 = 3(4) = 12.$$

Rule 3. If a constant c is subtracted from (added to) each value of the variable X_i, then

$$\sum_{i=1}^{n} (X_i - c) = \sum_{i=1}^{n} X_i - \sum_{i=1}^{n} c.$$

The sum of a variable minus (plus) a constant equals the sum of the variable minus (plus) the sum of the constant (nc). If $c = 2$, $X_1 = 4$, $X_2 = 7$, and $X_3 = 9$, then

$$\sum_{i=1}^{3} (X_i - c) = (4 - 2) + (7 - 2) + (9 - 2)$$

$$= \sum_{i=1}^{3} X_i - \sum_{i=1}^{3} c = (4 + 7 + 9) - 3(2)$$

$$= 20 - 6 = 14.$$

Rule 3 also applies to the sum of the differences in two variables. For example,

$$\sum_{i=1}^{n} (X_i - Y_i) = \sum_{i=1}^{n} X_i - \sum_{i=1}^{n} Y_i.$$

Rule 4. The sum of the squared values of variable X_i is

$$\sum_{i=1}^{n} X_i^2 = X_1^2 + X_2^2 + \cdots + X_n^2.$$

If $X_1 = 3$, $X_2 = 4$, and $X_3 = 2$, then

$$\sum_{i=1}^{3} X_i^2 = 3^2 + 4^2 + 2^2 = 9 + 16 + 4 = 29.$$

Note that

$$\sum_{i=1}^{n} X_i^2 \neq \left(\sum_{i=1}^{n} X_i \right)^2.$$

The former is a sum of squared values of a variable while the latter is the square of the sum of the values of a variable. For the above values of X_i,

$$\left(\sum_{i=1}^{3} X_i \right)^2 = (3 + 4 + 2)^2 = 9^2 = 81.$$

Rule 5. The sum of the products for paired observations of variable X_i and Y_i is

$$\sum_{i=1}^{n} X_i Y_i = X_1 Y_1 + X_2 Y_2 + \cdots + X_n Y_n.$$

If $X_1 = 2$, $X_2 = 1$, $X_3 = 4$, $Y_1 = 3$, $Y_2 = 5$, and $Y_3 = 10$, then

$$\sum_{i=1}^{3} X_i Y_i = (2)(3) + (1)(5) + (4)(10)$$

$$= 6 + 5 + 40 = 51.$$

Note that

$$\sum_{i=1}^{n} X_i Y_i \neq \left(\sum_{i=1}^{n} X_i \right) \left(\sum_{i=1}^{n} Y_i \right).$$

The first is a sum of products while the second is the product of two sums. For the values of X_i and Y_i above,

$$\left(\sum_{i=1}^{3} X_i \right) \left(\sum_{i=1}^{3} Y_i \right) = (2 + 1 + 4)(4 + 3 + 5) = (7)(12) = 84.$$

$$\text{Values of } P(X = x \mid n, \pi) = \frac{n!}{x!(n-x)!} \pi^x(1-\pi)^{n-x}$$

for $n = 1(1)20$, $x = 0(1)n$, and $\pi = 0.05(0.05)0.50$.*

symmetrical to .50 also

$\pi_1 = k/N$

n	x	0.05	0.10	0.15	0.20	0.25	0.30	0.35	0.40	0.45	0.50
1	0	0.9500	0.9000	0.8500	0.8000	0.7500	0.7000	0.6500	0.6000	0.5500	0.5000
	1	0.0500	0.1000	0.1500	0.2000	0.2500	0.3000	0.3500	0.4000	0.4500	0.5000
2	0	0.9025	0.8100	0.7225	0.6400	0.5625	0.4900	0.4225	0.3600	0.3025	0.2500
	1	0.0950	0.1800	0.2550	0.3200	0.3750	0.4200	0.4550	0.4800	0.4950	0.5000
	2	0.0025	0.0100	0.0225	0.0400	0.0625	0.0900	0.1225	0.1600	0.2025	0.2500
3	0	0.8574	0.7290	0.6141	0.5120	0.4219	0.3430	0.2746	0.2160	0.1664	0.1250
	1	0.1354	0.2430	0.3251	0.3840	0.4219	0.4410	0.4436	0.4320	0.4084	0.3750
	2	0.0071	0.0270	0.0574	0.0960	0.1406	0.1890	0.2389	0.2880	0.3341	0.3750
	3	0.0001	0.0010	0.0034	0.0080	0.0156	0.0270	0.0429	0.0640	0.0911	0.1250
4	0	0.8145	0.6561	0.5220	0.4096	0.3164	0.2401	0.1785	0.1296	0.0915	0.0625
	1	0.1715	0.2916	0.3685	0.4096	0.4219	0.4116	0.3845	0.3456	0.2995	0.2500
	2	0.0135	0.0486	0.0975	0.1536	0.2109	0.2646	0.3105	0.3456	0.3675	0.3750
	3	0.0005	0.0036	0.0115	0.0256	0.0469	0.0756	0.1115	0.1536	0.2005	0.2500
	4	0.0000	0.0001	0.0005	0.0016	0.0039	0.0081	0.0150	0.0256	0.0410	0.0625
5	0	0.7738	0.5905	0.4437	0.3277	0.2373	0.1681	0.1160	0.0778	0.0503	0.0312
	1	0.2036	0.3280	0.3915	0.4096	0.3955	0.3602	0.3124	0.2592	0.2059	0.1562
	2	0.0214	0.0729	0.1382	0.2048	0.2637	0.3087	0.3364	0.3456	0.3369	0.3125
	3	0.0011	0.0081	0.0244	0.0512	0.0879	0.1323	0.1811	0.2304	0.2757	0.3125
	4	0.0000	0.0004	0.0022	0.0064	0.0146	0.0284	0.0488	0.0768	0.1128	0.1562
	5	0.0000	0.0000	0.0001	0.0003	0.0010	0.0024	0.0053	0.0102	0.0185	0.0312
6	0	0.7351	0.5314	0.3771	0.2621	0.1780	0.1176	0.0754	0.0467	0.0277	0.0156
	1	0.2321	0.3543	0.3993	0.3932	0.3560	0.3025	0.2437	0.1866	0.1359	0.0938
	2	0.0305	0.0984	0.1762	0.2458	0.2966	0.3241	0.3280	0.3110	0.2780	0.2344
	3	0.0021	0.0146	0.0415	0.0819	0.1318	0.1852	0.2355	0.2765	0.3032	0.3125
	4	0.0001	0.0012	0.0055	0.0154	0.0330	0.0595	0.0951	0.1382	0.1861	0.2344
	5	0.0000	0.0001	0.0004	0.0015	0.0044	0.0102	0.0205	0.0369	0.0609	0.0938
	6	0.0000	0.0000	0.0000	0.0001	0.0002	0.0007	0.0018	0.0041	0.0083	0.0156
7	0	0.6983	0.4783	0.3206	0.2097	0.1335	0.0824	0.0490	0.0280	0.0152	0.0078
	1	0.2573	0.3720	0.3960	0.3670	0.3115	0.2471	0.1848	0.1306	0.0872	0.0547
	2	0.0406	0.1240	0.2097	0.2753	0.3115	0.3177	0.2985	0.2613	0.2140	0.1641
	3	0.0036	0.0230	0.0617	0.1147	0.1730	0.2269	0.2679	0.2903	0.2918	0.2734
	4	0.0002	0.0026	0.0109	0.0287	0.0577	0.0972	0.1442	0.1935	0.2388	0.2734
	5	0.0000	0.0002	0.0012	0.0043	0.0115	0.0250	0.0466	0.0774	0.1172	0.1641
	6	0.0000	0.0000	0.0001	0.0004	0.0013	0.0036	0.0084	0.0172	0.0320	0.0547
	7	0.0000	0.0000	0.0000	0.0000	0.0001	0.0002	0.0006	0.0016	0.0037	0.0078

* Values of $P(X = x)$ for $\pi > 0.5$ are obtained in the table by using $(1 - \pi)$ as the probability of success and $(n - x)$ as the number of successes. For example, $P(X = 3 \mid 5, 0.7) = P(X = 2 \mid 5, 0.3) = 0.3087$.

APPENDIX B: BINOMIAL PROBABILITY MASS FUNCTION (*continued*)

Values of $P(X = x \mid n, \pi) = \dfrac{n!}{x!(n-x)!} \pi^x (1-\pi)^{n-x}$

							π					
n	x	0.05	0.10	0.15	0.20	0.25	0.30	0.35	0.40	0.45	0.50
8	0	0.6634	0.4305	0.2725	0.1678	0.1001	0.0576	0.0319	0.0168	0.0084	0.0039
	1	0.2793	0.3826	0.3847	0.3355	0.2670	0.1977	0.1373	0.0896	0.0548	0.0312
	2	0.0515	0.1488	0.2376	0.2936	0.3115	0.2965	0.2587	0.2090	0.1569	0.1094
	3	0.0054	0.0331	0.0839	0.1468	0.2076	0.2541	0.2786	0.2787	0.2568	0.2188
	4	0.0004	0.0046	0.0185	0.0459	0.0865	0.1361	0.1875	0.2322	0.2627	0.2734
	5	0.0000	0.0004	0.0026	0.0092	0.0231	0.0467	0.0808	0.1239	0.1719	0.2188
	6	0.0000	0.0000	0.0002	0.0011	0.0038	0.0100	0.0217	0.0413	0.0703	0.1094
	7	0.0000	0.0000	0.0000	0.0001	0.0004	0.0012	0.0033	0.0079	0.0164	0.0312
	8	0.0000	0.0000	0.0000	0.0000	0.0000	0.0001	0.0002	0.0007	0.0017	0.0039
9	0	0.6302	0.3874	0.2316	0.1342	0.0751	0.0404	0.0207	0.0101	0.0046	0.0020
	1	0.2985	0.3874	0.3679	0.3020	0.2253	0.1556	0.1004	0.0605	0.0339	0.0176
	2	0.0629	0.1722	0.2597	0.3020	0.3003	0.2668	0.2162	0.1612	0.1110	0.0703
	3	0.0077	0.0446	0.1069	0.1762	0.2336	0.2668	0.2716	0.2508	0.2119	0.1641
	4	0.0006	0.0074	0.0283	0.0661	0.1168	0.1715	0.2194	0.2508	0.2600	0.2461
	5	0.0000	0.0008	0.0050	0.0165	0.0389	0.0735	0.1181	0.1672	0.2128	0.2461
	6	0.0000	0.0001	0.0006	0.0028	0.0087	0.0210	0.0424	0.0743	0.1160	0.1641
	7	0.0000	0.0000	0.0000	0.0003	0.0012	0.0039	0.0098	0.0212	0.0407	0.0703
	8	0.0000	0.0000	0.0000	0.0000	0.0001	0.0004	0.0013	0.0035	0.0083	0.0176
	9	0.0000	0.0000	0.0000	0.0000	0.0000	0.0000	0.0001	0.0003	0.0008	0.0020
10	0	0.5987	0.3487	0.1969	0.1074	0.0563	0.0282	0.0135	0.0060	0.0025	0.0010
	1	0.3151	0.3874	0.3474	0.2684	0.1877	0.1211	0.0725	0.0403	0.0207	0.0098
	2	0.0746	0.1937	0.2759	0.3020	0.2816	0.2335	0.1757	0.1209	0.0763	0.0439
	3	0.0105	0.0574	0.1298	0.2013	0.2503	0.2668	0.2522	0.2150	0.1665	0.1172
	4	0.0010	0.0112	0.0401	0.0881	0.1460	0.2001	0.2377	0.2508	0.2384	0.2051
	5	0.0001	0.0015	0.0085	0.0264	0.0584	0.1029	0.1536	0.2007	0.2340	0.2461
	6	0.0000	0.0001	0.0012	0.0055	0.0162	0.0368	0.0689	0.1115	0.1596	0.2051
	7	0.0000	0.0000	0.0001	0.0008	0.0031	0.0090	0.0212	0.0425	0.0746	0.1172
	8	0.0000	0.0000	0.0000	0.0001	0.0004	0.0014	0.0043	0.0106	0.0229	0.0439
	9	0.0000	0.0000	0.0000	0.0000	0.0000	0.0001	0.0005	0.0016	0.0042	0.0098
	10	0.0000	0.0000	0.0000	0.0000	0.0000	0.0000	0.0000	0.0001	0.0003	0.0010
11	0	0.5688	0.3138	0.1673	0.0859	0.0422	0.0198	0.0088	0.0036	0.0014	0.0005
	1	0.3293	0.3835	0.3248	0.2362	0.1549	0.0932	0.0518	0.0266	0.0125	0.0054
	2	0.0867	0.2131	0.2866	0.2953	0.2581	0.1998	0.1395	0.0887	0.0513	0.0269
	3	0.0137	0.0710	0.1517	0.2215	0.2581	0.2568	0.2254	0.1774	0.1259	0.0806
	4	0.0014	0.0158	0.0536	0.1107	0.1721	0.2201	0.2428	0.2365	0.2060	0.1611
	5	0.0001	0.0025	0.0132	0.0388	0.0803	0.1321	0.1830	0.2207	0.2360	0.2256
	6	0.0000	0.0003	0.0023	0.0097	0.0268	0.0566	0.0985	0.1471	0.1931	0.2256
	7	0.0000	0.0000	0.0003	0.0017	0.0064	0.0173	0.0379	0.0701	0.1128	0.1611
	8	0.0000	0.0000	0.0000	0.0002	0.0011	0.0037	0.0102	0.0234	0.0462	0.0806
	9	0.0000	0.0000	0.0000	0.0000	0.0001	0.0005	0.0018	0.0052	0.0126	0.0269
	10	0.0000	0.0000	0.0000	0.0000	0.0000	0.0000	0.0002	0.0007	0.0021	0.0054
	11	0.0000	0.0000	0.0000	0.0000	0.0000	0.0000	0.0000	0.0000	0.0002	0.0005

$$\text{Values of } P(X = x \mid n, \pi) = \frac{n!}{x!(n-x)!} \pi^x (1 - \pi)^{n-x}$$

						π					
n	x	0.05	0.10	0.15	0.20	0.25	0.30	0.35	0.40	0.45	0.50
12	0	0.5404	0.2824	0.1422	0.0687	0.0317	0.0138	0.0057	0.0022	0.0008	0.0002
	1	0.3413	0.3766	0.3012	0.2062	0.1267	0.0712	0.0368	0.0174	0.0075	0.0029
	2	0.0988	0.2301	0.2924	0.2835	0.2323	0.1678	0.1088	0.0639	0.0339	0.0161
	3	0.0173	0.0852	0.1720	0.2362	0.2581	0.2397	0.1954	0.1419	0.0923	0.0537
	4	0.0021	0.0213	0.0683	0.1329	0.1936	0.2311	0.2367	0.2128	0.1700	0.1208
	5	0.0002	0.0038	0.0193	0.0532	0.1032	0.1585	0.2039	0.2270	0.2225	0.1934
	6	0.0000	0.0005	0.0040	0.0155	0.0401	0.0792	0.1281	0.1766	0.2124	0.2256
	7	0.0000	0.0000	0.0006	0.0033	0.0115	0.0291	0.0591	0.1009	0.1489	0.1934
	8	0.0000	0.0000	0.0001	0.0005	0.0024	0.0078	0.0199	0.0420	0.0762	0.1208
	9	0.0000	0.0000	0.0000	0.0001	0.0004	0.0015	0.0048	0.0125	0.0277	0.0537
	10	0.0000	0.0000	0.0000	0.0000	0.0000	0.0002	0.0008	0.0025	0.0068	0.0161
	11	0.0000	0.0000	0.0000	0.0000	0.0000	0.0000	0.0001	0.0003	0.0010	0.0029
	12	0.0000	0.0000	0.0000	0.0000	0.0000	0.0000	0.0000	0.0000	0.0001	0.0002
13	0	0.5133	0.2542	0.1209	0.0550	0.0238	0.0097	0.0037	0.0013	0.0004	0.0001
	1	0.3512	0.3672	0.2774	0.1787	0.1029	0.0540	0.0259	0.0113	0.0045	0.0016
	2	0.1109	0.2448	0.2937	0.2680	0.2059	0.1388	0.0836	0.0453	0.0220	0.0095
	3	0.0214	0.0997	0.1900	0.2457	0.2517	0.2181	0.1651	0.1107	0.0660	0.0349
	4	0.0028	0.0277	0.0838	0.1535	0.2097	0.2337	0.2222	0.1845	0.1350	0.0873
	5	0.0003	0.0055	0.0266	0.0691	0.1258	0.1803	0.2154	0.2214	0.1989	0.1571
	6	0.0000	0.0008	0.0063	0.0230	0.0559	0.1030	0.1546	0.1968	0.2169	0.2095
	7	0.0000	0.0001	0.0011	0.0058	0.0186	0.0442	0.0833	0.1312	0.1775	0.2095
	8	0.0000	0.0000	0.0001	0.0011	0.0047	0.0142	0.0336	0.0656	0.1089	0.1571
	9	0.0000	0.0000	0.0000	0.0001	0.0009	0.0034	0.0101	0.0243	0.0495	0.0873
	10	0.0000	0.0000	0.0000	0.0000	0.0001	0.0006	0.0022	0.0065	0.0162	0.0349
	11	0.0000	0.0000	0.0000	0.0000	0.0000	0.0001	0.0003	0.0012	0.0036	0.0095
	12	0.0000	0.0000	0.0000	0.0000	0.0000	0.0000	0.0000	0.0001	0.0005	0.0016
	13	0.0000	0.0000	0.0000	0.0000	0.0000	0.0000	0.0000	0.0000	0.0000	0.0001
14	0	0.4877	0.2288	0.1028	0.0440	0.0178	0.0068	0.0024	0.0008	0.0002	0.0001
	1	0.3593	0.3559	0.2539	0.1539	0.0832	0.0407	0.0181	0.0073	0.0027	0.0009
	2	0.1229	0.2570	0.2912	0.2501	0.1802	0.1134	0.0634	0.0317	0.0141	0.0056
	3	0.0259	0.1142	0.2056	0.2501	0.2402	0.1943	0.1366	0.0845	0.0462	0.0222
	4	0.0037	0.0349	0.0998	0.1720	0.2202	0.2290	0.2022	0.1549	0.1040	0.0611
	5	0.0004	0.0078	0.0352	0.0860	0.1468	0.1963	0.2178	0.2066	0.1701	0.1222
	6	0.0000	0.0013	0.0093	0.0322	0.0734	0.1262	0.1759	0.2066	0.2088	0.1833
	7	0.0000	0.0002	0.0019	0.0092	0.0280	0.0618	0.1082	0.1574	0.1952	0.2095
	8	0.0000	0.0000	0.0003	0.0020	0.0082	0.0232	0.0510	0.0918	0.1398	0.1833
	9	0.0000	0.0000	0.0000	0.0003	0.0018	0.0066	0.0183	0.0408	0.0762	0.1222
	10	0.0000	0.0000	0.0000	0.0000	0.0003	0.0014	0.0049	0.0136	0.0312	0.0611
	11	0.0000	0.0000	0.0000	0.0000	0.0000	0.0002	0.0010	0.0033	0.0093	0.0222
	12	0.0000	0.0000	0.0000	0.0000	0.0000	0.0000	0.0001	0.0005	0.0019	0.0056
	13	0.0000	0.0000	0.0000	0.0000	0.0000	0.0000	0.0000	0.0001	0.0002	0.0009
	14	0.0000	0.0000	0.0000	0.0000	0.0000	0.0000	0.0000	0.0000	0.0000	0.0001

$$\text{Values of } P(X = x \mid n, \pi) = \frac{n!}{x!(n - x)!} \, \pi^x(1 - \pi)^{n - x}$$

π

n	x	0.05	0.10	0.15	0.20	0.25	0.30	0.35	0.40	0.45	0.50
15	0	0.4633	0.2059	0.0874	0.0352	0.0134	0.0047	0.0016	0.0005	0.0001	0.0000
	1	0.3658	0.3432	0.2312	0.1319	0.0668	0.0305	0.0126	0.0047	0.0016	0.0005
	2	0.1348	0.2669	0.2856	0.2309	0.1559	0.0916	0.0476	0.0219	0.0090	0.0032
	3	0.0307	0.1285	0.2184	0.2501	0.2252	0.1700	0.1110	0.0634	0.0318	0.0139
	4	0.0049	0.0428	0.1156	0.1876	0.2252	0.2186	0.1792	0.1268	0.0780	0.0417
	5	0.0006	0.0105	0.0449	0.1032	0.1651	0.2061	0.2123	0.1859	0.1404	0.0916
	6	0.0000	0.0019	0.0132	0.0430	0.0917	0.1472	0.1906	0.2066	0.1914	0.1527
	7	0.0000	0.0003	0.0030	0.0138	0.0393	0.0811	0.1319	0.1771	0.2013	0.1964
	8	0.0000	0.0000	0.0005	0.0035	0.0131	0.0348	0.0710	0.1181	0.1647	0.1964
	9	0.0000	0.0000	0.0001	0.0007	0.0034	0.0116	0.0298	0.0612	0.1048	0.1527
	10	0.0000	0.0000	0.0000	0.0001	0.0007	0.0030	0.0096	0.0245	0.0515	0.0916
	11	0.0000	0.0000	0.0000	0.0000	0.0001	0.0006	0.0024	0.0074	0.0191	0.0417
	12	0.0000	0.0000	0.0000	0.0000	0.0000	0.0001	0.0004	0.0016	0.0052	0.0139
	13	0.0000	0.0000	0.0000	0.0000	0.0000	0.0000	0.0001	0.0003	0.0010	0.0032
	14	0.0000	0.0000	0.0000	0.0000	0.0000	0.0000	0.0000	0.0000	0.0001	0.0005
	15	0.0000	0.0000	0.0000	0.0000	0.0000	0.0000	0.0000	0.0000	0.0000	0.0000
16	0	0.4401	0.1853	0.0743	0.0281	0.0100	0.0033	0.0010	0.0003	0.0001	0.0000
	1	0.3706	0.3294	0.2097	0.1126	0.0535	0.0228	0.0087	0.0030	0.0009	0.0002
	2	0.1463	0.2745	0.2775	0.2111	0.1336	0.0732	0.0353	0.0150	0.0056	0.0018
	3	0.0359	0.1423	0.2285	0.2463	0.2079	0.1465	0.0888	0.0468	0.0215	0.0085
	4	0.0061	0.0514	0.1311	0.2001	0.2252	0.2040	0.1553	0.1014	0.0572	0.0278
	5	0.0008	0.0137	0.0555	0.1201	0.1802	0.2099	0.2008	0.1623	0.1123	0.0667
	6	0.0001	0.0028	0.0180	0.0550	0.1101	0.1649	0.1982	0.1983	0.1684	0.1222
	7	0.0000	0.0004	0.0045	0.0197	0.0524	0.1010	0.1524	0.1889	0.1969	0.1746
	8	0.0000	0.0001	0.0009	0.0055	0.0197	0.0487	0.0923	0.1417	0.1812	0.1964
	9	0.0000	0.0000	0.0001	0.0012	0.0058	0.0185	0.0442	0.0840	0.1318	0.1746
	10	0.0000	0.0000	0.0000	0.0002	0.0014	0.0056	0.0167	0.0392	0.0755	0.1222
	11	0.0000	0.0000	0.0000	0.0000	0.0002	0.0013	0.0049	0.0142	0.0337	0.0667
	12	0.0000	0.0000	0.0000	0.0000	0.0000	0.0002	0.0011	0.0040	0.0115	0.0278
	13	0.0000	0.0000	0.0000	0.0000	0.0000	0.0000	0.0002	0.0008	0.0029	0.0085
	14	0.0000	0.0000	0.0000	0.0000	0.0000	0.0000	0.0000	0.0001	0.0005	0.0018
	15	0.0000	0.0000	0.0000	0.0000	0.0000	0.0000	0.0000	0.0000	0.0001	0.0002
	16	0.0000	0.0000	0.0000	0.0000	0.0000	0.0000	0.0000	0.0000	0.0000	0.0000
17	0	0.4181	0.1668	0.0631	0.0225	0.0075	0.0023	0.0007	0.0002	0.0000	0.0000
	1	0.3741	0.3150	0.1893	0.0957	0.0426	0.0169	0.0060	0.0019	0.0005	0.0001
	2	0.1575	0.2800	0.2673	0.1914	0.1136	0.0581	0.0260	0.0102	0.0035	0.0010
	3	0.0415	0.1556	0.2359	0.2393	0.1893	0.1245	0.0701	0.0341	0.0144	0.0052
	4	0.0076	0.0605	0.1457	0.2093	0.2209	0.1868	0.1320	0.0796	0.0411	0.0182
	5	0.0010	0.0175	0.0668	0.1361	0.1914	0.2081	0.1849	0.1379	0.0875	0.0472
	6	0.0001	0.0039	0.0236	0.0680	0.1276	0.1784	0.1991	0.1839	0.1432	0.0944
	7	0.0000	0.0007	0.0065	0.0267	0.0668	0.1201	0.1685	0.1927	0.1841	0.1484
	8	0.0000	0.0001	0.0014	0.0084	0.0279	0.0644	0.1134	0.1606	0.1883	0.1855
	9	0.0000	0.0000	0.0003	0.0021	0.0093	0.0276	0.0611	0.1070	0.1540	0.1855

$$\text{Values of } P(X = x \mid n, \pi) = \frac{n!}{x!(n-x)!} \, \pi^x (1 - \pi)^{n-x}$$

π

n	x	0.05	0.10	0.15	0.20	0.25	0.30	0.35	0.40	0.45	0.50
	10	0.0000	0.0000	0.0000	0.0004	0.0025	0.0095	0.0263	0.0571	0.1008	0.1484
	11	0.0000	0.0000	0.0000	0.0001	0.0005	0.0026	0.0090	0.0242	0.0525	0.0944
	12	0.0000	0.0000	0.0000	0.0000	0.0001	0.0006	0.0024	0.0081	0.0215	0.0472
	13	0.0000	0.0000	0.0000	0.0000	0.0000	0.0001	0.0005	0.0021	0.0068	0.0182
	14	0.0000	0.0000	0.0000	0.0000	0.0000	0.0000	0.0001	0.0004	0.0016	0.0052
	15	0.0000	0.0000	0.0000	0.0000	0.0000	0.0000	0.0000	0.0001	0.0003	0.0010
	16	0.0000	0.0000	0.0000	0.0000	0.0000	0.0000	0.0000	0.0000	0.0000	0.0001
	17	0.0000	0.0000	0.0000	0.0000	0.0000	0.0000	0.0000	0.0000	0.0000	0.0000
18	0	0.3972	0.1501	0.0536	0.0180	0.0056	0.0016	0.0004	0.0001	0.0000	0.0000
	1	0.3763	0.3002	0.1704	0.0811	0.0338	0.0126	0.0042	0.0012	0.0003	0.0001
	2	0.1683	0.2835	0.2556	0.1723	0.0958	0.0458	0.0190	0.0069	0.0022	0.0006
	3	0.0473	0.1680	0.2406	0.2297	0.1704	0.1046	0.0547	0.0246	0.0095	0.0031
	4	0.0093	0.0700	0.1592	0.2153	0.2130	0.1681	0.1104	0.0614	0.0291	0.0117
	5	0.0014	0.0218	0.0787	0.1507	0.1988	0.2017	0.1664	0.1146	0.0666	0.0327
	6	0.0002	0.0052	0.0301	0.0816	0.1436	0.1873	0.1941	0.1655	0.1181	0.0708
	7	0.0000	0.0010	0.0091	0.0350	0.0820	0.1376	0.1792	0.1892	0.1657	0.1214
	8	0.0000	0.0002	0.0022	0.0120	0.0376	0.0811	0.1327	0.1734	0.1864	0.1669
	9	0.0000	0.0000	0.0004	0.0033	0.0139	0.0386	0.0794	0.1284	0.1694	0.1855
	10	0.0000	0.0000	0.0001	0.0008	0.0042	0.0149	0.0385	0.0771	0.1248	0.1669
	11	0.0000	0.0000	0.0000	0.0001	0.0010	0.0046	0.0151	0.0374	0.0742	0.1214
	12	0.0000	0.0000	0.0000	0.0000	0.0002	0.0012	0.0047	0.0145	0.0354	0.0708
	13	0.0000	0.0000	0.0000	0.0000	0.0000	0.0002	0.0012	0.0045	0.0134	0.0327
	14	0.0000	0.0000	0.0000	0.0000	0.0000	0.0000	0.0002	0.0011	0.0039	0.0117
	15	0.0000	0.0000	0.0000	0.0000	0.0000	0.0000	0.0000	0.0002	0.0009	0.0031
	16	0.0000	0.0000	0.0000	0.0000	0.0000	0.0000	0.0000	0.0000	0.0001	0.0006
	17	0.0000	0.0000	0.0000	0.0000	0.0000	0.0000	0.0000	0.0000	0.0000	0.0001
	18	0.0000	0.0000	0.0000	0.0000	0.0000	0.0000	0.0000	0.0000	0.0000	0.0000
19	0	0.3774	0.1351	0.0456	0.0144	0.0042	0.0011	0.0003	0.0001	0.0000	0.0000
	1	0.3774	0.2852	0.1529	0.0685	0.0268	0.0093	0.0029	0.0008	0.0002	0.0000
	2	0.1787	0.2852	0.2428	0.1540	0.0803	0.0358	0.0138	0.0046	0.0013	0.0003
	3	0.0533	0.1796	0.2428	0.2182	0.1517	0.0869	0.0422	0.0175	0.0062	0.0018
	4	0.0112	0.0798	0.1714	0.2182	0.2023	0.1491	0.0909	0.0467	0.0203	0.0074
	5	0.0018	0.0266	0.0907	0.1636	0.2023	0.1916	0.1468	0.0933	0.0497	0.0222
	6	0.0002	0.0069	0.0374	0.0955	0.1574	0.1916	0.1844	0.1451	0.0949	0.0518
	7	0.0000	0.0014	0.0122	0.0443	0.0974	0.1525	0.1844	0.1797	0.1443	0.0961
	8	0.0000	0.0002	0.0032	0.0166	0.0487	0.0981	0.1489	0.1797	0.1771	0.1442
	9	0.0000	0.0000	0.0007	0.0051	0.0198	0.0514	0.0980	0.1464	0.1771	0.1762
	10	0.0000	0.0000	0.0001	0.0013	0.0066	0.0220	0.0528	0.0976	0.1449	0.1762
	11	0.0000	0.0000	0.0000	0.0003	0.0018	0.0077	0.0233	0.0532	0.0970	0.1442
	12	0.0000	0.0000	0.0000	0.0000	0.0004	0.0022	0.0083	0.0237	0.0529	0.0961
	13	0.0000	0.0000	0.0000	0.0000	0.0001	0.0005	0.0024	0.0085	0.0233	0.0518
	14	0.0000	0.0000	0.0000	0.0000	0.0000	0.0001	0.0006	0.0024	0.0082	0.0222

APPENDIX B: BINOMIAL PROBABILITY MASS FUNCTION (*concluded*)

$$\text{Values of } P(X = x \mid n, \pi) = \frac{n!}{x!(n - x)!} \pi^x(1 - \pi)^{n - x}$$

| | | | | | | | π | | | | | |
n	x	0.05	0.10	0.15	0.20	0.25	0.30	0.35	0.40	0.45	0.50
	15	0.0000	0.0000	0.0000	0.0000	0.0000	0.0000	0.0001	0.0005	0.0022	0.0074
	16	0.0000	0.0000	0.0000	0.0000	0.0000	0.0000	0.0000	0.0001	0.0005	0.0018
	17	0.0000	0.0000	0.0000	0.0000	0.0000	0.0000	0.0000	0.0000	0.0001	0.0003
	18	0.0000	0.0000	0.0000	0.0000	0.0000	0.0000	0.0000	0.0000	0.0000	0.0000
	19	0.0000	0.0000	0.0000	0.0000	0.0000	0.0000	0.0000	0.0000	0.0000	0.0000
20	0	0.3585	0.1216	0.0388	0.0115	0.0032	0.0008	0.0002	0.0000	0.0000	0.0000
	1	0.3774	0.2702	0.1368	0.0576	0.0211	0.0068	0.0020	0.0005	0.0001	0.0000
	2	0.1887	0.2852	0.2293	0.1369	0.0669	0.0278	0.0100	0.0031	0.0008	0.0002
	3	0.0596	0.1901	0.2428	0.2054	0.1339	0.0716	0.0323	0.0123	0.0040	0.0011
	4	0.0133	0.0898	0.1821	0.2182	0.1897	0.1304	0.0738	0.0350	0.0139	0.0046
	5	0.0022	0.0319	0.1028	0.1746	0.2023	0.1789	0.1272	0.0746	0.0365	0.0148
	6	0.0003	0.0089	0.0454	0.1091	0.1686	0.1916	0.1712	0.1244	0.0746	0.0370
	7	0.0000	0.0020	0.0160	0.0545	0.1124	0.1643	0.1844	0.1659	0.1221	0.0739
	8	0.0000	0.0004	0.0046	0.0222	0.0609	0.1144	0.1614	0.1797	0.1623	0.1201
	9	0.0000	0.0001	0.0011	0.0074	0.0271	0.0654	0.1158	0.1597	0.1771	0.1602
	10	0.0000	0.0000	0.0002	0.0020	0.0099	0.0308	0.0686	0.1171	0.1593	0.1762
	11	0.0000	0.0000	0.0000	0.0005	0.0030	0.0120	0.0336	0.0710	0.1185	0.1602
	12	0.0000	0.0000	0.0000	0.0001	0.0008	0.0039	0.0136	0.0355	0.0727	0.1201
	13	0.0000	0.0000	0.0000	0.0000	0.0002	0.0010	0.0045	0.0146	0.0366	0.0739
	14	0.0000	0.0000	0.0000	0.0000	0.0000	0.0002	0.0012	0.0049	0.0150	0.0370
	15	0.0000	0.0000	0.0000	0.0000	0.0000	0.0000	0.0003	0.0013	0.0049	0.0148
	16	0.0000	0.0000	0.0000	0.0000	0.0000	0.0000	0.0000	0.0003	0.0013	0.0046
	17	0.0000	0.0000	0.0000	0.0000	0.0000	0.0000	0.0000	0.0000	0.0002	0.0011
	18	0.0000	0.0000	0.0000	0.0000	0.0000	0.0000	0.0000	0.0000	0.0000	0.0002
	19	0.0000	0.0000	0.0000	0.0000	0.0000	0.0000	0.0000	0.0000	0.0000	0.0000
	20	0.0000	0.0000	0.0000	0.0000	0.0000	0.0000	0.0000	0.0000	0.0000	0.0000

Source: Harold J. Larson, *Statistics: An Introduction* (New York: John Wiley & Sons, Inc., 1975). Copyright © 1975 by John Wiley & Sons, Inc. Reprinted by permission of John Wiley & Sons, Inc.

APPENDIX C: COMBINATIONS AND PERMUTATIONS

$$\text{Values of } \frac{n!}{x!(n-x)!} \text{ for } n = 1(1)20 \text{ and } x = 0(1)10.*$$

This gives (1) the number of combinations of x out of n distinct items, $_nC_x$; and (2) the number of permutations of n items where x are of one kind and $n - x$ are of another kind, $_nP_{x, n-x}$.

n \ x	0	1	2	3	4	5	6	7	8	9	10
0	1										
1	1	1									
2	1	2	1								
3	1	3	3	1							
4	1	4	6	4	1						
5	1	5	10	10	5	1					
6	1	6	15	20	15	6	1				
7	1	7	21	35	35	21	7	1			
8	1	8	28	56	70	56	28	8	1		
9	1	9	36	84	126	126	84	36	9	1	
10	1	10	45	120	210	252	210	120	45	10	1
11	1	11	55	165	330	462	462	330	165	55	11
12	1	12	66	220	495	792	924	792	495	220	66
13	1	13	78	286	715	1 287	1 716	1 716	1 287	715	286
14	1	14	91	364	1 001	2 002	3 003	3 432	3 003	2 002	1 001
15	1	15	105	455	1 365	3 003	5 005	6 435	6 435	5 005	3 003
16	1	16	120	560	1 820	4 368	8 008	11 440	12 870	11 440	8 008
17	1	17	136	680	2 380	6 188	12 376	19 448	24 310	24 310	19 448
18	1	18	153	816	3 060	8 568	18 564	31 824	43 758	48 620	43 758
19	1	19	171	969	3 876	11 628	27 132	50 388	75 582	92 378	92 378
20	1	20	190	1 140	4 845	15 504	38 760	77 520	125 970	167 960	184 756

* If $x > 10$, use the relation $_nC_x = {_nC_{n-x}}$. For example, $_{20}C_{11} = {_{20}C_9} = 167{,}960$.

Source: Reprinted with permission from *C.R.C. Standard Mathematical Tables* (Cleveland, Ohio: Chemical Rubber Publishing Co., 12th ed., 1959), p. 388.

APPENDIX D: POISSON PROBABILITY MASS FUNCTION

$$\text{Values of } P(X = x \,|\, \lambda t) = \frac{(\lambda t)^x e^{-\lambda t}}{x!}$$

for $\lambda t = 0.005, 0.01(0.01)0.09, 0.1(0.1)10$. Values less than 0.00005 are shown as 0.0000.

λt

x	0.005	0.01	0.02	0.03	0.04	0.05	0.06	0.07	0.08	0.09
0	0.9950	0.9900	0.9802	0.9704	0.9608	0.9512	0.9418	0.9324	0.9231	0.9139
1	0.0050	0.0099	0.0192	0.0291	0.0384	0.0476	0.0565	0.0653	0.0738	0.0823
2	0.0000	0.0000	0.0002	0.0004	0.0008	0.0012	0.0017	0.0023	0.0030	0.0037
3	0.0000	0.0000	0.0000	0.0000	0.0000	0.0000	0.0000	0.0001	0.0001	0.0001

x	0.1	0.2	0.3	0.4	0.5	0.6	0.7	0.8	0.9	1.0
0	0.9048	0.8187	0.7408	0.6703	0.6065	0.5488	0.4966	0.4493	0.4066	0.3679
1	0.0905	0.1637	0.2222	0.2681	0.3033	0.3293	0.3476	0.3595	0.3659	0.3679
2	0.0045	0.0164	0.0333	0.0536	0.0758	0.0988	0.1217	0.1438	0.1647	0.1839
3	0.0002	0.0011	0.0033	0.0072	0.0126	0.0198	0.0284	0.0383	0.0494	0.0613
4	0.0000	0.0001	0.0002	0.0007	0.0016	0.0030	0.0050	0.0077	0.0111	0.0153
5	0.0000	0.0000	0.0000	0.0001	0.0002	0.0004	0.0007	0.0012	0.0020	0.0031
6	0.0000	0.0000	0.0000	0.0000	0.0000	0.0000	0.0001	0.0002	0.0003	0.0005
7	0.0000	0.0000	0.0000	0.0000	0.0000	0.0000	0.0000	0.0000	0.0000	0.0001

x	1.1	1.2	1.3	1.4	1.5	1.6	1.7	1.8	1.9	2.0
0	0.3329	0.3012	0.2725	0.2466	0.2231	0.2019	0.1827	0.1653	0.1496	0.1353
1	0.3662	0.3614	0.3543	0.3452	0.3347	0.3230	0.3106	0.2975	0.2842	0.2707
2	0.2014	0.2169	0.2303	0.2417	0.2510	0.2584	0.2640	0.2678	0.2700	0.2707
3	0.0738	0.0867	0.0998	0.1128	0.1255	0.1378	0.1496	0.1607	0.1710	0.1804
4	0.0203	0.0260	0.0324	0.0395	0.0471	0.0551	0.0636	0.0723	0.0812	0.0902
5	0.0045	0.0062	0.0084	0.0111	0.0141	0.0176	0.0216	0.0260	0.0309	0.0361
6	0.0008	0.0012	0.0018	0.0026	0.0035	0.0047	0.0061	0.0078	0.0098	0.0120
7	0.0001	0.0002	0.0003	0.0005	0.0008	0.0011	0.0015	0.0020	0.0027	0.0034
8	0.0000	0.0000	0.0001	0.0001	0.0001	0.0002	0.0003	0.0005	0.0006	0.0009
9	0.0000	0.0000	0.0000	0.0000	0.0000	0.0000	0.0001	0.0001	0.0001	0.0002

x	2.1	2.2	2.3	2.4	2.5	2.6	2.7	2.8	2.9	3.0
0	0.1225	0.1108	0.1003	0.0907	0.0821	0.0743	0.0672	0.0608	0.0050	0.0498
1	0.2572	0.2438	0.2306	0.2177	0.2052	0.1931	0.1815	0.1703	0.1596	0.1494
2	0.2700	0.2681	0.2652	0.2613	0.2565	0.2510	0.2450	0.2384	0.2314	0.2240
3	0.1890	0.1966	0.2033	0.2090	0.2138	0.2176	0.2205	0.2225	0.2237	0.2240
4	0.0992	0.1082	0.1169	0.1254	0.1336	0.1414	0.1488	0.1557	0.1622	0.1680
5	0.0417	0.0476	0.0538	0.0602	0.0668	0.0735	0.0804	0.0872	0.0940	0.1008
6	0.0146	0.0174	0.0206	0.0241	0.0278	0.0319	0.0362	0.0407	0.0455	0.0504
7	0.0044	0.0055	0.0068	0.0083	0.0099	0.0118	0.0139	0.0163	0.0188	0.0216
8	0.0011	0.0015	0.0019	0.0025	0.0031	0.0038	0.0047	0.0057	0.0068	0.0081
9	0.0003	0.0004	0.0005	0.0007	0.0009	0.0011	0.0014	0.0018	0.0022	0.0027
10	0.0001	0.0001	0.0001	0.0002	0.0002	0.0003	0.0004	0.0005	0.0006	0.0008
11	0.0000	0.0000	0.0000	0.0000	0.0000	0.0001	0.0001	0.0001	0.0002	0.0002
12	0.0000	0.0000	0.0000	0.0000	0.0000	0.0000	0.0000	0.0000	0.0000	0.0001

$$\text{Values of } P(X = x \mid \lambda t) = \frac{(\lambda t)^x e^{-\lambda t}}{x!}$$

λt

x	3.1	3.2	3.3	3.4	3.5	3.6	3.7	3.8	3.9	4.0
0	0.0450	0.0408	0.0369	0.0334	0.0302	0.0273	0.0247	0.0224	0.0202	0.0183
1	0.1397	0.1304	0.1217	0.1135	0.1057	0.0984	0.0915	0.0850	0.0789	0.0733
2	0.2165	0.2087	0.2008	0.1929	0.1850	0.1771	0.1692	0.1615	0.1539	0.1465
3	0.2237	0.2226	0.2209	0.2186	0.2158	0.2125	0.2087	0.2046	0.2001	0.1954
4	0.1734	0.1781	0.1823	0.1858	0.1888	0.1912	0.1931	0.1944	0.1951	0.1954
5	0.1075	0.1140	0.1203	0.1264	0.1322	0.1377	0.1429	0.1477	0.1522	0.1563
6	0.0555	0.0608	0.0662	0.0716	0.0771	0.0826	0.0881	0.0936	0.0989	0.1042
7	0.0246	0.0278	0.0312	0.0348	0.0385	0.0425	0.0466	0.0508	0.0551	0.0595
8	0.0095	0.0111	0.0129	0.0148	0.0169	0.0191	0.0215	0.0241	0.0269	0.0298
9	0.0033	0.0040	0.0047	0.0056	0.0066	0.0076	0.0089	0.0102	0.0116	0.0132
10	0.0010	0.0013	0.0016	0.0019	0.0023	0.0028	0.0033	0.0039	0.0045	0.0053
11	0.0003	0.0004	0.0005	0.0006	0.0007	0.0009	0.0011	0.0013	0.0016	0.0019
12	0.0001	0.0001	0.0001	0.0002	0.0002	0.0003	0.0003	0.0004	0.0005	0.0006
13	0.0000	0.0000	0.0000	0.0000	0.0001	0.0001	0.0001	0.0001	0.0002	0.0002
14	0.0000	0.0000	0.0000	0.0000	0.0000	0.0000	0.0000	0.0000	0.0000	0.0001

x	4.1	4.2	4.3	4.4	4.5	4.6	4.7	4.8	4.9	5.0
0	0.0166	0.0150	0.0136	0.0123	0.0111	0.0101	0.0091	0.0082	0.0074	0.0067
1	0.0679	0.0630	0.0583	0.0540	0.0500	0.0462	0.0427	0.0395	0.0365	0.0337
2	0.1393	0.1323	0.1254	0.1188	0.1125	0.1063	0.1005	0.0948	0.0894	0.0842
3	0.1904	0.1852	0.1798	0.1743	0.1687	0.1631	0.1574	0.1517	0.1460	0.1404
4	0.1951	0.1944	0.1933	0.1917	0.1898	0.1875	0.1849	0.1820	0.1789	0.1755
5	0.1600	0.1633	0.1662	0.1687	0.1708	0.1725	0.1738	0.1747	0.1753	0.1755
6	0.1093	0.1143	0.1191	0.1237	0.1281	0.1323	0.1362	0.1398	0.1432	0.1462
7	0.0640	0.0686	0.0732	0.0778	0.0824	0.0869	0.0914	0.0959	0.1002	0.1044
8	0.0328	0.0360	0.0393	0.0428	0.0463	0.0500	0.0537	0.0575	0.0614	0.0653
9	0.0150	0.0168	0.0188	0.0209	0.0232	0.0255	0.0280	0.0307	0.0334	0.0363
10	0.0061	0.0071	0.0081	0.0092	0.0104	0.0118	0.0132	0.0147	0.0164	0.0181
11	0.0023	0.0027	0.0032	0.0037	0.0043	0.0049	0.0056	0.0064	0.0073	0.0082
12	0.0008	0.0009	0.0011	0.0014	0.0016	0.0019	0.0022	0.0026	0.0030	0.0034
13	0.0002	0.0003	0.0004	0.0005	0.0006	0.0007	0.0008	0.0009	0.0011	0.0013
14	0.0001	0.0001	0.0001	0.0001	0.0002	0.0002	0.0003	0.0003	0.0004	0.0005
15	0.0000	0.0000	0.0000	0.0000	0.0001	0.0001	0.0001	0.0001	0.0001	0.0002

x	5.1	5.2	5.3	5.4	5.5	5.6	5.7	5.8	5.9	6.0
0	0.0061	0.0055	0.0050	0.0045	0.0041	0.0037	0.0033	0.0030	0.0027	0.0025
1	0.0311	0.0287	0.0265	0.0244	0.0225	0.0207	0.0191	0.0176	0.0162	0.0149
2	0.0793	0.0746	0.0701	0.0659	0.0618	0.0580	0.0544	0.0509	0.0477	0.0446
3	0.1348	0.1293	0.1239	0.1185	0.1133	0.1082	0.1033	0.0985	0.0938	0.0892
4	0.1719	0.1681	0.1641	0.1600	0.1558	0.1515	0.1472	0.1428	0.1383	0.1339

$$\text{Values of } P(X = x \mid \lambda t) = \frac{(\lambda t)^x e^{-\lambda t}}{x!}$$

λt

x	5.1	5.2	5.3	5.4	5.5	5.6	5.7	5.8	5.9	6.0
5	0.1753	0.1748	0.1740	0.1728	0.1714	0.1697	0.1678	0.1656	0.1632	0.1606
6	0.1490	0.1515	0.1537	0.1555	0.1571	0.1584	0.1594	0.1601	0.1605	0.1606
7	0.1086	0.1125	0.1163	0.1200	0.1234	0.1267	0.1298	0.1326	0.1353	0.1377
8	0.0692	0.0731	0.0771	0.0810	0.0849	0.0887	0.0925	0.0962	0.0998	0.1033
9	0.0392	0.0423	0.0454	0.0486	0.0519	0.0552	0.0586	0.0620	0.0654	0.0688
10	0.0200	0.0220	0.0241	0.0262	0.0285	0.0309	0.0334	0.0359	0.0386	0.0413
11	0.0093	0.0104	0.0116	0.0129	0.0143	0.0157	0.0173	0.0190	0.0207	0.0225
12	0.0039	0.0045	0.0051	0.0058	0.0065	0.0073	0.0082	0.0092	0.0102	0.0113
13	0.0015	0.0018	0.0021	0.0024	0.0028	0.0032	0.0036	0.0041	0.0046	0.0052
14	0.0006	0.0007	0.0008	0.0009	0.0011	0.0013	0.0015	0.0017	0.0019	0.0022
15	0.0002	0.0002	0.0003	0.0003	0.0004	0.0005	0.0006	0.0007	0.0008	0.0009
16	0.0001	0.0001	0.0001	0.0001	0.0001	0.0002	0.0002	0.0002	0.0003	0.0003
17	0.0000	0.0000	0.0000	0.0000	0.0000	0.0001	0.0001	0.0001	0.0001	0.0001

x	6.1	6.2	6.3	6.4	6.5	6.6	6.7	6.8	6.9	7.0
0	0.0022	0.0020	0.0018	0.0017	0.0015	0.0014	0.0012	0.0011	0.0010	0.0009
1	0.0137	0.0126	0.0116	0.0106	0.0098	0.0090	0.0082	0.0076	0.0070	0.0064
2	0.0417	0.0390	0.0364	0.0340	0.0318	0.0296	0.0276	0.0258	0.0240	0.0223
3	0.0848	0.0806	0.0765	0.0726	0.0688	0.0652	0.0617	0.0584	0.0552	0.0521
4	0.1294	0.1249	0.1205	0.1162	0.1118	0.1076	0.1034	0.0992	0.0952	0.0912
5	0.1579	0.1549	0.1519	0.1487	0.1454	0.1420	0.1385	0.1349	0.1314	0.1277
6	0.1605	0.1601	0.1595	0.1586	0.1575	0.1562	0.1546	0.1529	0.1511	0.1490
7	0.1399	0.1418	0.1435	0.1450	0.1462	0.1472	0.1480	0.1486	0.1489	0.1490
8	0.1066	0.1099	0.1130	0.1160	0.1188	0.1215	0.1240	0.1263	0.1284	0.1304
9	0.0723	0.0757	0.0791	0.0825	0.0858	0.0891	0.0923	0.0954	0.0985	0.1014
10	0.0441	0.0469	0.0498	0.0528	0.0558	0.0588	0.0618	0.0649	0.0679	0.0710
11	0.0245	0.0265	0.0285	0.0307	0.0330	0.0353	0.0377	0.0401	0.0426	0.0452
12	0.0124	0.0137	0.0150	0.0164	0.0179	0.0194	0.0210	0.0227	0.0245	0.0264
13	0.0058	0.0065	0.0073	0.0081	0.0089	0.0098	0.0108	0.0119	0.0130	0.0142
14	0.0025	0.0029	0.0033	0.0037	0.0041	0.0046	0.0052	0.0058	0.0064	0.0071
15	0.0010	0.0012	0.0014	0.0016	0.0018	0.0020	0.0023	0.0026	0.0029	0.0033
16	0.0004	0.0005	0.0005	0.0006	0.0007	0.0008	0.0010	0.0011	0.0013	0.0014
17	0.0001	0.0002	0.0002	0.0002	0.0003	0.0003	0.0004	0.0004	0.0005	0.0006
18	0.0000	0.0001	0.0001	0.0001	0.0001	0.0001	0.0001	0.0002	0.0002	0.0002
19	0.0000	0.0000	0.0000	0.0000	0.0000	0.0000	0.0000	0.0001	0.0001	0.0001

x	7.1	7.2	7.3	7.4	7.5	7.6	7.7	7.8	7.9	8.0
0	0.0008	0.0007	0.0007	0.0006	0.0006	0.0005	0.0005	0.0004	0.0004	0.0003
1	0.0059	0.0054	0.0049	0.0045	0.0041	0.0038	0.0035	0.0032	0.0029	0.0027
2	0.0208	0.0194	0.0180	0.0167	0.0156	0.0145	0.0134	0.0125	0.0116	0.0107
3	0.0492	0.0464	0.0438	0.0413	0.0389	0.0366	0.0345	0.0324	0.0305	0.0286
4	0.0874	0.0836	0.0799	0.0764	0.0729	0.0696	0.0663	0.0632	0.0602	0.0573

Values of $P(X = x \mid \lambda t) = \dfrac{(\lambda t)^x e^{-\lambda t}}{x!}$

λt

x	7.1	7.2	7.3	7.4	7.5	7.6	7.7	7.8	7.9	8.0
5	0.1241	0.1204	0.1167	0.1130	0.1094	0.1057	0.1021	0.0986	0.0951	0.0916
6	0.1468	0.1445	0.1420	0.1394	0.1367	0.1339	0.1311	0.1282	0.1252	0.1221
7	0.1489	0.1486	0.1481	0.1474	0.1465	0.1454	0.1442	0.1428	0.1413	0.1396
8	0.1321	0.1337	0.1351	0.1363	0.1373	0.1382	0.1388	0.1392	0.1395	0.1396
9	0.1042	0.1070	0.1096	0.1121	0.1144	0.1167	0.1187	0.1207	0.1224	0.1241
10	0.0740	0.0770	0.0800	0.0829	0.0858	0.0887	0.0914	0.0941	0.0967	0.0993
11	0.0478	0.0504	0.0531	0.0558	0.0585	0.0613	0.0640	0.0667	0.0695	0.0722
12	0.0283	0.0303	0.0323	0.0344	0.0366	0.0388	0.0411	0.0434	0.0457	0.0481
13	0.0154	0.0168	0.0181	0.0196	0.0211	0.0227	0.0243	0.0260	0.0278	0.0296
14	0.0078	0.0086	0.0095	0.0104	0.0113	0.0123	0.0134	0.0145	0.0157	0.0169
15	0.0037	0.0041	0.0046	0.0051	0.0057	0.0062	0.0069	0.0075	0.0083	0.0090
16	0.0016	0.0019	0.0021	0.0024	0.0026	0.0030	0.0033	0.0037	0.0041	0.0045
17	0.0007	0.0008	0.0009	0.0010	0.0012	0.0013	0.0015	0.0017	0.0019	0.0021
18	0.0003	0.0003	0.0004	0.0004	0.0005	0.0006	0.0006	0.0007	0.0008	0.0009
19	0.0001	0.0001	0.0001	0.0002	0.0002	0.0002	0.0003	0.0003	0.0003	0.0004
20	0.0000	0.0000	0.0001	0.0001	0.0001	0.0001	0.0001	0.0001	0.0001	0.0002
21	0.0000	0.0000	0.0000	0.0000	0.0000	0.0000	0.0000	0.0000	0.0001	0.0001

x	8.1	8.2	8.3	8.4	8.5	8.6	8.7	8.8	8.9	9.0
0	0.0003	0.0003	0.0002	0.0002	0.0002	0.0002	0.0002	0.0002	0.0001	0.0001
1	0.0025	0.0023	0.0021	0.0019	0.0017	0.0016	0.0014	0.0013	0.0012	0.0011
2	0.0100	0.0092	0.0086	0.0079	0.0074	0.0068	0.0063	0.0058	0.0054	0.0050
3	0.0269	0.0252	0.0237	0.0222	0.0208	0.0195	0.0183	0.0171	0.0160	0.0150
4	0.0544	0.0517	0.0491	0.0466	0.0443	0.0420	0.0398	0.0377	0.0357	0.0337
5	0.0882	0.0849	0.0816	0.0784	0.0752	0.0722	0.0692	0.0663	0.0635	0.0607
6	0.1191	0.1160	0.1128	0.1097	0.1066	0.1034	0.1003	0.0972	0.0941	0.0911
7	0.1378	0.1358	0.1338	0.1317	0.1294	0.1271	0.1247	0.1222	0.1197	0.1171
8	0.1395	0.1392	0.1388	0.1382	0.1375	0.1366	0.1356	0.1344	0.1332	0.1318
9	0.1256	0.1269	0.1280	0.1290	0.1299	0.1306	0.1311	0.1315	0.1317	0.1318
10	0.1017	0.1040	0.1063	0.1084	0.1104	0.1123	0.1140	0.1157	0.1172	0.1186
11	0.0749	0.0776	0.0802	0.0828	0.0853	0.0878	0.0902	0.0925	0.0948	0.0970
12	0.0505	0.0530	0.0555	0.0579	0.0604	0.0629	0.0654	0.0679	0.0703	0.0728
13	0.0315	0.0334	0.0354	0.0374	0.0395	0.0416	0.0438	0.0459	0.0481	0.0504
14	0.0182	0.0196	0.0210	0.0225	0.0240	0.0256	0.0272	0.0289	0.0306	0.0324
15	0.0098	0.0107	0.0116	0.0126	0.0136	0.0147	0.0158	0.0169	0.0182	0.0194
16	0.0050	0.0055	0.0060	0.0066	0.0072	0.0079	0.0086	0.0093	0.0101	0.0109
17	0.0024	0.0026	0.0029	0.0033	0.0036	0.0040	0.0044	0.0048	0.0053	0.0058
18	0.0011	0.0012	0.0014	0.0015	0.0017	0.0019	0.0021	0.0024	0.0026	0.0029
19	0.0005	0.0005	0.0006	0.0007	0.0008	0.0009	0.0010	0.0011	0.0012	0.0014
20	0.0002	0.0002	0.0002	0.0003	0.0003	0.0004	0.0004	0.0005	0.0005	0.0006
21	0.0001	0.0001	0.0001	0.0001	0.0001	0.0002	0.0002	0.0002	0.0002	0.0003
22	0.0000	0.0000	0.0000	0.0000	0.0001	0.0001	0.0001	0.0001	0.0001	0.0001

$$\text{Values of } P(X = x \,|\, \lambda t) = \frac{(\lambda t)^x e^{-\lambda t}}{x!}$$

x	9.1	9.2	9.3	9.4	9.5	9.6	9.7	9.8	9.9	10.0
0	0.0001	0.0001	0.0001	0.0001	0.0001	0.0001	0.0001	0.0001	0.0001	0.0000
1	0.0010	0.0009	0.0009	0.0008	0.0007	0.0007	0.0006	0.0005	0.0005	0.0005
2	0.0046	0.0043	0.0040	0.0037	0.0034	0.0031	0.0029	0.0027	0.0025	0.0023
3	0.0140	0.0131	0.0123	0.0115	0.0107	0.0100	0.0093	0.0087	0.0081	0.0076
4	0.0319	0.0302	0.0285	0.0269	0.0254	0.0240	0.0226	0.0213	0.0201	0.0189
5	0.0581	0.0555	0.0530	0.0506	0.0483	0.0460	0.0439	0.0418	0.0398	0.0378
6	0.0881	0.0851	0.0822	0.0793	0.0764	0.0736	0.0709	0.0682	0.0656	0.0631
7	0.1145	0.1118	0.1091	0.1064	0.1037	0.1010	0.0982	0.0955	0.0928	0.0901
8	0.1302	0.1286	0.1269	0.1251	0.1232	0.1212	0.1191	0.1170	0.1148	0.1126
9	0.1317	0.1315	0.1311	0.1306	0.1300	0.1293	0.1284	0.1274	0.1263	0.1251
10	0.1198	0.1210	0.1219	0.1228	0.1235	0.1241	0.1245	0.1249	0.1250	0.1251
11	0.0991	0.1012	0.1031	0.1049	0.1067	0.1083	0.1098	0.1112	0.1125	0.1137
12	0.0752	0.0776	0.0799	0.0822	0.0844	0.0866	0.0888	0.0908	0.0928	0.0948
13	0.0526	0.0549	0.0572	0.0594	0.0617	0.0640	0.0662	0.0685	0.0707	0.0729
14	0.0342	0.0361	0.0380	0.0399	0.0419	0.0439	0.0459	0.0479	0.0500	0.0521
15	0.0208	0.0221	0.0235	0.0250	0.0265	0.0281	0.0297	0.0313	0.0330	0.0347
16	0.0118	0.0127	0.0137	0.0147	0.0157	0.0168	0.0180	0.0192	0.0204	0.0217
17	0.0063	0.0069	0.0075	0.0081	0.0088	0.0095	0.0103	0.0111	0.0119	0.0128
18	0.0032	0.0035	0.0039	0.0042	0.0046	0.0051	0.0055	0.0060	0.0065	0.0071
19	0.0015	0.0017	0.0019	0.0021	0.0023	0.0026	0.0028	0.0031	0.0034	0.0037
20	0.0007	0.0008	0.0009	0.0010	0.0011	0.0012	0.0014	0.0015	0.0017	0.0019
21	0.0003	0.0003	0.0004	0.0004	0.0005	0.0006	0.0006	0.0007	0.0008	0.0009
22	0.0001	0.0001	0.0002	0.0002	0.0002	0.0002	0.0003	0.0003	0.0004	0.0004
23	0.0000	0.0001	0.0001	0.0001	0.0001	0.0001	0.0001	0.0001	0.0002	0.0002
24	0.0000	0.0000	0.0000	0.0000	0.0000	0.0000	0.0000	0.0001	0.0001	0.0001

Source: Robert Parsons, *Statistical Analysis: A Decision Making Approach*, 2nd ed. (New York: Harper & Row, Publishers, 1978). Table D, pp. 768–71. Reproduced with permission.

APPENDIX E: AREAS OF THE NORMAL DISTRIBUTION

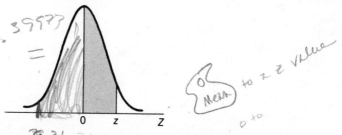

Each value in the table is the proportion of total area between the mean and $Z = (X - \mu)/\sigma$ standard deviations away from the mean. Values of Z are given to two decimal places as a row heading plus a column heading. For example, the value $Z = 1.25$ is located as the row heading 1.2 plus the column heading .05. The value at the intersection of this row and column indicates that the proportion of total area between the mean and 1.25 standard deviations away from the mean (for either $Z = +1.25$ or $Z = -1.25$) is 0.39435.

Z or $\dfrac{X - \mu}{\sigma}$	0.00	0.01	0.02	0.03	0.04	0.05	0.06	0.07	0.08	0.09
0.0	0.00000	0.00399	0.00798	0.01197	0.01595	0.01994	0.02392	0.02790	0.03188	0.03586
0.1	0.03983	0.04380	0.04776	0.05172	0.05567	0.05962	0.06356	0.06749	0.07142	0.07535
0.2	0.07926	0.08317	0.08706	0.09095	0.09483	0.09871	0.10257	0.10642	0.11026	0.11409
0.3	0.11791	0.12172	0.12552	0.12930	0.13307	0.13683	0.14058	0.14431	0.14803	0.15173
0.4	0.15542	0.15910	0.16276	0.16640	0.17003	0.17364	0.17724	0.18082	0.18439	0.18793
0.5	0.19146	0.19497	0.19847	0.20194	0.20540	0.20884	0.21226	0.21566	0.21904	0.22240
0.6	0.22575	0.22907	0.23237	0.23565	0.23891	0.24215	0.24537	0.24857	0.25175	0.25490
0.7	0.25804	0.26115	0.26424	0.26730	0.27035	0.27337	0.27637	0.27935	0.28230	0.28524
0.8	0.28814	0.29103	0.29389	0.29673	0.29955	0.30234	0.30511	0.30785	0.31057	0.31327
0.9	0.31594	0.31859	0.32121	0.32381	0.32639	0.32894	0.33147	0.33398	0.33646	0.33891
1.0	0.34134	0.34375	0.34614	0.34850	0.35083	0.35314	0.35543	0.35769	0.35993	0.36214
1.1	0.36433	0.36650	0.36864	0.37076	0.37286	0.37493	0.37698	0.37900	0.38100	0.38298
1.2	0.38493	0.38686	0.38877	0.39065	0.39251	0.39435	0.39617	0.39796	0.39973	0.40147
1.3	0.40320	0.40490	0.40658	0.40824	0.40988	0.41149	0.41309	0.41466	0.41621	0.41774
1.4	0.41924	0.42073	0.42220	0.42364	0.42507	0.42647	0.42786	0.42922	0.43056	0.43189
1.5	0.43319	0.43448	0.43574	0.43699	0.43822	0.43943	0.44062	0.44179	0.44295	0.44408
1.6	0.44520	0.44630	0.44738	0.44845	0.44950	0.45053	0.45154	0.45254	0.45352	0.45449
1.7	0.45543	0.45637	0.45728	0.45818	0.45907	0.45994	0.46080	0.46164	0.46246	0.46327
1.8	0.46407	0.46485	0.46562	0.46638	0.46712	0.46784	0.46856	0.46926	0.46995	0.47062
1.9	0.47128	0.47193	0.47257	0.47320	0.47381	0.47441	0.47500	0.47558	0.47615	0.47670
2.0	0.47725	0.47778	0.47831	0.47882	0.47932	0.47982	0.48030	0.48077	0.48124	0.48169
2.1	0.48214	0.48257	0.48300	0.48341	0.48382	0.48422	0.48461	0.48500	0.48537	0.48574
2.2	0.48610	0.48645	0.48679	0.48713	0.48745	0.48778	0.48809	0.48840	0.48870	0.48899
2.3	0.48928	0.48956	0.48983	0.49010	0.49036	0.49061	0.49086	0.49111	0.49134	0.49158
2.4	0.49180	0.49202	0.49224	0.49245	0.49266	0.49286	0.49305	0.49324	0.49343	0.49361

Z or $\dfrac{X-\mu}{\sigma}$	0.00	0.01	0.02	0.03	0.04	0.05	0.06	0.07	0.08	0.09
2.5	0.49379	0.49396	0.49413	0.49430	0.49446	0.49461	0.49477	0.49492	0.49506	0.49520
2.6	0.49534	0.49547	0.49560	0.49573	0.49585	0.49598	0.49609	0.49621	0.49632	0.49643
2.7	0.49653	0.49664	0.49674	0.49683	0.49693	0.49702	0.49711	0.49720	0.49728	0.49736
2.8	0.49744	0.49752	0.49760	0.49767	0.49774	0.49781	0.49788	0.49795	0.49801	0.49807
2.9	0.49813	0.49819	0.49825	0.49831	0.49836	0.49841	0.49846	0.49851	0.49856	0.49861
3.0	0.49865	0.49869	0.49874	0.49878	0.49882	0.49886	0.49889	0.49893	0.49897	0.49900
3.1	0.49903	0.49906	0.49910	0.49913	0.49916	0.49918	0.49921	0.49924	0.49926	0.49929
3.2	0.49931	0.49934	0.49936	0.49938	0.49940	0.49942	0.49944	0.49946	0.49948	0.49950
3.3	0.49952	0.49953	0.49955	0.49957	0.49958	0.49960	0.49961	0.49962	0.49964	0.49965
3.4	0.49966	0.49968	0.49969	0.49970	0.49971	0.49972	0.49973	0.49974	0.49975	0.49976
3.5	0.49977									
3.6	0.49984									
3.7	0.49989									
3.8	0.49993									
3.9	0.49995									
4.0	0.49997									

Source: John R. Stockton and Charles T. Clark, *Introduction to Business and Economic Statistics*, 5th ed. (Cincinnati, Ohio: South-Western Publishing Co., 1975), Appendix H, p. 569, Reproduced with permission.

APPENDIX F: VALUES OF THE UNIT NORMAL LOSS FUNCTION, $L(D)$

D	0.00	0.01	0.02	0.03	0.04	0.05	0.06	0.07	0.08	0.09
0.0	0.3989	0.3940	0.3890	0.3841	0.3793	0.3744	0.3697	0.3649	0.3602	0.3556
0.1	0.3509	0.3464	0.3418	0.3373	0.3328	0.3284	0.3240	0.3197	0.3154	0.3111
0.2	0.3069	0.3027	0.2986	0.2944	0.2904	0.2863	0.2824	0.2784	0.2745	0.2706
0.3	0.2668	0.2630	0.2592	0.2555	0.2518	0.2481	0.2445	0.2409	0.2374	0.2339
0.4	0.2304	0.2270	0.2236	0.2203	0.2169	0.2137	0.2104	0.2072	0.2040	0.2009
0.5	0.1978	0.1947	0.1917	0.1887	0.1857	0.1828	0.1799	0.1771	0.1742	0.1714
0.6	0.1687	0.1659	0.1633	0.1606	0.1580	0.1554	0.1528	0.1503	0.1478	0.1453
0.7	0.1429	0.1405	0.1381	0.1358	0.1334	0.1312	0.1289	0.1267	0.1245	0.1223
0.8	0.1202	0.1181	0.1160	0.1140	0.1120	0.1100	0.1080	0.1061	0.1042	0.1023
0.9	0.1004	0.09860	0.09680	0.09503	0.09328	0.09156	0.08986	0.08819	0.08654	0.08491
1.0	0.08332	0.08174	0.08019	0.07866	0.07716	0.07568	0.07422	0.07279	0.07138	0.06999
1.1	0.06862	0.06727	0.06595	0.06465	0.06336	0.06210	0.06086	0.05964	0.05844	0.05726
1.2	0.05610	0.05496	0.05384	0.05274	0.05165	0.05059	0.04954	0.04851	0.04750	0.04650
1.3	0.04553	0.04457	0.04363	0.04270	0.04179	0.04090	0.04002	0.03916	0.03831	0.03748
1.4	0.03667	0.03587	0.03508	0.03431	0.03356	0.03281	0.03208	0.03137	0.03067	0.02998
1.5	0.02931	0.02865	0.02800	0.02736	0.02674	0.02612	0.02552	0.02494	0.02436	0.02380
1.6	0.02324	0.02270	0.02217	0.02165	0.02114	0.02064	0.02015	0.01967	0.01920	0.01874
1.7	0.01829	0.01785	0.01742	0.01699	0.01658	0.01617	0.01578	0.01539	0.01501	0.01464
1.8	0.01428	0.01392	0.01357	0.01323	0.01290	0.01257	0.01226	0.01195	0.01164	0.01134
1.9	0.01105	0.01077	0.01049	0.01022	$0.0^2 9957$	$0.0^2 9698$	$0.0^2 9445$	$0.0^2 9198$	$0.0^2 8957$	$0.0^2 8721$
2.0	$0.0^2 8491$	$0.0^2 8266$	$0.0^2 8046$	$0.0^2 7832$	$0.0^2 7623$	$0.0^2 7418$	$0.0^2 7219$	$0.0^2 7024$	$0.0^2 6835$	$0.0^2 6649$
2.1	$0.0^2 6468$	$0.0^2 6292$	$0.0^2 6120$	$0.0^2 5952$	$0.0^2 5788$	$0.0^2 5628$	$0.0^2 5472$	$0.0^2 5320$	$0.0^2 5172$	$0.0^2 5028$
2.2	$0.0^2 4887$	$0.0^2 4750$	$0.0^2 4616$	$0.0^2 4486$	$0.0^2 4358$	$0.0^2 4235$	$0.0^2 4114$	$0.0^2 3996$	$0.0^2 3882$	$0.0^2 3770$
2.3	$0.0^2 3662$	$0.0^2 3556$	$0.0^2 3453$	$0.0^2 3352$	$0.0^2 3255$	$0.0^2 3159$	$0.0^2 3067$	$0.0^2 2977$	$0.0^2 2889$	$0.0^2 2804$
2.4	$0.0^2 2720$	$0.0^2 2640$	$0.0^2 2561$	$0.0^2 2484$	$0.0^2 2410$	$0.0^2 2337$	$0.0^2 2267$	$0.0^2 2199$	$0.0^2 2132$	$0.0^2 2067$
2.5	$0.0^2 2004$	$0.0^2 1943$	$0.0^2 1883$	$0.0^2 1826$	$0.0^2 1769$	$0.0^2 1715$	$0.0^2 1662$	$0.0^2 1610$	$0.0^2 1560$	$0.0^2 1511$
2.6	$0.0^2 1464$	$0.0^2 1418$	$0.0^2 1373$	$0.0^2 1330$	$0.0^2 1288$	$0.0^2 1247$	$0.0^2 1207$	$0.0^2 1169$	$0.0^2 1132$	$0.0^2 1095$
2.7	$0.0^2 1060$	$0.0^2 1026$	$0.0^3 9928$	$0.0^3 9607$	$0.0^3 9295$	$0.0^3 8992$	$0.0^3 8699$	$0.0^3 8414$	$0.0^3 8138$	$0.0^3 7870$
2.8	$0.0^3 7611$	$0.0^3 7359$	$0.0^3 7115$	$0.0^3 6879$	$0.0^3 6650$	$0.0^3 6428$	$0.0^3 6213$	$0.0^3 6004$	$0.0^3 5802$	$0.0^3 5606$
2.9	$0.0^3 5417$	$0.0^3 5233$	$0.0^3 5055$	$0.0^3 4883$	$0.0^3 4716$	$0.0^3 4555$	$0.0^3 4398$	$0.0^3 4247$	$0.0^3 4101$	$0.0^3 3959$
3.0	$0.0^3 3822$	$0.0^3 3689$	$0.0^3 3560$	$0.0^3 3436$	$0.0^3 3316$	$0.0^3 3199$	$0.0^3 3087$	$0.0^3 2978$	$0.0^3 2873$	$0.0^3 2771$
3.1	$0.0^3 2673$	$0.0^3 2577$	$0.0^3 2485$	$0.0^3 2396$	$0.0^3 2311$	$0.0^3 2227$	$0.0^3 2147$	$0.0^3 2070$	$0.0^3 1995$	$0.0^3 1922$
3.2	$0.0^3 1852$	$0.0^3 1785$	$0.0^3 1720$	$0.0^3 1657$	$0.0^3 1596$	$0.0^3 1537$	$0.0^3 1480$	$0.0^3 1426$	$0.0^3 1373$	$0.0^3 1322$
3.3	$0.0^3 1273$	$0.0^3 1225$	$0.0^3 1179$	$0.0^3 1135$	$0.0^3 1093$	$0.0^3 1051$	$0.0^3 1012$	$0.0^4 9734$	$0.0^4 9365$	$0.0^4 9009$
3.4	$0.0^4 8666$	$0.0^4 8335$	$0.0^4 8016$	$0.0^4 7709$	$0.0^4 7413$	$0.0^4 7127$	$0.0^4 6852$	$0.0^4 6587$	$0.0^4 6331$	$0.0^4 6085$
3.5	$0.0^4 5848$	$0.0^4 5620$	$0.0^4 5400$	$0.0^4 5188$	$0.0^4 4984$	$0.0^4 4788$	$0.0^4 4599$	$0.0^4 4417$	$0.0^4 4242$	$0.0^4 4073$
3.6	$0.0^4 3911$	$0.0^4 3755$	$0.0^4 3605$	$0.0^4 3460$	$0.0^4 3321$	$0.0^4 3188$	$0.0^4 3059$	$0.0^4 2935$	$0.0^4 2816$	$0.0^4 2702$
3.7	$0.0^4 2592$	$0.0^4 2486$	$0.0^4 2385$	$0.0^4 2287$	$0.0^4 2193$	$0.0^4 2103$	$0.0^4 2016$	$0.0^4 1933$	$0.0^4 1853$	$0.0^4 1776$
3.8	$0.0^4 1702$	$0.0^4 1632$	$0.0^4 1563$	$0.0^4 1498$	$0.0^4 1435$	$0.0^4 1375$	$0.0^4 1317$	$0.0^4 1262$	$0.0^4 1208$	$0.0^4 1157$
3.9	$0.0^4 1108$	$0.0^4 1061$	$0.0^4 1016$	$0.0^5 9723$	$0.0^5 9307$	$0.0^5 8908$	$0.0^5 8525$	$0.0^5 8158$	$0.0^5 7806$	$0.0^5 7469$
4.0	$0.0^5 7145$	$0.0^5 6835$	$0.0^5 6538$	$0.0^5 6253$	$0.0^5 5980$	$0.0^5 5718$	$0.0^5 5468$	$0.0^5 5227$	$0.0^5 4997$	$0.0^5 4777$
4.1	$0.0^5 4566$	$0.0^5 4364$	$0.0^5 4170$	$0.0^5 3985$	$0.0^5 3807$	$0.0^5 3637$	$0.0^5 3475$	$0.0^5 3319$	$0.0^5 3170$	$0.0^5 3027$
4.2	$0.0^5 2891$	$0.0^5 2760$	$0.0^5 2635$	$0.0^5 2516$	$0.0^5 2402$	$0.0^5 2292$	$0.0^5 2188$	$0.0^5 2088$	$0.0^5 1992$	$0.0^5 1901$
4.4	$0.0^5 1814$	$0.0^5 1730$	$0.0^5 1650$	$0.0^5 1574$	$0.0^5 1501$	$0.0^5 1431$	$0.0^5 1365$	$0.0^5 1301$	$0.0^5 1241$	$0.0^5 1183$
4.4	$0.0^5 1127$	$0.0^5 1074$	$0.0^5 1024$	$0.0^6 9756$	$0.0^6 9296$	$0.0^6 8857$	$0.0^6 8437$	$0.0^6 8037$	$0.0^6 7655$	$0.0^6 7290$
4.5	$0.0^6 6942$	$0.0^6 6610$	$0.0^6 6294$	$0.0^6 5992$	$0.0^6 5704$	$0.0^6 5429$	$0.0^6 5167$	$0.0^6 4917$	$0.0^6 4679$	$0.0^6 4452$
4.6	$0.0^6 4236$	$0.0^6 4029$	$0.0^6 3833$	$0.0^6 3645$	$0.0^6 3467$	$0.0^6 3297$	$0.0^6 3135$	$0.0^6 2981$	$0.0^6 2834$	$0.0^6 2694$
4.7	$0.0^6 2560$	$0.0^6 2433$	$0.0^6 2313$	$0.0^6 2197$	$0.0^6 2088$	$0.0^6 1984$	$0.0^6 1884$	$0.0^6 1790$	$0.0^6 1700$	$0.0^6 1615$
4.8	$0.0^6 1533$	$0.0^6 1456$	$0.0^6 1382$	$0.0^6 1312$	$0.0^6 1246$	$0.0^6 1182$	$0.0^6 1122$	$0.0^6 1065$	$0.0^6 1011$	$0.0^7 9588$
4.9	$0.0^7 9096$	$0.0^7 8629$	$0.0^7 8185$	$0.0^7 7763$	$0.0^7 7362$	$0.0^7 6982$	$0.0^7 6620$	$0.0^7 6276$	$0.0^7 5950$	$0.0^7 5640$

Source: Reproduced from Robert O. Schlaifer, *Probability and Statistics for Business Decisions*, published by McGraw-Hill Book Company, 1959, by specific permission from the copyright holder, the President and Fellows of Harvard College.

APPENDIX G: STUDENT'S *t* DISTRIBUTION

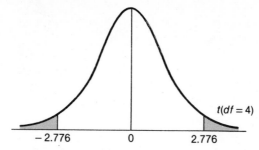

$t(df = 4)$

−2.776   0   2.776

The table provides values of the statistic $t = (\bar{X} - \mu)/\hat{\sigma}_{\bar{X}}$ for degrees of freedom df = 1(1)30, 40, 60, 120, ∞ and selected one- and two-tailed probabilities α. To illustrate, let df = 4. A one-tailed probability is $P(t > 2.776) = P(t < -2.776) = 0.025$. A two-tailed probability is $P(|t| > 2.776) = 0.05$.

					α					
	0.25	0.20	0.15	0.10	0.05	0.025	0.01	0.005	0.0005	One-Tailed
df	0.50	0.40	0.30	0.20	0.10	0.05	0.02	0.01	0.001	Two-Tailed
1	1.000	1.376	1.963	3.078	6.314	12.706	31.821	63.657	636.619	
2	0.816	1.061	1.386	1.886	2.920	4.303	6.965	9.925	31.598	
3	0.765	0.978	1.250	1.638	2.353	3.182	4.541	5.841	12.941	
4	0.741	0.941	1.190	1.533	2.132	2.776	3.747	4.604	8.610	
5	0.727	0.920	1.156	1.476	2.015	2.571	3.365	4.032	6.859	
6	0.718	0.906	1.134	1.440	1.943	2.447	3.143	3.707	5.959	
7	0.711	0.896	1.119	1.415	1.895	2.365	2.998	3.499	5.405	
8	0.706	0.889	1.108	1.397	1.860	2.306	2.896	3.355	5.041	
9	0.703	0.883	1.100	1.383	1.833	2.262	2.821	3.250	4.781	
10	0.700	0.879	1.093	1.372	1.812	2.228	2.764	3.169	4.587	
11	0.697	0.876	1.088	1.363	1.796	2.201	2.718	3.106	4.437	
12	0.695	0.873	1.083	1.356	1.782	2.179	2.681	3.055	4.318	
13	0.694	0.870	1.079	1.350	1.771	2.160	2.650	3.012	4.221	
14	0.692	0.868	1.076	1.345	1.761	2.145	2.624	2.977	4.140	
15	0.691	0.866	1.074	1.341	1.753	2.131	2.602	2.947	4.073	
16	0.690	0.865	1.071	1.337	1.746	2.120	2.583	2.921	4.015	
17	0.689	0.863	1.069	1.333	1.740	2.110	2.567	2.898	3.965	
18	0.688	0.862	1.067	1.330	1.734	2.101	2.552	2.878	3.922	
19	0.688	0.861	1.066	1.328	1.729	2.093	2.539	2.861	3.883	
20	0.687	0.860	1.064	1.325	1.725	2.086	2.528	2.845	3.850	
21	0.686	0.859	1.063	1.323	1.721	2.080	2.518	2.831	3.819	
22	0.686	0.858	1.061	1.321	1.717	2.074	2.508	2.819	3.792	
23	0.685	0.858	1.060	1.319	1.714	2.069	2.500	2.807	3.767	
24	0.685	0.857	1.059	1.318	1.711	2.064	2.492	2.797	3.745	
25	0.684	0.856	1.058	1.316	1.708	2.060	2.485	2.787	3.725	
26	0.684	0.856	1.058	1.315	1.706	2.056	2.479	2.779	3.707	
27	0.684	0.855	1.057	1.314	1.703	2.052	2.473	2.771	3.690	
28	0.683	0.855	1.056	1.313	1.701	2.048	2.467	2.763	3.674	
29	0.683	0.854	1.055	1.311	1.699	2.045	2.462	2.756	3.659	
30	0.683	0.854	1.055	1.310	1.697	2.042	2.457	2.750	3.646	
40	0.681	0.851	1.050	1.303	1.684	2.021	2.423	2.704	3.551	
60	0.679	0.848	1.046	1.296	1.671	2.000	2.390	2.660	3.460	
120	0.677	0.845	1.041	1.289	1.658	1.980	2.358	2.617	3.373	
∞	0.674	0.842	1.036	1.282	1.645	1.960	2.326	2.576	3.291	

Source: Abridged from Table III of R. A. Fisher and F. Yates, *Statistical Tables for Biological, Agricultural and Medical Research*, published by Longman Group Ltd., London, (previously published by Oliver & Boyd, Edinburgh), and by permission of the authors and publishers.

APPENDIX H: CRITICAL VALUES FOR THE CHI-SQUARE DISTRIBUTION

Example:
For $v = 10$ degrees of freedom:
$P(x^2 > 15.99) = 0.10$.

v	0.995	0.99	0.975	0.95	0.90	0.75	0.50	0.25	0.10	0.05	0.025	0.01	0.005
1	0.0^4393	0.0^3157	0.0^3982	0.0^239	0.0158	0.102	0.455	1.323	2.71	3.84	5.02	6.63	7.88
2	0.0100	0.0201	0.0506	0.103	0.211	0.575	1.386	2.77	4.61	5.99	7.38	9.21	10.60
3	0.0717	0.115	0.216	0.352	0.584	1.213	2.37	4.11	6.25	7.81	9.35	11.34	12.84
4	0.207	0.297	0.484	0.711	1.064	1.923	3.36	5.39	7.78	9.49	11.14	13.28	14.86
5	0.412	0.554	0.831	1.145	1.610	2.67	4.35	6.63	9.24	11.07	12.83	15.09	16.75
6	0.676	0.872	1.237	1.635	2.20	3.45	5.35	7.84	10.64	12.59	14.45	16.81	18.55
7	0.989	1.239	1.690	2.17	2.83	4.25	6.35	9.04	12.02	14.07	16.01	18.48	20.3
8	1.344	1.646	2.18	2.73	3.49	5.07	7.34	10.22	13.36	15.51	17.53	20.1	22.0
9	1.735	2.09	2.70	3.33	4.17	5.90	8.34	11.39	14.68	16.92	19.02	21.7	23.6
10	2.16	2.56	3.25	3.94	4.87	6.74	9.34	12.55	15.99	18.31	20.5	23.2	25.2
11	2.60	3.05	3.82	4.57	5.58	7.58	10.34	13.70	17.28	19.68	21.9	24.7	26.8
12	3.07	3.57	4.40	5.23	6.30	8.44	11.34	14.85	18.55	21.0	23.3	26.2	28.3
13	3.57	4.11	5.01	5.89	7.04	9.30	12.34	15.98	19.81	22.4	24.7	27.7	29.8
14	4.07	4.66	5.63	6.57	7.79	10.17	13.34	17.12	21.1	23.7	26.1	29.1	31.3
15	4.60	5.23	6.26	7.26	8.55	11.04	14.34	18.25	22.3	25.0	27.5	30.6	32.8
16	5.14	5.81	6.91	7.96	9.31	11.91	15.34	19.37	23.5	26.3	28.8	32.0	34.3
17	5.70	6.41	7.56	8.67	10.09	12.79	16.34	20.5	24.8	27.6	30.2	33.4	35.7
18	6.26	7.01	8.23	9.39	10.86	13.68	17.34	21.6	26.0	28.9	31.5	34.8	37.2
19	6.84	7.63	8.91	10.12	11.65	14.56	18.34	22.7	27.2	30.1	32.9	36.2	38.6
20	7.43	8.26	9.59	10.85	12.44	15.45	19.34	23.8	28.4	31.4	34.2	37.6	40.0
21	8.03	8.90	10.28	11.59	13.24	16.34	20.3	24.9	29.6	32.7	35.5	38.9	41.4
22	8.64	9.54	10.98	12.34	14.04	17.24	21.3	26.0	30.8	33.9	36.8	40.3	42.8
23	9.26	10.20	11.69	13.09	14.85	18.14	22.3	27.1	32.0	35.2	38.1	41.6	44.2
24	9.89	10.86	12.40	13.85	15.66	19.04	23.3	28.2	33.2	36.4	39.4	43.0	45.6
25	10.52	11.52	13.12	14.61	16.47	19.94	24.3	29.3	34.4	37.7	40.6	44.3	46.9
26	11.16	12.20	13.84	15.38	17.29	20.8	25.3	30.4	35.6	38.9	41.9	45.6	48.3
27	11.81	12.88	14.57	16.15	18.11	21.7	26.3	31.5	36.7	40.1	43.2	47.0	49.6
28	12.46	13.56	15.31	16.93	18.94	22.7	27.3	32.6	37.9	41.3	44.5	48.3	51.0
29	13.12	14.26	16.05	17.71	19.77	23.6	28.3	33.7	39.1	42.6	45.7	49.6	52.3
30	13.79	14.95	16.79	18.49	20.6	24.5	29.3	34.8	40.3	43.8	47.0	50.9	53.7
40	20.7	22.2	24.4	26.5	29.1	33.7	39.3	45.6	51.8	55.8	59.3	63.7	66.8
50	28.0	29.7	32.4	34.8	37.7	42.9	49.3	56.3	63.2	67.5	71.4	76.2	79.5
60	35.5	37.5	40.5	43.2	46.5	52.3	59.3	67.0	74.4	79.1	83.3	88.4	92.0
70	43.3	45.4	48.8	51.7	55.3	61.7	69.3	77.6	85.5	90.5	95.0	100.4	104.2
80	51.2	53.5	57.2	60.4	64.3	71.1	79.3	88.1	96.6	101.9	106.6	112.3	116.3
90	59.2	61.8	65.6	69.1	73.3	80.6	89.3	98.6	107.6	113.1	118.1	124.1	128.3
100	67.3	70.1	74.2	77.9	82.4	90.1	99.3	109.1	118.5	124.3	129.6	135.8	140.2

Source: Abridged from Catherine M. Thompson, "Table of Percentage Points of the χ^2 Distribution," *Biometrika*, vol. 32 (1941), pp. 187–91, and published here by permission of the author and the editor of *Biometrika*.

α is the right-tail probability.   Example: $P(F_{5,7} > 3.97) = 0.05$.

Denominator df	α	\[Numerator df\] 1	2	3	4	5	6	7	8	9
1	0.100	39.86	49.50	53.59	55.83	57.24	58.20	58.91	59.44	59.86
	0.050	161.4	199.5	215.7	224.6	230.2	234.0	236.8	238.9	240.5
	0.025	647.8	799.5	864.2	899.6	921.8	937.1	948.2	956.7	963.3
	0.010	4052	4999.5	5403	5625	5764	5859	5928	5982	6022
2	0.100	8.53	9.00	9.16	9.24	9.29	9.33	9.35	9.37	9.38
	0.050	18.51	19.00	19.16	19.25	19.30	19.33	19.35	19.37	19.38
	0.025	38.51	39.00	39.17	39.25	39.30	39.33	39.36	39.37	39.39
	0.010	98.50	99.00	99.17	99.25	99.30	99.33	99.36	99.37	99.39
3	0.100	5.54	5.46	5.39	5.34	5.31	5.28	5.27	5.25	5.24
	0.050	10.13	9.55	9.28	9.12	9.01	8.94	8.89	8.85	8.81
	0.025	17.44	16.04	15.44	15.10	14.88	14.73	14.62	14.54	14.47
	0.010	34.12	30.82	29.46	28.71	28.24	27.91	27.67	27.49	27.35
4	0.100	4.54	4.32	4.19	4.11	4.05	4.01	3.98	3.95	3.94
	0.050	7.71	6.94	6.59	6.39	6.26	6.16	6.09	6.04	6.00
	0.025	12.22	10.65	9.98	9.60	9.36	9.20	9.07	8.98	8.90
	0.010	21.20	18.00	16.69	15.98	15.52	15.21	14.98	14.80	14.66
5	0.100	4.06	3.78	3.62	3.52	3.45	3.40	3.37	3.34	3.32
	0.050	6.61	5.79	5.41	5.19	5.05	4.95	4.88	4.82	4.77
	0.025	10.01	8.43	7.76	7.39	7.15	6.98	6.85	6.76	6.68
	0.010	16.26	13.27	12.06	11.39	10.97	10.67	10.46	10.29	10.16
6	0.100	3.78	3.46	3.29	3.18	3.11	3.05	3.01	2.98	2.96
	0.050	5.99	5.14	4.76	4.53	4.39	4.28	4.21	4.15	4.10
	0.025	8.81	7.26	6.60	6.23	5.99	5.82	5.70	5.60	5.52
	0.010	13.75	10.92	9.78	9.15	8.75	8.47	8.26	8.10	7.98
7	0.100	3.59	3.26	3.07	2.96	2.88	2.83	2.78	2.75	2.72
	0.050	5.59	4.74	4.35	4.12	3.97	3.87	3.79	3.73	3.68
	0.025	8.07	6.54	5.89	5.52	5.29	5.12	4.99	4.90	4.82
	0.010	12.25	9.55	8.45	7.85	7.46	7.19	6.99	6.84	6.72
8	0.100	3.46	3.11	2.92	2.81	2.73	2.67	2.62	2.59	2.56
	0.050	5.32	4.46	4.07	3.84	3.69	3.58	3.50	3.44	3.39
	0.025	7.57	6.06	5.42	5.05	4.82	4.65	4.53	4.43	4.36
	0.010	11.26	8.65	7.59	7.01	6.63	6.37	6.18	6.03	5.91
9	0.100	3.36	3.01	2.81	2.69	2.61	2.55	2.51	2.47	2.44
	0.050	5.12	4.26	3.86	3.63	3.48	3.37	3.29	3.23	3.18
	0.025	7.21	5.71	5.08	4.72	4.48	4.32	4.20	4.10	4.03
	0.010	10.56	8.02	6.99	6.42	6.06	5.80	5.61	5.47	5.35
10	0.100	3.29	2.92	2.73	2.61	2.52	2.46	2.41	2.38	2.35
	0.050	4.96	4.10	3.71	3.48	3.33	3.22	3.14	3.07	3.02
	0.025	6.94	5.46	4.83	4.47	4.24	4.07	3.95	3.85	3.78
	0.010	10.04	7.56	6.55	5.99	5.64	5.39	5.20	5.06	4.94
11	0.100	3.23	2.86	2.66	2.54	2.45	2.39	2.34	2.30	2.27
	0.050	4.84	3.98	3.59	3.36	3.20	3.09	3.01	2.95	2.90
	0.025	6.72	5.26	4.63	4.28	4.04	3.88	3.76	3.66	3.59
	0.010	9.65	7.21	6.22	5.67	5.32	5.07	4.89	4.74	4.63
12	0.100	3.18	2.81	2.61	2.48	2.39	2.33	2.28	2.24	2.21
	0.050	4.75	3.89	3.49	3.26	3.11	3.00	2.91	2.85	2.80
	0.025	6.55	5.10	4.47	4.12	3.89	3.73	3.61	3.51	3.44
	0.010	9.33	6.93	5.95	5.41	5.06	4.82	4.64	4.50	4.39

Denominator df	α	\multicolumn{9}{c}{Numerator df}
		1	2	3	4	5	6	7	8	9
13	0.100	3.14	2.76	2.56	2.43	2.35	2.28	2.23	2.20	2.16
	0.050	4.67	3.81	3.41	3.18	3.03	2.92	2.83	2.77	2.71
	0.025	6.41	4.97	4.35	4.00	3.77	3.60	3.48	3.39	3.31
	0.010	9.07	6.70	5.74	5.21	4.86	4.62	4.44	4.30	4.19
14	0.100	3.10	2.73	2.52	2.39	2.31	2.24	2.19	2.15	2.12
	0.050	4.60	3.74	3.34	3.11	2.96	2.85	2.76	2.70	2.65
	0.025	6.30	4.86	4.24	3.89	3.66	3.50	3.38	3.29	3.21
	0.010	8.86	6.51	5.56	5.04	4.69	4.46	4.28	4.14	4.03
15	0.100	3.07	2.70	2.49	2.36	2.27	2.21	2.16	2.12	2.09
	0.050	4.54	3.68	3.29	3.06	2.90	2.79	2.71	2.64	2.59
	0.025	6.20	4.77	4.15	3.80	3.58	3.41	3.29	3.20	3.12
	0.010	8.68	6.36	5.42	4.89	4.56	4.32	4.14	4.00	3.89
16	0.100	3.05	2.67	2.46	2.33	2.24	2.18	2.13	2.09	2.06
	0.050	4.49	3.63	3.24	3.01	2.85	2.74	2.66	2.59	2.54
	0.025	6.12	4.69	4.08	3.73	3.50	3.34	3.22	3.12	3.05
	0.010	8.53	6.23	5.29	4.77	4.44	4.20	4.03	3.89	3.78
17	0.100	3.03	2.64	2.44	2.31	2.22	2.15	2.10	2.06	2.03
	0.050	4.45	3.59	3.20	2.96	2.81	2.70	2.61	2.55	2.49
	0.025	6.04	4.62	4.01	3.66	3.44	3.28	3.16	3.06	2.98
	0.010	8.40	6.11	5.18	4.67	4.34	4.10	3.93	3.79	3.68
18	0.100	3.01	2.62	2.42	2.29	2.20	2.13	2.08	2.04	2.00
	0.050	4.41	3.55	3.16	2.93	2.77	2.66	2.58	2.51	2.46
	0.025	5.98	4.56	3.95	3.61	3.38	3.22	3.10	3.01	2.93
	0.010	8.29	6.01	5.09	4.58	4.25	4.01	3.84	3.71	3.60
19	0.100	2.99	2.61	2.40	2.27	2.18	2.11	2.06	2.02	1.98
	0.050	4.38	3.52	3.13	2.90	2.74	2.63	2.54	2.48	2.42
	0.025	5.92	4.51	3.90	3.56	3.33	3.17	3.05	2.96	2.88
	0.010	8.18	5.93	5.01	4.50	4.17	3.94	3.77	3.63	3.52
20	0.100	2.97	2.59	2.38	2.25	2.16	2.09	2.04	2.00	1.96
	0.050	4.35	3.49	3.10	2.87	2.71	2.60	2.51	2.45	2.39
	0.025	5.87	4.46	3.86	3.51	3.29	3.13	3.01	2.91	2.84
	0.010	8.10	5.85	4.94	4.43	4.10	3.87	3.70	3.56	3.46
21	0.100	2.96	2.57	2.36	2.23	2.14	2.08	2.02	1.98	1.95
	0.050	4.32	3.47	3.07	2.84	2.68	2.57	2.49	2.42	2.37
	0.025	5.83	4.42	3.82	3.48	3.25	3.09	2.97	2.87	2.80
	0.010	8.02	5.78	4.87	4.37	4.04	3.81	3.64	3.51	3.40
22	0.100	2.95	2.56	2.35	2.22	2.13	2.06	2.01	1.97	1.93
	0.050	4.30	3.44	3.05	2.82	2.66	2.55	2.46	2.40	2.34
	0.025	5.79	4.38	3.78	3.44	3.22	3.05	2.93	2.84	2.76
	0.010	7.95	5.72	4.82	4.31	3.99	3.76	3.59	3.45	3.35
23	0.100	2.94	2.55	2.34	2.21	2.11	2.05	1.99	1.95	1.92
	0.050	4.28	3.42	3.03	2.80	2.64	2.53	2.44	2.37	2.32
	0.025	5.75	4.35	3.75	3.41	3.18	3.02	2.90	2.81	2.73
	0.010	7.88	5.66	4.76	4.26	3.94	3.71	3.54	3.41	3.30
24	0.100	2.93	2.54	2.33	2.19	2.10	2.04	1.98	1.94	1.91
	0.050	4.26	3.40	3.01	2.78	2.62	2.51	2.42	2.36	2.30
	0.025	5.72	4.32	3.72	3.38	3.15	2.99	2.87	2.78	2.70
	0.010	7.82	5.61	4.72	4.22	3.90	3.67	3.50	3.36	3.26

APPENDIX I: CRITICAL VALUES OF THE *F* DISTRIBUTION (*continued*)

Denominator df	α	\multicolumn{9}{c}{Numerator df}								
		1	2	3	4	5	6	7	8	9
25	0.100	2.92	2.53	2.32	2.18	2.09	2.02	1.97	1.93	1.89
	0.050	4.24	3.39	2.99	2.76	2.60	2.49	2.40	2.34	2.28
	0.025	5.69	4.29	3.69	3.35	3.13	2.97	2.85	2.75	2.68
	0.010	7.77	5.57	4.68	4.18	3.85	3.63	3.46	3.32	3.22
26	0.100	2.91	2.52	2.31	2.17	2.08	2.01	1.96	1.92	1.88
	0.050	4.23	3.37	2.98	2.74	2.59	2.47	2.39	2.32	2.27
	0.025	5.66	4.27	3.67	3.33	3.10	2.94	2.82	2.73	2.65
	0.010	7.72	5.53	4.64	4.14	3.82	3.59	3.42	3.29	3.18
27	0.100	2.90	2.51	2.30	2.17	2.07	2.00	1.95	1.91	1.87
	0.050	4.21	3.35	2.96	2.73	2.57	2.46	2.37	2.31	2.25
	0.025	5.63	4.24	3.65	3.31	3.08	2.92	2.80	2.71	2.63
	0.010	7.68	5.49	4.60	4.11	3.78	3.56	3.39	3.26	3.15
28	0.100	2.89	2.50	2.29	2.16	2.06	2.00	1.94	1.90	1.87
	0.050	4.20	3.34	2.95	2.71	2.56	2.45	2.36	2.29	2.24
	0.025	5.61	4.22	3.63	3.29	3.06	2.90	2.78	2.69	2.61
	0.010	7.64	5.45	4.57	4.07	3.75	3.53	3.36	3.23	3.12
29	0.100	2.89	2.50	2.28	2.15	2.06	1.99	1.93	1.89	1.86
	0.050	4.18	3.33	2.93	2.70	2.55	2.43	2.35	2.28	2.22
	0.025	5.59	4.20	3.61	3.27	3.04	2.88	2.76	2.67	2.59
	0.010	7.60	5.42	4.54	4.04	3.73	3.50	3.33	3.20	3.09
30	0.100	2.88	2.49	2.28	2.14	2.05	1.98	1.93	1.88	1.85
	0.050	4.17	3.32	2.92	2.69	2.53	2.42	2.33	2.27	2.21
	0.025	5.57	4.18	3.59	3.25	3.03	2.87	2.75	2.65	2.57
	0.010	7.56	5.39	4.51	4.02	3.70	3.47	3.30	3.17	3.07
40	0.100	2.84	2.44	2.23	2.09	2.00	1.93	1.87	1.83	1.79
	0.050	4.08	3.23	2.84	2.61	2.45	2.34	2.25	2.18	2.12
	0.025	5.42	4.05	3.46	3.13	2.90	2.74	2.62	2.53	2.45
	0.010	7.31	5.18	4.31	3.83	3.51	3.29	3.12	2.99	2.89
60	0.100	2.79	2.39	2.18	2.04	1.95	1.87	1.82	1.77	1.74
	0.050	4.00	3.15	2.76	2.53	2.37	2.25	2.17	2.10	2.04
	0.025	5.29	3.93	3.34	3.01	2.79	2.63	2.51	2.41	2.33
	0.010	7.08	4.98	4.13	3.65	3.34	3.12	2.95	2.82	2.72
120	0.100	2.75	2.35	2.13	1.99	1.90	1.82	1.77	1.72	1.68
	0.050	3.92	3.07	2.68	2.45	2.29	2.17	2.09	2.02	1.96
	0.025	5.15	3.80	3.23	2.89	2.67	2.52	2.39	2.30	2.22
	0.010	6.85	4.79	3.95	3.48	3.17	2.96	2.79	2.66	2.56
∞	0.100	2.71	2.30	2.08	1.94	1.85	1.77	1.72	1.67	1.63
	0.050	3.84	3.00	2.60	2.37	2.21	2.10	2.01	1.94	1.88
	0.025	5.02	3.69	3.12	2.79	2.57	2.41	2.29	2.19	2.11
	0.010	6.63	4.61	3.78	3.32	3.02	2.80	2.64	2.51	2.41

Denominator df	α	Numerator df 10	12	15	20	24	30	40	60	120	∞
1	0.100	60.19	60.71	61.22	61.74	62.00	62.26	62.53	62.79	63.06	63.33
	0.050	241.9	243.9	245.9	248.0	249.1	250.1	251.1	252.2	253.3	254.3
	0.025	968.6	976.7	984.9	993.1	997.2	1001	1006	1010	1014	1018
	0.010	6056	6106	6157	6209	6235	6261	6287	6313	6339	6366
2	0.100	9.39	9.41	9.42	9.44	9.45	9.46	9.47	9.47	9.48	9.49
	0.050	19.40	19.41	19.43	19.45	19.45	19.46	19.47	19.48	19.49	19.50
	0.025	39.40	39.41	39.43	39.45	39.46	39.46	39.47	39.48	39.49	39.50
	0.010	99.40	99.42	99.43	99.45	99.46	99.47	99.47	99.48	99.49	99.50
3	0.100	5.23	5.22	5.20	5.18	5.18	5.17	5.16	5.15	5.14	5.13
	0.050	8.79	8.74	8.70	8.66	8.64	8.62	8.59	8.57	8.55	8.53
	0.025	14.42	14.34	14.25	14.17	14.12	14.08	14.04	13.99	13.95	13.90
	0.010	27.23	27.05	26.87	26.69	26.60	26.50	26.41	26.32	26.22	26.13
4	0.100	3.92	3.90	3.87	3.84	3.83	3.82	3.80	3.79	3.78	3.76
	0.050	5.96	5.91	5.86	5.80	5.77	5.75	5.72	5.69	5.66	5.63
	0.025	8.84	8.75	8.66	8.56	8.51	8.46	8.41	8.36	8.31	8.26
	0.010	14.55	14.37	14.20	14.02	13.93	13.84	13.75	13.65	13.56	13.46
5	0.100	3.30	3.27	3.24	3.21	3.19	3.17	3.16	3.14	3.12	3.10
	0.050	4.74	4.68	4.62	4.56	4.53	4.50	4.46	4.43	4.40	4.36
	0.025	6.62	6.52	6.43	6.33	6.28	6.23	6.18	6.12	6.07	6.02
	0.010	10.05	9.89	9.72	9.55	9.47	9.38	9.29	9.20	9.11	9.02
6	0.100	2.94	2.90	2.87	2.84	2.82	2.80	2.78	2.76	2.74	2.72
	0.050	4.06	4.00	3.94	3.87	3.84	3.81	3.77	3.74	3.70	3.67
	0.025	5.46	5.37	5.27	5.17	5.12	5.07	5.01	4.96	4.90	4.85
	0.010	7.87	7.72	7.56	7.40	7.31	7.23	7.14	7.06	6.97	6.88
7	0.100	2.70	2.67	2.63	2.59	2.58	2.56	2.54	2.51	2.49	2.47
	0.050	3.64	3.57	3.51	3.44	3.41	3.38	3.34	3.30	3.27	3.23
	0.025	4.76	4.67	4.57	4.47	4.42	4.36	4.31	4.25	4.20	4.14
	0.010	6.62	6.47	6.31	6.16	6.07	5.99	5.91	5.82	5.74	5.65
8	0.100	2.54	2.50	2.46	2.42	2.40	2.38	2.36	2.34	2.32	2.29
	0.050	3.35	3.28	3.22	3.15	3.12	3.08	3.04	3.01	2.97	2.93
	0.025	4.30	4.20	4.10	4.00	3.95	3.89	3.84	3.78	3.73	3.67
	0.010	5.81	5.67	5.52	5.36	5.28	5.20	5.12	5.03	4.95	4.86
9	0.100	2.42	2.38	2.34	2.30	2.28	2.25	2.23	2.21	2.18	2.16
	0.050	3.14	3.07	3.01	2.94	2.90	2.86	2.83	2.79	2.75	2.71
	0.025	3.96	3.87	3.77	3.67	3.61	3.56	3.51	3.45	3.39	3.33
	0.010	5.26	5.11	4.96	4.81	4.73	4.65	4.57	4.48	4.40	4.31
10	0.100	2.32	2.28	2.24	2.20	2.18	2.16	2.13	2.11	2.08	2.06
	0.050	2.98	2.91	2.85	2.77	2.74	2.70	2.66	2.62	2.58	2.54
	0.025	3.72	3.62	3.52	3.42	3.37	3.31	3.26	3.20	3.14	3.08
	0.010	4.85	4.71	4.56	4.41	4.33	4.25	4.17	4.08	4.00	3.91
11	0.100	2.25	2.21	2.17	2.12	2.10	2.08	2.05	2.03	2.00	1.97
	0.050	2.85	2.79	2.72	2.65	2.61	2.57	2.53	2.49	2.45	2.40
	0.025	3.53	3.43	3.33	3.23	3.17	3.12	3.06	3.00	2.94	2.88
	0.010	4.54	4.40	4.25	4.10	4.02	3.94	3.86	3.78	3.69	3.60
12	0.100	2.19	2.15	2.10	2.06	2.04	2.01	1.99	1.96	1.93	1.90
	0.050	2.75	2.69	2.62	2.54	2.51	2.47	2.43	2.38	2.34	2.30
	0.025	3.37	3.28	3.18	3.07	3.02	2.96	2.91	2.85	2.79	2.72
	0.010	4.30	4.16	4.01	3.86	3.78	3.70	3.62	3.54	3.45	3.36

Denominator df	α	Numerator df									
		10	12	15	20	24	30	40	60	120	∞
13	0.100	2.14	2.10	2.05	2.01	1.98	1.96	1.93	1.90	1.88	1.85
	0.050	2.67	2.60	2.53	2.46	2.42	2.38	2.34	2.30	2.25	2.21
	0.025	3.25	3.15	3.05	2.95	2.89	2.84	2.78	2.72	2.66	2.60
	0.010	4.10	3.96	3.82	3.66	3.59	3.51	3.43	3.34	3.25	3.17
14	0.100	2.10	2.05	2.01	1.96	1.94	1.91	1.89	1.86	1.83	1.80
	0.050	2.60	2.53	2.46	2.39	2.35	2.31	2.27	2.22	2.18	2.13
	0.025	3.15	3.05	2.95	2.84	2.79	2.73	2.67	2.61	2.55	2.49
	0.010	3.94	3.80	3.66	3.51	3.43	3.35	3.27	3.18	3.09	3.00
15	0.100	2.06	2.02	1.97	1.92	1.90	1.87	1.85	1.82	1.79	1.76
	0.050	2.54	2.48	2.40	2.33	2.29	2.25	2.20	2.16	2.11	2.07
	0.025	3.06	2.96	2.86	2.76	2.70	2.64	2.59	2.52	2.46	2.40
	0.010	3.80	3.67	3.52	3.37	3.29	3.21	3.13	3.05	2.96	2.87
16	0.100	2.03	1.99	1.94	1.89	1.87	1.84	1.81	1.78	1.75	1.72
	0.050	2.49	2.42	2.35	2.28	2.24	2.19	2.15	2.11	2.06	2.01
	0.025	2.99	2.89	2.79	2.68	2.63	2.57	2.51	2.45	2.38	2.32
	0.010	3.69	3.55	3.41	3.26	3.18	3.10	3.02	2.93	2.84	2.75
17	0.100	2.00	1.96	1.91	1.86	1.84	1.81	1.78	1.75	1.72	1.69
	0.050	2.45	2.38	2.31	2.23	2.19	2.15	2.10	2.06	2.01	1.96
	0.025	2.92	2.82	2.72	2.62	2.56	2.50	2.44	2.38	2.32	2.25
	0.010	3.59	3.46	3.31	3.16	3.08	3.00	2.92	2.83	2.75	2.65
18	0.100	1.98	1.93	1.89	1.84	1.81	1.78	1.75	1.72	1.69	1.66
	0.050	2.41	2.34	2.27	2.19	2.15	2.11	2.06	2.02	1.97	1.92
	0.025	2.87	2.77	2.67	2.56	2.50	2.44	2.38	2.32	2.26	2.19
	0.010	3.51	3.37	3.23	3.08	3.00	2.92	2.84	2.75	2.66	2.57
19	0.100	1.96	1.91	1.86	1.81	1.79	1.76	1.73	1.70	1.67	1.63
	0.050	2.38	2.31	2.23	2.16	2.11	2.07	2.03	1.98	1.93	1.88
	0.025	2.82	2.72	2.62	2.51	2.45	2.39	2.33	2.27	2.20	2.13
	0.010	3.43	3.30	3.15	3.00	2.92	2.84	2.76	2.67	2.58	2.49
20	0.100	1.94	1.89	1.84	1.79	1.77	1.74	1.71	1.68	1.64	1.61
	0.050	2.35	2.28	2.20	2.12	2.08	2.04	1.99	1.95	1.90	1.84
	0.025	2.77	2.68	2.57	2.46	2.41	2.35	2.29	2.22	2.16	2.09
	0.010	3.37	3.23	3.09	2.94	2.86	2.78	2.69	2.61	2.52	2.42
21	0.100	1.92	1.87	1.83	1.78	1.75	1.72	1.69	1.66	1.62	1.59
	0.050	2.32	2.25	2.18	2.10	2.05	2.01	1.96	1.92	1.87	1.81
	0.025	2.73	2.64	2.53	2.42	2.37	2.31	2.25	2.18	2.11	2.04
	0.010	3.31	3.17	3.03	2.88	2.80	2.72	2.64	2.55	2.46	2.36
22	0.100	1.90	1.86	1.81	1.76	1.73	1.70	1.67	1.64	1.60	1.57
	0.050	2.30	2.23	2.15	2.07	2.03	1.98	1.94	1.89	1.84	1.78
	0.025	2.70	2.60	2.50	2.39	2.33	2.27	2.21	2.14	2.08	2.00
	0.010	3.26	3.12	2.98	2.83	2.75	2.67	2.58	2.50	2.40	2.31
23	0.100	1.89	1.84	1.80	1.74	1.72	1.69	1.66	1.62	1.59	1.55
	0.050	2.27	2.20	2.13	2.05	2.01	1.96	1.91	1.86	1.81	1.76
	0.025	2.67	2.57	2.47	2.36	2.30	2.24	2.18	2.11	2.04	1.97
	0.010	3.21	3.07	2.93	2.78	2.70	2.62	2.54	2.45	2.35	2.26
24	0.100	1.88	1.83	1.78	1.73	1.70	1.67	1.64	1.61	1.57	1.53
	0.050	2.25	2.18	2.11	2.03	1.98	1.94	1.89	1.84	1.79	1.73
	0.025	2.64	2.54	2.44	2.33	2.27	2.21	2.15	2.08	2.01	1.94
	0.010	3.17	3.03	2.89	2.74	2.66	2.58	2.49	2.40	2.31	2.21

APPENDIX I: CRITICAL VALUES OF THE F DISTRIBUTION (concluded)

Denominator df	α	Numerator df									
		10	12	15	20	24	30	40	60	120	∞
25	0.100	1.87	1.82	1.77	1.72	1.69	1.66	1.63	1.59	1.56	1.52
	0.050	2.24	2.16	2.09	2.01	1.96	1.92	1.87	1.82	1.77	1.71
	0.025	2.61	2.51	2.41	2.30	2.24	2.18	2.12	2.05	1.98	1.91
	0.010	3.13	2.99	2.85	2.70	2.62	2.54	2.45	2.36	2.27	2.17
26	0.100	1.86	1.81	1.76	1.71	1.68	1.65	1.61	1.58	1.54	1.50
	0.050	2.22	2.15	2.07	1.99	1.95	1.90	1.85	1.80	1.75	1.69
	0.025	2.59	2.49	2.39	2.28	2.22	2.16	2.09	2.03	1.95	1.88
	0.010	3.09	2.96	2.81	2.66	2.58	2.50	2.42	2.33	2.23	2.13
27	0.100	1.85	1.80	1.75	1.70	1.67	1.64	1.60	1.57	1.53	1.49
	0.050	2.20	2.13	2.06	1.97	1.93	1.88	1.84	1.79	1.73	1.67
	0.025	2.57	2.47	2.36	2.25	2.19	2.13	2.07	2.00	1.93	1.85
	0.010	3.06	2.93	2.78	2.63	2.55	2.47	2.38	2.29	2.20	2.10
28	0.100	1.84	1.79	1.74	1.69	1.66	1.63	1.59	1.56	1.52	1.48
	0.050	2.19	2.12	2.04	1.96	1.91	1.87	1.82	1.77	1.71	1.65
	0.025	2.55	2.45	2.34	2.23	2.17	2.11	2.05	1.98	1.91	1.83
	0.010	3.03	2.90	2.75	2.60	2.52	2.44	2.35	2.26	2.17	2.06
29	0.100	1.83	1.78	1.73	1.68	1.65	1.62	1.58	1.55	1.51	1.47
	0.050	2.18	2.10	2.03	1.94	1.90	1.85	1.81	1.75	1.70	1.64
	0.025	2.53	2.43	2.32	2.21	2.15	2.09	2.03	1.96	1.89	1.81
	0.010	3.00	2.87	2.73	2.57	2.49	2.41	2.33	2.23	2.14	2.03
30	0.100	1.82	1.77	1.72	1.67	1.64	1.61	1.57	1.54	1.50	1.46
	0.050	2.16	2.09	2.01	1.93	1.89	1.84	1.79	1.74	1.68	1.62
	0.025	2.51	2.41	2.31	2.20	2.14	2.07	2.01	1.94	1.87	1.79
	0.010	2.98	2.84	2.70	2.55	2.47	2.39	2.30	2.21	2.11	2.01
40	0.100	1.76	1.71	1.66	1.61	1.57	1.54	1.51	1.47	1.42	1.38
	0.050	2.08	2.00	1.92	1.84	1.79	1.74	1.69	1.64	1.58	1.51
	0.025	2.39	2.29	2.18	2.07	2.01	1.94	1.88	1.80	1.72	1.64
	0.010	2.80	2.66	2.52	2.37	2.29	2.20	2.11	2.02	1.92	1.80
60	0.100	1.71	1.66	1.60	1.54	1.51	1.48	1.44	1.40	1.35	1.29
	0.050	1.99	1.92	1.84	1.75	1.70	1.65	1.59	1.53	1.47	1.39
	0.025	2.27	2.17	2.06	1.94	1.88	1.82	1.74	1.67	1.58	1.48
	0.010	2.63	2.50	2.35	2.20	2.12	2.03	1.94	1.84	1.73	1.60
120	0.100	1.65	1.60	1.55	1.48	1.45	1.41	1.37	1.32	1.26	1.19
	0.050	1.91	1.83	1.75	1.66	1.61	1.55	1.50	1.43	1.35	1.25
	0.025	2.16	2.05	1.94	1.82	1.76	1.69	1.61	1.53	1.43	1.31
	0.010	2.47	2.34	2.19	2.03	1.95	1.86	1.76	1.66	1.53	1.38
∞	0.100	1.60	1.55	1.49	1.42	1.38	1.34	1.30	1.24	1.17	1.00
	0.050	1.83	1.75	1.67	1.57	1.52	1.46	1.39	1.32	1.22	1.00
	0.025	2.05	1.94	1.83	1.71	1.64	1.57	1.48	1.39	1.27	1.00
	0.010	2.32	2.18	2.04	1.88	1.79	1.70	1.59	1.47	1.32	1.00

Source: Abridged from M. Merrington and C. M. Thompson, "Tables of Percentage Points of the Inverted Beta (F) Distribution," Biometrika, vol. 33 (1943). Used by permission.

APPENDIX J: TABLE OF CRITICAL VALUES OF *T* IN THE WILCOXON SIGNED RANK TEST

	Level of significance			
n	*0.025*	*0.01*	*0.005*	*One-tailed*
	0.05	*0.02*	*0.01*	*Two-tailed*
6	0	—	—	
7	2	0	—	
8	4	2	0	
9	6	3	2	
10	8	5	3	
11	11	7	5	
12	14	10	7	
13	17	13	10	
14	21	16	13	
15	25	20	16	
16	30	24	20	
17	35	28	23	
18	40	33	28	
19	46	38	32	
20	52	43	38	
21	59	49	43	
22	66	56	49	
23	73	62	55	
24	81	69	61	
25	89	77	68	

Source: Adapted from Table I of F. Wilcoxon, *Some Rapid Approximate Statistical Procedures* (New York: American Cyanamid Company, 1949), p. 13, with the permission of the publisher.

APPENDIX K: CRITICAL VALUES OF T_L AND T_U FOR THE MANN-WHITNEY-WILCOXON TEST

A. $\alpha = .025$ one-tailed; $\alpha = .05$ two-tailed.

											n_1									
	3		4		5		6		7		8		9		10	
n_2	T_L	T_U	T_L	T_U	T_L	T_U	T_L	T_U	T_L	T_U	T_L	T_U	T_L	T_U	T_L	T_U
3	5	16	6	18	6	21	7	23	7	26	8	28	8	31	9	33
4	6	18	11	25	12	28	12	32	13	35	14	38	15	41	16	44
5	6	21	12	28	18	37	19	41	20	45	21	49	22	53	24	56
6	7	23	12	32	19	41	26	52	28	56	29	61	31	65	32	70
7	7	26	13	35	20	45	28	56	37	68	39	73	41	78	43	83
8	8	28	14	38	21	49	29	61	39	73	49	87	51	93	54	98
9	8	31	15	41	22	53	31	65	41	78	51	93	63	108	66	114
10	9	33	16	44	24	56	32	70	43	83	54	98	66	114	79	131

B. $\alpha = .05$ one-tailed; $\alpha = .10$ two-tailed.

											n_1									
	3		4		5		6		7		8		9		10	
n_2	T_L	T_U	T_L	T_U	T_L	T_U	T_L	T_U	T_L	T_U	T_L	T_U	T_L	T_U	T_L	T_U
3	6	15	7	17	7	20	8	22	9	24	9	27	10	29	11	31
4	7	17	12	24	13	27	14	30	15	33	16	36	17	39	18	42
5	7	20	13	27	19	36	20	40	22	43	24	46	25	50	26	54
6	8	22	14	30	20	40	28	50	30	54	32	58	33	63	35	67
7	9	24	15	33	22	43	30	54	39	66	41	71	43	76	46	80
8	9	27	16	36	24	46	32	58	41	71	52	84	54	90	57	95
9	10	29	17	39	25	50	33	63	43	76	54	90	66	105	69	111
10	11	31	18	42	26	54	35	67	46	80	57	95	69	111	83	127

Note: The test statistic, T_1, must be computed for n_1 and $n_1 \leq n_2$.

Source: From F. Wilcoxon and R. A. Wilcox, "Some Rapid Approximate Statistical Procedures," 1964, pp. 20–23. Reproduced with the permission of R. A. Wilcox and the Lederle Laboratories.

APPENDIX L: FOUR-PLACE LOGARITHMS

N	0	1	2	3	4	5	6	7	8	9
10	0000	0043	0086	0128	0170	0212	0253	0294	0334	0374
11	0414	0453	0492	0531	0569	0607	0645	0682	0719	0755
12	0792	0828	0864	0899	0934	0969	1004	1038	1072	1106
13	1139	1173	1206	1239	1271	1303	1335	1367	1399	1430
14	1461	1492	1523	1553	1584	1614	1644	1673	1703	1732
15	1761	1790	1818	1847	1875	1903	1931	1959	1987	2014
16	2041	2068	2095	2122	2148	2175	2201	2227	2253	2279
17	2304	2330	2355	2380	2405	2430	2455	2480	2504	2529
18	2553	2577	2601	2625	2648	2672	2695	2718	2742	2765
19	2788	2810	2833	2856	2878	2900	2923	2945	2967	2989
20	3010	3032	3054	3075	3096	3118	3139	3160	3181	3201
21	3222	3243	3263	3284	3304	3324	3345	3365	3385	3404
22	3424	3444	3464	3483	3502	3522	3541	3560	3579	3598
23	3617	3636	3655	3674	3692	3711	3729	3747	3766	3784
24	3802	3820	3838	3856	3874	3892	3909	3927	3945	3962
25	3979	3997	4014	4031	4048	4065	4082	4099	4116	4133
26	4150	4166	4183	4200	4216	4232	4249	4265	4281	4298
27	4314	4330	4346	4362	4378	4393	4409	4425	4440	4456
28	4472	4487	4502	4518	4533	4548	4564	4579	4594	4609
29	4624	4639	4654	4669	4683	4698	4713	4728	4742	4757
30	4771	4786	4800	4814	4829	4843	4857	4871	4886	4900
31	4914	4928	4942	4955	4969	4983	4997	5011	5024	5038
32	5051	5065	5079	5092	5105	5119	5132	5145	5159	5172
33	5185	5198	5211	5224	5237	5250	5263	5276	5289	5302
34	5315	5328	5340	5353	5366	5378	5391	5403	5416	5428
35	5441	5453	5465	5478	5490	5502	5514	5527	5539	5551
36	5563	5575	5587	5599	5611	5623	5635	5647	5658	5670
37	5682	5694	5705	5717	5729	5740	5752	5763	5775	5786
38	5798	5809	5821	5832	5843	5855	5866	5877	5888	5899
39	5911	5922	5933	5944	5955	5966	5977	5988	5999	6010
40	6021	6031	6042	6053	6064	6075	6085	6096	6107	6117
41	6128	6138	5149	6160	6170	6180	6191	6201	6212	6222
42	6232	6243	6253	6263	6274	6284	6294	6304	6314	6325
43	6336	6345	6355	6365	6375	6385	6395	6405	6415	6425
44	6435	6444	6454	6464	6474	6484	6493	6503	6513	6522
45	6532	6542	6551	6561	6571	6580	6590	6599	6609	6618
46	6628	6637	6646	6656	6665	6675	6684	6693	6702	6712
47	6721	6730	6739	6749	6758	6767	6776	6785	6794	6803
48	6812	6821	6830	6839	6848	6857	6866	6875	6884	6893
49	6902	6911	6920	6928	6937	6946	6955	6964	6972	6981
50	6990	6998	7007	7016	7024	7033	7042	7050	7059	7067
51	7076	7084	7093	7101	7110	7118	7126	7135	7143	7152
52	7160	7168	7177	7185	7193	7202	7210	7218	7226	7235
53	7243	7251	7259	7267	7275	7284	7292	7300	7308	7316
54	7324	7332	7340	7348	7356	7364	7372	7380	7388	7396
55	7404	7412	7419	7427	7435	7443	7451	7459	7466	7474
56	7482	7490	7497	7505	7513	7520	7528	7536	7543	7551
57	7559	7566	7574	7582	7589	7597	7604	7612	7619	7627
58	7634	7642	7649	7657	7664	7672	7679	7686	7694	7701
59	7709	7716	7723	7731	7738	7745	7752	7760	7767	7774

N	0	1	2	3	4	5	6	7	8	9
60	7782	7789	7796	7803	7810	7818	7825	7832	7839	7846
61	7853	7860	7868	7875	7882	7889	7896	7903	7910	7917
62	7924	7931	7938	7945	7952	7959	7966	7973	7980	7987
63	7993	8000	8007	8014	8021	8028	8035	8041	8048	8055
64	8062	8069	8075	8082	8089	8096	8102	8109	8116	8122
65	8129	8136	8142	8149	8156	8162	8169	8176	8182	8189
66	8195	8202	8209	8215	8222	8228	8235	8241	8248	8254
67	8261	8267	8274	8280	8287	8293	8299	8306	8312	8319
68	8325	8331	8338	8344	8351	8357	8363	8370	8376	8382
69	8388	8395	8401	8407	8414	8420	8426	8432	8439	8445
70	8451	8457	8463	8470	8476	8482	8488	8494	8500	8506
71	8513	8519	8525	8531	8537	8543	8549	8555	8561	8567
72	8573	8579	8585	8591	8597	8603	8609	8615	8621	8627
73	8633	8639	8645	8651	8657	8663	8669	8675	8681	8686
74	8692	8698	8704	8710	8716	8722	8727	8733	8739	8745
75	8751	8756	8762	8768	8774	8779	8785	8791	8797	8802
76	8808	8814	8820	8825	8831	8837	8842	8848	8854	8859
77	8865	8871	8876	8882	8887	8893	8899	8904	8910	8915
78	8921	8927	8932	8938	8943	8949	8954	8960	8965	8971
79	8976	8982	8987	8993	8998	9004	9009	9015	9020	9025
80	9031	9036	9042	9047	9053	9058	9063	9069	9074	9079
81	9085	9090	9096	9101	9106	9112	9117	9122	9128	9133
82	9138	9143	9149	9154	9159	9165	9170	9175	9180	9186
83	9191	9196	9201	9206	9212	9217	9222	9227	9232	9238
84	9243	9248	9253	9258	9263	9269	9274	9279	9284	9289
85	9294	9299	9304	9309	9315	9320	9325	9330	9335	9340
86	9345	9350	9355	9360	9365	9370	9375	9380	9385	9390
87	9395	9400	9405	9410	9415	9420	9425	9430	9435	9440
88	9445	9450	9455	9460	9465	9469	9474	9479	9484	9489
89	9494	9499	9504	9509	9513	9518	9523	9528	9533	9538
90	9542	9547	9552	9557	9562	9566	9571	9576	9581	9586
91	9590	9595	9600	9605	9609	9614	9619	9624	9628	9633
92	9638	9643	9647	9652	9657	9661	9666	9671	9675	9680
93	9685	9689	9694	9699	9703	9708	9713	9717	9722	9727
94	9731	9736	9741	9745	9750	9754	9759	9763	9768	9773
95	9777	9782	9786	9791	9795	9800	9805	9809	9814	9818
96	9823	9827	9832	9836	9841	9845	9850	9854	9859	9863
97	9868	9872	9877	9881	9886	9890	9894	9899	9903	9908
98	9912	9917	9921	9926	9930	9934	9939	9943	9948	9952
99	9956	9961	9965	9969	9974	9978	9983	9987	9991	9996

Source: Donald R. Plane and Edward B. Oppermann, *Statistics for Management Decisions* (Dallas, Texas: Business Publications, 1977), Appendix B, pp. 450–51. © 1977 by Business Publications, Inc.

ANSWERS TO ODD-NUMBERED QUESTIONS AND EXERCISES

CHAPTER 2:
DESCRIPTIVE
STATISTICS

Questions

A. True or False

1. **False.**

3. True.

5. False

7. True.

9. False.

B. Fill in

1. Variance.

3. Ogive.

5. Skewed.

7. Coefficient of variation.

9. Secondary.

11. 2.4.

Exercises

1.

Payment Due	Number of Customers
$ 5.00 and under $20.00	9
20.00 and under 35.00	9
35.00 and under 50.00	8
50.00 and under 65.00	9
65.00 and under 80.00	3
80.00 and under 95.00	2
	40

3. The histogram depends on the frequency distribution prepared in Exercise 2.

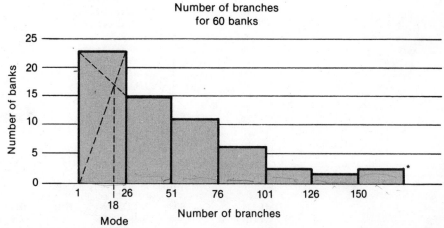

**Number of branches
for 60 banks**

* Two values of 188 and 215.

559

5. *a.* 32.4.   *b.* 28.8.

   c. 131.24.   *d.* 11.456.

7. *a.* $\bar{X} = 5.1$; $S = 2.234$.

   b. 7.75; ninety percent of the calls last less than 7.75 minutes.

9. *a.* 426.   *b.* 106.   *c.* $S = 4.60$.

11. *a.* 27.4.   *b.* 28.   *c.* No mode.   *d.* 10.04.   *e.* 3.17.

13. *a.* $276.74.   *b.* $64.94.   *c.* $52.52.

15. 30.

17. *a.* 6.

19. *a.* 1033.9.   *c.* $S_Y = S_X = $40,504.23$.

21. Broad Meadows; $V_{RR} = 22.4\%$; $V_{BM} = 26.74\%$.

23.

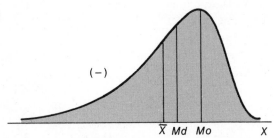

$(-)$

$\bar{X}$ Md Mo X

The extreme values are the small numerical values in a negatively skewed distribution.

25. *a.* The frequency distribution and histogram prepared by the MINITAB computer package are given below. Note that two classes have "0" frequencies. This situation should be avoided.

MIDDLE OF INTERVAL	NUMBER OF OBSERVATIONS	
0.	9	*********
20.	17	*****************
40.	12	************
60.	5	*****
80.	11	***********
100.	2	**
120.	1	*
140.	1	*
160.	0	
180.	1	*
200.	0	
220.	1	*

b. Descriptive measures determined with the SPSS computer package are:

VAR01   NUMBER OF BRANCH BANKS

MEAN	45.900	MEDIAN	32.167	STD DEV	43.192
VARIANCE	1865.546	SKEWNESS	1.798	RANGE	213.000
MINIMUM	2.000	MAXIMUM	215.000		

CHAPTER 3:
INDEX NUMBERS

Questions

A. True or False

1. True.
3. True.
5. False.
7. False.
9. True.

B. Fill in

1. Relative of aggregates.
3. 150.
5. Composite.
7. Average of relatives.

Exercises

1. *a.* 150. *b.* 151.9. *c.* 147.9. *d.* 147.9.

3. *a.* 115. *b.* 108.3. *c.* 121.5. *d.* 121.5.

5. 16.8 percent.

7.

Year	(a) Consumer Price Index (1967–100)	(b) U.S. Disposable Personal Income per Capita in 1967 dollars
1965	94.5	2578
1966	97.2	2679
1967	100.0	2749
1968	104.3	2824
1969	109.8	2851
1970	116.3	2903
1971	121.3	2972
1972	125.3	3067

c. 19 percent.

9. *a.* Real dollar personal income $= \dfrac{\text{Current dollar personal income}}{\text{CPI}} \cdot 100.$

b. 38.6 percent increase.

11. No; the decrease, in real dollars, is 10.8 percent.

13.

	Indexes (1975 = 100)	
	1976	*1977*
a. RA	122.0	165.0
b. AR	155.9	204.5

15. 133.1.

17. No; this is a 1.03 percent increase.

CHAPTER 4:
INTRODUCTION TO
PROBABILITY

Questions

A. True or False

1. False.

3. True.

5. False.

7. False.

9. False.

B. Fill in

1. $\frac{1}{4}$.

3. Mutually exclusive.

5. Equally likely.

7. 0.24.

9. Sample space.

Exercises

1. *a.* Not mutually exclusive.   *b.* Mutually exclusive.

   c. Not mutually exclusive.   *d.* Not mutually exclusive.

   e. Mutually exclusive.   *f.* Mutually exclusive.

3. *a.* Clubs, Diamonds, Hearts, and Spades.

   b. Yes; a single card cannot belong to more than one suit.

   c. 1.   *d.* "Clubs or Spades."

5. *a.* No.   *b.* No.   *c.* Yes.   *d.* No.

7. 56.   9. 65,780.

11. 36;

Sum x	$P(x)$
2	1/36
3	2/36
4	3/36
5	4/36
6	5/36
7	6/36
8	5/36
9	4/36
10	3/36
11	2/36
12	1/36
	1

13. 1:9.

15. *a.* 0.20; simple.   *b.* 0.80; simple.

     c. 0.10; joint.   *d.* 0.33; conditional.

     e. 1.00; conditional.   *f.* No; $P(\text{male}|\text{tennis}) \neq P(\text{male})$.

17. 0.90.

19. *a.* $\frac{4}{15}$.   *b.* $\frac{1}{15}$.   *c.* $\frac{8}{15}$.

     d.

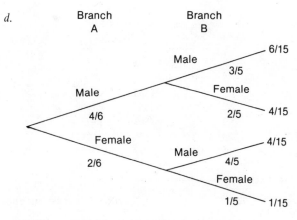

21. *a.* 0.55.   *b.* 0.60.   *c.* 0.45.   *d.* 0.30.

     e.

	Campus	*Off-Campus*	*Total*
Undergraduate	0.35	0.25	0.60
Graduate	0.10	0.30	0.40
Total	0.45	0.55	1.00

     f. No; $P(\text{campus housing}|\text{graduate}) \neq P(\text{campus housing})$.

23. *a.* 0.10.   *b.* 0.50.

25. $P(A|\text{accident}) = 0.324$; $P(B|\text{accident}) = 0.405$; $P(C|\text{accident}) = 0.270$.

CHAPTER 5:
PROBABILITY
DISTRIBUTIONS

Questions

A. True or False.

 1. False.

 3. False.

 5. True.

 7. False.

 9. False.

B. Fill in

1. Hypergeometric.

3. 7.

5. Standardized.

7. $\pi < 0.10$.

9. 0.

Exercises

1. $E(X) = 6.3$; $\sigma^2(X) = 0.81$.

3. *a.*

x	$P(x)$	$x \cdot P(x)$	$x^2 \cdot P(x)$
0	0	0	0
1	$\frac{1}{55}$	0.0182	0.0182
2	$\frac{4}{55}$	0.1455	0.2909
3	$\frac{9}{55}$	0.4909	1.4727
4	$\frac{16}{55}$	1.1636	4.6545
5	$\frac{25}{55}$	2.2727	11.3636
	1	4.0909	17.7999

b.

Probability mass function

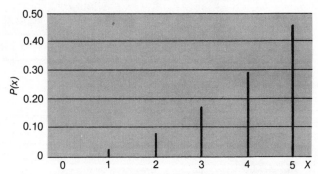

c. $E(X) = 4.0909$; $\sigma^2(X) = 1.0644$; $\sigma(X) = 1.0317$.

5. *a.* $P(X) = \frac{1}{10}$ for $x = 1, 2, \ldots, 10$, where the machine number is $Y = X - 1$.

    b. The production rate must be the same for all machines.

7. 0.0553; $E(X) = 11$; $\sigma^2(X) = 4.95$.

9. 0.0046; the proportion of accounts with errors may be larger than 10 percent.

11. **0.2684.**

13. *a.* 0.4696.   *b.* 0.0311.   *c.* $E(X) = 1$; $\sigma^2(X) = 0.6316$.

15. *a.* 0.4401.   *b.* 0.5599.

17. 0.0839.

19. *a.* 3.4.   *b.* 0.1469.

21. *a.* 0.5373.   *b.* 0.0183.

23. *a.* 0.2125.   *b.* 0.8743.

25. *a.*

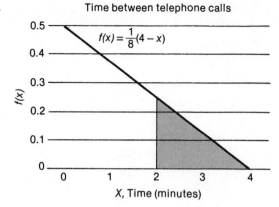

Time between telephone calls

$f(x) = \frac{1}{8}(4 - x)$

b. $\frac{1}{4}$, *c.* $\frac{1}{2}$.

27. *a.* $f(x) = 0.5$ for $2 \le X \le 4$.

b.

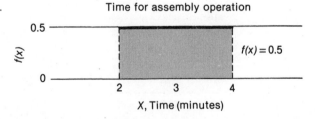

Time for assembly operation

$f(x) = 0.5$

c. $E(X) = 3; \sigma^2(X) = 0.333.$ *d.* 0.25.

29. *a.* 0.47725.   *b.* 0.97725.   *c.* 0.27337.   *d.* 0.39128.   *e.* 0.20289.

31. *a.* 0.34458.   *b.* (600)(0.00256) = 1.536, or 1 to 2.   *c.* 12.2

33. *a.* 0.15866.   *b.* 0.93319.   *c.* 20.56.

35. $\mu = 72.5, \sigma = 7.5.$ 37. 0.10565.

39. *a.* 0.04006.   *b.* 0.57048.

41. *a.* $E(Y) = 40; \sigma^2(Y) = 35.$ *b.* $E(R) = -20; \sigma^2(R) = 35.$
     c. $E(W) = 130; \sigma^2(W) = 590.$

CHAPTER 6:
SAMPLING AND
SAMPLING
DISTRIBUTIONS

Questions

A. True or False

1. True.

3. False.

5. True.

7. False.

9. False.

B. Fill in

1. π (the population proportion).

3. The standard error of the mean.

5. $n \geq 30$.

7. $\mu_1 = \mu_2$.

9. Variability or dispersion.

Exercises

3. *a.* $_6C_4 = 15$.

b.

X	$P(\bar{X})$
15	$0.0667 \approx 0.07$
15.5	0.07
16	0.13
16.5	0.13
17	0.20
17.5	0.13
18	0.13
18.5	0.07
19	0.07

c. $E(\bar{X}) = 17.0$; $\sigma_{\bar{X}}^2 = 1.2$.　　*d.* $\mu_X = 17.0$; $\sigma_X^2 = 11.67$.

e. $E(\bar{X}) = 17.0$; $\sigma_{\bar{X}}^2 = 1.168$ (difference from part *c* is due to rounding).

5. Regardless of sample size, $E(\bar{X}) = \mu_X = 17$. When $n = 1$, $\sigma_{\bar{X}}^2 = \sigma_X^2$; when $n = N$, $\sigma_{\bar{X}}^2 = 0$.

7. The sampling distributions in parts *a*, *b*, and *d* are normally distributed.

a. $E(\bar{X}) = 8,000$; $\sigma_{\bar{x}} = 200$.　　$E(\bar{X}) = 11,000$; $\sigma_{\bar{x}} = 200$.

b. They are the same because variances and sample sizes are equal and fpc is one.

c. $E(\bar{X}_1 - \bar{X}_2) = -3,000$; $\sigma_{\bar{X}_1 - \bar{X}_2} = 282.84$.

9. *a.* $\pi = 0.6$.

b.

p	$P(p)$
1.000	0.1
0.667	0.6
0.333	0.3

c. $E(p) = 0.6$; $\sigma_p^2 = 0.04$.

11. *a.* Normal with $E(\bar{X}) = 30$; $\sigma_{\bar{x}} = 0.5$.　*b.* 0.0228.　*c.* 30.64

.0275　29.36

13. Sampling distribution is normal with $p = 0.15$ and $\sigma_p = 0.046$.

a. $P(p \geq 0.167) = 0.3557$.　*b.* 0.015.

15. *a.* 0.1056.　*b.* 0.0062.　*c.* 0.3174.　*d.* 0.4649.

17. *a.* Normal with $E(\bar{X}_1 - \bar{X}_2) = -5$; $\sigma_{\bar{X}_1 - \bar{X}_2} = 0.93$.　*b.* Approximately zero.

19. The sampling distribution of $(\bar{X}_1 - \bar{X}_2)$ is normal with parameters $E(\bar{X}_1 - \bar{X}_2) = 12$; $\sigma_{\bar{X}_1 - \bar{X}_2} = 3.79$.

  a. 0.1446.   *b.* 0.5.   *c.* 0.1142.

CHAPTER 7:
STATISTICAL DECISION THEORY

Questions

A. True or False

1. False.
3. False.
5. True.
7. True.
9. True.

B. Fill in

1. States of nature.
3. EVPI.
5. Zero.
7. μ_b.
9. decrease.

Exercises

3. *a.* Premium = \$10,000.   *b.* $p = 0.9992$.

  c.

	Act	
Shipment	Insure	Do Not Insure
S_1: Safe	4000	0
S_2: Total Loss	$-4{,}996{,}000$	0

5. *a.* Opportunity loss table:

	Act		
State	A_1	A_2	A_3
S_1	2	0	5
S_2	0	3	5
S_3	0	8	15
S_4	18	15	0

$b.$ $A^* = A_3$ $c.$ $EVPI = 4.$

$d.$ $EMUU = EM(A_3) = 12.5.$ $e.$ $EMUC = 16.5.$

7. $a.$

S_i	$P(S_i \mid N)$
S_1	0.26
S_2	0.29
S_3	0.06
S_4	0.39

$b.$ $A^* = A_3; EVPI = EOL(A^*) = 3.65.$

9. $a.$ $EM(A_1) = 1.4; EM(A_2) = 0; A^* = A_1.$ $b.$ $EVPI = 1.2.$

11.
S_i	$P(S_i \mid N)$
S_1	0.100
S_2	0.375
S_3	0.525

$EM(A_1) = -0.85; EM(A_2) = 0; A^* = A_2.$

13. $a.$ $N(304.9, 8.87).$ $b.$ $N(3015, 4.96).$

15. $a.$ $EM(A_1) = 4500; EM(A_2) = 5000; A^* = A_2.$

$b.$ $\mu_b = 800.$ $c.$ $EVPI = 383.25.$

17. $a.$ $EVPI = 715.75.$ $b.$ $EVPI = 395.6$ $c.$ $EVPI = 1766.4.$

19. $a.$ Payoff $(A_1) = 0$; payoff $(A_2) = 1,200,000 - 5,000\mu.$

$b.$ $A^* = A_2$ $c.$ $EVPI = 39,560.$

21. $a.$ $N(140.05, 0.99).$ $b.$ $A^* = A_2; EVPI$ is approximately zero.

CHAPTER 8:
INTRODUCTION TO STATISTICAL INFERENCE

Questions

A. True or False

1. True.

3. False.

5. False.

7. True.

B. Fill in

1. Sample mean.

3. Level of significance.

5. One.

7. Unbiased.

9. 2.33.

Exercises

1. See chapter discussion.

3. $E(\hat{\sigma}^2) = \sigma^2 = 5.$

5. *a.* $618.

   b. $600.62 to $635.38. The annual mean expenditure is $600.62 to $635.38, with 92 percent confidence.

7. ~~245.~~ 49

9. *a.* $86,000.   *b.* 126.

11. *a.* 0.33 to 0.53.   *b.* 653.

13. 297.

15. Accept $H_0: \mu = 4.5$ with $Z = -1.5$ and p-value $= 0.13362$.
    No; the claim appears legitimate.

17. Reject $H_0: \mu = \$210$ with $Z = -2.4$ and p-value $= 0.00820$.
    Yes; the result does conflict with the insurer's claim.

19. *a.* 0.86638; this is $1 - \alpha$, $P(\text{accepting } H_0 | H_0 \text{ true})$.

    b. 0.13362; this is α, $P(\text{Type I error})$.

    c. 0.68525; this is β, $P(\text{Type II error} | \mu = 0.260)$.

    d. 0.50; this is $1 - \beta$, $P(\text{rejecting } H_0 | H_0 \text{ false})$ when $\mu = 0.235$.

21. Reject $H_0: \pi = 0.60$ with $Z = 2.71$ and p-value $= 0.00672$.
    No; the sample result does not support the campaign manager's belief.

23. Accept $H_0: \pi = 0.60$ with $Z = -1.30$ and p-value $= 0.0968$.
    No; these findings are not significantly less than 0.60.

CHAPTER 9: ADDITIONAL TOPICS IN STATISTICAL INFERENCE

Questions

A. True or False

1. False.

3. False.

5. True.

7. True.

B. Fill in

1. Normal.

3. 2.306.

5. Accepted.

7. 28.

9. 123.074; 156.926.

Exercises

1. $60.86 to $104.54.

3. 6.69 to 9.31.

5. Reject $H_0: \mu = 4500$ with $t = 3.11$ and $0.005 < $ p-value < 0.01. No; the average kilowatt hours used may be larger.

7. Accept $H_0: \mu = \$42,000$ with $t = -1.635$ and $0.05 < $ p-value < 0.10. No; the manager's claim appears correct.

9. *a.* Accept $H_0: \mu_1 - \mu_2 = 0$ with $Z = -1.10$ and p-value $= 0.27134$. Average verbal scores appear to be the same.

   b. Reject $H_0: \mu_1 - \mu_2 = 0$ with $Z = -5.42$ and p-value < 0.00006. Average mathematics scores appear to be different.

11. Accept $H_0: \mu_1 - \mu_2 = 0$ with $t = -1.12$ and $0.20 < $ p-value < 0.30. Average salaries in the two companies appear to be the same.

13. Reject $H_0: \mu_1 - \mu_2 = 0$ with $t = 2.667$ and $0.005 < $ p-value < 0.01. Yes; the chemical appears to improve paper life.

15. Reject $H_0: \pi_1 - \pi_2 = 0$ with $Z = 2.44$ and p-value $= 0.01468$. There appears to be a difference in east- and west-coast customer preference for unbranded products.

17. Accept $H_0: \pi_1 - \pi_2 = 0$ with $Z = -0.98$ and p-value $= 0.32708$.

19. The estimated difference in the two cities' average cost for appendectomies is, with 95 percent confidence, $21.15 to $68.85.

21. There is 95 percent confidence that the mean difference in tear force required for treated and untreated papers is within 0.0034 to 0.0286 grams.

CHAPTER 10:
THE CHI-SQUARE
DISTRIBUTION

Questions

A. True or False

1. True.

3. False.

5. True.

7. False.

B. Fill in

1. $v - 2$.

3. 10.

5. Contingency.

7. $n - 1$.

Exercises

1. Mean $= 10$; variance $= 20$; mode $= 8$.

3. *a.* 23.7.   *b.* 11.34.   *c.* 1.61.   *d.* 0.05.

5. *a.* Accept $H_0:\sigma^2 = 60$ since computed value $= 13.3$ and critical values are 8.91 and 32.9.

    b. 26.578 to 79.051

7. Accept $H_0: \sigma^2 = 16{,}000{,}000$ since computed value $= 4.9$ and critical values are 3.33 and 16.92.

9. Reject $H_0: p_j = 0.10$ for $j = 0, 1, \ldots, 9$ since computed value $= 30.13$ and critical value $= 21.7$.

11. Accept H_0: 20 percent is priced at a discount, 60 percent is standard, and 20 percent is premium; computed value $= 3.2$ and critical value $= 4.61$. Sample evidence does support executive's belief.

13. Accept H_0: there is no relationship between major and sex, or equivalently H_0: major is independent of sex; computed value $= 1.790$ and critical value $= 11.34$.

15. Accept H_0: employee opinion is independent of classification. Computed value $= 10.186$ and critical value $= 10.64$.

17. Accept H_0: vacancies follow a Poisson distribution with $\lambda = 0.5$. Computed value $= 0.39$ and, at the 0.05 level, critical value $= 3.84$.

CHAPTER 11:
THE *F*
DISTRIBUTION AND
ANALYSIS OF
VARIANCE

Questions

A. True or False

   1. True.

   3. False.

   5. False.

   7. True.

   9. True.

B. Fill in

   1. v_1 and v_2.

   3. H_0:all $\tau_j = 0$.

   5. *SSTR*, *SSB*, and *SSE*.

   7. Equal.

   9. Positively.

Exercises

1. *a.* (1) 6.63;   (2) 2.73;   (3) 0.21;   (4) 3.69.

5. Accept $H_0: \sigma_1^2 = \sigma_2^2$ since computed $F = 1.15$ and critical values are 0.31 and 3.18.

7. Accept $H_0: \sigma_1^2 = \sigma_2^2$ since computed $F = 0.79$ and critical values are 0.35 and 2.86.

9. *a.* $N = 12$.   *b.* $c = 4$.

    c. Accept H_0: all $\tau_j = 0$ since computed $F = 1.67$ and critical value is 4.07.

11.

Source of Variation	SS	v	MS	F
Treatments	30	2	15	5
Error.	36	12	3	
Total	66	14		

Reject $H_0: \mu_1 = \mu_2 = \mu_3$ since critical value $= 3.89$.

13. *a.* $SSTR = 463.45$; $SST = 587.78$; $F = 27.9$.
      Since the critical value $= 3.68$, reject H_0.

    b. 26 to 31.   *c.* -8.37 to -1.29.

15.

Source of Variation	SS	v	MS	F
Treatments	160	2	80	7.11
Blocks	160	4	40	
Error.	90	8	11.25	
Total	410	14		

    b. Since the critical value of $F = 4.46$, reject H_0: all $\tau_j = 0$.

17. *a.* $c = 4$.   *b.* $r = 5$.

    c. The computed $F = 3.01$, and the critical $F = 3.26$. Accept H_0: all $\beta_i = 0$; blocking was not effective.

    d. Computed $F = 6.02$; critical $F = 3.49$; reject H_0.

19. $SSTR = 13{,}444.8$; $SSB = 2146.2$; $SST = 18{,}217.2$; and computed $F = 20.48$. Reject H_0: all $\tau_j = 0$ since critical $F = 5.95$.

CHAPTER 12:
SIMPLE LINEAR REGRESSION

Questions

A. True or False

   1. False.

   3. False.

   5. False.

   7. True.

   9. False.

B. Fill in

   1. β_1.

   3. $\hat{\sigma}_\varepsilon^2$ (or $\hat{\sigma}_\varepsilon$ or SSE).

   5. Normal.

7. *t.*

9. *SSE.*

Exercises

3. *a.* Computed $t = -6.62$; critical values $= \pm 3.182$; reject H_0.

   b. -1.54 to -0.54.

5. $r^2 = 0.9346$.

7. *a.* 0.125.

   b. Computed $t = 9.8$; critical values $= \pm 2.686$; reject H_0.

9. *a.* $3.770 \leq E(Y_{120}) \leq 4.229$.

   b. $2.4 \leq Y_{100(\text{new})} \leq 3.6$.

   c. $\hat{Y} = -0.5$.

11. $5.444 \leq Y_{300(\text{new})} \leq 7.556$; reprimanded, it appears he is overspending.

13. *b.* $\hat{Y} = 2.40 + 0.053X$.

    c. Accept $H_0: \beta_1 = 0$; computed $t = 0.541$, critical values of $t = \pm 1.782$.

15. *b.* $\hat{Y} = -2.59 + 0.9855X$.

    c. Reject $H_0: \beta_1 = 0$; computed $t = 16.02$, critical values of $t = \pm 2.306$.

17. *a.* $\hat{Y} = 60.059 + 10.876X$.   *b.* 76.001.   *c.* 16.78.

    d. Reject $H_0: \beta_1 = 0$; computed $t = 2.65$, critical values $= \pm 2.11$.   *e.* 81.81.

CHAPTER 13: EXTENSIONS OF REGRESSION ANALYSIS

Questions

A. True or False.

1. True.

3. False.

5. True.

7. True.

9. False.

11. False.

B. Fill in

1. $4^2 = 16.0$.

3. Dummy variables.

5. Stepwise, forward selection, backward elimination.

7. *SST.*

9. Three.

Exercises

3. Reject H_0 since computed $F = 3.55$ and critical value of $F = 3.36$.

5. *a.* $SSR = 3,384$.   *b.* $MSR = 3,384$; $MSE = 14.5$.
   c. Reject H_0: the computed $F = 233.38$, the critical $F = 4.20$.

7. *a.* $SSR = 2,636$; $MSR = 1,318$; $MSE = 12.74$.
   b. Reject H_0: the computed $F = 103.45$, the critical $F = 3.35$.
   c. No.

9. *a.* $\hat{Y} = 2.0636$.

11. Note to reader: Rounding error is particularly important in multiple regression problems. We suggest that you do not round to less than five digits.
   a. Define $X_2 = 1$ if night class
                   0 if not night class.

Night Class	X_2
No	0
No	0
Yes	1
Yes	1
No	0
No	0
Yes	1
No	0
Yes	1
No	0
Yes	1

   b. $\hat{Y} = 4.4813 - 0.0186X_1 - 0.7106X_2$.
   c. Reject H_0; computed $F = 11.69$, critical $F = 4.46$.
   d. Reject H_0: computed $F = 14.65$, critical $F = 5.32$.
   e. $\hat{Y} = 3.2499$.

13. *a.* $SSE(X_1) = 6.0$; $SSE(X_1, X_2) = 2.2$.   *b.* $SSR(X_2 | X_1) = 3.8$.
   c. $SST(X_1) = SST(X_1, X_2) = 20.0$.
   d. Reject H_0; computed $F = 190.17$, critical $F = 3.202$.
   e. Reject H_0; computed $F = 81.19$; critical $F = 4.052$.
   f. $R^2 = 0.89$.

15. *a.* Reject H_0; computed $F = 19.72$, critical $F = 3.06$.
   b. For following parts, critical value of $t = \pm 2.131$.
      (1) Reject H_0: $\beta_1 = 0$; computed $t = 3.72$.
      (2) Reject H_0: $\beta_2 = 0$; computed $t = 5.09$.
      (3) Accept H_0: $\beta_3 = 0$; computed $t = 1.22$.
      (4) Reject H_0: $\beta_4 = 0$; computed $t = -3.75$.
   c. 0.798.   *d.* 6.2522.

CHAPTER 14:
TIME SERIES
ANALYSIS

Questions   A.   True or False

1.   False.

3.   True.

5.   False.

7.   False.

9.   True.

B.   Fill in

1.   94.3.

3.   1988.

5.   30,000.

Exercises   1.   The moving averages are:

Quarter		$\bar{y}_t$	Quarter		$\bar{y}_t$
1967	III	5.02	1972	I	4.20
	IV	5.02		II	4.26
				III	4.38
				IV	4.48
1968	I	5.04	1973	I	4.58
	II	5.08		II	4.86
	III	5.08		III	5.12
	IV	5.04		IV	5.30
1969	I	5.00	1974	I	5.54
	II	4.88		II	5.58
	III	4.68		III	5.20
	IV	4.50		IV	4.92
1970	I	4.32	1975	I	4.80
	II	4.08		II	4.64
	III	3.98		III	4.76
	IV	4.00		IV	5.12
1971	I	4.00	1976	I	5.30
	II	4.02		II	5.30
	III	4.14			
	IV	4.18			

3.   *a.* $\hat{Y} = 4117.2 + 53.4X$.   *b.* $\hat{Y}_{82} = 5131.8$; $\hat{Y}_{85} = 5452.2$.

c.

Year	Cyclical Relatives
1968	89.5
1969	101.3
1970	103.4
1971	99.2
1972	101.4
1973	108.6
1974	114.6
1975	88.5
1976	94.7
1977	98.7

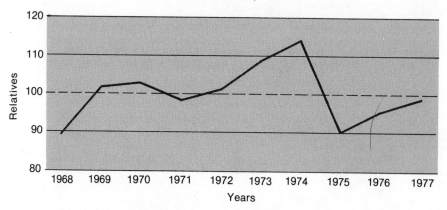

Cyclical relatives for
U.S. aluminum production

5. *a.* $\hat{Y} = 309.909 + 16.587X + 1.185X^2$.   *b.* 1983.

7. *a.* $\hat{Y} = 0.2226 + 0.0081X + 0.0005X^2$.   *b.* 0.7376 cents per gallon.

    c. All factors that potentially impact gasoline availability and demand. Changes in OPEC pricing; the interruption of oil exports (Iran, 1979), and domestic production rates are some of these factors.

9. *a.* A quadratic trend is best; $\hat{Y} = 536.806 - 9.624X + 0.260X^2$.

    b. No, this quadratic trend is decreasing by a decreasing amount annually, indicating that the rate of decline is slowing down.

    c. 451 firms.

11. The exponential trend is best; $\log \hat{Y} = 3.7853 + 0.0449X$. $\hat{Y}_{81} = \$23,400$ million.

13. *b.* The trend is linear; $\hat{Y} = 18,481.14 - 183.06X$.

    c. The trend in U.S. population under age 5 is 18,481.14 thousand in 1967 and decreases annually by 183.06 thousand.

    d. $\hat{Y}_{80} = 13,904.64$ thousand.

15. *a.* Quadratic: $\hat{Y} = 89.24 + 9.19X + 0.34X^2$.
      Exponential: $\log \hat{Y} = 1.9364 + 0.0436X$.

    b. $\sum(Y - \hat{Y})^2 = 1464.7$ for the quadratic and 1674.5 for the exponential, indicating that the quadratic provides a better fit.

    c. Quadratic: $\hat{Y}_{79} = 230.37$. Exponential: $\hat{Y}_{79} = 261$.

17. *a.* Seasonal indexes are:

January	73.41
February	88.46
March	102.54
April	113.29
May	116.51
June	94.08
July	74.32
August	73.33
September	104.75
October	127.28
November	130.14
December	101.89

   b. $219.32 million.

19. Quarterly forecasts are: (I) $101,250; (II) $128,750; (III) $152,500; (IV) $117,500.

21. 9750.

23. 1,500,000.

CHAPTER 15:
NONPARAMETRIC
METHODS

Questions

A. True or False

1. True.

3. True.

5. False.

7. False.

9. True.

B. Fill in

1. 11 (eleven).

3. One-way ANOVA.

5. 820.

7. Normal.

9. Nominal.

Exercises

1. *a.* 8.5.   *b.* 15.   *c.* 7.5.

3. *a.* $T_L = 66$.   *b.* $T_L = 14$ and $T_U = 30$.   *c.* $T_U = 43$.

5. Accept $H_0: M = 75$; $T_- = 14.5$ and critical value of $T = 4$.

7. Accept $H_0: M_D \geq 0$; $T_+ = 8$ and critical value of $T = 6$. There was not a significant increase in sales.

9. Reject $H_0: M_D \geq 0$; $T_+ = 1$ and critical value of $T = 4$. Yes, gas mileage is increased with radial tires.

11. Reject H_0: $M_1 - M_2 = 0$; computed $Z = 2.70$ and critical values are ± 1.96. Yes, there is a significant difference in assembly times.

13. Accept H_0: $M_1 - M_2 \leq 0$; computed value of $T_1 = 41$ and critical value $T_U = 45$. Joggers' QPA is not significantly greater than nonjoggers'.

15. Accept H_0: $M_1 = M_2 = M_3$; computed value of $H = 4.63$, critical value $= 5.99$. No significant differences in job satisfaction.

17. Accept H_0: no association; computed $t = -0.67$, critical values $= \pm 3.055$.

19. Accept H_0: interview rank and college rank are not associated; $r_s = 0.594$, computed $t = 2.09$, and critical values $= \pm 2.306$.

INDEX

Index

*This book has been set in 10 and 9 Times Roman,
leaded 2 points. Chapter numbers are 72 point Times
Roman Bold and chapter titles are 36 point Helvetica.
The size of the type page is 27 by 49 picas.*